Welcome to

McGraw Hill

ASVAB

Congratulations! You've chosen *McGraw Hill: ASVAB*. You probably know McGraw Hill from many of the textbooks you used in school or college. Now we're ready to help you reach your next goal—an exciting career in the U.S. Armed Forces.

This book gives you everything you need to succeed on the ASVAB. You'll get in-depth instruction and review of every topic tested, tips and strategies for every question type, and plenty of practice exams to boost your test-taking confidence. You'll also get some answers to your questions about enlistment and jobs in today's military. Additionally, in the following pages you'll find:

- **How to Get the Most Out of This Book:** A map of what you can find in this book.

- **What It Takes to Achieve Enlistment:** Learn the difference between selection and classification and how the military uses your ASVAB scores.

- **Your ASVAB Preparation Program:** Figure out what you know and don't know and decide on a study strategy.

- **Your ASVAB Study Schedule:** Be focused with your study time.

- **Test-Time Advice:** Follow this good advice.

- **Other Criteria for Enlistment:** Keep these factors in mind as you apply for the military.

ABOUT McGRAW HILL

This book has been created by a unit of McGraw Hill. McGraw Hill is a leading global provider of instructional assessment and reference materials in both print and digital form. McGraw Hill has offices in 33 countries and publishes in more than 65 languages. With a broad range of products and services—from traditional textbooks to the latest in online and multimedia learning—we engage, stimulate, and empower students and professionals of all ages, helping them meet the increasing challenges of the twenty-first-century knowledge economy.

Learn more. **Mc Graw Hill** **Do more.**

How to Get the Most Out of This Book

This book provides the material you need to score well on the ASVAB. It will teach you the knowledge and skills that are required for this difficult exam, including information about the types of questions the test includes. It also provides ample practice for you to refine the skills you are learning and then test yourself with full-length practice tests. To make the best use of this book, follow the following suggestions. Then develop your own personalized ASVAB study plan (see Chapter 6).

1 **If you are not sure which branch of the military interests you:**
- Read Chapter 1: "Deciding on Military Service."
- Go to Part 5 and learn about various military occupations.

2 **If you want to know about the ASVAB and why scoring high matters to you:**
- Read Chapter 2: "Introducing the ASVAB."
- Read Chapter 3: "ASVAB Scores and Score Reports."
- Read Chapter 4: "The ASVAB, the AFQT, and Military Entrance."
- Read Chapter 5: "Taking the CAT-ASVAB."

3 **If you want general study tips and guidelines for planning your study:**
- Review Chapter 6: "Follow These Suggestions to Ace the ASVAB."

4 **If you want to find out how well you might do right now on the real ASVAB:**
- Take the ASVAB Diagnostic Test Form in Part 2 and use it to identify your strengths and weaknesses.

5 **If you want to refresh your knowledge of the ASVAB subject areas:**
- Go to Part 3 of this book, which provides a complete review of typical topics tested.

6 **If you want to practice taking the ASVAB test:**
- Go to Part 4 of this book and take the ASVAB Practice Test Forms 1 through 3 under actual test conditions.

7 **If you want to know about some specific jobs and in the military:**
- Go to Part 5 of this book.

8 **If you are active duty and want to raise your score to change your military occupational specialty:**
- Go to Part 3 of this book to review the kinds of topics you may be tested on.
- Go to Part 2 to take a diagnostic test and Part 4 to take more practice items.

What It Takes to Achieve Enlistment

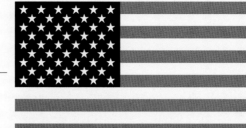

You can think of military enlistment as a two-step process: first selection and then classification. To be initially "selected" to enter the military at all, you need to do well on the ASVAB, in particular on four of the tests that make up the Armed Forces Qualifying Test (AFQT). So it is important to focus on those four tests: Arithmetic Reasoning, Word Knowledge, Paragraph Comprehension, and Mathematics Knowledge. You need to score well on those four tests in order to reach the minimum AFQT score to even be considered for enlistment. You should aim for higher than the minimum; the higher your score, the more job and training options you have. High scores may even have an enlistment bonus!

Once you meet the selection criteria, the second step is classification. That means a review of your scores in various combinations, called line scores, to see which jobs/training programs you are eligible for. Many of the line scores include one or more of the initial AFQT scores, so, again, concentrating on the four AFQT tests is very important.

ASVAB Factoids

The ASVAB is the most widely used multiple aptitude test in the world.

The ASVAB was introduced in 1968. Over 40 million examinees have taken various updated forms of the ASVAB since then.

The ASVAB is administered via both computer and paper and pencil. Roughly 90 percent of military applicants take the test via computer. Every year more high school or postsecondary school students are taking the computerized version.

The CAT-ASVAB was implemented after decades of extensive research and evaluation. It was the first large-scale adaptive test battery to be administered in a high-stakes setting.

The CAT-ASVAB generally takes less time to complete than the paper-and-pencil ASVAB.

Women compose about 20 percent of enlisted personnel in the military. The percentage varies by Service.

Your ASVAB Preparation Program

Take the Diagnostic Test

Start your ASVAB preparation program by taking the Diagnostic Test in Part 2 of this book. The steps outlined as follows will help you figure out your strengths and weaknesses. Be sure to read the directions for each test and work as fast as you can while being very careful with your answers.

Starting on page 116, you can begin to score each of the tests. Depending on how well you do, you may be happy with the results or realize that you need to buckle down and start to study hard. Here is an example of how to interpret your scores:

Part 4. Paragraph Comprehension

Item Number	Correct Answer	Mark X if You Picked the Correct Answer
1	A	X
2	B	
3	C	X
4	A	
5	D	X
6	B	
7	A	X
8	B	X
9	C	
10	A	X
11	C	X
12	A	
13	B	X
14	C	
15	D	X
Total Correct		**9**

Score Interpretation–Total Correct

15 14	This is pretty good work. Review the explanations of the answers you got incorrect. This test is a part of the AFQT, and you must do well.
13 12	You are doing pretty well. Review the explanations for the items you got incorrect. This test is a part of the AFQT, and you must do well.
11 10 9 8 7 6 5 4 3 2 1 0	You need to keep studying. Pay close attention to the explanations for each item, even for the ones you got correct. This will help you understand why the answer is correct. Spend time reviewing the Paragraph Comprehension and Word Knowledge reviews in Part 3 of this book. Keep reading books and newspapers.

If you scored 11 or lower on this test (the dark gray area for any of the tests in this book), you need to put some skin in the game and start studying. Fortunately, this ASVAB book shows you how to do better in two ways. First, for the Diagnostic Test and the other three full ASVAB tests, this ASVAB book gives you the explanation for the correct answers. For example, for the Paragraph Comprehension test, pages 132 and 133 give you an explanation of why each correct answer is correct. This type of information is provided for the Diagnostic Test and the other three practice tests. Be sure to read the explanations because they will help you do better on the real ASVAB.

Second, for the Paragraph Comprehension test, there is an entire review section that helps build your skills in this important ASVAB area. The review information for Paragraph Comprehension starts on page 227. This type of information is provided for each of the content areas as described in the next section.

If you faithfully follow this advice, you will know which tests you need to focus on with more time and energy. That's where you should put in more study time.

Review ASVAB Test Content

Starting on page 141, you will find an extensive review of all the ASVAB test topics starting with the General Science test. Each content area is thoroughly covered and contains an end-of-section quiz that will help you figure out if you learned the material in the review section. Where you answered incorrectly, go back to the section and study the material until you are sure you understand it.

This book provides almost 200 pages of content review, because you can't predict specifically what information will be asked on the ASVAB. Part 3 covers all ASVAB test topics and gives you the tools you need to answer ASVAB questions. For example, you will learn how to figure out vocabulary words, how to set up math problems to get the right answer, how to answer geometry items, how to decipher levers and pulleys, how to figure out electrical circuits, how to conquer angles and geometric shapes, how to answer auto and shop questions, and hundreds of other topics that you could be tested on.

Memorizing *isn't* always the best way to score high on the ASVAB. Be sure that you *understand* the concepts behind each question. For example, you could memorize the answer to a math question, but on the real ASVAB you probably won't find that same question. You will need to understand the steps in solving the problem, regardless of the numbers involved.

Take the Practice Tests

Once you have taken the Diagnostic Test, reviewed the test content that you needed to study, and taken the end-of-section quizzes, you are now ready to test yourself again. Go to page 337 to start the first test form in the book. Sit in a quiet location and prepare to spend about 2 to 2½ hours taking this test. Make sure you have a few pencils and a clock to show the time.

Carefully tear out page 339 and use it as your answer sheet. When you are ready to begin, read the instructions for each section and answer the questions to the best of your ability. When you have finished, go to pages 388 to 396 and compare your answers with the correct answers. Check the score interpretation chart for each subject test. If your number of correct answers is in the dark gray area, you need to review that test content in depth to raise your score. You will probably find some areas where you can improve. Go back to Part 3 of this book and spend some quality time studying. If your number of correct answers is in the light gray area, you're doing better, but you should still spend some time studying that content area to improve your score.

Here are some tips to help you prioritize your study time:

- If your scores are in the dark gray area on the AFQT tests—Arithmetic Reasoning, Word Knowledge, Paragraph Comprehension, or Mathematics Knowledge—you should spend most of your time on the AFQT subjects.

- If your scores on the AFQT tests are in the light gray areas but you have other scores in the dark gray areas, go back and review the content for tests where your scores were lowest.

- If all of your scores are in the light gray areas, focus again on the four AFQT tests.

- If it takes you more than 2½ hours to complete each test form, try to speed up your pace.

Plan Your Study Time

Let's be realistic. Maybe you were a whiz kid in high school and just need to brush up your knowledge and skills. Perhaps you totally blew off high school and just got your diploma by the skin of your teeth, or perhaps you have been away from school activities for a few years.

Regardless of your situation, the advice in the previous section on how to use this book is right on target. The difference may be in the amount of time you need to spend on getting yourself up to speed on the content of a particular test.

If you just need a refresher, you should probably spend about one to two hours per day for two weeks in focused study. If you really need to improve after starting from a low point, you should plan on one to three months of work, at least 6 to 8 hours per week.

Set up your study schedule accordingly. Use the chart in Figure 6.3 to see an example of what you might do, and then you can construct your own schedule with the chart in Figure 6.4.

Don't be a couch potato. Add some exercise into your schedule as it will keep your brain clear and your stress level down. This will also help you in the enlistment process, as you will need a decent level of fitness to be accepted into the military and to do well in your basic training. Keep in mind that many military applicants are rejected because they are not in shape!

Study Tips

Dos:

▶ Find a quiet place to study. Quiet means no TV, texting, phone, games, or other distractions.

▶ Take notes and summarize what you have learned as you move through the material.

▶ Highlight special facts or concepts you want to remember.

▶ Take a short, 10-minute break every 45 minutes.

▶ Think about the kinds of test questions you might be asked as you study the content review sections of this book.

▶ Eat healthy foods and eliminate fatty foods and alcohol and eliminate anything that may be interfering with your focus and learning.

▶ Stick to your study schedule but be sure to also include a decent amount of sleep each night.

Don'ts:

▶ Pretend you are studying, when you are actually daydreaming, gaming, scrolling though your feed, or multitasking. You are just wasting your time if you are not focused.

▶ Study with others unless they are very serious about studying for the ASVAB or they are serious about helping you study.

▶ Wait until the last few days before the real test to start studying.

▶ Get stressed out. Know and believe that if you study the material you will do well on the ASVAB.

Test-Time Advice

The Day Before the Test

- It's too late to cram now. You have followed a good study schedule and have prioritized your studies. Be proud of your hard work. It will pay off.

- Put this book down and just use the day to get some fresh air. Go for a run or a long walk, or relax with your favorite healthy hobby.

- Eat light, healthy meals and go to sleep a little earlier than usual. No alcohol, please. Try for a good eight hours of shut-eye.

The Day of the Test

- Get up early and take a short walk or run. A half-hour of exercise will do you wonders. It will clear the cobwebs from your brain and help you relax.

- Spend a few minutes skimming any important notes you may have taken—no more than a half-hour, though.

- Have a decent breakfast, but no sticky buns or heavy, fatty food. Think cereal, yogurt, fruit, eggs, veggies, milk, and coffee. You won't feel your best if you overload on goopy food.

- Do some deep-breathing exercises to relieve any stress.

- Get to the testing room with time to spare. No sense in adding more pressure on yourself by running late.

- Adjust your attitude upward. You have worked hard, you know the material, and you will do well. Be confident in your ability. Turn your knowledge into a powerhouse of confidence.

During the Test

- Show your stuff. If you have followed the study and review suggestions in this book, you will be able to go into the real test with a positive attitude. Feel confident that you've got the right stuff to succeed in the military.

- Breathe. Take a few moments to breathe deeply and relax if you get stressed during the test. You will feel better.

- Pace yourself. Budget enough time for each question so that you won't have to rush at the end of the section.

- Stay focused. Ignore the things going on around you that you cannot control.

- Use what you know. When you don't know the answer and need to guess, try to make an educated guess by eliminating answer choices that you know are wrong. The more you can eliminate, the better your chance of getting the answer right.

Other Criteria for Enlistment

Education

Success in any branch of the military depends on a good education, so a high school diploma and even some college credits is most desirable. Candidates with a GED can enlist, but some services may limit opportunities. It is very difficult to be considered a serious candidate without either a high school diploma or accepted alternative credential. Your minimum qualifying score on the AFQT can change depending on whether or not you have a regular high school diploma. In any case, staying in school is important for entering the military. The military does not want high school dropouts.

Physical Requirements

Because of the varying physical demands on service members in each branch, physical requirements vary greatly. These differences can vary even within each branch of the service. Generally speaking, potential service members should be in good physical condition, of appropriate weight, and be able to pass a standard physical screening prior to entry. Certain physical conditions can disqualify you, but talking with a recruiter will help you understand the specifics.

Behavior and Moral Character

The services do not want to enlist troublemakers, extremists, or persons with serious offenses. Each service takes a different approach to evaluating the severity and number of offenses on a candidate's record. The results of this evaluation may—or may not—disqualify candidates. A review is done on a case-by-case basis.

A word of caution: The military looks down on persons who lie about their record. The military conducts background checks; if you've provided false information, you'll be exposed as a liar and receive the appropriate punishment.

Age Requirements

Take a look at the following chart to see the age requirements for each of the services and the Coast Guard.

	Army	Marine Corps	Navy	Air Force	Space Force	Coast Guard
Active Duty	17–35	17–28	17–39	17–39	Contact your	17–32
Reserve	17–35	17–28	17–39	17–38	Air Force	17–40
Guard	17–35	N/A	N/A	17–39	recruiter for the	N/A
Service Academies	17–23	17–23	17–23	17–23	requirements.	17–22

Gender

Both men and women are eligible to enlist in the military. All jobs in the military are open to women. The percentage of women serving on active duty in the military has steadily increased over the years. Women play a very important role in today's military.

Citizenship

Properly documented noncitizens may enlist. However, each service has its own enlistment requirements for noncitizens. You should contact a recruiter for details about a specific situation. For enlistment purposes, the United States includes Guam, Puerto Rico, the U.S. Virgin Islands, the Northern Marianas Islands, American Samoa, the Federated States of Micronesia, the Marshall Islands, and Palau. There are some special programs for certain ethic and cultural groups who may want to enlist.

NOW GO ACE THE ASVAB! YOU CAN DO IT!
BEST WISHES FOR A SUCCESSFUL MILITARY CAREER

McGRAW HILL

ASVAB

McGRAW HILL
ASVAB

Armed Services
Vocational Aptitude Battery

Fifth Edition

Dr. Janet E. Wall
Former Manager
ASVAB Career Exploration Program
Department of Defense

New York   Chicago   San Francisco   Athens   London   Madrid   Mexico City
Milan   New Delhi   Singapore   Sydney   Toronto

1 2 3 4 5 6 7 8 9   LOV   25 24 23 22 21

ISBN   978-1-264-27746-9
MHID   1-264-27746-6

e-ISBN   978-1-264-27747-6
e-MHID   1-264-27747-4

McGraw Hill books are available at special quantity discounts to use as premiums and sales promotions, or for use in corporate training programs. To contact a representative please visit the Contact Us pages at www.mhprofessional.com.

The views expressed in this book are those of the author and do not reflect the official position of the Department of Defense or any other agency of the U.S. Government.

ASVAB is a registered trademark of the United States Department of Defense, which was not involved in the production of, and does not endorse, this product.

The Electronics Information, Auto and Shop Information, and Mechanical Comprehension portions of this book were contributed by David Tenenbaum.

McGraw Hill is committed to making our products accessible to all learners. To learn more about the available support and accommodations we offer, please contact us at accessibility@mheducation.com. We also participate in the Access Text Network (www.accesstext.org), and ATN members may submit requests through ATN.

CONTENTS

PART 4. Three ASVAB Practice Test Forms 335

PART 5. Jobs in Today's Military 563

Foreword

If you are interested in joining the U.S. Army, Navy, Marine Corps, Air Force, Space Force, or Coast Guard, you will have to take a test called the Armed Services Vocational Aptitude Battery, or ASVAB. It actually is a set of tests that measure aptitude in a wide range of skills. To be accepted for enlistment by any one of the military services, you must reach at least a minimal standard on a combination of the verbal and math tests. What you may not know is how important it is that your scores on the ASVAB tests be as high as possible. Why? You will have more job choices!

The strength of the Armed Forces of the United States is based on highly capable service members who can think clearly and logically, can operate the highly technical equipment, and can deal effectively with a broad variety of tasks. The ASVAB is required of all who want to enlist in any of the American military services to determine their qualifications for placement in one of the many military occupational specialties each service requires. Therefore, the higher the scores you achieve, the more military occupational specialties will be made available to you from which to choose.

If you are an active-duty military service member and want to change your military specialty, the score you enlisted with may not be high enough to qualify. Your general knowledge may be rusty, and you may benefit greatly from a refresher. This book will remind you of the ASVAB requirements.

It can provide a solid review of the material you may encounter on the test.

McGraw Hill ASVAB provides essential information and practice tests on the various areas measured by the ASVAB. A careful study of this book will help you increase your knowledge of critical information to score higher on the tests, and it also will familiarize you with the test itself to make you more confident when you take it.

The author of this book, Dr. Janet E. Wall, is a uniquely qualified expert on the ASVAB. As a former manager of the ASVAB Career Exploration Program, she has been a leader in the designing and providing content to the ASVAB program for several years as a senior official in the U.S. Department of Defense. Her educational and professional accomplishments in career development and assessment are nationally recognized. Her book will teach you what is necessary to score high on the ASVAB.

I am confident that you will find military service challenging and rewarding as a career or for one or more enlistments. Good luck on the ASVAB, and thank you for considering military service to our country.

Lieutenant General (U.S. Army, Retired) Robert G. Gard, Jr., PhD
Former Director, Human Resources Development, U.S. Army
Former President, National Defense University

From the Author

I am so glad you have chosen this book to help you rock the ASVAB. I am Dr. Janet Wall, and I very much want you to score high on the ASVAB and have a successful and satisfying enlistment in the military.

I have worked for the U.S. Department of Defense (DoD) for almost 15 years. In that time, I was responsible for several aspects of the Department of Defense overseas schools and later with the DoD group that created the ASVAB tests and the ASVAB Career Exploration Program. In those jobs, I wrote, edited, and/or reviewed test items in a variety of subject areas. I also led the development of the ASVAB Career Exploration Program used by more than a million youth each year. I even trained recruiters, counselors, and educators on the program!

For the last several years, I have been focused on career development—that is, getting people into good jobs that match their abilities, interests, and values. I have done this by consulting with several career guidance systems, publishing books, co-authoring an abilities assessment, and training career counselors to become more skilled at providing services to students and clients.

First and foremost, I consider myself to be a **teacher**. So, in this book, I try very hard to teach you the concepts that can help you score high on the ASVAB. No book will tell you the ASVAB test answers, but I have tried my very best to help you learn, think, and figure out the answers to the questions you may be issued on the test.

It is my most sincere wish that you score high on the ASAVB so that you have more control over the various jobs and opportunities the military can provide you. I want you to serve our country well, honorably, and to the best of your ability. Our country needs well qualified and dedicated men and women. I hope that is you!

Dr. Janet Wall, CCSP, JCTC, CDFI, NCDA Fellow

Introduction

If you are considering buying this book or have already done so, you must be interested in the Armed Services Vocational Aptitude Battery—the ASVAB. Without a doubt, this book is for you. It is packed with information and advice to help you do your best on the test. If you pay close attention to the advice you'll find here, you can get a much better ASVAB score.

Why should you pay attention to the advice? If you are interested in learning about careers and career planning, the ASVAB can help you know about yourself and how well you are suited for various occupations so that you can make better decisions. If your goal is to enter the military, you will want to perform your best on the ASVAB so that your options for military training programs and careers are the very best possible. In either case, this book will help you reach your goals. In either case, you need this book.

The contents of this book are based on the very latest reports and documents provided and sponsored by the U.S. Department of Defense (DoD) as of this writing. On occasion, DoD may make some small changes that are not reflected here.

Part 1 of this book will help you learn about the various branches of the military so that you can determine which opportunities are best for you. This information comes directly from the Services. It will show you how and why your ASVAB results matter to the military and matter to you. You will also learn about the differences between the paper-and-pencil ASVAB and the computer versions.

This section will also provide you with successful study skills and "blue-ribbon" test-taking strategies that you can use to reach your goals. These strategies are some of the most powerful and effective ones known today. By using the information in this section, you will learn how to plan your studying activities to fit your schedule.

Part 2 contains a full-length simulated ASVAB form that is just like the real thing! By taking this Diagnostic Test Form, you will find out exactly what to expect on the ASVAB. You will also learn which kinds of test questions you can answer with confidence, and which kinds give you difficulty. By pinpointing your ASVAB strengths and weaknesses, you will be able to focus your preparation on the test topics that you need to study most. This section will also provide you with the exact directions that you will hear from your test administrator and see in your test booklet if you are taking the paper-and-pencil version of the test. No other test preparation book has this information. At the end of the test, you will be able to see the answers to the items and explanations that will help you understand why the correct answer is correct and why the other answers are wrong. The information in these explanations can help you prepare for the real ASVAB.

Part 3 of this book will give you the most comprehensive review of the ASVAB test topics you are likely to encounter. No other book on the market and no existing website will provide you with as complete a review as this book does. These topic reviews are keyed to the ASVAB and can definitely help you improve your ASVAB scores, but they can also help you score high on other tests such as the SAT and ACT. You'll also find topic-based quizzes that can help you know your strengths and weaknesses.

Part 4 contains three additional full-length model ASVAB forms. Each one gives you the opportunity to practice taking a test that is just like the real ASVAB. Use these Practice Test Forms to strengthen your test-taking skills. Take each one under actual test conditions. Use a timer to help you keep track of the time limit for each section. Keep a record of your results and note the improvements you see from one test to the next in your scores and in the number of questions you are able to answer in the time allowed,

Practice is the key to higher ASVAB scores! In addition, read the answer explanations that follow each test form. These explanations offer you even more information about the tested area beyond the review sections. Between the subject area reviews and these explanations, you will certainly be in a position to score at the upper end of the ASVAB.

Part 5 gives you information on military occupations, including a description of the occupation, what people in the occupation do, helpful attributes, and services that offer that occupation.

HOW TO USE THIS BOOK

If you are not sure which branch of the military interests you:

- Read Chapter 1: "Deciding on Military Service."
- Go to Part 5 and learn about various military occupations.

If you want to know about the ASVAB and why it is important to the military:

- Read Chapter 2: "Introducing to the ASVAB."
- Read Chapter 3: "ASVAB Scores and Score Reports."
- Read Chapter 4: "The ASVAB, the AFQT, and Military Entrance."
- Read Chapter 5: "Taking the CAT-ASVAB."

If you want general study tips and guidelines for planning your study:

- Review Chapter 6: "Follow These Suggestions to Ace the ASVAB."

If you want to find out how well you might do right now on the real ASVAB:

- Take the ASVAB Diagnostic Test Form in Part 2 and use it to identify your strengths and weaknesses.

If you want to refresh your knowledge of the ASVAB test areas:

- Go to Part 3 of this book, which provides a complete review of every topic tested.

If you want to practice taking the ASVAB test:

- Go to Part 4 of this book and take ASVAB Practice Test Forms 1 through 3 under actual test conditions.

If you want to learn about some specific jobs in the military:

- Go to Part 5 of this book.

If you are active duty and want to raise your score to change your military occupational specialty:

- Go to Part 3 of this book to review the kinds of topics you may be tested on.
- Go to Part 2 to take a diagnostic test and Part 4 to take more practice items.

This book contains information that was current and accurate at the time of writing. On occasion the branches of the military change policies and requirements, so checking with a military recruiter will give you the most current data.

This book does not guarantee that it contains the actual ASVAB test questions. There are several forms of the paper-and-pencil version of the ASVAB and many hundreds of items on the computerized version of the test. This book, if you use it as we recommend, will certainly give you a major advantage in scoring high on the ASVAB. This will give you increased job and training opportunities that will help you in your military career and in civilian jobs.

Good luck in your efforts!

> *You don't always get what you wish for; you get what you worked for.*
> —DANIEL MILSTEIN

Part 1

ALL
ABOUT
THE
ASVAB

Chapter 1

Deciding on Military Service

YOUR GOALS FOR THIS CHAPTER:

- **Find out about opportunities in the armed services.**

- **Learn about enlistment, training, advancement, and education programs in each service branch.**

- **Get information about military pay and benefits.**

Your choice to join the military is a personal one. You may wish to join because it is an honorable family tradition. You may wish to join because the military offers you the best training and education opportunities available to you or anyone. You may wish to join because the military is the largest employer of young people just starting their careers. You may wish to join because there is a strong family history and tradition of serving your country in uniform. You may wish to join because you have a deep desire to give back to your country what it has given you, your family, and your friends.

Regardless of your motivation, the decision to join the military is a serious and important one that will influence your life in many ways.

One of your first decisions will be to identify the branch of the armed services you wish to join. Your decision will determine what you do, where you live, what training you will receive, and your future opportunities.

The following pages give you an overview of the five, including the Army, Navy, Marine Corps, Air Force, and Space Force. The Coast Guard is not part of the military but is under the direction of the Department of Homeland Security.

Since policies and programs can change, your best information comes from your recruiter. Always ask them the questions that are on your mind.

Here are some questions you can ask.

QUESTION TO ASK A RECRUITER

Before you meet with a recruiter, have some specific questions in mind.

Here are some questions to think about.

General Questions

- How is your military branch different from the others?
- What is the recruiting process like from beginning to end?
- Why should I join the (service)?
- What's the Delayed Entry Program?

Basic Training

- What really happens in basic training?
- What's the balance of classroom and physical training?
- What kind of condition do you have to be in at the start?
- What are the physical standards candidates must meet?
- What are training and drill instructors like today?
- What percentage of people who start basic training complete it?
- Can two friends go through basic training at the same time?

The First Enlistment Term

- How long does the first term last? Do you have programs of different lengths?
- Can an entrant choose the military job they want? How is the job assignment made?
- Are there any jobs with enlistment bonuses?
- Can a trainee choose to serve overseas?
- How much does a new recruit earn, and what are the benefits?
- How often are service members promoted?

Education

- What kind of training comes after Basic Training?
- How good are your military job-training schools?
- How does the training relate to civilian jobs?
- What are all the ways a service member can earn college credits during enlistment?
- What are your tuition-support programs? How does a recruit qualify for them?

OPPORTUNITIES IN THE ARMY

The U.S. Army is the oldest of the Services having been established in 1796. It is mostly a land-based Service, but it does have responsibilities in air and water. Its purpose is "to fight and win our Nation's wars, by providing prompt, sustained land dominance, across the full range of military operations and the spectrum of conflict, in support combatant commanders." Personnel work all across the globe and are the major ground-based offensive and defensive force of the United States.

Specifically, its duties are as follow:

- Preserving the peace and security and providing for the defense of the United States, the commonwealths, and possessions and any areas occupied by the United States
- Supporting the national policies
- Implementing the national objectives
- Overcoming any nations responsible for aggressive acts that imperil the peace and security of the United States

The Army is divided into the Regular Army, which is the full-time active-duty component; the Army Reserve; and the Army National Guard. Except for Washington, D.C., the National Guard is under the authority of each state governor. The DC National Guard is under the authority of the president of the United States, although there is a movement to change that.

The Army is divided into several functional areas. Some examples include Infantry, Corps of Engineers, Field Artillery, Air Defense, Aviation, Cyber Force, Special Forces, Armor, Signal Corps, Judge Advocate General, Military Police, Intelligence, Finance and Comptroller, Psychological Operations, Medical, and others. Those who enlist in the Army will find hundreds of challenging career opportunities that can offer a lifetime of security and excitement to them and their families.

Enlistment

You can enlist in the Army for three, four, five, or six years. You must be between 17 and 35 years old, an American citizen or registered alien, and in good health and physical condition. To determine what careers are best suited for, you must take the Armed Services Vocational Aptitude Battery (ASVAB). The ASVAB is offered at most high schools and at military enlistment processing sites.

In most cases, qualified applicants can be guaranteed their choice of training or duty assignment. There are often combinations of guarantees that are particularly attractive to those who are qualified. For those who wish to be guaranteed a specific school, a particular area of assignment, or both, the Army offers the Future Soldier Training Program (FSTP). An applicant for the FSTP can reserve a school or an assignment choice as much as one year in advance of entry into active duty. Other enlistment programs include the Army Civilian Acquired Skills Program, which gives recognition to those skills acquired through civilian training or experience. This program allows enlisted members with previously acquired training to be promoted more quickly than they ordinarily would be. In some cases, the Army also offers enlistment bonuses.

Enlistment programs and options vary from time to time. Local Army recruiters always have the latest information and are ready to answer inquiries without obligation.

Training

Initial Army training is provided in two phases: basic training and advanced individual training (job training).

Basic Training Basic training is a rigorous orientation for men and women entering the Army. Basic training transforms new enlistees from civilians into soldiers. During basic training, new soldiers gain the discipline, spirit, pride, knowledge, and physical conditioning necessary to perform Army duties. Army basic training is given at Fort Benning, Georgia; Fort Jackson, South Carolina; Fort Leonard Wood, Missouri; Fort Sill, Oklahoma; and Fort Knox, Kentucky.

Upon reporting for basic training, new soldiers are assigned to a training company and are issued uniforms and equipment. They are introduced to their training leaders, otherwise known as drill sergeants. Drill sergeants are experienced noncommissioned officers who direct soldiers' training to ensure that they are successful.

Army basic training stresses teamwork. Soldiers are trained in groups known as squads or platoons. These groups range from 9 to approximately 80 soldiers; they are small enough that individual soldiers can be recognized for their special abilities. Such groups tend to become closely knit teams and develop group pride and camaraderie during the 10 weeks of rigorous training they experience together.

Basic training is conducted on a demanding schedule, but individual soldiers progress at the rate they can handle best. Soldiers attend a variety of classes that include first aid, CPR, land navigation, physical fitness, self-defense, and military drills. All training emphasizes teamwork and Army values and, therefore, includes classes in human relations. These classes help trainees from different backgrounds learn to work closely together. Only limited personal time is available during basic training, but there is plenty of time for receiving and answering mail, for personal care, and for attending religious services.

Advanced Individual Training After basic training, Army soldiers go directly to advanced individual training in the occupational field that they have chosen and qualified for, where they learn a specific Army job. Advanced individual training schools are located at many Army bases throughout the country.

The Army offers skills training in a wide range of career fields, including programs maintenance, administration, electronics, healthcare, construction, and combat specialty occupations, to name a few.

Advanced individual training students generally attend traditional classes very similar to those in a high school or college. These classes are supplemented with demonstrations by highly qualified instructors and by practical exercises that use hands-on training, Army equipment, or Army procedures in a way that prepares students for their jobs. Many soldiers also receive on-the-job training, learning job skills by working at a job with other soldiers under the guidance of qualified instructors.

Advancement

Every job in the Army has a career path leading to increased pay and responsibility with well-defined promotion criteria. After six months of service, new soldiers advance to Private (E-2). The next step in the promotion ladder is Private First Class (E-3), which occurs after the twelfth month. Promotion to Corporal or Specialist (E-4) occurs after established time-in-grade and time-in-service requirements are met. These times vary, but soldiers can ordinarily expect to reach the level of corporal within their first three years of service.

Starting with grade E-5, promotions to Sergeant through Sergeant Major are made on a competitive basis. At each grade, there are minimum periods of time in service and time in grade that must be met before a soldier can be considered for promotion. In some cases, there also are educational requirements that must be met for promotion.

The Army offers a number of ways to advance beyond enlisted status as either a warrant officer or a commissioned officer. These programs usually are reserved for the best-qualified soldiers. Warrant officers perform duties similar to those of commissioned Army officers. Many warrant officers are directly appointed from the enlisted grades as vacancies occur. These opportunities usually exist in the technical fields, especially those involving maintenance of equipment. Other opportunities are available in Army administration, intelligence, and law enforcement.

Unique among the armed forces is the Army's Warrant Officer Aviator Program. Qualified personnel may enlist for Warrant Officer Candidate School and, upon completion, receive flight training and appointment as Army Warrant Officer Aviators. Enlisted soldiers may also compete for a limited number of selections to attend Officer Candidate School (OCS) or the U.S. Military Academy. Upon graduating from OCS or the academy, soldiers receive officer commissions.

Education Programs

For enlisted personnel, the Army has a well-defined system for progressive service school training. Soldiers are often able to volunteer for this schooling; in some cases, they are selected on a competitive basis.

As a soldier progresses in his or her career, advanced technical training opportunities are offered. These courses include, but are not limited to, advanced noncommissioned officer courses at the E-6 grade level, senior leadership at the E-7 grade, and the Sergeants Major Academy at the E-8 and E-9 levels.

Civilian education is stressed as a means to improve both the soldier's work performance and their preparedness for life in a technical and competitive society.

The Army Continuing Education System (ACES) provides assistance with tuition, counseling, academic services, and vocational-technical services at no cost. The ACES mission is to promote lifelong learning, readiness, and resilience through flexible and relevant education programs, services, and systems in support of the Total Army Family.

ACES provides Soldiers, Family Members, and Civilians services and information on the following:

Education and Career Counseling, Testing, Assessment, and Evaluation Services, Leadership Development, Tuition Assistance, Career Management, Transition Assistance, Basic Skills Education Program, Credentialing Assistance

OPPORTUNITIES IN THE NAVY

The Navy plays an important role in helping maintain the freedom of the seas. It defends the right of our country and its allies to travel and trade freely on the world's oceans and helps protect our country and national interests overseas during times of international conflict through power projection ashore. Navy sea and air power make it possible for our country to use the oceans when and where our national interests require it.

The U.S. Navy works closely with the U.S. Marine Corps and the Coast Guard. Although the Marine Corps functions as a separate branch of the military, it is administratively part of the U.S. Navy. The Coast Guard is administratively under the Department of Homeland Security, but it is under the administration of the U.S. Navy in times of war. All these branches are sea-based, so they must cooperate together in the protection of the United States.

Sailors work in various qualification areas, such as Surface Warfare, Aviation Warfare, Information Dominance Warfare, Naval Aircrew, Special Warfare, Seabee Warfare, Submarine Warfare, or Expeditionary Warfare. Enlisted members are said to be "rated," meaning that they have been trained and have earned "rating" similar to what is called a military occupational specialty.

These sailors operate globally and are responsible for driving, operating, and maintaining a variety of ships and boats. Some examples include aircraft carriers, amphibious warfare vessels, cruisers, destroyers, frigates, and littoral combat ships, mine countermeasures ships, patrol boats, and submarines. Some Navy personnel are trained to fly aircraft.

Enlistment

To qualify for enlistment in Navy programs, men and women must be between the ages of 17 and 39. Because most Navy programs require enlistees to be high school graduates, future sailors are allowed to choose a career and select a shipping date while completing their senior year in high school.

Initial enlistment in the Navy usually is for four years. However, five- and six-year enlistments are also available for men and women, depending on the programs they select.

After Navy people go through the enlistment process at a Military Entrance Processing Station, they are placed in the Delayed Entry Program (DEP) for a minimum of 90 days. While in DEP, recruits work on their Personal Qualification Standards (PQS) while they await their shipping date. DEP allows enlistees to finish high school or college, take care of personal business, or just relax before reporting for duty.

There is even an opportunity to earn advancement while in DEP, which translates into a higher pay grade when entering basic training. There is extra pay in the Navy for sea duty, submarine duty, demolition duty, diving duty, work as a crew member of an aviation team, and other jobs that require special training. Signing bonuses are available for those who enter the nuclear field or other highly technical fields.

Training

The Navy is known for the excellent training it provides. The Navy provides both recruit training and job training. The first assignment for every Navy enlistee is recruit training. It is a challenging period of transition from civilian to Navy life. It provides the discipline, knowledge, and physical conditioning necessary to continue serving in the Navy.

The Navy's recruit training command, or "boot camp," is located in Great Lakes, Illinois.

Physical fitness training includes push-ups, sit-ups, distance running, water survival, and swimming instruction. Recruits are tested for physical fitness at the beginning and end of recruit training.

Recruits are given classroom training covering more than 30 subjects, including aircraft and ship familiarization, career incentives, decision making, time management, military drill, Navy mission and organization, military customs and courtesies, and the chain of command. Recruits are also given hands-on training including basic deck seamanship, firefighting, firearms familiarization, and damage control.

After graduation from boot camp, many new sailors go directly to the technical school ("A" school) they signed up for at the Military Entrance Processing Station. The Navy has more than 72 from which enlistees may choose. Navy "A" schools are located on military bases throughout the United States, including Great Lakes, Illinois; San Diego, California; Newport, Rhode Island; and Pensacola, Florida.

Training programs range in length from a few weeks to many months, depending on the complexity of the subject. Those who complete recruit training and are still undecided about what career path they want to take in the Navy can begin an on-the-job apprenticeship training program.

Advancement

Like other branches of the armed services, the Navy has nine enlisted pay grades, from E-1 to E-9. A new enlistee entering the Navy is an E-1 (Seaman deciding on military service recruit). After nine months in the Navy, the E-1 normally is eligible for advancement to E-2 (Seaman Apprentice).

Navy promotions are based on (1) job performance, (2) competitive examination grades, (3) recommendations of supervisors, (4) length of service, and (5) time in present level of work. It is impossible to predict exactly when promotions will occur; however, every job in the Navy has a defined career path leading to supervisory positions.

People with highly developed skills in certain critical occupations may enter the Navy at advanced

pay grades. Some people qualify for one of the specialized technical training programs in the electronics, computer, or nuclear field, where advancement is often rapid.

Enlisted petty officer ratings (E-4 through E-9) are not to be confused with Navy commissioned officer rankings. More and more Navy enlisted personnel enter with some college courses or degrees. Most Navy commissioned officers have college degrees.

Education Programs

The Navy believes that the more education people receive, the better equipped they are to perform their jobs and fulfill personal goals. Navy College Offices are available worldwide and assist in providing opportunities for enlisted members to continue college or technical classes throughout their Navy careers. Through the Navy College, enlisted members can pursue all levels of education and training, from high school equivalency to vocational certificate to college degree(s), regardless of their location and duty station. Navy College offers on-duty and off-duty study to provide a complete package of educational benefits to Navy people. Enlisted members can enroll in any combination of Navy College programs and keep adding credits toward a civilian college degree or vocational certificate of their choice.

OPPORTUNITIES IN THE AIR FORCE

The mission of the Air Force is to defend the United States through control and exploitation of air and space. The Air Force flies and maintains aircraft, such as long-range bombers, supersonic fighters, Airborne Warning and Control System (AWACS) aircraft, and many others, whenever and wherever necessary, to protect the interests of America and American allies. Some pilot aircraft—everything from helicopters to the space shuttle. Many others do the jobs that support the Air Force's flying mission; they may work as firefighters, aircraft mechanics, security police, or air traffic controllers, or in many other Air Force career fields.

Enlistment

Applicants for enlistment in the Air Force must be in good health, possess good moral character, and make the minimum scores on the Armed Services Vocational Aptitude Battery (ASVAB) required for Air Force enlistment. They must also be at least 18 years of age.

Prior to taking the oath of enlistment, qualified applicants may be guaranteed either to receive training in a specific skill or to be assigned within a selected aptitude area. The Guaranteed Training Enlistment Program guarantees training and initial assignment in a specific job skill. The Aptitude Area Program guarantees classification into one of four aptitude areas (mechanical, administrative, general, or electronic); specific skills within these aptitude areas are selected during basic training.

Jobs in the Air Force are listed according to the Air Force Specialty Code (AFSC). For enlisted members, these jobs include a variety of fields, including computer specialties, mechanic specialties, aircrew, communication systems, avionics technicians, civil engineering, hospitality, mail operations, security forces, search and rescue specialties, and others. Air Force jobs for enlisted members are thought of as entry level, so the Air Force provides the technical training.

Entry-level airmen are considered the ranks or pay grades of E-1 through E-4. Those at the ranks of E-5 through E-9 and considered noncommissioned officers (NCOs) with E-7 through E-9 are called senior non-commissioned officers. Senior airmen who have completed Airman Leadership school are authorized to be supervisors.

As of this writing, some personnel are being transferred to the new Space Force in support of its mission.

Training

The Air Force provides two kinds of training to all enlistees: basic training and job training. Selected candidates can also pursue a management training program, explained later in this section.

Basic Training All Air Force basic military training (BMT) is conducted at Joint Base San Antonio–Lackland, Texas. BMT teaches enlistees how to adjust to military life, both physically and mentally, and promotes pride in being a member of the Air Force. It lasts approximately eight and a half weeks and consists of academic instruction, confidence courses, physical conditioning, and marksmanship training. Trainees who enlist with an aptitude-area guarantee receive orientation and individual counseling to help them choose a job specialty that is compatible with Air Force needs and with their aptitudes, education, civilian experience, and desires. After graduation from BMT, recruits receive job training in their assigned specialty.

Job Training All BMT graduates can go directly to one of the Technical Training Centers Department of Defense for formal, in-residence training. In-residence job training is conducted at Keesler AFB, Biloxi, Mississippi; Joint Base San Antonio–Lackland, San Antonio, Texas; Sheppard AFB, Wichita Falls, Texas; Goodfellow AFB, San Angelo, Texas; and several other locations nationwide. In formal classes and practice sessions, airmen learn the basic skills needed for the first assignment in their specialty.

Air Force training does not end with graduation from basic training and technical training school. After three months at their first permanent duty station, airmen begin on-the-job training (OJT). OJT is a two-part program consisting of self-study and supervised job performance. Airmen enroll in skill-related correspondence courses to gain broad knowledge of their Air Force job, and they study technical orders and directives to learn the specific tasks they must perform. They also work daily with their trainers and supervisors, who coach them during hands-on task performance. Through OJT, they develop the job skills needed to progress from apprentice airmen to skilled noncommissioned officers (NCOs). Airmen also complete advanced training and supplemental formal courses throughout their careers to increase their skills in using specific equipment or techniques.

Advancement

Typically, Airman Basic (pay grade E-1) is the initial enlisted grade. However, there are several programs available that may qualify individuals for enlistment at a higher initial grade.

Every job in the Air Force has a defined career path leading to supervisory positions. Airman Basic enlistees are normally promoted to Airman (E-2) upon completion of six months of service and to Airman First Class (E-3) after 16 months of service. Promotion to Senior Airman (E-4) usually occurs at the three-year point. However, some airmen qualify for accelerated promotion. Local Air Force recruiters have all the details on qualifications for accelerated promotions and advanced enlistment grades.

Promotions to the higher enlisted grades of Staff Sergeant (E-5), Technical Sergeant (E-6), Master Sergeant (E-7), and Senior and Chief Master Sergeant (E-8 and E-9) are competitive. Eligible airmen compete with others worldwide in the same grade and skill, based on test scores, performance ratings, decorations, and time in service and grade.

Education Programs

The Air Force has many education programs to help men and women pursue their educational goals while serving in the Air Force that can now award bachelor's degrees. These programs are in addition to veterans' educational benefits set up by the federal government for members of all services. All Air Force bases have education service centers, where trained counselors help airmen decide on a program or combination of programs and help them enroll.

OPPORTUNITIES IN THE MARINE CORPS

The U.S. Marine Corps was created November 10, 1775, by a resolution of the Continental Congress. Since then, the Marine corps has grown to be one of the most elite fighting forces in the world. Marines operate on various installations on land and on ships. These are found all around the world. To perform the many duties of the Marine Corps, officers and enlisted Marines fly planes and helicopters; operate radar equipment; drive armored vehicles; gather intelligence; survey and map territory; maintain and repair radios, computers, jeeps, trucks, tanks, and aircraft; and perform duties in 42 occupational fields with more than 220 different Military Occupational Specialties (MOS) open to entry-level Marines. The Marine Corps training programs offer practical, challenging, and progressive skill development. The Marine Corps stresses professional education for all ranks and emphasizes the development of mental strength as well as traditional physical prowess. In this way, the Marine Corps provides the nation with a modern, well-armed force that is both tough and smart.

Marines operate on various installations on land and on ships. These are found all around the world.

They also are called upon to support a variety of non-naval tasks including the following:

- Providing music for state functions at the White House (Marine Corps Band, The President's Own)
- Guarding Presidential retreats such as Camp David
- Providing transport for the President and Vice-President, as well as for Cabinet members and other VIPs
- Securing U.S. embassies and consulates around the world

Enlistment

Marine Corps active-duty enlistment terms are for four, five, or six years, depending on the type of enlistment program. Men and women enlisting in the Marine Corps must meet exacting physical, mental, and moral standards. Applicants must be between ages 17 and 28, American citizens or registered aliens, and in good health to ensure they can meet the rigorous physical training demands.

Applicants for enlistment can be guaranteed training and duty assignment with a wide variety of options, depending on their level of education and the qualifications they possess. All occupational fields are open to men and women. In addition to regular enlistment, the Marine Corps offers special enlistment programs.

Delayed Entry Program Enlistment in the Marine Corps Delayed Entry Program (DEP) allows applicants to postpose their initial active duty training for up to a full year. Enlisting in the DEP has two principal benefits. First, highly desirable enlistment programs available in limited numbers, such as computer and aviation specialties, can be reserved while the enlistee completes high school, a college semester, or other obligation. Second, DEP members meet regularly with recruiters and peers to mentally and physically prepare them for recruit training.

Training

Marine Corps training occurs in three sections: recruit training, combat training, and job training.

Recruit Training Upon completing the enlistment process, all applicants enter Marine Corps recruit training. They attend either at Parris Island, South Carolina, or San Diego, California. Recruit training is rigorous, demanding, and challenging. The overall goal of recruit training is to instill in recruits the military skills, knowledge, discipline, teamwork, pride, and self-confidence necessary to perform as U.S. Marines.

In the first several days at the recruit depot, a recruit is assigned to a platoon, receives a basic issue of uniforms and equipment, and is given a physical. Each platoon is led by a team of three or more drill instructors. A typical training day for recruits begins with reveille at 0500 (5 a.m.), continues with close order drill, physical training, and classes, then ends with taps at 2100 (9 p.m.).

Combat and Job Training Upon graduation from recruit training, each Marine takes a short vacation, then reports to combat training. Marines in combat arms job fields attend the School of Infantry, while Marines in other career fields attend the shorter Marine Combat Training. Upon graduation, Marines report to Military Occupational Specialty (MOS) training at their formal schools or to their unit for on-the-job training. The Marine Corps sends students to more than 200 basic formal schools and more than 300 advanced formal schools. The length of formal school training varies from a few weeks to more than a year, depending on the level of technical expertise and knowledge required to become proficient in a particular job skill. Emphasis is placed on practical application of newly acquired skills, and many times as soon as classroom instruction is complete, students are placed in an actual work environment to obtain experience and develop confidence.

After completion of entry-level MOS training, most Marines receive orders to operational units of the Fleet Marine Force. Marines assigned to more technical MOSs may require more advanced training and attend follow-on schools before their first assignments. After attaining some fleet experience, additional training in advanced or specialized subjects is available to Marines in many MOSs. Job performance requirements in many MOSs are comparable to requirements needed to earn a Certificate of Apprenticeship through the U.S. Department of Labor, and some Marines earn their certificates before completing their Marine Corps service.

Advancement

Advancement is directly linked to an individual's development as a Marine, of which performance in the Marine's MOS is one factor among many.

Each Marine is evaluated based on a "whole-Marine" concept. Physical fitness, rifle qualification, military and civilian education, work performance, time in grade and service, and other factors all contribute an individual's promotion prospects. Marines are in competition with peers of the same rank and MOS, ensuring the best and brightest rise to the top. Promotion becomes increasingly competitive as Marines advance in rank. In addition to minimum time-in-grade requirements, educational and other prerequisites factor into the promotion system. For example, to be selected for promotion to sergeant, a corporal must first have attended a formal Corporals' Course; for promotion to staff sergeant, a sergeant must have attended a Sergeants' Course; and so on. Unit leaders ensure their Marines have opportunities to attend appropriate formal Primary Military Education courses. Promotions to the Staff Noncommissioned Officer (SNCO) ranks are determined by promotion boards.

The Meritorious Promotion System is used to recognize Marines who demonstrate outstanding performance and professional development. Marines recommended for meritorious promotion are carefully screened for accelerated advancement before their peers. Qualified enlisted Marines can also compete for and be accepted into the officer corps through several different programs. Competition is keen, and only the best-qualified Marines are accepted.

Education Programs

All Marines on active duty are encouraged to continue their education by taking advantage of service schools and Marine Corps–funded off-duty courses at civilian colleges. The Marine Corps has developed an extensive professional military education program to provide Marine leaders with the skill, knowledge, understanding, and confidence that will better enable them to make sound military decisions. Several educational assistance programs are available to enlisted Marines.

OPPORTUNITIES IN THE COAST GUARD

The Coast Guard is one of the uniformed services, but it is part of the Department of Homeland Security, not the Department of Defense. It is a unique branch of the military responsible for an array of maritime duties, from ensuring safe and lawful commerce to performing rescue missions in severe conditions. If war is declared, the Coast Guard reports to the U.S. Department of Defense.

The role of the Coast Guard is to defend America's borders and protect the maritime environment. It is the smallest of all the services, with the possible exception of the Space Force, which is still evolving. Interestingly, it is considered to be the 12th largest naval force in the world.

The Coast Guard is a military, multimission, maritime force offering a unique blend of military, law enforcement, humanitarian, regulatory, and diplomatic capabilities. These capabilities underpin its three broad roles: maritime safety, maritime security, and maritime stewardship. There are 11 missions that are interwoven within these roles.

Missions

The Coast Guard is assigned 11 official missions.

Port and Waterway Security Along with search and rescue, port and waterway security is the Coast Guard's primary homeland security mission. Coast Guard members protect marine resources and maritime commerce, as well as those who live, work, or recreate on the water.

Port and waterway security also involves prevention of terrorist attacks and response when terrorist acts do occur. Counterterrorism preparedness and response operations all fall within the scope of port and waterway security.

Drug Interdiction The Coast Guard is the nation's first line of defense against drug smugglers seeking to bring illegal substances into the United States. The Coast Guard coordinates closely with other federal agencies and countries within a vast six-million-square-mile region to disrupt and deter the flow of illegal drugs. Coast Guard drug interdiction accounts for more than half of all U.S. government seizures of cocaine each year.

Aids to Navigation One important mission entrusted to the Coast Guard is the care and maintenance of maritime aids to navigation. Much like drivers need stoplights, street signs, and universally accepted driving rules, boaters also need equivalent nautical "rules of the road."

The Coast Guard is responsible for ensuring this network of signs, symbols, buoys, markers, lighthouses, and regulations is up to date and functioning properly so recreational and commercial boaters can safely navigate the maritime environment.

Search and Rescue Search and rescue (SAR) is one of the Coast Guard's oldest missions. Warding off the loss of life, personal injury, and property damage by helping boaters in distress has always been a top Coast Guard priority. Coast Guard SAR response involves multimission stations, cutters, aircraft, and boats linked by communications networks.

Living Marine Resources The nation's waterways and marine ecosystems are vital to the country's economy and health. Ensuring America enjoys a rich, diverse, and sustainable ocean environment is an important Coast Guard mission. This includes ensuring the country's protected marine species are provided the protection necessary to help their populations recover to healthy, sustainable levels.

Marine Safety While search and rescue is one of the Coast Guard's most well-known missions, crews do much more than save mariners in peril.

Promoting safe boating practices is a key objective to help prevent an incident at sea. The Coast Guard investigates maritime accidents, merchant vessels, offshore drilling units, and marine facilities. Additionally, the Coast Guard is responsible for licensing mariners, documenting U.S. flagged vessels, and implementing a variety of safety programs.

Despite best efforts, mariners sometimes find themselves in harm's way. When they do, the Coast Guard has a proud tradition of immediate response to save lives and property in peril. To be part of its search and rescue team takes more than physical ability; it also requires that special desire and bravery with which heroes are born.

Defense Readiness In our post-9/11 society, national security interests can no longer be defined solely in terms of direct military threats to America and its allies. The Coast Guard's role in national defense and antiterrorism is a cornerstone of homeland security efforts to protect the country from the ever-present threat of terrorism.

The Coast Guard has four major national defense missions: maritime intercept operations, deployed port operations/security and defense, peacetime engagement, and environmental defense operations. These missions are essential military tasks assigned to the Coast Guard as a component of joint and combined forces in peacetime, crisis, and war.

Migrant Interdiction Thousands of people try to enter this country illegally every year by sea, many via highly dangerous and illegal smuggling operations. Intercepting these offenders at sea means they can be safely returned to their country of origin without the costly processes required if they had successfully entered the United States.

As the United States' primary maritime law enforcement agency, the Coast Guard enforces immigration laws at sea. The Coast Guard conducts patrols and coordinates with federal agencies and foreign countries to detain undocumented migrants at sea and prohibit entry via maritime routes to the United States and its territories.

Marine Environmental Protection Protecting the delicate ecosystem of our oceans is a vital Coast Guard mission. The Coast Guard works with a variety of groups and organizations to ensure the livelihood of endangered marine species.

Through the Marine Environmental Protection program, the Coast Guard develops and enforces regulations to avert the introduction of invasive species into the maritime environment, stop unauthorized ocean dumping, and prevent oil and chemical spills.

Ice Operations Frigid, subzero temperatures heighten the dangers for any operation. Adding hazardous icy waters and icebergs makes for treacherous conditions for maritime commerce. To facilitate safe maritime commerce in icy waters and to protect communities in emergency situations, the Coast Guard conducts ice-breaking operations in the Great Lakes and Northeast regions.

Law Enforcement Preventing illegal foreign fishing vessels from encroaching on the Exclusive Economic Zone is a priority for the Coast Guard. Protecting the integrity of the nation's maritime borders and ensuring the health of U.S. fisheries is a vital part of the Coast Guard mission.

The Coast Guard also enforces international agreements to suppress illegal, unreported, and unregulated fishing activity in international waters.

Training

As with any military service, the journey begins at basic training.

Basic training is tough. Recruits are challenged every day, both mentally and physically. They are pushed, tested, and worked harder than they ever thought possible, but they graduate confident and proud.

Recruits arrive at Sexton Hall in Cape May, New Jersey, and the first days of basic training are spent getting oriented, receiving uniforms, getting haircuts (women may pin their hair up within regulation standards), and filling out entry forms.

Here, they meet their company commander (CC)— their mentor, instructor, leader, coach, and guide through basic training. The company commander's job is to motivate, teach self-discipline, and teach how to obey orders. From the CC, recruits learn commitment and service as a productive Coast Guard member.

Academic Training In addition to being subject to physical demands, recruits are challenged in the classroom. The intense academic program at basic training is designed to provide them with entry-level skills and knowledge needed to succeed in the junior member field. In class, they receive training in everything from military justice, customs, and ethics to Coast Guard history.

Moreover, many classes are hands-on. During training, they participate in a variety of practical instruction, including small arms training, seamanship, firefighting, and damage control. They fire at the basic pistol course, learn how to handle lines, practice helm commands, and participate in safety and rescue training. Additionally, all throughout basic training, they are coached through a myriad of team-building training exercises.

Around the fourth week of training, recruits are asked to request a geographic location for their first assignment and the type of unit they wish to serve, whether ashore or afloat. They receive their first assignment orders about a week later.

After graduation from basic training, recruits are ready to serve.

Advancement

Upon completing eight weeks of basic training, recruits are promoted to seaman or fireman (E-2). Normally, they then proceed to their first unit and learn the skills needed to move up to seaman or fireman (E-3). While at their first unit, they are exposed to many of the different ratings available in the Coast Guard and can narrow down the career path they want to pursue. For aviation-specific rates, the USCG Aviation Technical Training Center in Elizabeth City, North Carolina, incorporated the airmen training program into the "A" school curriculum.

Recruits who already know what they want to do in the Coast Guard can ask their recruiter about the guaranteed "A" school program. With guaranteed "A" school, they can go directly from basic training to their rating's training program. However, not all ratings participate in this program. Coast Guard training provides members with the highest level of readiness because lives and mission success depend on how the Corps performs its duties.

Enlisted personnel continue to receive higher-level training in their field, which leads to more responsibility, potentially a higher ranking, and greater pay.

To become a petty officer in the Coast Guard, individuals must either graduate from "A" school or complete the corresponding "striker" program (on-the-job training). Not all ratings have striker programs.

By demonstrating practical skills on the job, recruits can advance within their rating. They also have to successfully complete leadership training, pass the appropriate end-of-course test, and compete in a national service-wide exam.

The highest rank an enlisted member can attain is Master Chief Petty Officer (E-9). If they want to keep going, as a four-year enlistee in the Coast Guard, if they obtain the rank of E-5 or higher and have a minimum of 30 college credits, they can apply for Officer Candidate School. To get in, applicants compete against civilians who have a four-year college degree and other members of the Coast Guard in their position.

It is also possible to move up the enlisted ladder to warrant officer and later compete in the Warrant-to-Lieutenant program.

Those who join as officers participate in a different training program.

OPPORTUNITIES IN THE SPACE FORCE

Space has become essential to our security and prosperity—so much so that we need a branch of our military dedicated to its defense, just like we have branches of the military dedicated to protecting and securing the air, land, and sea.

Access to space is vital to national defense. Space affects almost every part of our daily lives and is fundamental to our economic system. For example, satellites not only power the GPS technology that we use daily but also allow us to surf the web and call our friends, enable first responders to communicate with each other in times of crisis, time-stamp transactions in the world financial market, and even allow us to use credit cards at gas pumps.

As a result of this need, the U.S. Space Force has become the newest branch of the military. It was established at the end of 2019 and is administratively within the U.S. Air Force. This is much like the relationship between the Marine Corps and the Navy. The staff of the Space Force will be responsible to organize, train, and equip its members. As of now, the Air Force will provide a significant amount of support functions to the Space Force, including logistics, base operating support, civilian personnel management, business systems, information technology support, and other services.

Its mission is to organize, train, and equip space forces to protect the U.S. and allied interests in space and to provide space capabilities to all forces. The responsibilities include developing Guardians, acquiring military space systems and equipment, and supporting combatant commands. Guardians will launch rockets, keep satellites safe and operational, and develop technology to defend our way of life on Earth through our interests in space.

Since it is a new branch of the military, the Space Force is still in transition. Its military personnel are being transferred from what was once the Air Force Space Command and are assigned to the Space Force. Although they are still Air Force Airmen, eventually, as new members have been recruited into the Space Force, they are called Space Force Guardians. In addition, the Department of Defense will be consolidating space-related activities and people from the space-related activities and people from the other services to the Space Force. Some personnel from the other services, not just the Air Force, will be detailed or transferred to the Space Force. It will take time to sort all this out. Military personnel from the other services can apply for a transfer to the Space Force.

As the Space Force evolves and matures, it will be recruiting specialized members into the Space Force directly. As of this writing, there are about 16,000 military and civilian personnel assigned to the Space Force. Currently, more than 2,500 Guardians have operated almost 100 spacecraft. This also will change over time.

The Space Force locations include Buckley, Colorado; Los Angeles, California; Patrick, Florida; Peterson, Colorado; Schriever, Colorado; and Vandenberg, California.

Right now, if you are interested in the Space Force, you would look to the Air Force for recruiting purposes. Be watchful over the years as this may change.

Typical occupational areas include Space Systems Operations, Cyber Systems Operations, Knowledge Operations Management, Cyber Security, Cable and Antenna Systems, Cyber Transport Systems, Fusion Analyst, Target Analyst, Client Systems, and others.

Having strong math, analytical, and information technology skills will be helpful if you want to enter the Space Force.

MILITARY PAY AND BENEFITS

The most important components of compensation in the military are pay and allowances. There are various types of pay. *Basic pay* is received by all and is the main component of an individual's salary. Basic pay increases about 2 to 3 percent each year.

There are other compensations for specific qualifications or events. For example, there are *special pays* for aviators and parachutists; special pays are also paid for dangerous or hardship duties.

Allowances are the second most important element of military pay. Allowances are monies provided for specific needs, such as food or housing. Monetary allowances are provided when the government does not provide for that specific need.

For example, not all military members and their families live in housing provided by the government. Those who live in government housing do not receive full housing allowances. Those who do not live in government housing receive allowances to assist them in obtaining commercial housing.

The most common allowances are the Basic Allowance for Subsistence (BAS) and Basic Allowance for Housing (BAH). A majority of the force receives both of these allowances and, in many cases, these allowances are a significant portion of the member's total pay.

Active-duty service members receive free medical and dental care. Spouses and dependent children of an active-duty service member can enroll in military health care (a small enrollment fee and annual deductible may apply).

Many allowances are not taxable, which is an additional benefit to military personnel. The charts on page 17 show the most recent pay for enlisted military members.

MILITARY MYTHS AND REALITIES

There are many myths about joining the military. The table explains some of the realities.

DIVERSITY IN THE MILITARY

Over the past several years, the U.S. military has taken steps to build a more diverse and inclusive force.

The current Secretary of Defense stated that the "American military works best when it represents all the American people." This is more than just having token minorities but having "the moral courage to include other perspectives and ideas into our decision making—perspectives that are based on lived experience. It's that experience and the professionalism and commitment of our people that has always been our decisive advantage."

The services have ended some restrictions for women in combat roles. Women and racial and ethnic minorities are still underrepresented in parts of the military, especially in higher ranks and pay grades. This is not unlike the situation in the civilian world of work. This, however, is changing.

Although the policies have wavered over time, the military has been receptive to openly gay individuals. It is thought to be the largest employer of transgender individuals. These, like other policies, change periodically, depending on the current administration and Congress.

Women in the Military

As of January 2016, all military occupations and positions are open to women. Although women

Myths	Realities
The Military is a roadblock to a higher education.	• Qualified servicemembers can receive thousands of dollars in tuition benefits. • The Military operates over 300 schools, teaching more than 10,000 courses. • The Military offers retired personnel up to $100/month reimbursement for tutorial assistance. • The new GI Bill improves those benefits.
People in the Military are not compensated as well as private-sector workers.	• Military pay is comparable to, and in some cases better than, its civilian counterpart. • The services offer significant signing bonuses. • After 20 years of service, retired personnel can potentially receive military retirement pay for life.
You don't need to finish high school to join the Military.	• You must have a high school diploma or equivalent to enlist. • A GED may be accepted with special approval.
Military training and jobs have little relation to the civilian world.	• About 85% of military jobs have direct civilian counterparts. • Veterans of the U.S. Armed Forces are less likely to be unemployed than nonveterans.
Women have a hard time achieving success in the Military.	• All jobs in the Military are open to women. • Women account for around 20% of the U.S. Military.

have served in combat prior to this date, about 10 percent of military jobs were closed to females. These included infantry, armor, reconnaissance, and some special operations units.

About 20 percent of the military enlisted personnel are women. This varies by Service. Part of the challenge is that, on average, men are generally stronger than women and many jobs in the military require strength. That said, there are women who are stronger than some men. The services have been looking at the physical requirements of military jobs to determine what is necessary to successfully accomplish these tasks.

If you are a female military applicant, you need to know that you will be assigned to positions based on ability, not gender, but that does not mean that there will be equal participation by men and women

in each occupation. For jobs that require a certain level of physical strength, you will have to meet the requirement.

For more information, visit these recruiting websites:

- www.goarmy.com
- www.navy.com
- www.airforce.com
- www.marines.com
- www.uscg.mil
- www.airforce.com/spaceforce

What If the Recruiter Won't Send Me to Take the ASVAB?

A recruiter's primary job is to ensure that applicants meet all necessary qualifications. Before an applicant is allowed to test, the recruiter will conduct an interview looking for disqualifying factors such as too young or old, too many dependents, a medical problem, drug usage, or criminal history. If a recruiter has determined that you are not qualified to enlist, he or she will not send you to be tested. Sometimes, recruiters have applicants take a short pre-screening test to get an estimate of how they would perform on the AFQT portion of the ASVAB. Based on these results, the recruiter may choose not to spend time and resources to send you for full ASVAB testing. The only way to know for sure is to ask your recruiter.

Estimated Pay Effective 2021 in Dollars				
Enlisted Monthly Pay Chart for Years of Service Ranging from Less than 2 to over 6				
	Less than 2	**Over 2**	**Over 3**	**Over 4**	**Over 6**
E-9					
E-8					
E-7	3,294	3,595	3,733	3,915	4,058
E-6	2,849	3,136	3,274	3,409	3,549
E-5	2,610	2,786	2,921	3,058	3,273
E-4	2,393	2,516	2,652	2,787	2,905
E-3	2,161	2,296	2,436	2,436	2,436
E-2	2,055	2,055	2,055	2,055	2,055
E-1	1,833	1,833	1,833	1,833	1,833
E-1 Less than 4 months	1,695				

Enlisted Monthly Pay Chart for Years of Service Ranging from 8 to over 16				
	Over 8	**Over 10**	**Over 12**	**Over 14**	**Over 16**
E-9		5,789	5,920	6,086	6,280
E-8	4,739	4,949	5,078	5,234	5,402
E-7	4,302	4,440	4,685	4,889	5,027
E-6	3,864	3,988	4,226	4,298	4,352
E-5	3,498	3,682	3,704	3,704	3,704
E-4	2,905	2,905	2,905	2,905	2,905
E-3	2,436	2,436	2,436	2,436	2,436
E-2	2,055	2,055	2,055	2,055	2,055
E-1	1,833	1,833	1,833	1,833	1,833

Enlisted Monthly Pay Chart for Years of Service Ranging from 18 to over 26				
	Over 18	**Over 20**	**Over 22**	**Over 24**	**Over 26**
E-9	6,477	6,791	7,057	7,336	7,764
E-8	5,706	5,860	6,123	6,268	6,626
E-7	5,175	5,232	5,425	5,528	5,921
E-6	4,413	4,413	4,413	4,413	4,413
E-5	3,704	3,704	3,704	3,704	3,704
E-4	2,905	2,905	2,905	2,905	2,905
E-3	2,436	2,436	2,436	2,436	2,436
E-2	2,055	2,055	2,055	2,055	2,055
E-1	1,833	1,833	1,833	1,833	1,833

For the most updated information, go to https://www.dfas.mil/MilitaryMembers/payentitlements/Pay-Tables/

Chapter 2

Introducing the ASVAB

YOUR GOALS FOR THIS CHAPTER:

- Identify the two major formats of the ASVAB.

- Find out what areas the ASVAB tests.

- Get important information about where you can take the ASVAB.

- Learn about the policy for retaking the ASVAB.

WHAT IS THE ASVAB?

You learned in the preceding chapter that everyone seeking to enlist in any branch of the U.S. Armed Forces must take the Armed Services Vocational Aptitude Battery (ASVAB). The education level of military personnel is a major concern, and the military does not take just anyone who wants to join. The ASVAB is one tool that the military uses to measure the abilities of potential recruits. The ASVAB is also given to high school students to help them explore their aptitudes and interests for different careers. Results from the high school assessment can be used for military entrance.

The ASVAB is actually a group of individual aptitude tests. The tests are listed in the following charts. Each test measures something that is important for military entrance or for acceptance into training programs for certain military jobs.

Once you are accepted into the military, your ASVAB scores are used to qualify you for various military occupations. The higher your scores, the more choices you will have for training in different occupations.

Different Formats of the ASVAB

The ASVAB comes in two formats. Persons who take the ASVAB in many schools and in certain other locations in the country are typically given a paper-and-pencil test battery. This format of the test has 225 items. Testing time,

including administration time and instructions, about three hours. The test taker reads the questions in a test booklet and answers them by filling in bubbles on a machine-readable answer sheet. The sheets are taken to a scoring location, and the results are returned to the school and to recruiters. There are four separate forms of the paper-and-pencil version of the test. The chart on page 20 shows the subtests that make up the paper-and-pencil ASVAB.

In recent years, the Department of Defense has implemented a computer format of the test. Individuals who take this form of the ASVAB sit in a room with computers and answer the questions using the keyboard or mouse. One of the special characteristics of the computer format is that the test is adapted to the ability level of each individual. The feature is called computer adaptive testing (CAT), so this version of the ASVAB is called CAT-ASVAB.

The CAT-ASVAB uses fewer items than the paper-and-pencil version and can take less time to

The Paper-and-Pencil ASVAB

Subtest	Minutes	Questions	Description
General Science	11	25	Measures knowledge of physical, earth and space, and biological sciences
Arithmetic Reasoning	36	30	Measures ability to solve basic arithmetic word problems
Word Knowledge	11	35	Measures ability to select the correct meaning of words presented through synonyms
Paragraph Comprehension	13	15	Measures ability to obtain information from written material
Mathematics Knowledge	24	25	Measures knowledge of high school mathematics concepts and applications
Electronics Information	9	20	Tests knowledge of electrical current, circuits, devices, and electronic systems
Auto and Shop Information	11	25	Measures knowledge of automotive maintenance and repair and wood and metal shop practices
Mechanical Comprehension	19	25	Measures knowledge of the principles of mechanical of mechanical devices, structural support, and properties of materials
Assembling Objects	15	25	Measures ability to comprehend the sizes and shapes of different objects shown in a series of pictures

complete. Because the items are tailored to your ability level, you will not receive many easy items or many items that are way too difficult for you. Items are selected based on whether or not you got the previous answers correct. The items that are given to you are drawn from a very large pool of items, and no two people get the exact same test. The chart at the top of page 21 shows the subtests that make up the CAT-ASVAB.

It doesn't matter which format of the ASVAB you take because you will end up with the same military enlistment score.

WHO TAKES THE ASVAB?

More than a million people take the ASVAB each year, making it the most popular aptitude test in the country. The ASVAB can be taken by students in grades 10, 11, and 12 and those in postsecondary schools. It is used in about 12,000 schools across the country. Many students take the ASVAB in order to help them identify their strengths and weakness and to help them seek out and explore careers and jobs.

Scores are acceptable for use in the military enlistment process if the scores are no more than two years old. If you took the ASVAB more than two years ago, you must take the test again for purposes of enlisting in the military.

Thousands of people take the ASVAB at government locations for the purpose of enlisting in the military. If you take the ASVAB at a government location, you must be 17 years of age or older for your scores to count for enlistment purposes.

WHERE CAN YOU TAKE THE ASVAB?

There are several places where you can take the ASVAB: at your school, at a Military Entrance Processing Station (MEPS), or at a mobile examining team site (MET). Since its inception, more than 40 million people have taken the ASVAB.

Your School

About 800,000 students take the ASVAB at their school every year. If you are a student at a high school or postsecondary school, it is very likely that the ASVAB is offered at your school at least once a year. It is offered at more than 12,000 schools across the United States. There is no charge to students for taking the ASVAB.

Watch for school announcements that mention testing dates and times. Keep your eyes open for announcements on the bulletin board. Visit your career center and ask about the ASVAB testing dates scheduled for your school. Ask your school

THE CAT-ASVAB

Subtest	Minutes	Questions	Description
General Science	10	15	Measures knowledge of physical, earth and space, and biological sciences
Arithmetic Reasoning	55	15	Measures ability to solve basic arithmetic word problems
Word Knowledge	9	15	Measures ability to select the correct meaning of words presented through synonyms
Paragraph Comprehension	27	10	Measures ability to obtain information from written material
Mathematics Knowledge	23	15	Measures knowledge of high school mathematics concepts and applications
Electronics Information	10	15	Tests knowledge of electrical current, circuit, devices, and electronic systems
Auto Information	7	10	Measures knowledge of automotive maintenance and repair
Shop Information	6	10	Measures knowledge of wood and metal shop practices
Mechanical Comprehension	22	15	Measures knowledge of the principles of mechanical devices, structural support, and properties of materials
Assembling Objects	17	15	Measures the ability to interpret diagrams showing special relationships and how objects are related and connected

counselor when the ASVAB will be offered at your school or a nearby school. Your school counselor has received information about the ASVAB from a local representative and should be able to tell you where and when the ASVAB will be offered. If the ASVAB is not offered at your school, your counselor can arrange to include you in a testing session at a nearby school.

The ASVAB can be taken by students in grades 10 through 12 and also by students at the postsecondary level. Scores at the tenth-grade level cannot be used for military entrance, but taking the test then is a good idea because it can give you an idea of how you will do on an ASVAB test that counts for military enlistment. It is good practice for other tests that you will take during your lifetime as well.

Scores from ASVAB tests at grades 11 and 12 and the postsecondary level can be used for military entrance for up to two years. If you took the test in eleventh or twelfth grade and you think that you could score higher, you may wish to retake the ASVAB at a MEPS to see if you can exceed your high school scores. Following the advice in this book will help you score higher.

If you take the ASVAB at a school, you will probably take it with a group of students. The administration procedures should be professionally delivered by competent government test administrators. This is important because you need to perform your very best on this test and any other test you may take.

The MEPS

You can take the ASVAB test at a local Military Entrance Processing Station (MEPS). There are 65 MEPS located all across the United States. (See the list of MEPS in this chapter.) At the MEPS, you will take a computer version of the test. This test will seem different from the paper-and-pencil version, as it will have fewer items, but those items will be tailored by the computer to your level of ability. This ASVAB test is called CAT-ASVAB. CAT means computer adaptive test. The word *adaptive* means that the test is tailored to or adapts to your particular ability level.

Don't worry about taking the CAT-ASVAB, as it will give you the same scores as the paper-and-pencil version.

Later on in this book, you will learn more about how the CAT-ASVAB works and what you should expect.

MET Sites

There are approximately 500 mobile examining team (MET) sites across the country. They have been set up to qualify military applicants at locations that may be remote or distant from MEPS. If you live a long way away from the nearest MEPS, you can take the ASVAB at a MET site. That way, you and the recruiter can determine whether you are qualified by aptitude to enter the military without spending the time and money it would take you to travel to a MEPS.

At a MET site you will receive the paper-and-pencil version of the ASVAB or the iCAT, which is delivered via the Internet.

WHO ADMINISTERS THE ASVAB?

The answer to this question depends on where you take the test battery. If you take the ASVAB at a school or at a MET site, you will have a trained civilian test administrator from the Department of Defense or other agencies. These are individuals who are contracted by the federal government to adhere to the strict timing and directions of ASVAB administration.

If you take the ASVAB at a MEPS, you will have a military administrator who will help you get started on the CAT-ASVAB. The directions will be self-explanatory, and you will determine the pace of the test. It is likely that you will finish the CAT-ASVAB in less time than it would take you to finish the paper-and-pencil instrument.

CONTACT INFORMATION FOR THE MILITARY ENTRANCE PROCESSING STATIONS

ALABAMA
Montgomery
Maxwell Air Force Base—Gunter Annex
705 McDonald Street
Building 1512
Montgomery, AL 36114-3110
Pʜᴏɴᴇ: (334) 223-2800
Fᴀx: (334) 416-5034
www.mepcom.army.mil/Units/Eastern-Sector/
    8th-Battalion/Montgomery

ALASKA
Anchorage
1717 "C" Street
Anchorage, AK 99501
Pʜᴏɴᴇ: (907) 274-9142
Fᴀx: (907) 274-7268
www.mepcom.army.mil/Units/Western-Sector/
    5th-Battalion/Anchorage

ARIZONA
Phoenix
2800 N. Central Ave
Suite 400
Phoenix, AZ 85004-1007
Pʜᴏɴᴇ: (602) 586-2580
Fᴀx: (602) 258-8206
www.mepcom.army.mil/Units/Western-Sector/
    7th-Battalion/Phoenix

ARKANSAS
Little Rock
1520 Riverfront Drive
Little Rock, AR 72202-1724
Pʜᴏɴᴇ: (501) 263-3410
Fᴀx: (501) 663-5863
www.mepcom.army.mil/Units/Western-Sector/
    11th-Battalion/Little-Rock

CALIFORNIA
Los Angeles
1776 Grand Ave
El Segundo, CA 90245
Pʜᴏɴᴇ: (310) 955-9600
Fᴀx: (310) 640-9754
www.mepcom.army.mil/Units/Western-Sector/
    7th-Battalion/Los-Angeles

Sacramento
1651 Alhambra Blvd.
Suite 100
Sacramento, CA 95816
Pʜᴏɴᴇ: (916) 309-3000
Fᴀx: (916) 455-9012
www.mepcom.army.mil/Units/Western-Sector/
    7th-Battalion/Sacramento

San Diego

4181 Ruffin Rd
Suite B
San Diego, CA 92123

PHONE: (858) 609-2300
FAX: (858) 874-0415
www.mepcom.army.mil/Units/Western-Sector/
 7th-Battalion/San-Diego

San Jose

546 Vernon Avenue
Mountain View, CA 94043

PHONE: (650) 429-2313
FAX: (650) 603-8225
www.mepcom.army.mil/Units/Western-Sector/
 7th-Battalion/San-Jose

COLORADO
Denver

721 19th Street
Suite 275
Denver, CO 80202

PHONE: (720) 462-4222
FAX: (303) 623-2210
www.mepcom.army.mil/Units/Western-Sector/
 3rd-Battalion/Denver

FLORIDA
Jacksonville

7178 Baymeadows Way
Jacksonville, FL 32256

PHONE: (904) 632-7300
FAX: (904) 737-5339
www.mepcom.army.mil/Units/Eastern-Sector/
 10th-Battalion/Jacksonville

Miami

7789 NW 48th Street
Suite 150
Miami, FL 33166

PHONE: (305) 908-6400
FAX: (305) 629-8923
www.mepcom.army.mil/Units/Eastern-Sector/
 10th-Battalion/Miami

Tampa

3520 West Waters Avenue
Tampa, FL 33614-2716

PHONE: (813) 462-3100
FAX: (813) 932-8763
www.mepcom.army.mil/Units/Eastern-Sector/
 10th-Battalion/Tampa

GEORGIA
Atlanta

1500 Hood Avenue, Building 720
Forest Park, GA 30297-5000

PHONE: (470) 346-6900
FAX: (404) 469-5367
www.mepcom.army.mil/Units/Eastern-Sector/
 10th-Battalion/Atlanta

HAWAII
Honolulu

490 Central Avenue
Pearl Harbor, HI 96860

PHONE: (808) 664-5900
FAX: (808) 664-3235
www.mepcom.army.mil/Units/Western-Sector/
 5th-Battalion/Honolulu

IDAHO
Boise

550 West Fort Street
MSC 044
Boise, ID 83724-0101

PHONE: (208) 605-3777
FAX: (208) 334-1580
www.mepcom.army.mil/Units/Western-Sector/
 3rd-Battalion/Boise

ILLINOIS
Chicago

8700 W. Bryn Mawr Ave
Suite 200
Chicago, IL 60631

PHONE: (773) 272-9756
FAX: (773) 714-8150
www.mepcom.army.mil/Units/Eastern-Sector/
 6th-Battalion/Chicago

INDIANA
Indianapolis

5541 Herbert Lord Drive
Indianapolis, IN 46216

PHONE: (463) 203-4159
FAX: (317) 554-0541
www.mepcom.army.mil/Units/Eastern-Sector/
   6th-Battalion/Indianapolis

IOWA
Des Moines

7105 NW 70th Avenue
Building S-71
Johnston, IA 50131

PHONE: (515) 348-3272
FAX: (515) 224-4906
www.mepcom.army.mil/Units/Western-Sector/
   1st-Battalion/Des-Moines

KENTUCKY
Louisville

600 Dr. Martin Luther King, Jr. Place
Room 477
Louisville, KY 40202

PHONE: (502) 540-8350
FAX: (502) 582-6566
www.mepcom.army.mil/Units/Eastern-Sector/
   8th-Battalion/Louisville

LOUISIANA
New Orleans

Belle Chasse Naval Air Station
400 Russell Avenue
New Orleans, LA 70143-5077

PHONE: (504) 799-3180
FAX: (504) 678-8925
www.mepcom.army.mil/Units/Western-Sector/
   11th-Battalion/New-Orleans

Shreveport

2715 Alkay Drive
Shreveport, LA 71118-2509

PHONE: (318) 216-4915
FAX: (318) 671-6097
www.mepcom.army.mil/Units/Western-Sector/
   11th-Battalion/Shreveport

MAINE
Portland

510 Congress Street
3rd Floor
Portland, ME 04101-3403

PHONE: (207) 560-2300
FAX: (207) 775-2947
www.mepcom.army.mil/Units/Eastern-Sector/
   2nd-Battalion/Portland-ME

MARYLAND
Baltimore

850 Chisholm Avenue
Fort Meade, MD 20755

PHONE: (410) 874-6350
FAX: (301) 677-0440
www.mepcom.army.mil/Units/Eastern-Sector/
   12th-Battalion/Baltimore

MASSACHUSETTS
Boston

Barnes Building
495 Summer Street
4th Floor
Boston, MA 02210

PHONE: (857) 338-2100
FAX: (617) 753-3997
www.mepcom.army.mil/Units/Eastern-Sector/
   2nd-Battalion/Boston

Springfield

551 Airlift Drive
Westover ARB
Chicopee, MA 01022-1519

PHONE: (413) 377-3333
FAX: (413) 593-9485
www.mepcom.army.mil/Units/Eastern-Sector/
   2nd-Battalion/Springfield

MICHIGAN
Detroit

1172 Kirts Blvd.
Troy, MI 48084-4846

PHONE: (248) 729-4200
FAX: (248) 244-9352
www.mepcom.army.mil/Units/Eastern-Sector/
   6th-Battalion/Detroit

Lansing
120 East Jolly Road
Lansing, MI 48910

PHONE: (517) 318-9888
FAX: (517) 877-9160
www.mepcom.army.mil/Units/Eastern-Sector/
  6th-Battalion/Lansing

MINNESOTA
Minneapolis
1 Federal Drive
Suite 3201
Fort Snelling, MN 55111

PHONE: (612) 217-7878
FAX: (612) 725-1749
www.mepcom.army.mil/Units/Western-Sector/
  1st-Battalion/Minneapolis

MISSISSIPPI
Jackson
McCoy Federal Bldg.
100 W. Capitol
Jackson, MS 39269

PHONE: (601) 863-2201
FAX: (601) 532-5708
www.mepcom.army.mil/Units/Eastern-Sector/
  8th-Battalion/Jackson

MISSOURI
Kansas City
10316 NW Prairie View Road
Kansas City, MO 64153-1350

PHONE: (816) 235-3400
FAX: (816) 891-8258
www.mepcom.army.mil/Units/Western-Sector/
  11th-Battalion/Kansas-City

St. Louis
Robert A. Young Federal Building
1222 Spruce Street
St. Louis, MO 63103-2816

PHONE: (314) 410-2200
FAX: (314) 331-5699
www.mepcom.army.mil/Units/Western-Sector/
  11th-Battalion/St-Louis

MONTANA
Butte
22 West Park Street
Butte, MT 59701

PHONE: (406) 221-3650
FAX: (406) 782-7797
www.mepcom.army.mil/Units/Western-Sector/
  3rd-Battalion/Butte

NEBRASKA
Omaha
4245 S. 121 St. Plaza
Omaha, NE 68137

PHONE: (531) 213-3939
FAX: (402) 733-7660
www.mepcom.army.mil/Units/Western-Sector/
  1st-Battalion/Omaha

NEW JERSEY
Fort Dix
Building 5645 Texas Avenue
Fort Dix, NJ 08640

PHONE: (609) 316-3660
FAX: (609) 562-5207
www.mepcom.army.mil/Units/Eastern-Sector/
  2nd-Battalion/Fort-Dix

NEW MEXICO
Albuquerque
10500 Copper Ave. NE
Suite J
Albuquerque, NM 87123

PHONE: (505) 404-3500
FAX: (505) 246-8861
www.mepcom.army.mil/Units/Western-Sector/
  3rd-Battalion/Albuquerque

NEW YORK
Albany
Leo W. O'Brien Federal Building
North Pearl Street & Clinton Avenue
Albany, NY 12207

PHONE: (518) 649-9888
FAX: (518) 320-9869
http://www.mepcom.army.mil/Units/
  Eastern-Sector/4th-Battalion/Albany/

Buffalo

2024 Entrance Ave
Building 799
Niagara Falls ARS, NY 14304-5000

Pʜᴏɴᴇ: (463) 203-4382
Fᴀx: (716) 501-9027
www.mepcom.army.mil/Units/Eastern-Sector/
    4th-Battalion/Buffalo

New York

Fort Hamilton Military Community
116 White Avenue
Brooklyn, NY 11252-4705

Pʜᴏɴᴇ: (929) 417-2777
Fᴀx: (718) 765-7338
www.mepcom.army.mil/Units/Eastern-Sector/
    2nd-Battalion/New-York

Syracuse

6001 East Malloy Road
Building 710
Syracuse, NY 13211-2100

Pʜᴏɴᴇ: (315) 468-7800
Fᴀx: (315) 455-7807
www.mepcom.army.mil/Units/Eastern-Sector/
    4th-Battalion/Syracuse

NORTH CAROLINA

Charlotte

3545 Whitehall Park
Suite 200
Charlotte, NC 28273

Pʜᴏɴᴇ: (980) 985-3200
Fᴀx: (704) 504-1335
www.mepcom.army.mil/Units/Eastern-Sector/
    12th-Battalion/Charlotte

Raleigh

2625 Appliance Court
Raleigh, NC 27604

Pʜᴏɴᴇ: (984) 328-8000
Fᴀx: (919) 755-1303
www.mepcom.army.mil/Units/Eastern-Sector/
    12th-Battalion/Raleigh

NORTH DAKOTA

Fargo

225 Fourth Avenue North Suite 210
Fargo, ND 58102

Pʜᴏɴᴇ: (701) 219-7377
Fᴀx: (701) 234-0597
www.mepcom.army.mil/Units/Western-Sector/
    1st-Battalion/Fargo

OHIO

Cleveland

20637 Emerald Parkway Drive
Cleveland, Ohio 44135-6023

Pʜᴏɴᴇ: (216) 430-8100
Fᴀx: (216) 430-8080
www.mepcom.army.mil/Units/Eastern-Sector/
    4th-Battalion/Cleveland

Columbus

775 Taylor Road
Gahanna, OH 43230

Pʜᴏɴᴇ: (614) 490-3200
Fᴀx: (614) 856-9065
www.mepcom.army.mil/Units/Eastern-Sector/
    6th-Battalion/Columbus

OKLAHOMA

Oklahoma City

301 Northwest 6th Street
Suite 150
Oklahoma City, OK 73102

Pʜᴏɴᴇ: (405) 416-6525
Fᴀx: (405) 609-8639
www.mepcom.army.mil/Units/Western-Sector/
    11th-Battalion/Oklahoma-City

OREGON

Portland

7545 NE Ambassador Place
Portland, OR 97220-1367

Pʜᴏɴᴇ: (917) 978-4859
Fᴀx: (503) 528-1640
www.mepcom.army.mil/Units/Western-Sector/
    5th-Battalion/Portland-OR

PENNSYLVANIA
Harrisburg
4641 Westport Drive
Mechanicsburg, PA 17055

PHONE: (717) 550-7700
FAX: (717) 691-8039
www.mepcom.army.mil/Units/Eastern-Sector/
    4th-Battalion/Harrisburg

Pittsburgh
William S. Moorehead Federal Building
1000 Liberty Avenue
Suite 1917
Pittsburgh, PA 15222-4101

PHONE: (412) 209-3725
FAX: (412) 281-5464
www.mepcom.army.mil/Units/Eastern-Sector/
    4th-Battalion/Pittsburgh

PUERTO RICO
San Juan
Millennium Park Plaza
15 Second St., Suite 310
Guaynabo, PR 00968-1741

PHONE: (305) 908-4120
FAX: (787) 277-7507
www.mepcom.army.mil/Units/Eastern-Sector/
    10th-Battalion/San-Juan

SOUTH CAROLINA
Fort Jackson
2435 Marion Avenue
Fort Jackson, SC 29207

PHONE: (803) 740-2801
FAX: (803) 751-5744
www.mepcom.army.mil/Units/Eastern-Sector/
    12th-Battalion/Fort-Jackson

SOUTH DAKOTA
Sioux Falls
2801 South Kiwanis Avenue
Suite 200
Sioux Falls, SD 57105

PHONE: (605) 305-5300
FAX: (605) 332-8412
www.mepcom.army.mil/Units/Western-Sector/
    1st-Battalion/Sioux-Falls

TENNESSEE
Knoxville
710 Locust St.
Room 600
Knoxville, TN 37902

PHONE: (865) 291-9400
FAX: (865) 531-8741
www.mepcom.army.mil/Units/Eastern-Sector/
    8th-Battalion/Knoxville

Memphis
1980 Nonconnah Blvd. 3rd Floor
Memphis, TN 38132

PHONE: (901) 291-3900
FAX: (901) 396-8124
www.mepcom.army.mil/Units/Eastern-Sector/
    8th-Battalion/Memphis

Nashville
20 Bridgestone Park
Nashville, TN 37214-2428

PHONE: (615) 928-4994
FAX: (615) 833-2570
www.mepcom.army.mil/Units/Eastern-Sector/
    8th-Battalion/Nashville

TEXAS
Amarillo
1100 South Fillmore
Suite 100
Amarillo, TX 79101

PHONE: (806) 290-9401
FAX: (806) 374-9332
www.mepcom.army.mil/Units/Western-Sector/
    9th-Battalion/Amarillo

Dallas
Federal Building
207 South Houston Street
Suite 400
Dallas, TX 75202

PHONE: (972) 367-2900
FAX: (214) 655-3213
www.mepcom.army.mil/Units/Western-Sector/
    9th-Battalion/Dallas

El Paso

6380 Morgan Avenue
Suite E
El Paso TX 79906-4611

Phone: (915) 995-3200
Fax: (915) 568-4477
www.mepcom.army.mil/Units/Western-Sector/
9th-Battalion/El-Paso

Houston

701 San Jacinto Street
P.O. Box 52309
Houston, TX 77052-2309

Phone: (346) 272-5770
Fax: (713) 718-4228
www.mepcom.army.mil/Units/Western-Sector/
9th-Battalion/Houston

San Antonio

1950 Stanley Road
Suite 103
Fort Sam Houston, TX 78234-2712

Phone: (726) 444-3600
Fax: (210) 295-9151
www.mepcom.army.mil/Units/Western-Sector/
9th-Battalion/San-Antonio

UTAH
Salt Lake City

546 West Amelia Earhart Drive
Suite 130
Salt Lake City, UT 84116

Phone: (385) 707-8478
Fax: (801) 975-3715
www.mepcom.army.mil/Units/Western-Sector/
3rd-Battalion/Salt-Lake-City

VIRGINIA
Fort Lee

2011 Mahone Avenue
Ft. Lee, VA 23801-1707

Phone: (804) 518-3480
Fax: (804) 765-4190
www.mepcom.army.mil/Units/Eastern-Sector/
12th-Battalion/Fort-Lee

WASHINGTON
Seattle

4735 East Marginal Way South
Suite 161, Box 16
Seattle, WA 98134-2388

Phone: (206) 701-5105
Fax: (206) 766-6430
www.mepcom.army.mil/Units/Western-Sector/
5th-Battalion/Seattle

Spokane

8570 West Highway 2
Spokane, WA 99224

Phone: (509) 456-4641
Fax: (509) 747-1988
www.mepcom.army.mil/Units/Western-Sector/
5th-Battalion/Spokane

WEST VIRGINIA
Beckley

409 Wood Mountain Road
Glen Jean, WV 25846

Phone: (304) 469-5460
Fax: (304) 465-3194
www.mepcom.army.mil/Units/Eastern-Sector/
12th-Battalion/Beckley

WISCONSIN
Milwaukee

11050 West Liberty Drive
Milwaukee, WI 53224

Phone: (414) 214-3200
Fax: (414) 359-1390
www.mepcom.army.mil/Units/Eastern-Sector/
6th-Battalion/Milwaukee

CAN YOU RETAKE THE ASVAB?

If you have taken the ASVAB within the past two years, you can retake the test as long as you follow certain rules. If you are taking the ASVAB for enlistment purposes, your most recent valid score is the one that will be considered.

The rules about retaking the ASVAB are as follows. First, you must take the entire test battery—that is, all subtests, not just one. Military applicants who have taken an initial ASVAB—student or enlistment—can retest after one calendar month has elapsed. For example, if you first took the test on February 3, the earliest you could retake it would be March 3. If you wished to retake the test a second time, you would have to wait until April 3. After that, you would need to wait at least six months before you could take the test again. In other words, if you first took the test on February 3, took a retest on March 3, and took a second retest on April 3, you would have to wait until October 3 before you could take a third retest (see the Sample Retesting Schedule chart on this page).

Retesting with the same version of the ASVAB that was used on any previous test is strictly prohibited for at least six months. If an applicant is retested with the same test version within a six-month period, the retest score will be invalidated and the previous valid test score will stand as the score of record. However, if the condition is the result of a MEPS or test administrator procedural or administrative error, the MEPS commander may authorize an immediate retest using a different ASVAB version. This is not such an issue with the CAT-ASVAB.

Applicants who are dismissed for cheating or disruptive behavior will have their test invalidated, and are not permitted to retest for six months from the date of the invalid test.

If you are taking the ASVAB in order to enlist in the military and your Armed Forces Qualification Test (AFQT) score (see Chapter 4) on your most recent test is 20 points or more higher than your score on an ASVAB you took less than six months previously, you may be required to complete a confirmation test.

Retaking the ASVAB for Enlistment If You Took It in High School

If you took the test in high school, should you take it again if you want to join the military? Remember that you can retake the test every six months and that your most recent valid score will be the one used for military enlistment purposes. That being the case, it is a good idea to take the test again if you think you can get a better score. Use the information in this book to help you score higher.

Retaking the ASVAB If You Are on Active Duty

Low ASVAB scores on initial tests do not have to be permanent, barring access to certain schools and other opportunities. Active-duty military personnel whose original ASVAB scores were low can retest to raise their scores and improve their eligibility for some programs. Higher ASVAB scores can help active-duty personnel change to more technical ratings and can improve eligibility for a class A school.

However, enlisted personnel are allowed to retake the ASVAB only once, and that test score becomes permanent, even if it is lower than the original score. As a result, if you are already in the military and wish to retake the ASVAB, you must be certain that you have made the improvements necessary to raise your score. If you want to retake the test, you must prove that you have improved your abilities

Sample Retesting Schedule

3 Feb.	3 Mar.	3 Apr.	3 May	3 June	3 July	3 Aug.	3 Sept.	3 Oct.
Initial test	First retest	Second retest						Third retest

enough—through training, practical experience, and schools—to expect a higher ASVAB score. See the review sections in this book to increase your score.

Improvements may be gained in a number of ways: functional skills training, completing study at public or private institutions, participation in training courses, study at academic skills learning centers, or using this test preparation book. Other proven ways to achieve higher ASVAB scores include command programs to enhance basic academic skills, attending boot camp, military experience, and increased maturity.

Chapter 3

ASVAB Scores and Score Reports

YOUR GOALS FOR THIS CHAPTER:

- **Find out what information will be on your score report.**
- **Understand the difference between standard scores and percentile scores.**
- **Review sample ASVAB score reports.**

Once you take the ASVAB, the score report you will receive depends on where you took the test. The amount of information you will receive also depends on whether you took the paper-and-pencil version or the computer-adaptive version of the ASVAB.

IF YOU TAKE THE ASVAB AT A SCHOOL

If you take the ASVAB at a school, a score report with be sent to your school. Your score report will include a number of different scores.

Subtest Scores and Career Exploration Scores

Your score report will include scores for each of the subtests in the test battery. These are General Science, Arithmetic Reasoning, Word Knowledge, Paragraph Comprehension, Mathematics Knowledge, Electronics Information, Auto and Shop Information, and Mechanical Comprehension. Mechanical Comprehension.

Your score report will also show three Career Exploration Scores: Verbal Skills, Math Skills, and Science and Technical Skills. These are composite scores combining your scores on several ASVAB subtests. The ASVAB Career Exploration Scores are a good indicator of the kinds of tasks that test takers do well and the kinds of tasks that they may find difficult.

- *Verbal Skills* is a general measure of the vocabulary and reading skills covered in the Word Knowledge and Paragraph Comprehension tests. People with high scores tend to do well in tasks that require good vocabulary or reading skills, while people with low scores have more difficulty with such tasks.

- *Math Skills* is a general measure of the mathematics skills covered in the Mathematics Knowledge and Arithmetic Reasoning tests. People with high scores tend to do well in tasks that require knowledge of mathematics, while people with low scores have more difficulty with these kinds of tasks.

- *Science and Technical Skills* is a general measure of science and technical skills, which are covered in the General Science, Electronics Information, and Mechanical Comprehension tests. People with high scores tend to do well in tasks that require scientific thinking or technical skills, while people with low scores have more difficulty with such tasks.

Standard Scores and Percentile Scores

Each of the scores just listed is reported in two ways: as standard scores and as percentile scores. Each standard score is calculated by applying statistical methods to the your raw score. This method produces a numerical score with a short range of possible statistical error above and below it. Most student ASVAB takers achieve standard scores between 30 and 70. This means that a standard score of 50 is an average score, a score of 60 is an above-average score, and a score of 30 is below average. The score report shows the numerical standard scores, which are estimates of your true skill level in that area. According to the report, if you took the test again, your new score would probably be similar to, but not necessarily exactly like, your initial score.

The percentile scores on the ASVAB score report indicate how well the you did in relation to others in the same grade. For each ASVAB test and composite (called a Career Exploration Score), students receive a same grade/same sex, same grade/opposite sex, and same grade/combined sex percentile score. For example, if you are a female eleventh grader, you would get percentile scores showing how well you did compared to other females in the eleventh grade, males in the eleventh grade, and all eleventh graders. For example, if you scored a percentile of 65 on Math Skills, that means you scored as well as or better than 65 out of 100 eleventh-grade females in Math Skills. When compared to males and all eleventh-grade students, your percentile score may be different.

Because the experiences of males and females differ, they can perform somewhat differently on tests. On the more technically oriented tests, such as Electronics Information, the mean performance of males is higher than that of females. This does not mean that women cannot learn this information or that they should be discouraged from considering occupations in related areas. Typically, this difference occurs because more males than females have had exposure to electronics principles. As a result, it is fairer to report how students do compared to members of their own sex, but also to let them know how they compare to members of the opposite sex on tests that might be important to them. For example, a female student might be interested in a career in mechanics, surveying, or civil engineering. Knowing how she scores relative to both her own sex and the opposite sex is useful information. In the past, these career fields have traditionally been dominated by males. Since she will be competing with males, it is important for her to know how she stands relative to males. The same is true for males interested in occupations traditionally dominated by females.

The AFQT is a general measure of trainability and a predictor of on-the-job performance. Chapter 4 explains how the AFQT is used in military recruiting.

Figure 3.1 shows a sample score report from an ASVAB test taken at a school.

IF YOU TAKE THE PAPER-AND-PENCIL ASVAB AT A MILITARY FACILITY

If you take the ASVAB at a military facility called a MET site, you may be given a paper-and-pencil version of the test. You won't receive as much information about your results as those who take the ASVAB at a high school or postsecondary school. At at MET site you also may be given an Internet-delivered version of the CATASVAB, the iCAT.

Figure 3.2 shows the only information you will receive: the AFQT percentile. This is an unofficial score calculated by the local test administrator.

IF YOU TAKE THE CAT-ASVAB AT A MEPS

If you take the CAT-ASVAB at a MEPS, you will receive more scoring information than if you took the test at a MET site but less than if you took the ASVAB at a high school. The CAT-ASVAB report provides your scores on every subtest of the ASVAB, your AFQT, and your general scores for qualifying for services occupations. The score report in Figure 3.3 is a sample of the CAT-ASVAB score report.

ASVAB SUMMARY RESULTS

Print No. 0005

Student
12th Gr Female (Form: 23G)
SSN: XXX-XX-9999
Test Date: Jul 11, 2005
Old Dominion Hs
Hometown DC

ASVAB Results

	Percentile Scores			12th Grade Standard Scores
	12th Grade Females	12th Grade Males	12th Grade Students	
Career Exploration Scores				
Verbal Skills	97	95	96	65
Math Skills	22	17	19	42
Science and Technical Skills	81	48	64	53
ASVAB Tests				
General Science	91	81	86	61
Arithmetic Reasoning	43	30	37	47
Word Knowledge	98	95	96	66
Paragraph Comprehension	92	91	91	62
Mathematics Knowledge	14	12	13	37
Electronics Information	13	10	11	38
Auto and Shop Information	53	21	37	45
Mechanical Comprehension	95	76	85	59

Military Entrance Score (AFQT) 57

12th Grade Standard Score Bands

EXPLANATION OF YOUR ASVAB PERCENTILE SCORES

Your ASVAB results are reported as percentile scores in the three highlighted columns to the left of the graph. Percentile scores show how you compare to other students - males and females, and for all students - in your grade. For example, a percentile score of 65 for an 11th grade female would mean she scored the same or better than 65 out of every 100 females in the 11th grade.

For purposes of career planning, knowing your relative standing in these comparison groups is important. Being male or female does not limit your career or educational choices. There are noticeable differences in how men and women score in some areas. Viewing your scores in light of your relative standing both to men and women may encourage you to explore areas that you might otherwise overlook.

You can use the Career Exploration Scores to evaluate your knowledge and skills in three general areas (Verbal, Math, and Science and Technical Skills). You can use the ASVAB Test Scores to gather information on specific skill areas. *Together, these scores provide a snapshot of your current knowledge and skills.* This information will help you develop and review your career goals and plans.

EXPLANATION OF YOUR ASVAB STANDARD SCORES

Your ASVAB results are reported as standard scores in the above graph. Your score on each test is identified by the "X" in the corresponding bar graph. You should view these scores as *estimates* of your true skill level in that area. If you took the test again, you probably would receive a somewhat different score. Many things, such as how you were feeling during testing, contribute to this difference. This difference is shown with gray score bands in the graph of your results. Your standard scores are based on the ASVAB tests and composites based on your grade level.

The score bands provide a way to identify some of your strengths. Overlapping score bands mean your true skill level is similar in both areas, so the real difference between specific scores might not be meaningful. If the score bands do not overlap, you probably are stronger in the area that has the higher score band.

The ASVAB is an aptitude test. It is neither an absolute measure of your skills and abilities nor a perfect predictor of your success or failure. A high score does not guarantee success, and a low score does not guarantee failure, in a future educational program or occupation. For example, if you have never worked with shop equipment or cars, you may not be familiar with the terms and concepts assessed by the Auto and Shop Information test. Taking a course or obtaining a part-time job in this area would increase your knowledge and improve your score if you were to take it again.

USING ASVAB RESULTS IN CAREER EXPLORATION

Your career and educational plans may change over time as you gain more experience and learn more about your interests. *Exploring Careers: The ASVAB Career Exploration Guide* can help you learn more about yourself and the world of work, to identify and explore potential goals, and develop an effective strategy to realize your goals. The *Guide* will help you identify occupations in line with your interests and skills. As you explore potentially satisfying careers, you will develop your career exploration and planning skills.

Meanwhile, your ASVAB results can help you in making well-informed choices about future high school courses.

We encourage you to discuss your ASVAB results with a teacher, counselor, parent, family member or other interested adult. These individuals can help you to view your ASVAB results in light of other important information, such as your interests, school grades, motivation, and personal goals.

USE OF INFORMATION

Personal identity information (name, social security number, street address, and telephone number) and test scores will not be released to any agency outside of the Department of Defense (DoD), the Armed Forces, the Coast Guard, and your school. Your school or local school system can determine any further release of information. The DoD will use your scores for recruiting and research purposes for up to two years. After that the information will be used by the DoD for research purposes only.

MILITARY ENTRANCE SCORES

The **Military Entrance Score** (also called AFQT, which stands for the Armed Forces Qualification Test) is the score used to determine your qualifications for entry into any branch of the United States Armed Forces or the Coast Guard. The Military Entrance Score predicts in a general way how well you might do in training and on the job in military occupations. Your score reflects your standing compared to American men and women 18 to 23 years of age.

Use Access Code: 123456789X
(for online Occu-find and FYI)
Access Code expires: Jul 15, 2007

Explore career possibilities by using your Access Code at

www.asvabprogram.com

SEE YOUR COUNSELOR FOR FURTHER INFORMATION

DD FORM 1304-5(S)

Figure 3.1. Sample score report from an ASVAB taken at a school

ASVAB SCORE AND TEST DESCRIPTIONS

Verbal Skills is a general measure of language and reading skills which combines the Word Knowledge and Paragraph Comprehension tests. People with high scores tend to do well in tasks that require good language or reading skills, while people with low scores have more difficulty with such tasks.

Math Skills is a general measure of mathematics skills which combines the Mathematics Knowledge and Arithmetic Reasoning tests. People with high scores tend to do well in tasks that require a knowledge of mathematics, while people with low scores have more difficulty with these kinds of tasks.

Science and Technical Skills is a general measure of science and technical skills which combines the General Science, Electronics Information, and Mechanical Comprehension tests. People with high scores tend to do well in tasks that require scientific thinking or technical skills, while people with low scores have more difficulty with such tasks.

General Science (GS) tests the ability to answer questions on a variety of science topics drawn from courses taught in most high schools. The life science items cover botany, zoology, anatomy and physiology, and ecology. The earth and space science items are based on astronomy, geology, meteorology, and oceanography. The physical science items measure force and motion mechanics, energy, fluids, atomic structure, and chemistry.

Arithmetic Reasoning (AR) tests the ability to solve basic arithmetic problems one encounters in everyday life. One-step and multi-step word problems require addition, subtraction, multiplication, and division, and choosing the correct order of operations when more than one step is necessary. The items include operations with whole numbers, operations with rational numbers, ratio and proportion, interest and percentage, and measurement. Arithmetic reasoning is one factor that helps characterize mathematics comprehension and it also assesses logical thinking.

Word Knowledge (WK) tests the ability to understand the meaning of words through synonyms - words having the same or nearly the same meaning as other words. The test is a measure of one component of reading comprehension since vocabulary is one of many factors that characterize reading comprehension.

Paragraph Comprehension (PC) tests the ability to obtain information from written material. Students read different types of passages of varying lengths and respond to questions based on information presented in each passage. Concepts include identifying stated and reworded facts, determining a sequence of events, drawing conclusions, identifying main ideas, determining the author's purpose and tone, and identifying style and technique.

Mathematics Knowledge (MK) tests the ability to solve problems by applying knowledge of mathematical concepts and applications. The problems focus on concepts and algorithms and involve number theory, numeration, algebraic operations and equations, geometry and measurement, and probability. Mathematics knowledge is one factor that characterizes mathematics comprehension; it also assesses logical thinking.

Electronics Information (EI) tests understanding of electrical current, circuits, devices, and systems. Electronics information topics include electrical circuits, electrical and electronic systems, electrical currents, electrical tools, symbols, devices, and materials.

Auto and Shop Information (AS) tests aptitude for automotive maintenance and repair and wood and metal shop practices. The test covers several areas commonly included in most high school auto and shop courses such as automotive components, automotive systems, automotive tools, troubleshooting and repair, shop tools, building materials, and building and construction procedures.

Mechanical Comprehension (MC) tests comprehension of the principles of mechanical devices and properties of materials. Mechanical comprehension topics include simple machines, compound machines, mechanical motion, fluid dynamics, properties of materials, and structural support.

Military Entrance Score (AFQT) is the score used if an individual decides to enter any of the armed services. See your local recruiter for details.

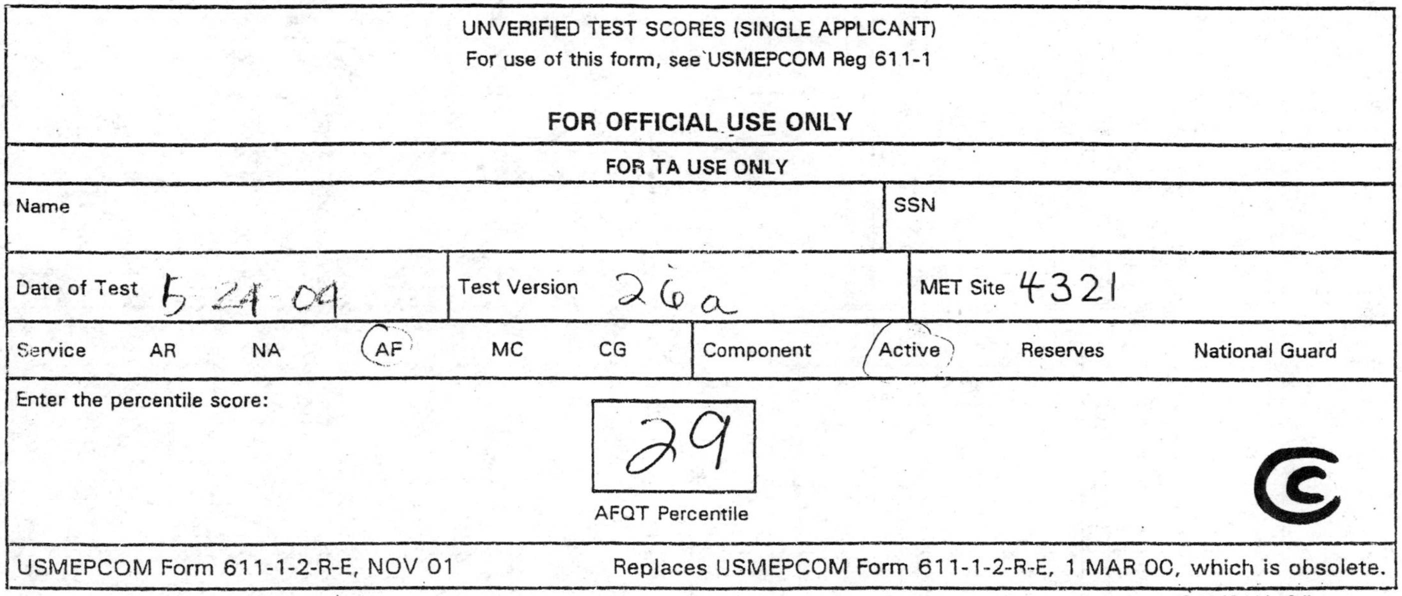

UNVERIFIED TEST SCORES (SINGLE APPLICANT)
For use of this form, see USMEPCOM Reg 611-1

FOR OFFICIAL USE ONLY

FOR TA USE ONLY

Name	SSN

Date of Test 5-24-09	Test Version 26a	MET Site 4321

Service AR NA (AF) MC CG	Component (Active) Reserves National Guard

Enter the percentile score:

29

AFQT Percentile

USMEPCOM Form 611-1-2-R-E, NOV 01 — Replaces USMEPCOM Form 611-1-2-R-E, 1 MAR 00, which is obsolete.

34813(11–01)

Figure 3.2. Sample score report from an ASVAB taken at a MET site. This is an unofficial score calculated by the local test administrator

FOR OFFICIAL USE ONLY

UNVERIFIED WINDOWS CAT-ASVAB TEST SCORE REPORT

Testing Site ID: 987654   Service: DNR

Testing Session: Date 2022/09/24   Starting Time: 05:00

Applicant: Name: Jay Jones   SSN: 123-45-6789

Test Form: 04D   Test Type: Initial

	GS	AR	WK	PC	MK	EI	AS	MC	AO	VE
Standard Scores:	56	57	56	35	46	37	37	48	40	50

COMPOSITE SCORE:

Army:	GT	CL	CO	EL	FA	GM	MM	OF	SC	ST
	108	101	092	092	-94	091	085	093	093	097

Air Force:	M	A	G	E
	38	40	59	47

Navy/CG:	GT	EL	BEE	ENG	MEC	MEC2	NUC	OPS	HM	ADM
	107	196	205	083	142	145	201	193	152	096

Marine:	MM	GT	EL	CL
	088	104	098	096

AFQT Percentile Score: 49

Figure 3.3. Sample CAT-ASVAB score report

Chapter 4

The ASVAB, the AFQT, and Military Entrance

YOUR GOALS FOR THIS CHAPTER:

- **Learn how ASVAB AFQT scores are used to determine eligibility for enlistment.**
- **Find out about other educational requirements for military recruits.**
- **Learn what happens during the rest of the military entrance process.**

When you take the ASVAB in order to enlist in the U.S. Armed Forces, your ASVAB scores will be used in several ways. In the previous chapter, you learned about an ASVAB score called the Armed Forces Qualifying Test (AFQT). You will need to achieve a certain AFQT score in order to be eligible for initial enlistment. Later, your ASVAB scores will be used to qualify you for various occupations in the military. The higher your scores, the more choices you will have for training in various occupations. So you want to do your best on the ASVAB so that you will have the widest choice of training programs and job responsibilities.

AFQT SCORES AND ELIGIBILITY FOR ENLISTMENT

The AFQT is a composite of your scores on four ASVAB tests. The AFQT is the primary score that indicates the training potential of persons who wish to enlist in the military. Your AFQT score determines your eligibility for military service. It also is able to predict how well you will perform on the job.

AFQT scores are reported as percentile scores. The percentile shows how your score compares to the scores of other test takers, and it also determines where you fit into what are called AFQT categories.

Persons in categories I and II are above average in trainability. Category III recruits tend to be average in trainability, and category IV recruits are below average in trainability. Applicants scoring in category V are forbidden by law to be accepted into the military. In addition, military applicants in category IV who do not have a high school diploma are generally forbidden to enter the military.

The services use these categories to set recruiting goals. Recruiters prefer recruiting people in the highest categories because it is shown that they are easier to train in the various occupations needed by the military. It is also more economical for the Department of Defense to recruit smarter individuals because they learn what is needed much more quickly and they perform better on the job.

Armed Forces Qualification Test (AFQT) Categories and Corresponding Percentile Score Ranges

AFQT Category	Percentile Score Range
I	93–99
II	65–92
IIIA	50–64
IIIB	31–49
IV	10–30
V	1–9

The services are required to enlist at least 60 percent of recruits from AFQT categories I through IIIA, and usually no more than 4 percent of the recruits can come from category IV. To give you an idea of the percentages of recruits in each AFQT category, look at Table 4.1. You can see that very few recruits come from category IV and none come from category V.

ADDITIONAL EDUCATIONAL REQUIREMENTS

The amount of education you have obtained also matters in your eligibility to enter military service. The Department of Defense uses a three-tier classification of educational credentials. The three tiers are

Tier 1. Regular high school graduates, adult diploma holders, and nongraduates with at least 15 hours of college credit.

Tier 2. Alternative credential holders, including those with a General Education Development (GED) certificate of high school equivalency.

Tier 3. Those with no educational credentials.

The reason for this classification system is that there is a strong relationship between educational

Table 4.1. Non-Prior Service (NPS) Active Component Enlisted Accessions, by Armed Forces Qualification Test (AFQT) Category, Service, and Gender with Civilian Comparision Group

	AFQT Category					I + II + IIIA
Gender	I	II	IIIA	IIIB	IV	Subtotal
Army						
Males	6.04	32.51	23.74	35.55	2.09	62.29
Females	3.15	25.31	24.54	45.17	1.71	53.00
Total	**5.52**	**31.21**	**23.89**	**37.29**	**2.02**	**60.61**
Navy						
Males	9.59	41.52	23.34	25.44	0	74.45
Females	4.12	30.80	27.73	37.29	0.01	62.64
Total	**8.26**	**38.91**	**24.41**	**28.33**	**0.01**	**71.57**
Marine Corps						
Males	4.66	37.25	27.52	30.53	0	69.43
Females	3.64	33.33	31.34	31.68	0	68.32
Total	**4.55**	**38.84**	**27.92**	**30.65**	**0**	**69.31**
Air Force						
Males	8.88	48.29	25.59	15.91	0	82.76
Females	4.53	39.83	30.64	23.96	0	75.00
Total	**7.79**	**46.17**	**26.85**	**17.93**	**0**	**80.82**
Coast Guard						
Males	7.35	39.83	20.95	20.01	2.04	68.13
Females	5.11	29.40	22.42	26.84	2.65	56.93
Total	**6.83**	**37.43**	**21.29**	**21.58**	**2.18**	**65.55**
Total DoD						
Males	7.04	38.36	24.80	28.70	0.81	70.19
Females	3.83	31.38	27.70	36.16	0.61	62.92
Total	**6.41**	**37.00**	**25.36**	**30.15**	**0.77**	**68.78**
Representative Group of U.S. 18–23 Yr-Old Civilians						
Males	8.12	28.97	15.32	18.38	19.63	52.41
Females	7.60	26.35	15.77	19.18	21.87	49.72
Total	**7.86**	**27.66**	**15.54**	**18.78**	**20.75**	**51.07**

credentials and successful completion of the first term of military service. That is, if you have a good AFQT and good educational credentials, you are likely to complete your required service and not drop out partway through (an outcome that would waste the taxpayer dollars spent to train you).

The services are required to ensure that at least 90 percent of first-time recruits are high school graduates. Services often set even higher educational standards, sometimes requiring nearly 100 percent of the recruits in their enlistment pool to be high school graduates. If you don't have a high school diploma, you will need a very high AFQT score.

There are different AFQT scoring standards for individuals in each tier. Generally, tier 3 applicants must have higher AFQT test scores than tier 2 applicants, who must have higher test scores than tier 1 individuals. The Air Force and Marine Corps follow these differential standards, requiring different minimum test scores for each tier. The Army and Navy require applicants with alternative credentials (tier 2) and those with no credentials (tier 3) to meet the same AFQT standards, which are more stringent than those for high school graduates (tier 1). Your best information comes from your recruiter on these issues.

So if you want to enlist in the military, your chances are much better if you have graduated from high school with a traditional diploma and score high on the AFQT.

BEYOND THE ASVAB
Physical Condition and Moral Character Requirements

If you achieve a satisfactory ASVAB score and continue the application process, you will be scheduled for a physical examination and background review at one of the 65 Military Entrance Processing Stations (MEPS). Physical examinations are important because everyone entering the armed forces must be in good health to endure the challenges of basic training and military service. Any physical disqualifications that appear during your MEPS physical exam may bar you from entry into the military. You will have to remove your outer clothing during parts of the examination.

The examination determines your fitness for military service. It includes measurement of blood pressure, pulse, visual acuity, and hearing; blood testing and urinalysis; drug and HIV testing; medical history; and possibly tests of strength and endurance. If you have a fixable or temporary medical

problem, you may be required to get treatment before proceeding. It is possible but very difficult to obtain a waiver of certain disqualifying medical conditions and be allowed to enlist. Check with your recruiter for specifics because they change from time to time.

Females will be provided a drape or gown. Your visit with the physician will be in a private room. Underclothing is required during your physical. A female attendant will accompany you when you must remove your clothing. You will also be given a pregnancy test.

> Obesity is the leading medical reason for not qualifying for the military.

To enter one of the services, you must meet rigorous moral character standards. You will be screened by the recruiter and will undergo an interview covering your background. You may undergo a financial credit check; a computerized search for a criminal record may also be conducted. Some types of criminal activity are clearly disqualifying; other cases require a waiver. The service to which you applied will examine your circumstances and make an individual determination of qualification. Since it has been shown that applicants with existing financial problems generally do not overcome those problems on junior enlisted pay, a credit history may be part of the decision to allow you to enlist or not.

Particularly due to the events of January 6, 2021, military applicants will be scrutinized more carefully for behavior resembling or supporting extremist activities. Be assured that you will be observed for any indications of extremism during you enlistment period. You will also receive training on how to avoid becoming the target of extremist groups that may want to recruit you to their cause.

Personality Testing

The Services have begun using a personality assessment called Tailored Adaptive Personality Assessment System (TAPAS), which is said to measure recruit potential.

The idea is to identify which personality types are attracted to and will do well in certain military jobs.

The assessment measures leadership, resilience, ingenuity, selflessness, commitment to serve, and propensity to attrit.

The goal is to find the "best fit" of people to various military careers. It has been found that those who are a "best fit" tend to outperform in the military, as opposed to others with the same ASVAB score.

Military Occupational Counseling

If your ASVAB scores, educational credentials, physical fitness, and moral character qualify you for entry and you wish to proceed with the process, you will meet with a service classification counselor at the MEPS to discuss options for enlistment. The counselor has the record of your qualifications and computerized information on available service training/skill openings, schedules, and enlistment incentives.

A recruit can sign up for a specific skill or for a broad occupational area (such as the mechanical or electronics area). In the Army, most recruits enter for specific skill training; the others are placed in a military occupational specialty during basic training. A large percentage of Air Force recruits enter for a specific skill, while the rest sign up for an occupational area and are classified into a specific skill while in basic training. In the Navy, many recruits enlist for a specific skill, while the rest go directly to the fleet after basic training, classified into airman, fireman, or seaman programs. A significant proportion of Marine Corps enlistees enter with a guaranteed occupational area and are assigned a specific skill within that area after recruit training; the rest enlist with either a specific job guarantee or assignment to a job after recruit training.

Your counselor will discuss your interests with you and will explain what the service has to offer. Typically, the counselor will describe a number of different occupations to you. In general, the higher your test scores, the more choices you will have. The counselor may suggest incentives to encourage you to choose hard-to-fill occupational specialties. You are free to accept or reject these offers. Many applicants do not decide immediately, but take time to discuss options with their family and

---

MEPS Visit Reminders

1. Discuss any childhood medical problems with your parents and bring documentation with you.
2. Bring your Social Security card, birth certificate, and driver's license.
3. Remove all piercings and earrings.
4. Profanity and offensive wording or pictures on clothing are not tolerated.
5. Hats are not permitted inside the MEPS.
6. If you wear either eyeglasses or contacts, bring them along with your prescription and lens case.
7. Bathe or shower the night before your examination.
8. Wear underclothes.
9. Get a good night's sleep before taking the ASVAB.
10. Wear neat, moderate, comfortable clothing.
11. Don't bring stereo headphones, watches, jewelry, excessive cash, or any other valuables.
12. Ask your recruiter for a list of recommended personal items to bring to basic training.
13. Processing starts early at the MEPS—You must report on time.

What Happens Next?

When you accept an offer and sign a contract, you will need to choose between two options. One option is to proceed directly to a recruit training center within a month of signing the contract. Most people choose the second option, which is to enter the Delayed Entry Program (DEP). This allows you up to a year before you need to report for duty. During this time, you can continue your education, obtain your high school diploma, take advantage of a supervised exercise program, and in general become acclimated to the military. The length of time in the DEP depends on the training opportunities for the occupation you have selected.

friends; others decide not to enlist. The services do not discriminate based on race, religion, or gender. In fact, about 20 percent of active duty and reserves are female.

If you are selected, you will be one of the many thousands of individuals who have met the standards and elected to serve and protect our country by joining the Armed Forces of the United States.

*A dream doesn't become reality through magic;
it takes sweat, determination and hard work.*
—COLIN POWELL

Chapter 5

Taking the CAT-ASVAB

YOUR GOALS FOR THIS CHAPTER:

- **Learn how the CAT-ASVAB works.**

- **Find out the pros and cons of taking the CAT-ASVAB.**

- **Preview the directions for the computer-based test.**

- **Find out strategies that can raise your CAT-ASVAB score.**

HOW THE CAT-ASVAB WORKS

The CAT-ASVAB is the computer-based version of the test that is offered to potential recruits at the military facilities called MEPS. The CAT-ASVAB subtests measure the same abilities as the paper-and-pencil ASVAB subtests. One difference is that the paper-and-pencil ASVAB's Automotive and Shop Information subtest is broken into two separate subtests in CAT-ASVAB. In addition, the tests are adaptive.

With a group-administered paper-and-pencil test, all examinees answer the same questions in the same order. An adaptive test automatically tailors questions to the ability level of the individual examinee. All examinees start with a question of medium difficulty. If you answer the question correctly, you are given a question that is more difficult. If you answer the question incorrectly, you are given a question that is easier. This pattern continues until the test is complete. Therefore, you only answer questions that are appropriate for your ability level, not wasting time answering questions that are too easy or too difficult.

If you take this version of the test, you will be seated in front of a computer and monitor in a room with others. The test comes up on the screen. As you answer the test questions, the program records your answer, scores it, and then calculates your ability level. Based on that information, the next item is selected. After you have completed the test, the AFQT and composite scores will be calculated.

The following chart shows the general idea of how the items are selected for you.

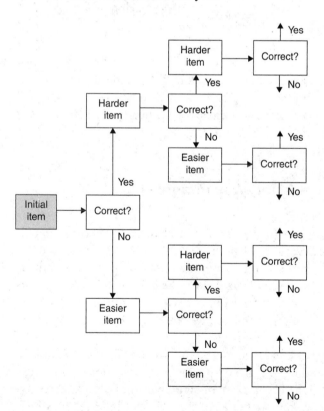

So you can see that the CAT-ASVAB is tailored to you and your ability level. The items come from a huge pool of items that range in difficulty from very easy to very hard. After each item is administered, information is collected and evaluated, and the item best suited to you is selected to be administered next. This selection process provides a more accurate score with fewer items than the paper-and-pencil version. On the whole, the CAT-ASVAB takes less time than the paper-and-pencil version.

The adaptive item selection process of the CAT-ASVAB results in higher levels of test-score precision and shorter test lengths than the paper-and-pencil ASVAB. The figure below illustrates that shorter test lengths can be achieved on the CAT-ASVAB by tailoring the test to the ability level of each examinee.

Test Type	Examinee Ability	Item Difficulty			Test Length
		Easy	Medium Hard	Hard	
P&P	All				30
CAT	Low				15
	Medium				15
	High				15

It is important for you to know that no two test takers get the same items, so you need a good background in the areas tested to do well. Chapters 7 through 18 in this book will certainly help you achieve a high score.

Possible Advantages

One advantage of taking the CAT-ASVAB is that it takes less time than the paper-and-pencil version. Also, the test can be scored immediately, and scoring errors are reduced because of the automation. The test can be administered with minimal advance notice, while administration of the paper-and-pencil ASVAB needs to be scheduled well in advance.

Unlike scores on the paper-and-pencil ASVAB, the CAT-ASVAB score you get is not based solely on the number of items you answered correctly. Some people have claimed that the CAT-ASVAB is easier and can give you better scores because you get more "points" for answering more difficult items. In truth, you will receive the same score regardless of which version of the ASVAB you take.

Possible Disadvantages

Unlike the paper-and-pencil ASVAB, the CAT-ASVAB does not give you an opportunity to go back to questions you previously answered to change your answer or think about the question again. You also cannot skip questions and go on to other questions that you might know. If you get items wrong, the computer will give you easier items, but your score will be lower than if you answered a few difficult items.

CAT-ASVAB Directions

When you arrive to take the CAT-ASVAB, you will be escorted to a waiting room. You will complete various forms, if you haven't already done so. Your Social Security number will be verified. There you will get a briefing similar to the one that follows.

Briefing

[*Your test administrator will read aloud to you directions similar to the following.*]

Welcome, I am [administrator's name], and I will be administering your test today. First of all, has anyone here taken the Armed Services Vocational Aptitude Battery at any time in the past, either in a high school or at another testing site, and not indicated this on the USMEPCOM Form 680-3AE you provided? It is extremely important to identify this because your test will be checked against a nationwide computer file for armed forces applicants. If it is discovered that you were previously tested, but did not tell us, the results from today's test may not be valid for enlistment.

[*If anyone raises their hand, the administrator will check that person's USMEPCOM Form 680-3AE to ensure that they have marked the retest box and entered the previous forms.*]

It is important that you are physically fit to take this test. Is there anybody here who doesn't feel well enough to take the examination?

[*If anyone says that they don't feel well, the administrator will remove that person from the session group, inform the service (if available), and indicate the reason for removal on USMEPCOM Form 680-3AE.*]

The test you are about to take is administered by computer. Instructions for taking the test are on the computer, and the guidelines are very easy to follow. If you need assistance during any part of the test, press the HELP key, raise your hand, and I'll assist you.

The use of calculators, crib sheets, or other devices designed to assist in testing is not permitted. No talking is allowed while in the testing room. Use the scratch paper and pencil on the side of your computer for any figuring you need to do while taking the test. If you need more paper or another pencil, press the red HELP key and then raise your hand.

After completing the examination, give your scratch paper to me, and then you will be released. If you are staying at the hotel, wait at the front desk for transportation.

Does anyone have any questions?

What Happens Next?

After the briefing by the test administrator, you will be brought into the testing room and assigned a seat.

The CAT-ASVAB is designed so that individuals with very little or no computer experience can take the exam.

The computer program will give you directions on how to answer the items. It will also tell you about time limits and give you the chance to do some practice items to be sure that you know how to navigate each test.

The directions are quite simple. If you need help during the session, you need only press the HELP key. The administrator will come by to assist you. You should do this only if you are experiencing technical problems with the test or if you are in need of more scratch paper or another pencil to use for your calculation.

Recent Changes

Recently the Department of Defense has been implementing the iCAT, a CAT ASVAB that is delivered via the Internet. Using the i-CAT will be similar to the computer version and will get you the same score.

Page 21 shows you the number of items and allotted time to complete the CAT-ASVAB items. On occasion, you will be given some extra items and extra time to complete them.

These are tryout questions that do not count toward your score. Tryout questions help the Department of Defense determine if the questions are suitable for inclusion in future CAT-ASVAB tests.

These will always be at the end of the test.

SPECIAL STRATEGIES FOR TAKING THE CAT-ASVAB

Except where noted, the strategies for taking the CAT-ASVAB are the same as for the paper-and-pencil test. Here are some special things to keep in mind:

- To raise your scores, try your very best to answer every item correctly. Take your time, especially with the early items in each test. If you get the early items correct, the computer will give you more difficult questions, and these carry more value. If you get the early items wrong, the computer will give you easier items, and even if you answer these correctly, your score will be lower than if you answered the same number of more difficult items. Unlike on the paper-and-pencil tests, guessing is not necessarily a good strategy.

- Focus hard and give the test your full attention.

- Pay attention to the directions for each test. These will be on the computer screen.

- Try hard to finish all the items. For every item you do not finish, there is a penalty. The more unfinished items, the higher the penalty. Also, the penalty is high if you answer the last few items on each test incorrectly. For most people, there is ample time to complete all the items.

THE PiCAT

Applicants who hope to join the military can now take the Armed Services Vocational Aptitude Battery pretest from any computer with Internet access. It is called the PiCAT, which stands for the Pre-screening, internet-delivered, Computer Adaptive Test. The PiCAT is an unproctored version of the full ASVAB that provides recruiters with the ability to determine whether you are qualified before sending applicants to a MEPS or MET site for additional testing.

In order to take the PiCAT, you need an authorization code from a recruiter, which will provide you with access to the test.

You must start the test within 72 hours of receiving the code and will have 24 hours to complete the test once you have started. The recruiter will have immediate access to your scores once the test has

---

CAT-ASVAB Myths

Myth: The CAT-ASVAB is harder than the paper-and-pencil ASVAB.

Reality: The CAT-ASVAB may seem harder to many people because the items are selected based on a person's previous answers. For example, a person with high ability will be able to skip over the easier questions that would be found on the paper-and-pencil version of the test.

A lot of research has been done to ensure that the score you would get on the paper-and-pencil or computer adaptive versions will be the same.

Myth: The paper-and-pencil ASVAB is easier than the CAT-ASVAB.

Reality: For persons of lower ability, the paper-and-pencil version may seem easier because the questions are generally arranged from easy to hard within each subtest. As a result, you will receive many easier items to answer before you are challenged with the more difficult ones.

On the CAT-ASVAB, once your general ability level is determined, you will be getting more difficult items to answer to see how well you do. The easier items are skipped. The final score will be the same, regardless of the version of the ASVAB you are given.

---

been completed. The test can only be taken once and will not be given to anyone who has previously taken the ASVAB or Student ASVAB.

You need to achieve a minimum AFQT on the first five sections in order to complete the test. If you do not achieve the minimum AFQT score, you will not be permitted to proceed to a proctored verification test and will have to take the full ASVAB to enlist.

Once you complete the PiCAT, you will have to take a short verification test at the MEPS. This test confirms that you took the PiCAT without assistance and validates the PiCAT score for use as the official test score of record. If confirmed, you will enlist under that PiCAT score. If not confirmed, you will have to take a full-length ASVAB. Current retest rules will apply.

You cannot use any resources such as books, the Internet, or people when taking the PiCAT. The test shows the recruiter your ability to pass the real ASVAB. If your PiCAT scores differ greatly from a real ASVAB, there may be a strong suspicion that you did not take the test as directed. To say the least, cheating is not tolerated by the military and consequences will be swift.

PiCAT can count as the official ASVAB. You can retest after a verified PiCAT is given; however, retest rules will apply, meaning applicants need to wait 30 days after completing the verification test to retest. You could be randomly chosen to take the ASVAB, even if you passed the PiCAT.

It is a good idea for you to take the PiCAT because you will gain familiarity with the ASVAB test and the types of questions that will be asked. For a lot of people who experience test anxiety, an added benefit is that the PiCAT is not timed.

Chapter 6

Follow These Suggestions to Ace the ASVAB

YOUR GOALS FOR THIS CHAPTER:

- **Create your own ASVAB study plan.**

- **Learn effective study techniques.**

- **Find out ways to reduce test anxiety.**

- **Learn test-taking pointers that can raise your score.**

SEVEN STEPS TO ACING THE ASVAB

There are several steps you can take to give yourself a leading edge on acing the ASVAB. If you follow these simple rules and stick to them during the time leading up to the ASVAB, you will do well. You need to be disciplined and persistent. Merely hoping that you will do well won't cut it. Merely flipping through this book won't do it. You need to work hard, stay consistent, and focus like a laser to achieve your goal.

1. Be Prepared

Don't count on luck, your good looks, or your charming personality to get you high scores on the ASVAB. The best way to do well on the ASVAB is to study hard and do well in your courses at school. Even if you did well in school, focusing on the contents of this book will help you perform better on the ASVAB. Since the main criterion for military entrance is the AFQT, you need to do especially well on the math and verbal tests. Reviewing the contents of this book should help boost your score in those areas.

The other ASVAB tests are used for entry into certain occupational fields and training programs. Doing well on these tests gives you a better chance of getting into the career field of your choice.

You have had years of opportunity to learn math, reading, and vocabulary, but you may or may not have taken courses in these other areas. For example, not everyone takes automotive courses or shop courses in school, so it will be very helpful to review the material in this book and take the practice exams. Not everyone has had formal training in the skills of mechanical comprehension. Spending time on learning the content of these areas is well worth your while in getting good jobs in the military.

2. Set Up a Study Plan and Don't Procrastinate

Give yourself three to five months of study and review before taking the ASVAB. You may need to spend more or less time studying, depending on how strong your abilities are in each content area. For example, maybe your math skills are just fine, but your knowledge of auto and shop concepts or general science needs a lot of work. Adjust the schedule to meet your own needs.

A sample 20-week study plan is shown in Figure 6.1. It is followed by Figure 6.2, a blank study plan for your own use.

Tasks begun well, likely have good finishes.

—SOPHOCLES
496–406 BC

Figure 6.1 Sample study plan

Week 1	Select a place to study and gather all necessary books and materials. Develop a weekly study plan.
Week 2	Take the diagnostic test in this book and identify your strengths and weaknesses. Determine which areas need the most work.
Week 3	Review the math skills section in this ASVAB book.
Week 4	Review math skills and take the ASVAB math tests in Practice Test 1 in this book.
Week 5	Review the vocabulary lists and meanings in this book and take the Word Knowledge test in Practice Test 1 in this book.
Week 6	Read some books and magazines. Identify the main characters and the main idea, and make inferences about what would happen next for 50 different paragraphs. Take the Paragraph Comprehension test in Practice Test 1 in this book.
Week 7	Review science concepts and take the General Science test in Practice Test 1 in this book.
Week 8	Continue to review math and science concepts.
Week 9	Work on review sections in this ASVAB book for Auto and Shop, Electrical Information, Mechanical Comprehension, and Assembling Objects.
Week 10	Work on review sections in this ASVAB book for Auto and Shop, Electrical Information, Mechanical Comprehension, and Assembling Objects. Take the Auto and Shop, Electrical Information, Mechanical Comprehension, and Assembling Objects tests in Practice Test 1 in this book.
Week 11	Go back and review the math and verbal concepts.
Week 12	Take Practice Test 2 and identify your strengths and weaknesses.
Week 13	Review all concepts.
Week 14	Review all concepts.
Week 15	Take Practice Test 3. Decide in what areas you need to continue your study efforts.
Week 16	Keep reviewing the content, focusing on your weakest areas. Concentrate on the AFQT tests.
Week 17	Review notes.
Week 18	Review subject reviews in this ASVAB book.
Week 19	Review.
Week 20	Review.

Figure 6.2   Study plan—blank

Week 1	
Week 2	
Week 3	
Week 4	
Week 5	
Week 6	
Week 7	
Week 8	
Week 9	
Week 10	
Week 11	
Week 12	
Week 13	
Week 14	
Week 15	
Week 16	
Week 17	
Week 18	
Week 20	

3. Set Up a Weekly Study Schedule

If you want to do well on the ASVAB, it is very important that you make studying for it a priority in your life. You should spend 4 to 6 hours per week on your studies for this test; more if you need it. Setting up a weekly schedule and sticking to it is very important.

In Figure 6.3 you'll see a sample weekly study schedule. It is followed by Figure 6.4, a blank form that you can use to make your own schedule.

- *When can you study?* Before completing the plan, you need to review your typical week to determine when you can make time to study. If you go to school or have a job, your time for study may be limited to evenings and weekends.

- *What time of day is best?* Do you work better in the mornings, afternoons, or evenings? Schedule your study time when you are most likely to be efficient in your study. Use the chart at the bottom of this page to determine if you are a morning or an evening person.

- *How long can you study?* You know yourself better than anyone. Are you able to study well for short or long periods? Can you focus for one hour and then get restless, or can you work undistracted and focused for several hours? Block out the times that seem ideal for you. Adjust the time as you learn what's best for you.

- *It's not just the clock time.* Don't fool yourself into thinking that just because the clock has indicated that three hours have passed, you have spent three hours studying. Make sure that you are spending quality time on your studies.

- *Make room for exercise.* It's very important that you schedule in some exercise, as that will help reduce the stress associated with working hard to achieve a goal. It will also keep you healthier so that your study schedule is not interrupted by illness.

- *Write down your study schedule and stick to it.* Posting your schedule for your friends and family to see can help them understand that you need to use your time for study. This could help eliminate distractions. If you are really serious about doing well on the ASVAB, you may need to give up some nonessential activities, like dates, movies, and hanging out with your friends. Don't worry. It's only temporary to reach your goal.

Be realistic. Don't plan study periods during the week if it is unlikely that you will follow through. In the beginning, you may schedule one or two study periods and then increase your commitment to study as you get focused and organized. Your success depends on you.

Tips for Managing Your Time

- State your goals.
- Make up a weekly list of things to accomplish.
- Make up a "to-do" list for tomorrow and set priorities.
- Keep track on paper of time spent studying and what you accomplished.
- Reward yourself for finishing items on your "to-do" list.

Are You a Morning or Evening Person?

Morning Person	Evening Person
I often wake up before the alarm in the mornings.	When I hear the alarm, I feel awful and hit the snooze button.
When I wake up, I am ready to go.	When it's 10 a.m., I am just starting to feel awake.
In the mornings, my mind is clear and alert.	I'm still a little groggy into the afternoon.
The best time to talk with me is in the mornings.	Don't talk to me in the morning, I only feel human in the evenings.
I fade about 10 or 11 p.m.	I'm starting to feel sharp at about 8 p.m.

Figure 6.3   Sample weekly planning schedule

	Monday	Tuesday	Wednesday	Thursday	Friday	Saturday	Sunday
8–9 a.m.	School/Work	School/Work	School/Work	School/Work	School/Work	Errands	Personal Time
9–10 a.m.	↓	↓	↓	↓	↓	↓	↓
10–11 a.m.							ASVAB study
11 a.m.–12 p.m.	↓	↓	↓	↓	↓	↓	↓
12–1 p.m.	Lunch	Lunch	Lunch	Lunch	Lunch		Lunch
1–2 p.m.	School/Work	School/Work	School/Work	School/Work	School/Work	Lunch	
2–3 p.m.						ASVAB study	Fitness activity
3–4 p.m.						↓	↓
4–5 p.m.	↓	↓	↓	↓	↓		
5–6 p.m.	Fitness activity	ASVAB study	Fitness activity	ASVAB study	TGIF		
6–7 p.m.	↓	↓	↓	↓		Movie and Dinner	
7–8 p.m.	Dinner	Dinner	Dinner	Dinner	Dinner		Dinner
8–9 p.m.							

Figure 6.4 Blank weekly planning schedule

	Monday	Tuesday	Wednesday	Thursday	Friday	Saturday	Sunday
8–9 a.m.							
9–10 a.m.							
10–11 a.m.							
11 a.m.–12 p.m.							
12–1 p.m.							
1–2 p.m.							
2–3 p.m.							
3–4 p.m.							
4–5 p.m.							
5–6 p.m.							
6–7 p.m.							
7–8 p.m.							
8–9 p.m.							

4. Pick a Location Where You Study Best

Some people work best in a quiet room with the door closed and no distractions. Others prefer to flop on the couch, have music playing in the background, put their feet up, and hit the books. Think hard about what works best for you. Some people think they are studying just because they have a book in their hands, but they aren't working effectively because they are really listening to the music, texting, or paying attention to what's on the TV. Be honest and smart about where you study. Your best bet is to find a quiet place that is well lit and has no distractions. Get real about what's best for you. Multitasking? It's a myth. Study without distractions to get the best results.

Best Study Setting	
Quiet	At a desk
Good lighting	No distractions
No phones	No excuses
No TV or music	

Check out your local library. Many people live in busy and active households, making it very difficult to find a quiet place to focus on their studies. Maybe your local library is the best place if you can't find a quiet place at home, school, or work. The library is well lit, and it should reduce any eyestrain, making it easier for you to concentrate on your studies and allowing you to study longer without tiring.

5. Get Support from Friends and Family

Tell your family and friends about your goals and your study schedule. This will show them that you are serious about your plan, and they will be more likely to give you the support and space that you need. It is important to have some cheerleaders to help you stay on your course. Post your schedule where your family can see it so that they can be supportive.

You may want to engage your family and friends in helping you. They can use flash cards to test you on what you know and can talk through some challenging concepts until you have those ideas down cold.

Since you will be spending a lot of time studying, ask your family if they would be willing to take on some of your household chores to give you more time to accomplish your goals. Do what you can to make your fair contribution, though.

When you have finished your study schedule, be sure to thank everyone who helped you achieve your goals. They probably had to make some sacrifices when you were studying.

Caution: Studying with Others
If you are taking the ASVAB with a friend or friends, it could be useful to study together. Friends may be able to help you with certain subject areas, and you may be able to help them with others. Be careful, though, because you might end up spending more time socializing than studying.

6. Know the Test

Good test preparation includes knowing what to expect on the test. What kinds of questions will you get? What test question format should you anticipate? Reviewing this book will go a long way toward helping you become comfortable with the types of test questions you will be seeing on the ASVAB.

7. Work at It

Everyone needs to work on succeeding and performing well. You are not alone. The smartest and most successful people that you know or have heard about haven't come by their successes easily. Everyone has to work at being successful. You can be successful as well, if you work at it. The great thing is that it is your choice. You can create your own destiny.

Steps to Acing the ASVAB!

1. Be prepared.
2. Set up a study plan and don't procrastinate.
3. Set up a weekly study schedule.
4. Pick a location where you study best.
5. Get support from family and friends.
6. Know the test.
7. Work at it—work hard.

TAKING CARE OF YOUR BODY AS WELL AS YOUR BRAIN

There are three areas that you need to attend to while you are studying: your stress level, exercise, and what you eat. Concentrating on these areas will help to facilitate the learning that you will undertake over the next several weeks.

Reduce Stress

When you are working toward an important goal, there is a certain level of stress that comes with the territory. If there is enough stress for a sustained period, eventually this can hurt your ability to focus and may even damage your health. You need to implement some stress-reducing actions on a regular basis.

It may sound silly, but staying quiet and doing some deep breathing exercises have been shown to reduce stress. If you have scheduled a two- to three-hour block of study time, do this for a couple minutes every hour or so.

Also, try visualizing some wonderful, idyllic places that are particularly soothing to you, like waves on a beach, a gentle spring shower, or a snow-capped mountain range, as you breathe deeply. This should help relax you.

If you are trying to sleep and are too wound up from working and studying, and you are in bed and the room is dark, start to breath deeply while relaxing your toes, then moving to your calves, your thighs, your hips, your shoulders, your arms, your jaw, your head, and so on. See the suggestions for

---

Relaxation Exercise

While in bed or in a dark, quiet place, lie still with your eyes closed. Let your body and mind slow down for a few moments. Stay still and quiet. Breathe deeply. Be aware of your breathing.

As you stay still and quiet, focus on your toes. Clench them and then let them relax. Do this two or three times. As you relax your toes, think about how they would feel if they turned into liquid. Let your toes turn to liquid. Release the tension. Then do this same exercise with your ankles, your calves, your knees, your thighs, your buttocks, your stomach, your chest, your back, your hands, your forearms, your biceps, your shoulders, your neck, your jaw, your cheeks, your eyes, and your forehead. Let them all turn to liquid as you move from your toes to your head.

Once your muscles are relaxed, let your mind float and dissolve. Let your thoughts flow away. Breathe deeply as you reduce your stress.

---

the relaxation exercise. There are several phone apps that have stress-reducing audio.

Exercises like those mentioned should help to reduce your stress level and should help you get to sleep faster and sleep better. With a good sleep, you will be more enthusiastic and better able to face the challenges of the next day and the next ASVAB study schedule. It will also help you retain more of what you study. Seven to nine hours of sleep is generally recommended.

Get Physical Exercise

Be sure to incorporate some vigorous physical exercise into your weekly routine. Doing some sort of aerobic activity, like running, power walking, or cycling, will help reduce your stress level and will also keep you physically stronger to contend with your study schedule. Don't skip your weight-training exercises, either. Physical activity has been shown to reduce stress and keep you healthier.

Plus, if you are really interested in joining and performing well in the military, you need to be physically fit, so why not start now to incorporate exercise into your weekly schedule? Start transforming that body fat into muscle. Do you need to lose extra pounds? Drop them now.

Eat and Act Right

It's very important for your heath that you maintain a good weight and eat well. Eating low-fat, healthy meals will keep your weight down, which will, in turn, help you feel more energetic, healthy, and focused on your task.

Abusing your body in any way will only hurt your efforts, as it will make you less alert and less focused. Clearly, if you are abusing drugs or alcohol, your entrance into the military will be problematic, at best. Get a grip on those behaviors, as they are likely to reduce your chances of qualifying for military enlistment. Shed those bad habits now!

Although you won't be disqualified from the military for smoking, it is a bad habit that damages your body and brain functioning. Smoking also reduces your athletic ability, a very important aspect of being successful in the military. Offload the smoking problem now, and if you haven't succumbed, don't start.

HOW TO STUDY

There are several ways that you can approach studying. The methods you select depend on how you have studied in the past and what feels most comfortable and productive to you. You may find many of the techniques listed here to be useful as you work on your preparation for the ASVAB.

> The secret to getting ahead is getting started.

Prioritize

Start with the most difficult subjects first. That way, you will give your freshest and most focused attention to the subject areas that are hardest for you.

Use Dead Time

Any time there is some space in your schedule, review your notes or lists.

Highlight Important Sections

As you review the contents in this book or any textbook that you use, it is helpful to use a highlighter to focus on important concepts. Another good strategy is to take notes on the concepts that seem important to understand and remember.

Create Concept Maps

Concept maps are somewhat like a visual outline. They help you keep things straight and show the relationships between and among various ideas.

The concept map at the bottom of this page begins to describe the various types of rocks on the Earth.

Read and Summarize

As you read a section of material, stop and think about what you read for a moment. In a notebook, write down what you read in your own words. Do this for each section. The more active and interactive you are with the subject material, the more you will remember. It's been shown that if you physically write something down with a pen or pencil on paper, you will retain it longer.

Develop Flash Cards

For things that you just need to memorize, develop flash cards with the question on one side and the answer on the other. Using flash cards, you can study while waiting for the bus, or your family and friends can test you. Just writing things down on the flash cards will help you retain the information.

Use the PQRST Method

Preview, question, read, summarize, and test yourself.

Preview. Skim the material to see what it generally contains.

Question. As you work through a section of material, think about what types of questions could be asked on a test. Write down the questions and the answers.

Read. Read sections of material carefully and actively.

Summarize. Summarize what you've read. Jot down notes, diagrams, key words, and processes.

Test. Recite and review what you have read and written down in your notes immediately after you have read and summarized.

THE DAY BEFORE THE TEST

Don't Cram

It's too late anyway. Studying for the ASVAB is not something you do the day before the test. At this point, it's too late to stuff information into your brain. If you paid attention to the advice in this book,

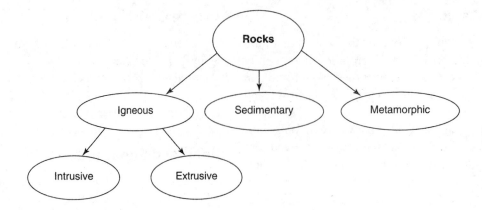

you have undertaken a steady pace of study that is directed toward success on the test.

Exercise Your Body

If you have studied hard and followed your study plan, you should take the day before the test to pay attention to your personal well-being. Take a run or a long walk, play tennis or basketball, go rock climbing, or have a picnic. Do whatever physical activity gives you joy. Relax and feel good. Drink a good amount of water.

Revel in the fact that you have worked hard and will have the opportunity to show what you know tomorrow. Be excited that your hard work will finally have a payoff as you take the ASVAB. You've worked hard, and now you will obtain your reward.

Have a Healthy Dinner

Eat a decent, low-fat meal. Chicken or fish and a good supply of vegetables would be appropriate. Don't drink any alcoholic beverages or high-sugar or caffeine drinks.

Sleep Well

You have worked hard for the past several months. If you have followed the advice in this book, you should go to bed feeling confident and relaxed, and with a smile on your face. You are going to rock the ASVAB! Go to sleep early and set your alarm well in advance of the time you need to be at the testing center. Setting a backup alarm is a good idea. If you have trouble sleeping, do some relaxation exercises.

THE DAY OF THE TEST
Eat a Good Breakfast

Have a modest breakfast, but lay off the butter, sausage, bacon, and syrup. Eat lightly, choosing foods that will give you energy and focus. Cereal and milk with some fruit would be good, as would oatmeal. A cup of coffee might sharpen your edge. Eating too much or eating foods that are too fatty can make you feel, think, and act sluggish, and you may not perform your best. Your best bet is to be eating healthy meals all along. One morning of eating right won't help you much. By all means, don't go to the test with an empty stomach.

Don't Drink a Lot of Fluid

Certainly don't go into the test dehydrated, but drinking lots of coffee and water before the test may cause you to be uncomfortable. This could distract you from performing your best.

Be Early

If you are taking the test at school, you probably need to arrive at your usual time. If you are taking

---

Reducing Test Anxiety

1. Be prepared. If you have studied hard, there is no reason that you can't show what you know.
2. Banish all your excuses for not studying.
3. Don't procrastinate.
4. Stay with an organized study routine that sets aside time each week to study.
5. Reduce as much stress from other aspects of your life as possible.
6. Stay in good physical condition by exercising and eating right.
7. Visualize your success. Imagine yourself receiving great ASVAB scores.

---

the test at a MEPS or a MET site, it's a good idea to arrive a little early and get comfortable with the testing room.

Relax and Glance at Your Notes

A little stress before you start the test is a good thing, as it will help you stay mentally sharp. An excess of stress can cause test anxiety, a condition that is not helpful to your performance. You have studied hard and done all you can do to prepare, so be confident in yourself and relax. If you feel you need to keep reviewing, scan your notes one more time, but then set them aside. If you follow the advice in this book, you will approach the test as an opportunity to show them what you know. Take a deep breath and begin.

Read or Listen to the Directions

The ASVAB has very specific directions for each test within the test battery. Read the directions carefully and listen intently as the directions are read to you by the test administrator or provided to you on the computer screen. The directions will tell you what you need to do and how you need to approach the test items. You will be given a specified amount of time for each subtest. Pay attention to the time schedule, especially for the paper-and-pencil version of the ASVAB. There will probably be a clock in the room, but bring a watch so that you can manage your time.

The directions that you will be given for both the paper-and-pencil version and the computer version are provided elsewhere in this book, so you should have little in the way of surprises.

Before each test, you will be given some practice items. Be sure to pay attention to those items and the directions. If there is something you don't understand, ask for clarification.

Test Administration (Paper-and-Pencil ASVAB)

The Room You will probably be tested in a room that has a desk or a table that you can use to open the test booklet and record your answers on the answer sheet. There will be a clock visible to you to help you with the timing of the test. The test administrator will indicate the time the test will end on a sign or on the board. The room should be free from noise and visual distractions. The lighting should be sufficient for you to see the questions easily. An audible electronic timer will be used for timing the various subtests to let you know when to stop answering questions.

There will be a test administrator for every 25 persons. If there are more than 25 people, there should be one proctor for every 25 people.

The test directions will be read to you verbatim and in English. You will be given scratch paper to work with.

Cheating—Don't Even Think About It If you are taking the paper-and-pencil version along with other people, individuals will be given one of several versions of the ASVAB. So copying from others is likely to give you the wrong answers. You will also be seated in such a way that you cannot readily observe another person's answer sheet, and if you can see the answer sheet, the test version and thus the answers are going to be different.

Use of unauthorized assistance such as calculators, slide rules, other mechanical devices, or crib sheets and going forward or backward to other subtests are considered cheating. If you are caught cheating, your test results will be invalidated and you will not be able to retake the ASVAB for six months.

Given that moral character is important for the military, starting off your career with cheating is, well, *not your best decision, to say the least.*

Test Administration (CAT-ASVAB)

You will likely be in a room with multiple computers. Each person gets his or her own set of questions based on how they answer initial questions.

See chapter 5 for some specifics on how the CAT-ASVAB works.

TEST-TAKING POINTERS

The review sections of this book have some test-taking pointers that are specific to the subject matter. There are also some general test-taking strategies for multiple-choice questions that are good to keep in mind.

- Read each question carefully; then try to answer it without looking at the possible answers given. If you find your answer among those provided, you probably have a correct answer.

- Read all the answers before deciding on the correct answer. In many instances, there could be more than one correct answer, but there will be one that is clearly the best.

- If there are two answers that are opposites, it is likely that one of them will be the correct answer.

- Try to eliminate answers that are clearly wrong and think through the possible remaining answers.

- Carefully consider answers that have qualifiers (*usually, mostly, most often, generally*). These tend to be correct.

- Pay attention to words like *always, never, all, none,* and *only*. They mean what they say.

- If you skip a question or want to go back and revisit it if you have time, jot the item number down on the scratch paper that you have been given. (This applies only to the paper-and-pencil version of the ASVAB.)

- For the paper-and-pencil version, if you have time, go back and review your answers, but remember that your first answer is usually correct. Change the answer only if you are very sure that it is wrong.

- Answer every item. On the paper-and-pencil version, there is no penalty for guessing. On the CAT-ASVAB, try to answer every question correctly. Your score is higher if you answer harder questions.

- A penalty procedure is applied to all examinees who do not complete the test before time runs out. In almost all cases, it is not necessary to apply the penalty as there is ample time to answer all the questions.

- Be sure the answer you have selected is the one you want to select, as you aren't allowed to return to a question once you have answered it.

- If time is running short, try to read and legitimately answer the questions, rather than filling in random guesses for the remaining items, as the CAT-ASVAB applies a relatively large penalty when several incorrect answers are provided toward the end of a subtest.

Perhaps the most valuable result of all education is the ability to make yourself do the things that you have to do, when it ought to be done, whether you like it or not.
—THOMAS HUXLEY

Believe you can and you are half way there.
—THEODORE ROOSEVELT

Part 2

ASVAB
DIAGNOSTIC
TEST
FORM

ASVAB Diagnostic Test Form

YOUR GOALS FOR THIS TEST:

- **Take a complete practice ASVAB under actual test conditions.**

- **Review the actual test directions that you will hear on test day.**

- **See a sample ASVAB answer sheet and learn how to mark your answers.**

- **Check your test answers and read explanations for every question.**

- **Use your results to identify your strengths and weaknesses.**

Now that you have learned all about the ASVAB and the important role it can play in your future, it is time to try out your test-taking skills. The following pages present a full-length practice ASVAB that will give you a very good idea of what the actual test is like. You will see samples of every type of ASVAB question, and you will get to know all the different topics that the real test covers. You will also read the actual test directions that you will hear on test day, and you will mark your answers on a copy of the actual ASVAB answer sheet used by the Department of Defense.

Try to take this diagnostic test under actual test conditions. Find a quiet place to work, and set aside a period of approximately $2^1/_2$ hours when you will not be disturbed. Work on only one section at a time, and use your watch or a timer to keep track of the time limits for each section. Mark your answers on the answer sheet, just as you will when you take the real exam.

At the end of the test, you'll find an answer key and explanations for every question. After you check your answers against the key, you can complete a chart that will show you how you did on the test and what test topics you might need to study more. Then review the explanations, paying particular attention to the ones for the questions that you answered incorrectly.

Once you have worked your way through this diagnostic test, you will know how ready you are right now for the actual ASVAB. You will find out which test subjects and which types of question are easy for you, and which ones give you trouble. You will also find out whether you are able to work fast enough to finish each section within the time allowed or whether you need to improve your test-taking speed. With the knowledge you gain, you'll be able to plan an ASVAB preparation program that fits your needs. Later on, when you read through the ASVAB subject reviews in Part 3 of this book, you'll know which topics and question types to focus on in order to get your best score.

TEST DIRECTIONS

IF YOU ARE TAKING THE PAPER-AND-PENCIL VERSION OF THE ASVAB, THE
TEST ADMINISTRATOR WILL READ THE FOLLOWING ALOUD TO YOU.

I am [Administrator's name], your test administrator for today. I am here to administer the Armed
Services Vocational Aptitude Test Battery, also known as the ASVAB. This is a battery of tests
designed to assess various aspects of school and career aptitudes. The scores you obtain will be
provided to your school guidance counselor to help you formulate your future plans. The scores
can be useful to you only if they accurately reflect your abilities; therefore, it is important for
you to do your best on these tests. Your score will have no effect on your school grades. If you
are ill or too tired to take the test, please raise your hand at this time.

At this time, please clear your desk. Each of you will be given a test booklet, an answer
sheet, scratch paper, and two pencils. Unauthorized assistance, such as calculators, slide rules,
other electronic or mechanical devices, crib sheets, or talking during the test, is not permitted.
Examinees who do not abide by these requirements will be dismissed from the test session, and
their tests will be invalidated. *Do not open* your test booklet until told to do so.

Now turn off any electronic devices you might have with you (i.e., cellular phones, pagers,
watch alarms, etc.). This will prevent disturbing other examinees in the room.

No breaks are permitted during the test. If it is necessary to use the restroom, please raise your
hand. I will collect your test materials and will give them back to you when you return. You may
not work on the same portion of the test once you return. Only one male and one female student
may be out of the room at the same time.

You should now have in front of you a test booklet, answer sheet, scratch paper, and two
pencils. If you are missing one of these items, please raise your hand now. Again, I must remind
you, do not open your test booklet until told to do so.

In marking your answer sheet, use only the pencils that have been provided to you. If both your
pencils break, please hold one up over your head and a replacement will be brought to you. Do not
make any stray marks on the answer sheet; do not fold, crease, or tear it. Damaged answer sheets
cannot be automatically scored. Please listen carefully to all directions. You will perform better if
you understand what you are doing. If the directions are not perfectly clear, raise your hand.

Please take your scratch paper and print your name and today's date in the upper right-hand
corner.

You should have a three-page answer sheet that is fastened together. (See the example on
pages 67, 68, and 69.) Do not separate them. With the fastened edge on your left, you will see
that on the upper center portion of the answer sheet there is a black printed seven-digit sequence
number. Find that number now. That same number should also be printed in the upper center
portion of pages 2 and 3. Check now to make sure that the sequence number is identical on all
pages of the answer sheet. Also check to see that page 2 has a "Signature" block and that page 3
has parts 1 through 9. If you find anything different from that, please raise your hand and you
will be provided with another answer sheet.

Turn to page 2 of the answer sheet. Essentially, the statement you are about to read tells you
what purposes your scores will serve: counseling, research, military recruiting activities, etc. *If
you fail to sign this statement acknowledging the intended uses and release of the test scores,
your test will not be processed and no score will be available for your use.* Read the statement

now. When you have finished reading the statement, please sign it and record today's date. A copy of this statement is available to you after the test session if you so desire.

Return to page 1 and turn your answer sheet sideways so you can read the section labeled "Student Name" at the upper left side. Please print your last name, one letter per box, starting with the first box on the left. Skip a box and print your first name, then skip another box and print your middle initial if you have one. It is all right if you run out of space; just print as much of your first name as possible. Below the boxes in which you printed your name are columns of letters. In each column, completely blacken the space corresponding to the same letter that you printed in the name boxes.

In block 2, titled "Home Mailing Address," enter as much of your street address as will fit in the boxes provided. If your mailing address is general delivery, enter "GENDEL." If you receive mail on a mail route, enter "ROUTE," skip a box, and enter your route number, letter, or name. If you have a post office box, enter "BOX," skip a space, and enter the box number. Then, as before, completely blacken in the appropriate spaces corresponding to the letters or numbers below the letters or numbers you have printed.

In block 3, titled "Home City," print the name of the city in which you live. Completely blacken in the corresponding spaces below.

In block 4, titled "ST" for state, print the two-letter abbreviation for your state, which is _____, and completely blacken in the corresponding spaces below.

In block 5, titled "Zip Code," enter the first five numbers of your home address zip code and completely blacken in the corresponding spaces below.

In block 6, titled "Area Code/Phone Number," enter the three-digit area code, then your seven-digit phone number and completely blacken in the corresponding spaces below. If you do not have a phone, leave this block blank.

In block 7, titled "School Code," enter the nine-digit number _____ and completely blacken in the corresponding paces below.

In block 8, print the name of this school.

In block 9, titled "Education Level," completely blacken in the space that corresponds to your present school grade. For example, if you are a senior, blacken in the circle next to 12. If you are a junior, blacken in the circle next to 11. Scores cannot be provided for you if you leave this block blank.

In block 10, titled "Sex," indicate your gender by completely blackening in the space next to "Male" or "Female."

In block 11, titled "Intentions," completely blacken in the space corresponding to your plans after completing high school. If you are not sure of your plans at this time, completely blacken in the space next to "Undecided."

In block 12, titled "Test Version," enter the test version number corresponding to the test version number printed on the front of your test booklet and completely blacken in the spaces below. The version number should either be 23A, 23B, 24A, or 24B. If you are unsure of your test version, please raise your hand now and someone will assist you.

Please leave block 13, titled "SS ASVAB," blank.

Are there any questions on filling out any of the blocks on page 1?

Now turn to page 2. On the line provided in the upper left-hand corner, print your last name, first name, and middle initial. Also enter your social security number on the line indicated as SSN.

In block 14, titled "Social Security Number," this is no longer required.

Please leave block 15, titled "TA IA Number," blank.

In block 16, titled "Code," place your school's distribution code and completely blacken in the corresponding spaces below the boxes.

In block 17, titled "Date of Birth," enter the year and date of your birth.

Leave block 18, titled "SP Studies," blank.

In block 19, titled "Test Booklet Number," enter the seven-digit number printed on the upper left corner of the cover of your test booklet. Completely blacken the corresponding spaces below. If your test booklet number has six digits, place a zero in the first box.

In block 20, titled "Racial Category," please blacken any of the appropriate responses to indicate the racial category or categories of which you consider yourself a member. If you do not wish to respond, please leave block 20 blank.

In block 21, titled "Ethnic Category," please blacken the appropriate response to indicate the ethnic category of which you consider yourself a member. If you do not wish to respond, please leave block 21 blank.

At this time, remove page 3 of the answer sheet. Pages 1 and 2 will now be collected from you.

On page 3, on the line provided in the upper left-hand corner, print your last name, first name, and middle initial. Also, enter your social security number on the line indicated as SSN.

Please open your test booklet to page 1 and read the instructions silently while I read them aloud.

GENERAL DIRECTIONS

DO NOT WRITE YOUR NAME OR MAKE ANY MARKS in this booklet. Mark your answers on the separate answer sheet. Use the scratch paper which was given to you for any figuring you need to do. Return this scratch paper with your other papers when you finish the test.

If you need another pencil while taking this test, hold your pencil above your head. A proctor will bring you another one.

This booklet contains 9 tests. Each test has its own instructions and time limit. When you finish a test you may check your work in that test ONLY. Do not go on to the next test until the examiner tells you to do so. Do not turn back to a previous test at any time.

For each question, be sure to pick the BEST ONE of the possible answers listed. When you have decided which one of the choices given is the best answer to the question, blacken the space on your answer sheet that has the same number and letter as your choice. Mark only in the answer space. BE CAREFUL NOT TO MAKE ANY STRAY MARKS ON YOUR ANSWER SHEET. Each test has a separate section on the answer sheet. Be sure you mark your answers for each test in the section that belongs to that test.

Here is an example of correct marking on an answer sheet.

S1 A triangle has

    A. 2 sides
    B. 3 sides
    C. 4 sides
    D. 5 sides

PRACTICE

S1 Ⓐ ● Ⓒ Ⓓ
S2 Ⓐ Ⓑ Ⓒ Ⓓ
S3 Ⓐ Ⓑ Ⓒ Ⓓ

The correct answer to Sample Question S1 is B.

Next to the item, note how space B opposite number S1 has been blackened. Your marks should look just like this and be placed in the space with the same number and letter as the correct answer to the question. Remember, there is only ONE BEST ANSWER for each question. If you are not sure of the answer, make the BEST GUESS you can. If you want to change your answer, COMPLETELY ERASE your first answer mark.

Answer as many questions as possible. Do not spend too much time on any one question. Work QUICKLY, but work ACCURATELY. DO NOT TURN THE PAGE UNTIL TOLD TO DO SO. Are there any questions?

ASVAB DIAGNOSTIC TEST FORM

ANSWER SHEET

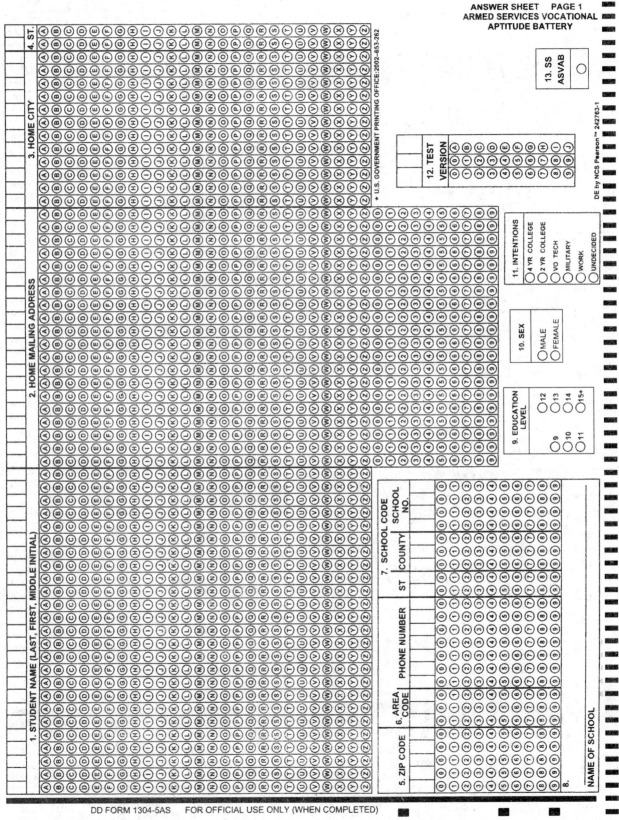

ASVAB DIAGNOSTIC TEST FORM

ANSWER SHEET

ANSWER SHEET   PAGE 2
ARMED SERVICES VOCATIONAL
APTITUDE BATTERY

16. CODE

15. TA ID NUMBER

14. SOCIAL SECURITY NO.

19. TEST BOOKLET NUMBER

18. SP STUDIES

17. DATE OF BIRTH

YEAR   MO   DAY

21. ETHNIC CATEGORY
- HISPANIC OR LATINO
- NOT HISPANIC OR LATINO

20. RACIAL CATEGORY
- AMERICAN INDIAN / ALASKA NATIVE
- ASIAN
- BLACK / AFRICAN AMERICAN
- NATIVE HAWAIIAN / OTHER PACIFIC ISLANDER
- WHITE

LAST

FIRST

MI

SSN

STUDENT TESTING PROGRAM
PRIVACY ACT STATEMENT
Armed Forces Vocational Aptitude Battery

1.   **AUTHORITY:** 10 USC 133 and 3013; and E.O. 9397

2.   **PRINCIPAL PURPOSE(S):** To compute and furnish test score products for career/vocational guidance and group assessment of aptitude test performance; for up to 2 years, to establish eligibility for enlistment (only for students at the eleventh grade or higher and only with the expressed permission of the school); for marketing evaluation, assessment of manpower trends and characteristics; and for related statistical studies and reports.

3.   **ROUTINE USE(S):** None

4.   **DISCLOSURE:** Voluntary. However, if you do not provide the requested information, your test will not be scored or otherwise processed.

SIGNATURE

DATE

ASVAB DIAGNOSTIC TEST FORM

ANSWER SHEET

PART 1. GENERAL SCIENCE

THE TEST ADMINISTRATOR WILL READ THE FOLLOWING ALOUD TO YOU.

Turn to Part 1 and read the directions for General Science silently while I read them aloud. This test has questions about science. Pick the BEST answer for each question and then blacken the space on your separate answer sheet that has the same number and letter as your choice. Here are three practice questions.

S1. A rose is a kind of
   A. animal.
   B. bird.
   C. flower.
   D. fish.

S2. A cat is a
   A. plant.
   B. mammal.
   C. reptile.
   D. mineral.

S3. The earth revolves around the
   A. sun.
   B. moon.
   C. meteorite.
   D. Mars.

Now look at the section of your answer sheet labeled "PRACTICE." Notice that answer space C has been marked for question S1. Now do practice questions S2 and S3 by yourself. Find the correct answer to the question, then mark the space on your answer sheet that has the same letter as the answer you picked. Do this now.

You should have marked B for question S2 and A for question S3. If you make any mistakes, erase your mark carefully and blacken the correct answer space. Do this now.

Your score on this test will be based on the number of questions you answer correctly. You should try to answer every question. If you finish before time is called, go back and check your work in this part <u>ONLY</u>. Now find the section of your answer sheet that is marked PART 1. When you are told to begin, start with question number 1 in Part 1 of your test booklet and answer space number 1 in Part 1 on your separate answer sheet. DO NOT TURN THE PAGE UNTIL TOLD TO DO SO. This part has 25 questions in it. You will have 11 minutes to complete it. Most of you should finish in this amount of time. Are there any questions?

Turn the page and begin.

1. The basic unit of life is
   A. the cell.
   B. nucleus.
   C. the carbon molecule.
   D. oxygen.

2. When a plant undergoes photosynthesis, one of the by-products is
   A. carbon dioxide.
   B. light.
   C. ozone.
   D. oxygen.

3. Why is the nucleus so important to a living cell?
   A. It controls the production and elimination of cell waste.
   B. It controls most of the activities of the cell.
   C. It controls photosynthesis.
   D. It uses oxygen to make carbon dioxide.

4. Which of the following do not have organelles?
   A. Muscle cells
   B. Protozoa
   C. Nerve cells
   D. Bacteria

5. Plants bend toward a source of light on account of
   A. geotropism.
   B. thigmotropism.
   C. chemotropism.
   D. phototropism.

6. Which of the following terms can be used to describe a bird?
   A. Invertebrate, omnivore
   B. Vertebrate, warm-blooded
   C. Endoskeleton, cold-blooded
   D. Crustacean, endoskeleton

7. Why are decomposers important to an ecosystem?
   A. They make nutrients available for plants to use.
   B. They create soil for farming.
   C. They stabilize the soil.
   D. They enhance the habitat with specialized particles important to life.

8. Which of the following is the part of the skeletal system that protects the spinal cord?
   A. Marrow
   B. Cartilage
   C. Ligaments
   D. Vertebrae

9. A bruise on your skin is the result of which of the following?
   A. Excess strain on the muscle
   B. Broken capillaries
   C. Overactive sweat glands
   D. Swelling

10. The function that moves food from one part of the digestive system to another is called
    A. respiration.
    B. digestion.
    C. mechanical decomposition.
    D. peristalsis.

11. Which of the following words are all associated with the nervous system?
    A. Nephrons, urethra, bladder
    B. Antigens, T cells, cilia
    C. Hypothalamus, pons, cortex
    D. Lymph nodes, spleen, tonsils

12. Which of the following is the name for the space between two nerve cells where an electric charge transmits information?
    A. Nerve cell gap
    B. Synapse
    C. Axon
    D. Dendrite

13. Which of the following states is likely to have the greatest biodiversity?
    A. North Dakota
    B. Illinois
    C. Nebraska
    D. Florida

14. Which of the following best represents what happens to ultraviolet radiation from the sun when it reaches Earth's atmosphere?
    A. It changes into chlorofluorocarbons.
    B. Some of it is absorbed by the ozone layer.
    C. All of it reaches Earth's surface.
    D. It is deflected back into space.

15. Which of the following properties of a sound wave affects how loud or soft the sound is?
    A. Crest
    B. Trough
    C. Frequency
    D. Amplitude

16. The major force causing sea-floor spreading, continental drift, and the collision of continental plates is
    A. subduction of Earth's core.
    B. creation of oceanic ridges.
    C. compression of continents.
    D. convection currents in the mantle.

17. When two tectonic plates converge and one rises above the other, which of the following are likely to form?
    A. Volcanoes
    B. Hot spots
    C. Tsunamis
    D. Rifts

18. Breaking waves or breakers that occur near the shoreline are caused by which of the following conditions?
    A. Earthquakes that occur far out in the ocean
    B. Friction between the wave trough and the ocean bottom
    C. The expansion of the wavelength of the wave
    D. Increasing tidal forces

19. How long does it take for Earth to make one revolution around the sun?
    A. 1 year
    B. 1 month
    C. 1 week
    D. 1 day

20. An eclipse of the sun occurs under which of the following conditions?
    A. Earth is between the sun and the moon.
    B. Earth is rotating faster than the moon.
    C. Earth, the sun, and the moon are perpendicular to each other.
    D. The moon is between the sun and Earth.

21. Earth's greenhouse effect is caused by which of the following conditions?
    A. Changing rainfall patterns from global warming
    B. Heat from the sun that is trapped by atmospheric gases
    C. The increase in the thickness of the ozone layer
    D. The number of plants in tropical rainforests trapping CO_2

22. Which of the following can hold the most moisture?
    A. Cold air
    B. Moving air
    C. Stable air
    D. Warm air

23. When a prevailing wind reaches a mountain range, what happens to the air?
    A. It is blocked and starts to move backward.
    B. It rises and cools, causing condensation.
    C. It expands and heats, resulting in rain.
    D. It expands and heats, resulting in a dry, desert-like area.

24. The process of erosion is linked to the formation of which of the following types of rocks?
    A. Igneous
    B. Extrusive
    C. Sedimentary
    D. Metamorphic

25. Scientists believe that the universe is expanding because of which of the following indicators?
    A. The difference in time it takes for light from various stars to reach us
    B. The red shift in the spectra of stars
    C. The difference in chemical composition of various stars
    D. The explosion of supernovas

STOP! DO NOT TURN THIS PAGE UNTIL TIME IS UP FOR THIS TEST. IF YOU FINISH BEFORE TIME IS UP, CHECK OVER YOUR WORK ON THIS TEST ONLY.

PART 2. ARITHMETIC REASONING

THE TEST ADMINISTRATOR WILL READ THE FOLLOWING ALOUD TO YOU.

Turn to Part 2 and read the directions for Arithmetic Reasoning silently while I read them aloud.

This is a test of arithmetic word problems. Each question is followed by four possible answers. Decide which answer is CORRECT, and then blacken the space on your answer sheet that has the same number and letter as your choice. Use your scratch paper for any figuring you wish to do.

Here is a sample question. DO NOT MARK your answer sheet for this or any further sample questions.

S1 A student buys a sandwich for 80 cents, milk for 20 cents, and pie for 30 cents. How much did the meal cost?

    A. $1.00
    B. $1.20
    C. $1.30
    D. $1.40

The total cost is $1.30; therefore, C is the right answer. Your score on this test will be based on the number of questions you answer correctly. You should try to answer every question. DO NOT SPEND TOO MUCH TIME on any one question. If you finish before time is called, go back and check your work in this part ONLY.

Now find the section of your answer sheet that is marked PART 2. When you are told to begin, start with question number 1 in Part 2 of your test booklet and answer space number 1 in Part 2 on your separate answer sheet.

DO NOT TURN THIS PAGE UNTIL TOLD TO DO SO. You will have 36 minutes for the 30 questions. Are there any questions?

Begin.

1. Lisa went to the grocery store and bought a package of hot dogs for $3.54, a package of buns for $1.79, a container of mustard for $2.10, a jar of relish for $2.16, and a bag of charcoal for $5.15. What was her total bill?
   A. $12.64
   B. $14.74
   C. $16.15
   D. $16.74

2. Lashawn bought a blouse for $17.16, a skirt for $38.19, and a purse for $35.14. If she hands the clerk a $100 bill, what is her change?
   A. $1.53
   B. $9.51
   C. $10.67
   D. $14.15

3. Cassidy answered 38 questions correctly on a test with 45 items. To the nearest percent, what percent did she answer correctly?
   A. 72%
   B. 78%
   C. 84%
   D. 89%

4. Metro's test scores for Social Studies are 81, 76, 91, 73, and 79. What is her average score?
   A. 80
   B. 87
   C. 91
   D. 93

5. A box of chocolates has 45 pieces. If 9 pieces have nuts in them, what percentage are without nuts?
   A. 20%
   B. 45%
   C. 79%
   D. 80%

6. In the Metro Bicycle Club, 6 of 48 members are females. What is the ratio of females to all club members?
   A. $\dfrac{3}{16}$
   B. $\dfrac{1}{15}$
   C. $\dfrac{1}{4}$
   D. $\dfrac{1}{8}$

7. A teacher has assigned 192 pages of reading. If Vanessa starts reading on Monday and reads 48 pages each day, on what day will she complete the assignment?
   A. Wednesday
   B. Thursday
   C. Friday
   D. Saturday

8. Samuel is saving coins in a jar so that he can purchase a software program. He has saved 94 nickels, 44 quarters, and 112 dimes. If he reaches into the jar to pick one coin at random, what is the probability that he will pick a dime?
   A. $\dfrac{1}{2}$
   B. $\dfrac{2}{3}$
   C. $\dfrac{56}{125}$
   D. $\dfrac{112}{138}$

9. Rene deposits $4,000 in a savings account that earns 4% simple interest per year. How much interest will she earn after 3 years?
   A. $112
   B. $350
   C. $380
   D. $480

10. Leah flies 5,200 miles in 11 hours. What is the average speed of her airplane?
    A. 472.73 miles/hour
    B. 499.17 miles/hour
    C. 512.33 miles/hour
    D. 587.43 miles/hour

11. The Kent High School Booster's Club has 250 members. In 90 days it hopes to have a membership of 376 by initiating a Web-based membership campaign. On average, how many new members does it need to sign up each day to reach that goal?
    A. 1
    B. 1.4
    C. 1.5
    D. 2.2

12. LeBron has a rectangular garden that is 3 feet by 7 feet. If he needs 2 ounces of fertilizer per square foot for the garden to flourish, how many ounces must he use?
    A. 21 oz
    B. 42 oz
    C. 44 oz
    D. 56 oz

13. The local electronics store is holding a 20%-off sale on all its products. Marla selects 2 CDs at the original price of $16.99 each. What is the price of the two CDs after the 20% discount?
    A. $15.99
    B. $22.50
    C. $25.78
    D. $27.18

14. The Berkeley Department Store is running a special back-to-school sale on shoes in September. A pair of shoes costs $67.50. In October the price increases to $87.75. What percent increase is the price in October?
    A. 15%
    B. 20%
    C. 24%
    D. 30%

15. Two numbers together add to 300. One number is twice the size of the other. What are the two numbers?
    A. 25; 50
    B. 50; 100
    C. 75; 150
    D. 100; 200

16. Logan is making a picture frame for a photo of his family. The photo is 8 inches by 10 inches. If the wood for the frame costs 50 cents an inch, how much will the wood for the frame cost?
    A. $9.00
    B. $12.00
    C. $15.00
    D. $18.00

17. The distance across the circular mouth of a volcano running through its center is 75 yards. What is the distance around its mouth?
    A. 145.5 yd
    B. 215.5 yd
    C. 235.5 yd
    D. 315.5 yd

18. Sue spent one-third of her life in Kansas. If Sue is 39, how many years did she spend in Kansas?
    A. 9
    B. 12
    C. 13
    D. 15

19. Brian shoots hoops with Dave, who is two years younger than Brian. If Brian is 17, how old is Dave?
    A. 15
    B. 17
    C. 19
    D. 21

20. Al has three times as much money as Bill. Bill has two times as much money as Charlie. If Charlie has $5.89, how much money does Al have?
    A. $11.78
    B. $19.79
    C. $29.24
    D. $35.34

21. The car owned by Marisa gets an average of 16 miles per gallon of gas on the highway. If she recently took a trip of 1,344 miles, how many gallons of gas did she use?
    A. 34 gallons
    B. 57 gallons
    C. 78 gallons
    D. 84 gallons

22. A rectangular swimming pool is 6 feet deep, 28 feet long, and 12 feet wide. What is the volume of the pool?
    A. 1,655 ft^3
    B. 1,799 ft^3
    C. 2,016 ft^3
    D. 2,546 ft^3

23. Damian wants to put a fence around his rectangular garden. The garden is 6 feet by 15 feet. How much fencing must he purchase?
    A. 30 feet
    B. 35 feet
    C. 42 feet
    D. 47 feet

24. In a recent election, Joe received 32% of the votes, Samantha received 14% of the votes, Sally received 42% of the votes, and Cal received 12% of the votes. If 5,000 votes were cast, how many votes did Samantha receive?
    A. 300
    B. 450
    C. 535
    D. 700

25. A circular pool is 3.5 meters deep and has a radius of 9 meters. What is the volume?
    A. 450.2 m^3
    B. 575.9 m^3
    C. 890.2 m^3
    D. 974.4 m^3

26. The diameter of a planet is 8,000 miles. What is its circumference?
    A. 19,412 miles
    B. 20,889 miles
    C. 25,120 miles
    D. 28,475 miles

27. Will has $4,355.19 in his checking account. He writes checks for $1,204.90 and $890.99. How much is left in Will's account?
    A. $1,189.65
    B. $2,259.30
    C. $2,867.30
    D. $2,989.50

28. A club collected $585.00. If 75% of that came from membership dues, how much money came from the dues?
    A. $413.15
    B. $438.75
    C. $465.75
    D. $525.05

29. Brandon buys a computer originally priced at $750.00, software programs originally priced at $398.75, a printer originally priced at $149.98, and a case of paper originally priced at $24.99. All items are on sale for 20% off. How much will Danny save because of the sale?
    A. $124.94
    B. $196.74
    C. $236.24
    D. $264.74

30. A ladder is placed against a building. If the ladder makes a 55° angle with the ground, what is the measure of the angle that the ladder makes with the building?
    A. 35°
    B. 40°
    C. 55°
    D. 90°

STOP! DO NOT TURN THIS PAGE UNTIL TIME IS UP FOR THIS TEST. IF YOU FINISH BEFORE TIME IS UP, CHECK OVER YOUR WORK ON THIS TEST ONLY.

PART 3. WORD KNOWLEDGE

THE TEST ADMINISTRATOR WILL READ THE FOLLOWING ALOUD TO YOU.

Now turn to Part 3 and read the directions for Word Knowledge silently while I read them aloud.

This is a test of your knowledge of work meanings. These questions consist of a sentence or phrase with a word or phrase underlined. From the four choices given, you are to decide which one MEANS THE SAME OR MOST NEARLY THE SAME as the underlined word or phrase. Once you have made your choice, mark the space on your answer sheet that has the same number and letter as your choice.

Look at the sample question.

S1 The weather in this geographic area tends to be <u>moderate</u>.

   A. Severe
   B. Warm
   C. Mild
   D. Windy

The correct answer is "mild," which is choice C. Therefore, you would have blackened in space C on your answer sheet.

Your score on this test will be based on the number of questions you answer correctly. You should try to answer every question. DO NOT SPEND TOO MUCH TIME on any one question. If you finish before time is called, go back and check your work in this part <u>ONLY</u>.

Now find the section of your answer sheet that is marked PART 3. When you are told to begin, start with question number 1 in Part 3 of your test booklet and answer space number 1 in Part 3 on your separate answer sheet.

DO NOT TURN THE PAGE UNTIL TOLD TO DO SO. You will have 11 minutes to complete the 35 questions in this part. Are there any questions?

Begin.

1. <u>Exempt</u> most nearly means
   A. empty.
   B. exit.
   C. excuse.
   D. anoint.

2. <u>Recruit</u> most nearly means
   A. react.
   B. enlist.
   C. permit.
   D. recur.

3. The decision was <u>unanimous</u>.
   A. Questioned
   B. Undisputed
   C. Selective
   D. Trivial

4. <u>Temperamental</u> most nearly means
   A. benevolent.
   B. lazy.
   C. volatile.
   D. demure.

5. He chose the <u>economical</u> solution.
   A. Costly
   B. Logical
   C. Comical
   D. Thrifty

6. The <u>promotion</u> made Darrell happy.
   A. Notice
   B. Gift
   C. Elevation
   D. Letter

7. Her behavior was <u>unpredictable</u>.
   A. Consistent
   B. Random
   C. Dependable
   D. Regular

8. <u>Deny</u> most nearly means
   A. state.
   B. avoid.
   C. refuse.
   D. open.

9. Making that decision at that time was <u>irresponsible</u>.
   A. Hapless
   B. Giddy
   C. Easy
   D. Reckless

10. <u>Persistent</u> most nearly means
    A. gentle.
    B. unrelenting.
    C. resigned.
    D. manageable.

11. The opponent was more <u>dominant</u>.
    A. Helpful
    B. Overriding
    C. Understanding
    D. Joyous

12. Marco was more <u>conscientious</u> than his friend.
    A. Painstaking
    B. Popular
    C. Friendly
    D. Handsome

13. <u>Diversion</u> most nearly means
    A. distraction.
    B. containment.
    C. pressure.
    D. supply.

14. The <u>assumption</u> was that she was correct.
    A. Belief
    B. Decision
    C. Rumor
    D. Guess

15. <u>Profess</u> most nearly means
 A. work.
 B. claim.
 C. disguise.
 D. accuse.

16. The substance of their argument was <u>trivial</u>.
 A. Sinister
 B. Inconsequential
 C. Sad
 D. Demanding

17. His attitude was <u>optimistic</u>.
 A. Grouchy
 B. Angry
 C. Unusual
 D. Positive

18. <u>Variable</u> most nearly means
 A. different.
 B. gloomy.
 C. changeable.
 D. bifurcate.

19. His words needed <u>clarification</u>.
 A. Expansion
 B. Explanation
 C. Elimination
 D. Recognition

20. The view was <u>panoramic</u>.
 A. Beautiful
 B. Unrealistic
 C. Stylistic
 D. Expansive

21. <u>Compassionate</u> most nearly means
 A. unwieldy.
 B. modern.
 C. concerned.
 D. restrictive.

22. <u>Contingent</u> most nearly means
 A. rapid.
 B. dependent.
 C. difficult.
 D. mysterious.

23. <u>Velocity</u> most nearly means
 A. variety.
 B. veracity.
 C. swiftness.
 D. overflow.

24. Her house was <u>immaculate</u>.
 A. Cleaned
 B. Spotless
 C. Large
 D. Fancy

25. The <u>visibility</u> that day was very poor.
 A. Clarity
 B. Test
 C. Observation
 D. Dichotomy

26. The wind was so strong that the boat <u>capsized</u>.
 A. Overturned
 B. Accelerated
 C. Tilted
 D. Docked

27. The elected official <u>abdicated</u> responsibility for the situation.
 A. Welcomed
 B. Cheered
 C. Transitioned
 D. Abandoned

28. Her skills made her <u>eligible</u> for the position.
 A. Qualified
 B. Eager
 C. Adequate
 D. Thankful

29. Buoyant most nearly means
   A. floating.
   B. happy.
   C. significant.
   D. acrobatic.

30. The pond was teeming with fish.
   A. Empty
   B. Opaque
   C. Crowded
   D. Vigorous

31. Bestow most nearly means
   A. remove.
   B. have.
   C. give.
   D. question.

32. Meander most nearly means
   A. twist.
   B. reverse.
   C. exceed.
   D. renew.

33. Infiltrate most nearly means
   A. interfere.
   B. seek.
   C. enter.
   D. penetrate.

34. His science grade plummeted after that difficult exam.
   A. Soared
   B. Plunged
   C. Increased
   D. Failed

35. The monument was defiled by the visitors.
   A. Honored
   B. Desecrated
   C. Cleaned
   D. Respected

STOP! DO NOT TURN THIS PAGE UNTIL TIME IS UP FOR THIS TEST. IF YOU FINISH BEFORE TIME IS UP, CHECK OVER YOUR WORK ON THIS TEST ONLY.

PART 4. PARAGRAPH COMPREHENSION

THE TEST ADMINISTRATOR WILL READ THE FOLLOWING ALOUD TO YOU.

Turn to Part 4 and read the directions for Paragraph Comprehension silently while I read them aloud.

This is a test of your ability to understand what you read. In this section you will find one or more paragraphs of reading material followed by incomplete statements or questions. You are to read the paragraph and select one of four lettered choices which BEST completes the statement or answers the question. When you have selected your answer, blacken the space on your answer sheet that has the same number and letter as your answer.

Your score on this test will be based on the number of questions you answer correctly. You should try to answer every question. DO NOT SPEND TOO MUCH TIME on any one question. If you finish before time is called, go back and check your work in this part ONLY.

Now find the section of your answer sheet that is marked PART 4. When you are told to begin, start with question number 1 in Part 4 of your test booklet and answer space number 1 in Part 4 on your separate answer sheet.

DO NOT TURN THE PAGE UNTIL TOLD TO DO SO. You will have 13 minutes to complete the 15 questions in this part. Are there any questions?

Begin.

The evidence is growing and is more convincing than ever! People of all ages who are generally inactive can improve their health and well-being by becoming active at a moderate intensity on a regular basis.

Regular physical activity substantially reduces the risk of dying of coronary heart disease, the nation's leading cause of death, and decreases the risk of stroke, colon cancer, diabetes, and high blood pressure. It also helps to control weight; contributes to healthy bones, muscles, and joints; reduces falls among older adults; helps to relieve the pain of arthritis; reduces symptoms of anxiety and depression; and is associated with fewer hospitalizations, physician visits, and medications. Moreover, physical activity need not be strenuous to be beneficial; people of all ages benefit from participating in regular, moderate-intensity physical activity, such as 30 minutes of brisk walking five or more times a week.

Despite the proven benefits of physical activity, more than 50 percent of American adults do not get enough physical activity to provide health benefits. Twenty-five percent of adults are not active at all in their leisure time. Activity decreases with age and is less common among women than men and among those with lower income and less education. Furthermore, there are racial and ethnic differences in physical activity rates, particularly among women. Insufficient physical activity is not limited to adults. More than a third of young people in grades 9 to 12 do not regularly engage in vigorous-intensity physical activity. Daily participation in high school physical education classes dropped from 42 percent in 2001 to 32 percent in 2020.

Physical activity can bring you many health benefits. People who enjoy participating in moderate-intensity or vigorous-intensity physical activity on a regular basis benefit by lowering their risk of developing coronary heart disease, stroke, non-insulin-dependent (type 2) diabetes mellitus, high blood pressure, and colon cancer by 30 to 50 percent. Additionally, active people have lower premature death rates than people who are the least active.

1. Which of the following best represents the main idea of this passage?
   A. Even a moderate amount of exercise has health benefits.
   B. People who do not exercise need to be convinced to do so.
   C. Exercises cures heart disease and decreases stroke.
   D. Strenuous exercise is better than moderate exercise.

2. Which of the following can be inferred from this passage?
   A. More exercise is better.
   B. If you exercise, you are likely to improve your health.
   C. Exercise eliminates heart disease and strokes.
   D. People who exercise have no weight problems.

3. According to the passage, which of the following is <u>not</u> a benefit of regular exercise?
   A. Lowering blood pressure
   B. Reducing colon cancer
   C. Reducing fatty acids
   D. Reducing coronary heart disease

4. According to the passage, which of the following is true?
   A. Twenty-five percent of adults do no exercise at all in their leisure time.
   B. People are living longer because of exercise.
   C. Older people tend to exercise more than younger people.
   D. People who have been hospitalized tend not to exercise.

5. About what percent of Americans do not get enough exercise to obtain health benefits?
   A. 25 percent
   B. 32 percent
   C. 42 percent
   D. 50 percent

During the first decade of the nineteenth century, the geographic image of western North America began to change dramatically. Based on the observations of the explorers Meriwether Lewis and George Rogers Clark, information gathered from native people, and Clark's own cartographic imagination, this image evolved from an almost empty interior with a hypothetical single mountain range serving as a western continental divide to an intricate one showing a labyrinth of mountains and rivers. A continent that had once seemed empty and simple was now becoming full and complex.

The Lewis and Clark expedition established the precedent for army exploration in the West. Major Stephen H. Long's Scientific Expedition (1819–1820) advanced that tradition, this time centering attention on the central and southern Great Plains and the Front Range of the Rockies. For the first time, an American exploring party included professional scientists (a zoologist and a botanist) and two skilled artists. While not every future American expedition took along such skilled observers, the pattern was set for increasingly scientific exploration.

It would take another 50 years after Lewis and Clark to complete the cartographic image of the West we know today. Other explorers and mapmakers followed, each revealing new geographic and scientific details about specific parts of the western landscape. But this revealing process was not a simple one. New knowledge did not automatically replace old ideas; some old notions—especially about river passages across the West—persisted well into the nineteenth century. In the decades after Lewis and Clark, the company of western explorers expanded to include fur traders, missionaries, and government topographers, culminating in the 1850s with the Army's Corps of Topographical Engineers surveying the southwestern and northwestern boundaries of the United States as well as the potential routes for a transcontinental railroad. By the time of the Civil War, an ocean-to-ocean American empire with borders clearly defined was a fact of continental life.

6. Which of the following best summarizes the passage?
   A. Western North America has a single mountain range that was unknown before the early nineteenth century.
   B. Nineteenth-century explorers found western North America to be geographically more complex than they had originally thought.
   C. Western North America was barren and mostly empty before the early nineteenth century.
   D. Lewis and Clark discovered many rivers in western North America.

7. In the passage the word *labyrinth* most nearly means
   A. jumble.
   B. line.
   C. concentration.
   D. beautiful design.

8. According to the passage, which event set the pattern for scientific exploration in the West?
   A. The exploration of the Army's Corps of Topographical Engineers
   B. Long's Scientific Expedition
   C. The exploration of Lewis and Clark
   D. The discovery of a major mountain range in the West

9. According to the passage, which of the following is true?
   A. Lewis and Clark explored the Front Range of the Rockies, helping the fur traders to expand their business in that area.
   B. Fur traders boosted the economy of what is now the western part of the United States.
   C. Native people provided Lewis and Clark with information about the topography of what is now the western part of the United States.
   D. Mapmakers helped plot the route for the transcontinental railroad.

10. According to the passage, what group of people were most instrumental in setting the pattern for more scientifically oriented exploration of the West?
    A. The military
    B. Fur traders
    C. Mapmakers
    D. Missionaries

Heroes were an important part of Greek mythology, but the characteristics that Greeks admired in a hero are not necessarily identical to those we admire today. Greek heroes are not always what modern readers might think of as "good role models." Their actions may strike us as morally dubious.

For example, consider the encounter between the legendary Greek hero Odysseus and the Cyclops. The Cyclops was one of a race of giants who lived by themselves on a remote, rarely visited island. The name Cyclops means "round eye," because these giants had only one eye in the middle of their forehead. They lived in caves, tended flocks of sheep, and ate the produce of their fields; they were shepherds.

Odysseus visited the island as part of his exploration to look for supplies. He brought with him a flask of wine. Although he was regarded as an intruder by the Cyclops, he helped himself to the giant's supplies without permission. The Cyclops became very angry. To ease the anger, Odysseus served him some wine. The Cyclops enjoyed the wine and asked for more. Later, when the Cyclops was in a wine-induced stupor, Odysseus attacked him in the eye. Later Odysseus bragged to his comrades about blinding the one-eyed creature.

This does not mean the Greeks admired thievery and bragging, however. What they admired about Odysseus, in this instance, was his capacity for quick thinking. Odysseus was also known for pulling off great feats with panache and self-confidence.

Not all Greek heroes were admired for the same reasons. Some, such as Odysseus, were admired for their resourcefulness and intelligence, whereas others, such as Herakles, were known for their strength and courage. Some were not particularly resourceful but depended on help to accomplish their tasks.

Whether or not a given action or quality was admired depended upon its ultimate results. Being headstrong might succeed in one instance but lead to failure in another. The Greeks held the characters in their legends accountable for their actions, and a hero might be punished as well as rewarded.

11. In this passage, what is the meaning of the word *dubious*?
    A. Crass
    B. Wrong
    C. Doubtful
    D. On the high ground

12. Based on this passage, why did the Greeks admire Odysseus?
    A. He could react quickly to a situation.
    B. He was a good role model.
    C. He attacked when the Cyclops was incapacitated.
    D. He was a hero in Greek mythology.

13. According to the passage, what was Herakles known for?
    A. Kindness
    B. Bravery
    C. Resourcefulness
    D. Failure

14. What is the major idea in this passage?
    A. Odysseus was a murderer.
    B. Odysseus was a thief.
    C. The Greeks judged their legendary heroes by the results they achieved.
    D. The Greeks admired each of their legendary heroes for the same reasons.

15. What was the major occupation of the Cyclops?
    A. Sailor
    B. Farmer
    C. Weaver
    D. Shepherd

STOP! DO NOT TURN THIS PAGE UNTIL TIME IS UP FOR THIS TEST. IF YOU FINISH BEFORE TIME IS UP, CHECK OVER YOUR WORK ON THIS TEST ONLY.

PART 5. MATHEMATICS KNOWLEDGE

THE TEST ADMINISTRATOR WILL READ THE FOLLOWING ALOUD TO YOU.

Now turn to Part 5 and read the directions for Mathematics Knowledge silently while I read them aloud.

This is a test of your ability to solve general mathematics problems. You are to select the correct response from the choices given. Then mark the space on your answer sheet that has the same number and letter as your choice. Use your scratch paper to do any figuring you wish to do.

Your score on this test will be based on the number of questions you answer correctly. You should try to answer every question. DO NOT SPEND TOO MUCH TIME on any one question. If you finish before time is called, go back and check your work in this part ONLY.

Now find the section of your answer sheet that is marked PART 5. When you are told to begin, start with question number 1 in Part 5 of your test booklet and answer space number 1 in Part 5 on your separate answer sheet.

DO NOT TURN THE PAGE UNTIL TOLD TO DO SO. You will have 24 minutes to complete the 25 questions in this part. Are there any questions?

Begin.

1. $x + 9 = 13$

 $x = ?$
 A. 3
 B. 4
 C. 5
 D. 13

2. $3^2 + 4(5 - 2) =$
 A. 21
 B. 36
 C. 39
 D. 40

3. $1\frac{1}{4} \times 2\frac{1}{2} =$
 A. 2
 B. $2\frac{1}{2}$
 C. $2\frac{3}{4}$
 D. $3\frac{1}{8}$

4. The square root of 64 =
 A. 4
 B. −8
 C. 9
 D. −9

5. What is the value of the following expression if $x = 2$ and $y = 4$?

 $2x^3y^2$

 A. 52
 B. 106
 C. 256
 D. 512

6. Add:

 $3x^2 + 3xy + 4y + 2y^2$
 $3x^2 + xy + 9y + y^2$
 $-5x^2 - 2xy - 13y + y^2$
 $-3x^2 + 3xy + y - 4y^2$

 A. $2x^2 + 4xy + y + y^2$
 B. $-2x^2 + 4xy + y + y^2$
 C. $2x^2 + 5xy + 2y + 2y^2$
 D. $-2x^2 + 5xy + y$

7. Factor the following expression: $y^2 - 16y + 48$
 A. $(y - 1)(y + 48)$
 B. $(y - 4)(y - 12)$
 C. $(y - 4)(y + 4)$
 D. $(y + 1)(y - 16)$

8. Solve for x.

 $12x + 6 = 8x + 10$

 A. $x = 1$
 B. $x = 2$
 C. $x = 4$
 D. $x = 6$

9. Solve for the two unknowns.

 $3y + 3x = 24$
 $6y + 3x = 39$

 A. $y = 3; x = 5$
 B. $y = 4; x = 6$
 C. $y = 5; x = 3$
 D. $y = 6; x = 2$

10. $\dfrac{x^{12}}{x^4} =$

 A. x^{-8}
 B. x^8
 C. x^{18}
 D. x^{-16}

11. Solve for y.

 $y^2 + 7y = -10$

 A. $y = -5; y = -2$
 B. $y = 1; y = -10$
 C. $y = -2; y = 6$
 D. $y = 5; y = -1$

12. Multiply:

$$\frac{6y}{11} \times \frac{2}{5x} =$$

A. $6y + 5x = 22$
B. $12y + 55$
C. $\dfrac{12y}{55x}$
D. $\dfrac{30xy}{22}$

13. Divide:

$$\frac{6x}{11} \div \frac{2}{5y}$$

A. $\dfrac{12x}{55y}$
B. $\dfrac{15xy}{11}$
C. $\dfrac{12}{55xy}$
D. $30xy$

14. $\sqrt[3]{27} =$
A. 3
B. 4
C. 5
D. 2.25

15. Solve for g.

$$\frac{g}{h} = a$$

A. $\dfrac{ga}{h}$
B. $g = \dfrac{a}{h}$
C. $g = ah$
D. $g = a \div h$

16. Which of the following is an equilateral triangle?

A.
B.
C.
D.

17. What is the perimeter of the following rectangle?

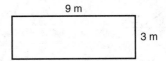

A. 12 m
B. 18 m
C. 24 m
D. 12 m²

18. What is the circumference of the following circle?

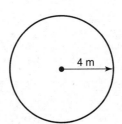

A. 6.56 m
B. 12.56 m
C. 25.12 m
D. 35.12 m

19. In the following equilateral triangle, what is the measure of ∠1?

   A. 30°
   B. 45°
   C. 60°
   D. 90°

20. If ∠1 is 34°, what is the measure of ∠2?

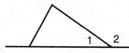

   A. 56°
   B. 110°
   C. 146°
   D. 180°

21. In the following right triangle, what is the length of side *AB*?

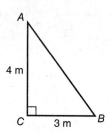

   A. 3 m
   B. 5 m
   C. 7 m
   D. 12 m

22. What is the area of the following triangle?

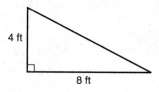

   A. 8 ft²
   B. 12 ft²
   C. 15 ft²
   D. 16 ft²

23. What is the diameter of the following circle?

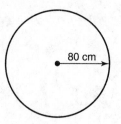

   A. 40 cm
   B. 80 cm
   C. 160 cm
   D. 640 cm

24. If lines *A* and *B* are parallel and are intersected by line *C* and ∠2 is 30°, what is the measure of ∠7?

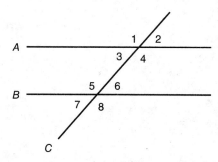

   A. 30°
   B. 50°
   C. 112°
   D. 130°

25. What is the volume of a cylinder with a height of 15 m and a radius of 5 m?
   A. 235.50 m³
   B. 1,177.50 m³
   C. 1,280.25 m³
   D. 1,340.25 m³

STOP! DO NOT TURN THIS PAGE UNTIL TIME IS UP FOR THIS TEST. IF YOU FINISH BEFORE TIME IS UP, CHECK OVER YOUR WORK ON THIS TEST ONLY.

PART 6. ELECTRONICS INFORMATION

THE TEST ADMINISTRATOR WILL READ THE FOLLOWING ALOUD TO YOU.

Now turn to Part 6 of your test booklet and read the directions for Electronics Information silently while I read them aloud.

This is a test of your knowledge of electrical, radio, and electronics information. You are to select the correct response from the choices given and then mark the space on your answer sheet that has the same number and letter as your choice.

Your score on this test will be based on the number of questions you answer correctly. You should try to answer every question. DO NOT SPEND TOO MUCH TIME on any one question. If you finish before time is called, go back and check your work in this part ONLY.

Now find the section of your answer sheet that is marked PART 6. When you are told to begin, start with question number 1 in Part 6 of your test booklet and answer space number 1 in Part 6 on your separate answer sheet.

DO NOT TURN THE PAGE UNTIL TOLD TO DO SO. You will have 9 minutes to complete the 20 questions in this part. Are there any questions?

Begin.

1. One hertz is defined as
   A. one ampere per second.
   B. one cycle per second.
   C. one volt per second.
   D. one coulomb per second.

2. One advantage of alternating current is that it is easy to
   A. store.
   B. convert into chemical energy.
   C. transport through wire.
   D. transmit through air.

3. Electric current can travel only through a
   A. wire.
   B. circuit.
   C. battery.
   D. load.

4. Copper is used in most electric wires because it is
   A. a semiconductor.
   B. not likely to heat up during normal usage.
   C. a good resistor and inexpensive.
   D. a good conductor and inexpensive.

5. Amperes are a measure of
   A. the number of electrons moving through a conductor.
   B. electrical pressure.
   C. a material's ability to store electric current.
   D. the total resistance of a series circuit.

6. Ohms are a measure of
   A. capacitance.
   B. resistance.
   C. current.
   D. cycles per second.

7. If amperes = volts/ohms, then volts =
   A. 1/(amperes × ohms).
   B. ohms/amperes.
   C. amperes/ohms.
   D. amperes × ohms.

8. A circuit with 20 amperes has a load of 12 ohms. What is the voltage?
   A. 60
   B. 120
   C. 240
   D. 24

9. If the resistance of a circuit is 0, the circuit is
   A. an open circuit.
   B. a short circuit.
   C. a superconductor circuit.
   D. a semiconductor circuit.

10. If a transformer raises the voltage, it will
    A. raise the resistance.
    B. reduce the resistance.
    C. raise the amperage.
    D. reduce the amperage.

11. Which of these devices depends on the close relationship between electricity and magnetism?
    A. Transformer
    B. Electromagnet
    C. Electric motor
    D. All of the above

12. If you are soldering two wires, you need solder,
    A. heat, and flux.
    B. and heat.
    C. and flux.
    D. flux, and electrical tape.

13. When connecting wires in the house, the bare copper wire is sometimes replaced by the
    A. the black wire.
    B. the white wire.
    C. the hot wire.
    D. conduit.

14. To hook up an electric heater, you need
    A. three supply wires.
    B. four supply wires.
    C. heavy supply wires.
    D. no supply wires.

15. In most home wiring, the hot wire is
    A. black.
    B. orange.
    C. green.
    D. white.

16. To control a light from each end of a hallway, you would install
    A. two single-pole switches.
    B. one four-way switch.
    C. one three-way switch.
    D. two three-way switches.

17. At 120 volts, a 15-ampere circuit will carry _____ power than a 20-ampere circuit.
    A. more
    B. less
    C. first less, then more
    D. first more, then less

18. To connect a battery properly, you must
    A. observe proper polarity.
    B. get the right voltage.
    C. both A and B.
    D. be certain the battery is chemical-free.

19. When you are making an electrical connection to a large battery,
    A. never tighten the terminal.
    B. tighten the terminal and then back off one turn.
    C. use salt water to prevent corrosion.
    D. tighten the terminal securely.

20. If you need to reduce the voltage in a certain part of a circuit, you could use a
    A. capacitor.
    B. transistor.
    C. resistor.
    D. inductor.

STOP! DO NOT TURN THIS PAGE UNTIL TIME IS UP FOR THIS TEST. IF YOU FINISH BEFORE TIME IS UP, CHECK OVER YOUR WORK ON THIS TEST ONLY.

PART 7. AUTO AND SHOP INFORMATION

THE TEST ADMINISTRATOR WILL READ THE FOLLOWING ALOUD TO YOU.

Now turn to Part 7 and read the directions for Auto and Shop Information silently while I read them aloud.

This test has questions about automobiles, shop practices, and the use of tools. Pick the BEST answer for each question and then mark the space on your answer sheet that has the same number and letter as your choice.

Your score on this test will be based on the number of questions you answer correctly. You should try to answer every question. DO NOT SPEND TOO MUCH TIME on any one question. If you finish before time is called, go back and check your work in this part ONLY.

Now find the section of your answer sheet that is marked PART 7. When you are told to begin, start with question number 1 in Part 7 of your test booklet and answer space number 1 in Part 7 on your separate answer sheet.

DO NOT TURN THE PAGE UNTIL TOLD TO DO SO. You will have 11 minutes to complete the 25 questions in this part. Are there any questions?

Begin.

1. If you hear a loud clanking noise from the rear of a rear-wheel-drive, front-engine car, you should look for trouble at
   A. the differential or driveshaft.
   B. the transmission.
   C. the engine.
   D. the clutch.

2. If one engine cylinder does not work right, you should check for problems in the
   A. drive train.
   B. pollution controls.
   C. ignition or fuel system.
   D. clutch.

3. If tire pressure gets low, what will happen?
   A. The tire will lose its shape and create extra friction, reducing gas mileage and possibly overheating the tire.
   B. The tire will not absorb energy from the engine.
   C. Vehicle weight will be reduced.
   D. The tire will get better traction on the highway.

4. If an engine with conventional ignition runs rough, what might you adjust on the breaker points, shown as part B of this diagram?
   A. The gap
   B. The timing
   C. Composition
   D. Lubrication

5. One important difference between front-wheel drive and rear-wheel drive is that in front-wheel drive,
   A. the layshaft moves the driveshaft.
   B. the engine must be rotated 180° from normal position.
   C. the driving wheels also steer.
   D. the rear wheels must also drive.

6. Which is the correct sequence in a four-cycle internal-combustion engine?
   A. Intake, compression, power, rest
   B. Reduction, power, compression, exhaust
   C. Intake, compression, power, exhaust
   D. Exhaust, compression, intake, power

7. If you suspect that not enough gasoline is entering the engine, what might you check first?
   A. Fuel system
   B. Ignition system
   C. Timing belt
   D. Engine control module

8. When a manual transmission is in a "direct-drive" gear, one revolution of the engine produces
   A. two revolutions of the driveshaft.
   B. one revolution of the driveshaft.
   C. one-half revolution of the driveshaft.
   D. two revolutions of the clutch.

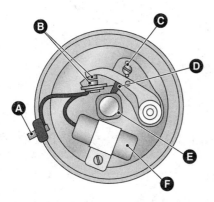

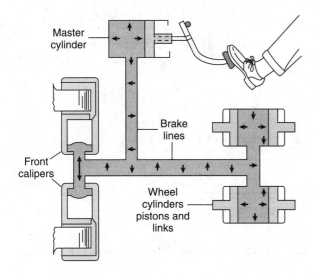

9. In the diagram above, the arrows indicate
   A. viscosity.
   B. brake fluid.
   C. filters.
   D. temperature.

10. What must happen before the starter motor is engaged?
    A. The brakes must be adjusted.
    B. The starter relay must be closed.
    C. The gas tank must be full.
    D. The oil pump must be automatically primed.

11. If the exhaust valve does not open,
    A. the fuel–air mix cannot enter the engine.
    B. the car will not pollute.
    C. the engine will run a bit rough.
    D. burned gas cannot leave the cylinder, and the engine will not run.

12. If the differential ratio is 3 to 1,
    A. one turn of the driveshaft produces three turns of the wheels.
    B. three turns of the jack shaft produce one turn of the wheels.
    C. the ring gear is probably disengaged.
    D. three turns of the driveshaft produce one turn of the wheels.

13. When choosing a slot screwdriver,
    A. use a blade that slips easily into the slot.
    B. make the blade as wide as the screw head.
    C. always use the shortest possible handle.
    D. file out the screw to fit the screwdriver.

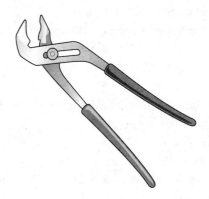

14. The tool shown above is a(n)
    A. locking pliers (Vise-Grips).
    B. arc-joint pliers (Channel Locks).
    C. monkey wrench.
    D. side-cutting electrician's pliers.

15. A plywood-cutting blade for a circular saw
    A. has a few big teeth because plywood is so much harder than regular wood.
    B. cannot be used with particle board.
    C. must not be too sharp or else it will burn the wood.
    D. has many small teeth to reduce splintering.

16. Which of the following is a poor use for a claw hammer?
    A. Pulling nails
    B. Hammering nails
    C. Driving screws
    D. Driving wood chisels

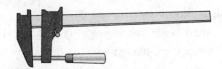

17. The tool shown above is used for
    A. clamping parts for gluing.
    B. tightening large pipe fittings.
    C. pressing a drill into the chuck.
    D. pressing glass into a window sash.

18. It helps to hit a larger chisel with a larger hammer because the hammer is
    A. heavier.
    B. more massive.
    C. bigger, and therefore less likely to miss the chisel.
    D. slower-moving.

19. Plywood is better than regular wood for a sub-floor mainly because
    A. it is more weatherproof.
    B. it is lighter.
    C. it is thinner.
    D. it is stronger.

20. Which tool might replace a hacksaw for some jobs?
    A. A circular saw with a carbide blade
    B. A keyhole saw
    C. A coping saw
    D. An oxyacetylene cutting torch

21. The tool shown above is a
    A. tin snip.
    B. magna-shear.
    C. bolt cutter.
    D. duckbill wrench.

22. Which of these tools does not start to clamp until you turn it?
    A. Screwdriver
    B. Locking pliers
    C. Pipe wrench
    D. Vise

23. What is not an advantage of a sharp drill bit?
    A. It creates a smaller amount of chips.
    B. It cuts with less pressure.
    C. It cuts cooler.
    D. It cuts faster.

24. Which of the following would not loosen a frozen nut?
    A. Heating it with a torch
    B. Pounding it with a hammer
    C. Soaking it in penetrating oil
    D. Dousing it in ice water

25. A plumb line makes what kind of angle where it intersects a level line?
    A. Acute
    B. Obtuse
    C. Square
    D. Straight

STOP! DO NOT TURN THIS PAGE UNTIL TIME IS UP FOR THIS TEST. IF YOU FINISH BEFORE TIME IS UP, CHECK OVER YOUR WORK ON THIS TEST ONLY.

PART 8. MECHANICAL COMPREHENSION

THE TEST ADMINISTRATOR WILL READ THE FOLLOWING ALOUD TO YOU.

Now turn to Part 8 and read the directions for Mechanical Comprehension silently while I read them aloud.

This test has questions about mechanical and physical principles. Study the picture and decide which answer is CORRECT and then mark the space on your separate answer sheet that has the same number and letter as your choice.

Your score on this test will be based on the number of questions you answer correctly. You should try to answer every question. DO NOT SPEND TOO MUCH TIME on any one question. If you finish before time is called, go back and check your work in this part ONLY.

Now find the section of your answer sheet that is marked PART 8. When you are told to begin, start with question number 1 in Part 8 of your test booklet and answer space number 1 in Part 8 on your separate answer sheet.

DO NOT TURN THE PAGE UNTIL TOLD TO DO SO. You will have 19 minutes to complete the 25 questions in this part. Are there any questions?

Begin.

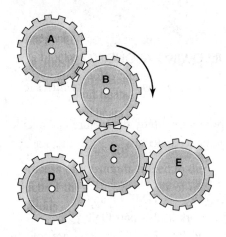

1. The diagram shows five gears. If gear B turns as shown, then the gears turning in the same direction are
   A. A and C.
   B. A and D.
   C. C and D.
   D. D and E.

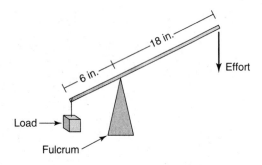

2. What is the theoretical mechanical advantage in using this lever?
   A. 1
   B. 2
   C. 3
   D. 4

3. A wrench is used to turn a bolt that has 20 threads per inch. After 15 complete turns of the wrench, the bolt will have moved
   A. ¼ inch.
   B. ½ inch.
   C. ¾ inch.
   D. 1 inch.

4. A gear and pinion have a ratio of 5 to 1. If the gear is rotating at a speed of 150 revolutions per minute (rpm), the speed of the pinion is most nearly
   A. 750 rpm.
   B. 300 rpm.
   C. 150 rpm.
   D. 30 rpm.

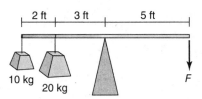

5. In the diagram, water is flowing from left to right. What is true of the water after it enters the 3-inch pipe?
   A. Its pressure decreases, but its speed increases.
   B. Its pressure and speed both decrease.
   C. Its pressure and speed both increase.
   D. Its pressure increases, but its speed decreases.

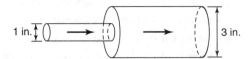

6. What is the force F needed to balance this lever?
   A. 20 kg
   B. 22 kg
   C. 25 kg
   D. 30 kg

7. Gear A, with 48 teeth, meshes with gear B, with 12 teeth. For every rotation that gear A makes, how many rotations does gear B make?
   A. 1
   B. 2
   C. 4
   D. 8

8. Pushing a heavy concrete block up an inclined plane will be easier if you
   A. raise the angle of the plane.
   B. turn the block upside down.
   C. heat the block with an electric coil.
   D. put the block on a wheelbarrow.

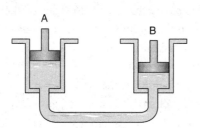

9. These pipes and cylinders are part of a hydraulic mechanism. If the piston in cylinder A moves,
   A. the piston in cylinder B will move.
   B. the fluid in cylinder A will not move.
   C. the piston in cylinder B will not move.
   D. the fluid in cylinder A will move, but the fluid in cylinder B will not move.

10. A 3 ft × 6 ft tank is filled with 3,600 lb of water. What is the pressure on the bottom, in lb/ft²?
    A. 100
    B. 200
    C. 400
    D. 800

11. Which cannot bend without breaking?

    A.

    B.

    C.

    D.

12. A knife blade is an example of which kind of simple machine?
    A. Pulley
    B. Wheel and axle
    C. Wedge
    D. Gear

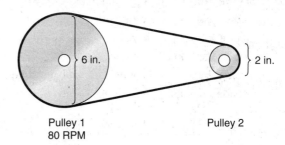

Pulley 1
80 RPM

Pulley 2

13. In the pulley (sheave) system shown, pulley 1 is rotating at 80 rpm. How fast is pulley 2 rotating?
    A. 180 rpm
    B. 240 rpm
    C. 300 rpm
    D. 360 rpm

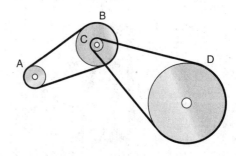

14. In the pulley system shown, which pulley is rotating the slowest?
    A. A
    B. B
    C. C
    D. D

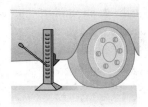 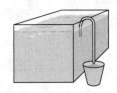

15. Operating this car jack will take less effort if you
    A. add weight to the other side of the car.
    B. get a longer jack handle.
    C. open the hood of the car.
    D. replace the jack handle with one just like it made of wood.

16. Which material is best for the floor of a fireplace?
    A. Wood
    B. Plastic
    C. Stone
    D. Glass

17. What effort is required to lift the load?
    A. 40 lb
    B. 50 lb
    C. 100 lb
    D. 200 lb

18. What makes the water flow out of the tank and through the siphon hose into the bucket?
    A. Air pressure
    B. Water pressure
    C. The weight of the water
    D. Magnetism

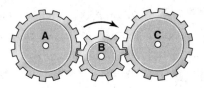

19. In the diagram, gear B is an idler gear. If gear B is rotating clockwise,
    A. gear A is rotating clockwise, but gear C is rotating counterclockwise.
    B. gear A and gear C are rotating clockwise.
    C. gear A and gear C are rotating counterclockwise.
    D. gear A is rotating counterclockwise, but gear C is rotating clockwise.

20. A wrench uses which simple machine to turn a bolt?
    A. Lever
    B. Inclined plane
    C. Pulley
    D. Wheel and axle

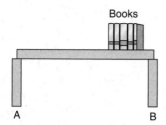

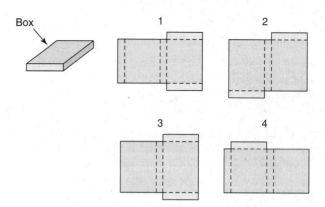

21. The diagram shows a box of books on a shelf supported by two posts. Which post bears the greater part of the weight?
    A. Post A
    B. Post B
    C. Each post bears the same amount of weight.
    D. The shelf distributes the weight equally.

24. Which flat cardboard pattern can be folded along the dotted lines to form the complete, totally enclosed box shown?
    A. 1
    B. 2
    C. 3
    D. 4

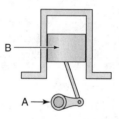

22. Water is flowing into the tank from the upper faucet at the rate of 240 gallons per hour and draining out of the tank through the lower faucet at a rate of 1 gallon per minute. How many more gallons of water will be in the tank after 6 minutes?
    A. 14
    B. 16
    C. 18
    D. 22

25. After the crank (A) rotates 9.5 times, where will the piston (B) be?
    A. In the same place as shown
    B. At the top of the cylinder
    C. At the bottom of the cylinder
    D. B or C

23. The diagram at the right shows a rotating disk. When the disk rotates, which point travels farthest?
    A. Point A
    B. Point B
    C. Point C
    D. Points B and C both travel the same distance.

STOP! DO NOT TURN THIS PAGE UNTIL TIME IS UP FOR THIS TEST. IF YOU FINISH BEFORE TIME IS UP, CHECK OVER YOUR WORK ON THIS TEST ONLY.

PART 9. ASSEMBLING OBJECTS

THE TEST ADMINISTRATOR WILL READ THE FOLLOWING ALOUD TO YOU.

Now turn to Part 9 and read the directions for Assembling Objects silently while I read them aloud.

This test has questions that will measure your spatial ability. Study the diagram, decide which answer is CORRECT, and then mark the space on your separate answer sheet that has the same number and letter as your choice.

Your score on this test will be based on the number of questions you answer correctly. You should try to answer every question. DO NOT SPEND TOO MUCH TIME on any one question. If you finish before time is called, go back and check your work in this part ONLY.

Now find the section of your answer sheet that is marked PART 9. When you are told to begin, start with question 1 in Part 9 of your test booklet and answer space number 1 in Part 9 on your separate answer sheet.

DO NOT TURN THE PAGE UNTIL TOLD TO DO SO. You will have 15 minutes to complete the 25 questions in this part. Are there any questions?

Begin.

For Questions 1–18, which figure best shows how the objects in the left box will appear if they are fit together?

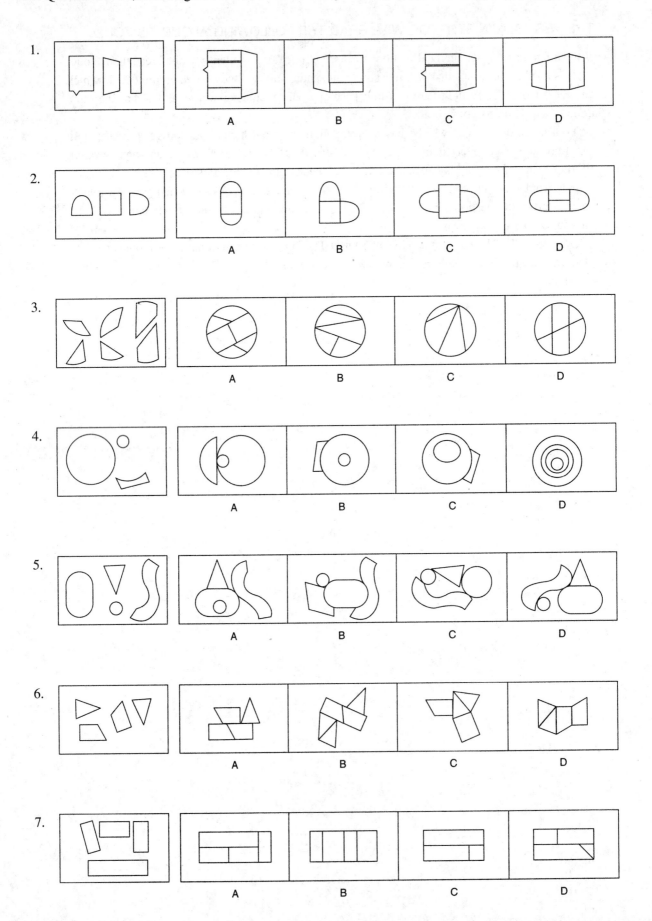

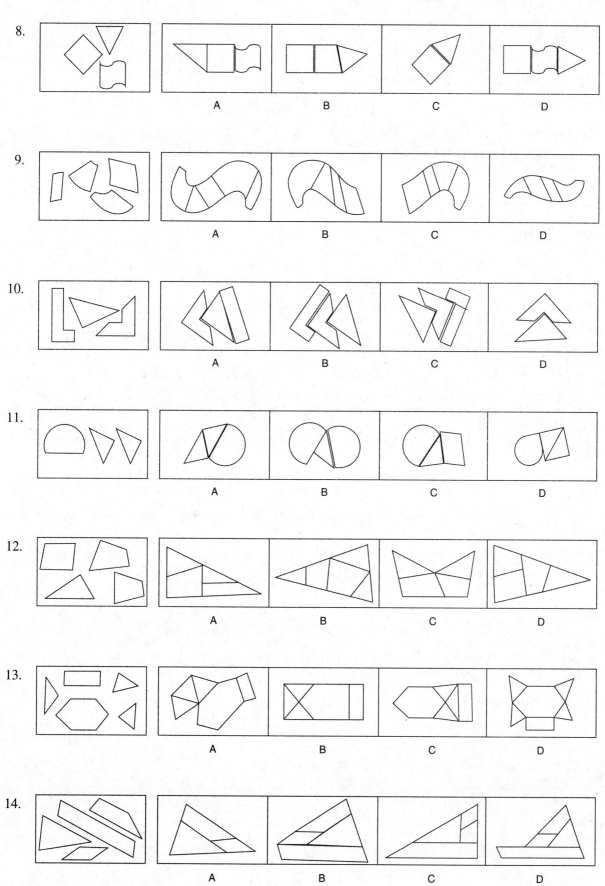

8.

9.

10.

11.

12.

13.

14.

15.

16.

17.

18.

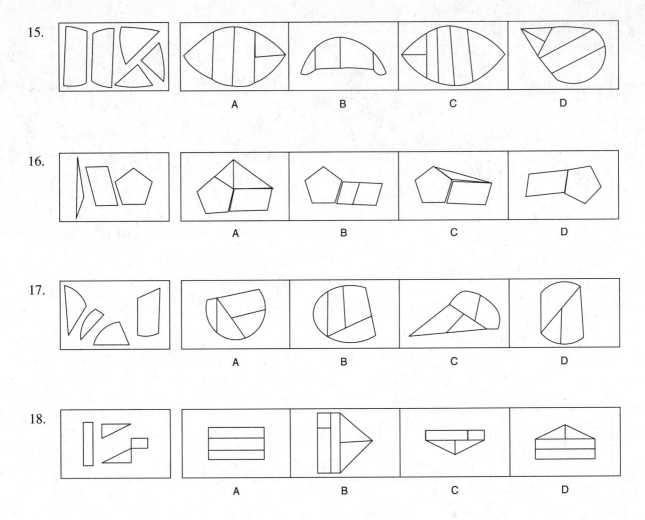

For Questions 19–25, which figure best shows how the objects in the left box will touch if the letters for each object are matched?

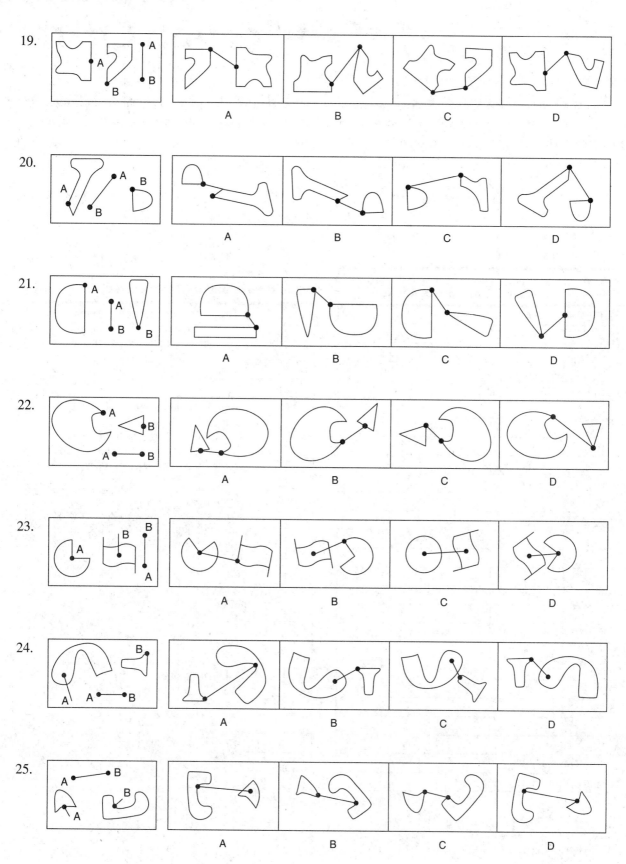

Answer Keys and Self-Scoring Charts

The following answer keys show the correct answers for each part of the diagnostic test that you just took. For each question, compare your answer to the correct answer. Mark an X in the column to the right if you got the item correct. Then total the number correct for each part of the test. Find that number in the corresponding chart to the right of the answer key. See the suggestions listed for your performance.

Part 1. General Science

Item Number	Correct Answer	Mark X if You Picked the Correct Answer
1	A	
2	D	
3	B	
4	D	
5	D	
6	B	
7	A	
8	D	
9	B	
10	D	
11	C	
12	B	
13	D	
14	B	
15	D	
16	D	
17	A	
18	B	
19	A	
20	D	
21	B	
22	D	
23	B	
24	C	
25	B	
Total Correct		

Score Interpretation–Total Correct

25	This is pretty good work. Review the explanations of the answers you got incorrect.
24	
23	
22	
21	
20	You are doing pretty well. Review the explanations for the items you answered incorrectly. If you have time, review the other explanations and you will learn even more.
19	
18	
17	
16	
15	You need to keep studying. Pay close attention to the explanations for each item, even for the ones you got correct.
14	
13	
12	
11	
10	Spend time working on the General Science review in Part 3 of this book.
9	
8	
7	
6	If you have copies of your school science textbooks, it would be a good idea to study those.
5	
4	
3	
2	
1	
0	

Part 2. Arithmetic Reasoning

Item Number	Correct Answer	Mark X if You Picked the Correct Answer
1	B	
2	B	
3	C	
4	A	
5	D	
6	D	
7	B	
8	C	
9	D	
10	A	
11	B	
12	B	
13	D	
14	D	
15	D	
16	D	
17	C	
18	C	
19	A	
20	D	
21	D	
22	C	
23	C	
24	D	
25	C	
26	C	
27	B	
28	B	
29	D	
30	A	
Total Correct		

Score Interpretation–Total Correct

Score	Interpretation
30	This is pretty good work. Review the explanations of the answers you got incorrect. This test is a part of the AFQT, and you must do well.
29	
28	
27	
26	
25	You are doing pretty well. Review the explanations for the items you answered incorrectly. If you have time, review the other explanations and you will learn even more.
24	
23	
22	
21	
20	You need to keep studying. Pay close attention to the explanations for each item, even for the ones you got correct. Spend time working on the Arithmetic Reasoning review in Part 3 of this book. Keep working and reworking problems until you are comfortable with the processes.
19	
18	
17	
16	
15	
14	
13	
12	
11	
10	
9	
8	
7	
6	
5	
4	
3	
2	
1	
0	

Part 3. Word Knowledge

Item Number	Correct Answer	Mark X if You Picked the Correct Answer
1	C	
2	B	
3	B	
4	C	
5	D	
6	C	
7	B	
8	C	
9	D	
10	B	
11	B	
12	A	
13	A	
14	A	
15	B	
16	B	
17	D	
18	C	
19	B	
20	D	
21	C	
22	B	
23	C	
24	B	
25	A	
26	A	
27	D	
28	A	
29	A	
30	C	
31	C	
32	A	
33	D	
34	B	
35	B	
Total Correct		

Score Interpretation–Total Correct

35	This is pretty good work. Review
34	the explanations of the answers you
33	got incorrect. This is an important
32	test that contributes to your AFQT.
31	
30	You are doing pretty well. Review
29	the explanations for the items you
28	answered incorrectly. If you have
27	time, review the other explana-
	tions and you will learn even
26	more.
25	You need to keep studying. Pay
24	close attention to the explanations
23	for each item, even for the ones
	you got correct.
22	
21	
20	Spend time working on the Word
	Knowledge review in Part 3 of this
19	book.
18	
17	
16	Keep reading and identifying
	words you don't know.
15	
14	
13	
12	
11	
10	
9	
8	
7	
6	
5	
4	
3	
2	
1	
0	

Part 4. Paragraph Comprehension

Item Number	Correct Answer	Mark X if You Picked the Correct Answer
1	A	
2	B	
3	C	
4	A	
5	D	
6	B	
7	A	
8	B	
9	C	
10	A	
11	C	
12	A	
13	B	
14	C	
15	D	
Total Correct		

Score Interpretation–Total Correct

15 14	This is pretty good work. Review the explanations of the answers you got incorrect. This test is a part of the AFQT, and you must do well.
13 12	You are doing pretty well. Review the explanations for the items you got incorrect. This test is a part of the AFQT, and you must do well.
11 10 9 8 7 6 5 4 3 2 1 0	You need to keep studying. Pay close attention to the explanations for each item, even for the ones you got correct. This will help you understand why the answer is correct. Spend time reviewing the Paragraph Comprehension and Word Knowledge reviews in Part 3 of this book. Keep reading books and newspapers.

Part 5. Mathematics Knowledge

Item Number	Correct Answer	Mark X if You Picked the Correct Answer
1	B	
2	A	
3	D	
4	B	
5	C	
6	D	
7	B	
8	A	
9	C	
10	B	
11	A	
12	C	
13	B	
14	A	
15	C	
16	D	
17	C	
18	C	
19	C	
20	C	
21	B	
22	D	
23	C	
24	A	
25	B	
Total Correct		

Score Interpretation–Total Correct

25	This is pretty good work. Review the explanations of the answers you got incorrect.
24	
23	
22	You are doing pretty well. Review the explanations for the items you answered incorrectly. If you have time, review the other explanations and you will learn even more.
21	
20	
19	This test is part of the AFQT. You must perform well on this test.
18	
17	You need to keep studying. Pay close attention to the explanations for each item, even for the ones you got correct.
16	
15	
14	
13	Spend time working on the Mathematics Knowledge reviews in Part 3 of this book, and work and rework the problems until you fully understand the processes.
12	
11	
10	
9	
8	
7	
6	
5	
4	
3	
2	
1	
0	

Part 6. Electronics Information

Item Number	Correct Answer	Mark X if You Picked the Correct Answer
1	B	
2	C	
3	B	
4	D	
5	A	
6	B	
7	D	
8	C	
9	B	
10	D	
11	D	
12	A	
13	D	
14	C	
15	A	
16	D	
17	B	
18	C	
19	D	
20	C	
Total Correct		

Score Interpretation–Total Correct

20	This is pretty good work. Review the explanations of the answers you got incorrect.
19	
18	
17	You are doing pretty well. Review the explanations for the items you answered incorrectly. If you have time, review the other explanations and you will learn even more.
16	
15	
14	
13	
12	You need to keep studying. Pay close attention to the explanations for each item, even for the ones you got correct.
11	
10	
9	
8	Spend time working on the Electronics Information review in Part 3 of this book.
7	
6	
5	Your school electronics or physics books may have some helpful information as well.
4	
3	
2	
1	
0	

Part 7. Auto and Shop Information

Item Number	Correct Answer	Mark X if You Picked the Correct Answer
1	A	
2	C	
3	A	
4	A	
5	C	
6	C	
7	A	
8	B	
9	B	
10	B	
11	D	
12	D	
13	B	
14	B	
15	D	
16	C	
17	A	
18	B	
19	D	
20	D	
21	A	
22	C	
23	A	
24	D	
25	C	
Total Correct		

Score Interpretation—Total Correct

25	This is pretty good work. Review the explanations of the answers you got incorrect.
24	
23	
22	
21	
20	You are doing pretty well. Review the explanations for the items you answered incorrectly. If you have time, review the other explanations and you will learn even more.
19	
18	
17	
16	
15	
14	You need to keep studying. Pay close attention to the explanations for each item, even for the ones you got correct.
13	
12	
11	
10	Spend time working on the Auto and Shop Information review in Part 3 of this book.
9	
8	
7	
6	
5	
4	
3	
2	
1	
0	

Part 8. Mechanical Comprehension

Item Number	Correct Answer	Mark X if You Picked the Correct Answer
1	D	
2	C	
3	C	
4	A	
5	D	
6	B	
7	C	
8	D	
9	A	
10	B	
11	A	
12	C	
13	B	
14	D	
15	B	
16	C	
17	B	
18	A	
19	C	
20	D	
21	B	
22	C	
23	B	
24	A	
25	A	
Total Correct		

Score Interpretation–Total Correct

25	This is pretty good work. Review the explanations of the answers you got incorrect.
24	
23	
22	
21	
20	You are doing pretty well. Review the explanations for the items you answered incorrectly. If you have time, review the other explanations and you will learn even more.
19	
18	
17	
16	
15	You need to keep studying. Pay close attention to the explanations for each item, even for the ones you got correct.
14	
13	
12	
11	Spend time working on the Mechanical Comprehension review in Part 3 of this book
10	
9	
8	
7	
6	
5	
4	
3	
2	
1	
0	

Part 9. Assembling Objects

Item Number	Correct Answer	Mark X if You Picked the Correct Answer
1	C	
2	B	
3	D	
4	B	
5	A	
6	D	
7	A	
8	D	
9	C	
10	B	
11	A	
12	D	
13	C	
14	B	
15	A	
16	C	
17	A	
18	C	
19	D	
20	B	
21	C	
22	A	
23	D	
24	B	
25	D	
Total Correct		

Score Interpretation–Total Correct

Score	Interpretation
25	This is pretty good work. Review the explanations of the answers you got incorrect.
24	
23	
22	
21	
20	You are doing pretty well. Review the explanations for the items you answered incorrectly. If you have time, review the other explanations and you will learn even more.
19	
18	
17	
16	
15	Keep studying. Pay close attention to the explanations for each item, even for the ones you got correct.
14	
13	
12	
11	Spend time working on the Assembling Objects review in Part 3 of this book.
10	
9	
8	
7	
6	
5	
4	
3	
2	
1	
0	

Answers and Explanations

PART 1. GENERAL SCIENCE

1. **A.** The nucleus is an important part of the cell. Although carbon and oxygen are important to life, the basic unit of life is the cell. Choice A is the correct answer.

2. **D.** During the process of photosynthesis, a plant uses carbon dioxide and water. Through the interaction of chlorophyll and light, glucose (a sugar) and oxygen are created. Carbon dioxide (choice A) is not a by-product of photosynthesis, but rather the raw material used by the plant, and light (choice B) acts to initiate the process of photosynthesis as it interacts with chlorophyll. Ozone is found high in the atmosphere and is not created by plants. Choice D, oxygen, is the correct answer.

3. **B.** The nucleus is important because it controls most of the activities of the cell. Choice B is the correct answer.

4. **D.** Muscle cells (choice A) and nerve cells (choice C) all have organelles. Protozoa (choice B), which are one-celled organisms, also have organelles. Bacteria (choice D) do not. Choice D is the correct answer.

5. **D.** Geotropism refers to plant growth responding to the force of gravity. Thigmotropism is when a plant moves or grows in response to touch or contact. Chemotropism is when a plant responds to chemicals. Phototropism is when a plant responds to light. D is the correct answer.

6. **B.** Birds are vertebrates that are warm-blooded. Choice B is the correct answer.

7. **A.** Decomposers break down tissue and release nutrients and carbon dioxide into the ecosystem. Choice A is the correct answer.

8. **D.** Marrow (choice A) is the soft tissue within the bone and does not protect the spinal cord. Cartilage (choice B) is a smooth, slippery covering on the ends of bones that acts as a shock absorber. Its purpose is not to protect the spinal cord. Ligaments (choice C) are one of the ways in which bones are joined and move together. They do not protect the spinal cord. The vertebrae, found in your back, do indeed protect the spinal cord. Choice D is the correct answer.

9. **B.** A bruise on your skin may cause a bit of swelling (choice D), but the bruise is not related to your sweat glands (choice C). Excess strain on your muscles (choice A) may cause some discomfort, but that is not likely to break your capillaries. Bruising is caused by some traumatic event that breaks the capillaries. Choice B is the correct answer.

10. **D.** Peristalsis is the muscular contraction of the digestive system that moves food from one part of the system to the next. For example, peristalsis is responsible for moving food from the stomach to the large intestine.

11. **C.** The hypothalamus, pons, and cortex (choice C) are all parts of the brain and thus of the nervous system. Nephrons, urethra, and bladder (choice A) are part of the excretory system. Antigens, T cells, and cilia (choice B) are all part of the immune system. Lymph nodes, spleen, and tonsils (choice D) are part of the lymphatic system. Choice C is the correct answer.

12. **B.** Dendrites (choice D) and axons (choice C) are parts of the nerve cell. The nerve cell gap (choice A) is not the correct answer because it is not related to the nervous system. The space between two nerve cells where an electric charge transmits information is called a synapse. Choice B is the correct answer.

13. **D.** Biodiversity tends to increase with warmer climate. Florida is the warmest state of those listed, so choice D is the correct answer.

14. **B.** Ultraviolet light is absorbed, at least in part, by the ozone layer in the atmosphere. Some of it passes through to Earth's surface. With the thinning of the ozone layer as a result of the introduction of chlorofluorocarbons into the atmosphere, more ultraviolet light is reaching Earth's surface, with possibly damaging effects on humans. Choice B is the correct answer.

15. **D.** The amplitude of a sound wave is responsible for the loudness of the sound. Choice D is the correct answer.

16. **D.** All the alternatives have something to do with sea-floor spreading, but only one is the cause. Subduction (choice A) is a downward force or movement that is a result of sea-floor spreading, not the cause. The same can be said for creation of the ridges (choice B) and for the compression of continents (choice C). They are both results, not causes. All these features are produced by the convection currents in the mantle, which cause the uprising and subduction of the crust. Choice D is the correct answer.

17. **A.** A rift (choice D) is a valley in the ocean floor formed by tectonic plates that are separating. A tsunami (choice C) is a wall of water that can overwhelm a shoreline. It is created when an earthquake creates an ocean wave. Hot spots (choice B) are undersea fissures that are spouting magma from the sea floor. The Hawaiian Islands have been and continue to be created from hot spots. When two tectonic plates converge and one rises above the other, it is common to find areas that experience earthquakes and volcanic activity. Choice A, volcanoes, is the correct answer.

18. **B.** When a wave is created in the ocean by persistent surface winds, the wave moves in the direction of the winds. When that wave approaches land, it moves through shallow water and the trough (bottom) of the wave starts to scrape the ocean bottom. The scraping causes friction, and that makes the trough of the wave move more slowly than the crest (top) of the wave. Eventually the crest just topples over, creating a breaker. Choice B is the correct answer.

19. **A.** The definition of a year is the time it takes for Earth to make one revolution around the sun. Choice A, 1 year, is the correct answer.

20. **D.** A solar eclipse is created when the moon blocks the light from the sun from reaching Earth by casting a shadow on Earth's surface. The shadow covers only a very small area, so a total solar eclipse is visible only on a limited part of Earth's surface at any one time. The shadow has two parts, a darker part called the umbra, where a total eclipse occurs, and a lighter part called the penumbra, where a partial eclipse can be seen. In order for a solar eclipse to occur, the moon has to be between the sun and Earth. Choice D is the correct answer.

21. **B.** When solar radiation reaches Earth's atmosphere, the radiation is either reflected, absorbed by Earth's atmosphere, or trapped by Earth's atmosphere. When it is trapped by the atmosphere, it causes the atmosphere and Earth's surface to heat up. Choice B is the correct answer.

22. **D.** Warm air can hold more moisture than cold air. Air masses formed over the tropical oceans can hold significant amounts of moisture, while air masses formed over cold oceans or land masses are generally drier. Whether the air is moving or stable does not much affect the humidity in the air. Choice D, warm air, is the correct answer.

23. **B.** When a prevailing wind containing moisture reaches a mountain range, it rises and cools, causing condensation in the form of rain or snow. The side of the mountain range on which this happens is called the windward side. Much of the moisture in the air is deposited on the windward side of a mountain range. When the air reaches the tops of the mountains, it starts to spill down the other side. As it descends, it heats up. The side of the mountain range on which this happens is called

the leeward side. Often you will find deserts on the leeward side of a mountain range. The climate is definitely drier on the leeward side. Choice B is the correct answer.

24. **C.** Igneous rocks (choice A) are formed from molten material that eventually cools. Extrusive rocks (choice B) are igneous rocks that have formed on or above Earth's surface. Metamorphic rocks (choice D) are created by high temperature and pressure. Sedimentary rocks (choice C) are formed by rock that has been broken up into pieces and cemented together. The pieces can be somewhat large, as in a conglomerate, or small, such as in shale or limestone. Choice C is the correct answer.

25. **B.** When objects move away from a viewer, the light waves they produce lengthen, shifting toward the red end of the spectrum. Scientists have observed that light from distant stars shifts toward the red end of the spectrum. This supports the conclusion that the universe is expanding. Choice B is the correct answer.

PART 2. ARITHMETIC REASONING

1. **B.** This is a simple addition problem. Add the amounts to get the total of $14.74.

2. **B.** This problem has two steps. First you need to add the amount spent. Then you need to subtract that amount from the $100 bill. The total amount spent is $90.49, so, subtracting that from $100, the change is $9.51.

3. **C.** In order to calculate the percent, you need to divide the total number of items into the number answered correctly. In this case, you would divide 45 into 38 and get 0.8444. To change that to a percent, move the decimal point two places to the right and add the percent sign to get 84.44%. Since the question asks for the nearest percent, the answer would be 84%.

4. **A.** To find the average, add up the numbers and divide by the number of numbers. In this problem,

the numbers add to 400. Dividing that by 5 gives the correct result of 80 as the average.

5. **D.** This problem needs to be completed in two steps. First, you need to determine what percent of chocolates have nuts, then you subtract that from 100% to determine what percent do not have nuts. So, dividing 9 by 45 results in the percent that have nuts, which is 20%. Subtracting that number from 100% gives the correct answer: 80% of the chocolates do not have nuts.

6. **D.** To find the ratio of one number to another, create a fraction to show the relationship. In this problem, there are 6 female club members out of a total of 48 members. So the fraction that shows the relationship is $\frac{6}{48}$. This fraction needs to be simplified to $\frac{1}{8}$, the correct answer. You might see a ratio written as 1:8 on the ASVAB test. $\frac{1}{8}$ and 1:8 mean the same thing.

7. **B.** If Vanessa has 192 pages of reading and reads 48 pages each day, on Monday she has read 48 pages, on Tuesday she has read 2(48) or 96 pages, on Wednesday she has ready 3(48) or 144 pages, and on Thursday she has read 4(48) or 192 pages. Thursday is the correct answer. Another way to do this is to divide 192 by 48 to get 4. If Monday is 1, Tuesday is 2, Wednesday is 3, and Thursday is 4. Again, Thursday is the correct answer.

8. **C.** Probability = number of favorable outcomes/ number of possible outcomes. In this problem, a favorable outcome would be selecting a dime. There are 112 dimes or favorable outcomes. The number of possible outcomes is the sum of all the coins, or $94 + 44 + 112 = 250$. So Probability = $\frac{112}{250}$. Simplified, this is $\frac{56}{125}$.

9. **D.** Simple interest problems use the formula $I = prt$, where I is the amount of interest, p is the principal or the amount saved or invested, r is the rate of interest, and t is the amount of time that the interest is accruing. In this problem, you are asked to find the interest, given that the principal is $4,000, the rate of interest is 4%, and

the amount of time is 3 years. Substituting this information into the formula, you get

$$I = 4,000(0.04)(3) = \$480 \text{ of interest}$$

10. **A.** Set this up as a proportion.

$$\frac{5,200 \text{ miles}}{11 \text{ hours}} = \frac{x \text{ miles}}{1 \text{ hour}}$$

Cross-multiply to get

$$11x = 5,200$$
$$x = 472.73 \text{ miles per hour}$$

11. **B.** The Booster's Club has 250 members and wants to reach 376 members in 90 days. Subtracting 250 from 376, the club needs to attract 126 new members. To do this in 90 days, the club must attract $126 \div 90 =$ an average of 1.4 members each day.

12. **B.** Here you need to calculate the number of square feet in the garden and then determine how many ounces of fertilizer are needed. The area of a rectangular garden is calculated by multiplying the length $\times$ the width. So in this problem the area is 21 ft^2. If you need 2 oz of fertilizer for every square foot, then you need to multiply 2 by the number of square feet.

$$21 \times 2 = 42 \text{ oz}$$

13. **D.** Calculate the cost of the two CDs:

$$\$16.99 \times 2 = \$33.98$$

Reduce that amount by 20% by multiplying the total cost by 0.20:

$$0.20 \times 33.98 = \$6.80$$

Subtract $6.80 from the original cost:

$$\$33.98 - \$6.80 = \$27.18$$

A faster way to do this (and speed is important on the ASVAB) is to multiply the original cost by 80% or 0.80 because that is the cost of the items after the discount.

$$0.80 \times \$33.98 = \$27.18$$

14. **D.** Use the following formula:

$$\text{Percent of change} = \frac{\text{amount of change}}{\text{starting point}}$$

In this problem, the starting point is $67.50. The amount of change is the difference between the starting price and the ending price of $87.75.

$$\$87.75 - \$67.50 = \$20.25$$

Substitute the information into the formula:

$$\text{Percent of change} = \frac{\$20.25}{\$67.50} = 0.3$$

Change that number to a percent by moving the decimal place two places to the right and adding the percent sign.

$$0.3 = 30\%$$

15. **D.** Set this up as an equation. A number plus 2 times that number is 300.

$$x + 2x = 300$$

Solve for x.

$$3x = 300$$
$$x = 100$$

So one number is 100 and the other is twice that, or 200.

16. **D.** To answer this question, you need to know the perimeter of the frame. For a rectangle, the perimeter is $2l + 2w$. If $w = 8$ and $l = 10$, then the perimeter $= 2(8) + 2(10) = 36$ inches. At $0.50 per inch, multiply $36 \times \$0.50$ to get the price of the wood.

$$36 \times \$0.50 = \$18.00$$

17. **C.** To find the circumference of a circle, use the formula $C = 2\pi r$. In this problem, you are given the diameter. You need to use the radius, which is $\frac{1}{2}d$, or 37.5 yards. Substitute the information into the formula.

$$C = 2\pi(37.5)$$
$$C = 2(3.14)(37.5) = 235.5 \text{ yd}$$

Note that you can skip a step if you remember the formula $C = \pi d$. Substitute the information you have.

$$C = \pi(75) = 235.5 \text{ yd}$$

18. **C.** In this problem, you need to find 1/3 of 39. Dividing 39 by 3 gives you 13.

19. **A.** Create an equation to solve this problem.

$$\text{Dave} = \text{Brian} - 2.$$

You are told that Brian is 17. Substitute that information into the formula.

$$\text{Dave} = 17 - 2$$
$$\text{Dave} = 15$$

20. **D.** Create equations to solve this problem.

$$\text{Al} = 3(\text{Bill})$$
$$\text{Bill} = 2(\text{Charlie})$$
$$\text{Charlie} = \$5.89$$

Solve for Al. Substitute what you know into the formulas.

$$\text{Bill} = 2(\$5.89)$$
$$\text{Bill} = \$11.78$$

To find Al, substitute what you know about Bill.

$$\text{Al} = 3(\$11.78)$$
$$\text{Al} = \$35.34$$

21. **D.** In this problem, you need to calculate the number of 16-mpg units there are in 1,344. Do this by dividing 1,344 by 16.

$$1,344 \div 16 = 84 \text{ gallons}$$

22. **C.** To solve this problem, use the formula $V = lwh$. You have been given the dimensions of $l = 28$ ft, $w = 12$ ft, and $h = 6$ ft. Substitute that information into the formula.

$$V = (28)(12)(6)$$
$$V = 2016 \text{ ft}^3$$

23. **C.** You need to calculate the perimeter of the garden. The perimeter is the sum of all the sides,

or use the formula $P = 2l + 2w$. You are given the dimensions of 6 feet and 15 feet. Substitute that information into the formula.

$$P = 2(6) + 2(15)$$
$$P = 42 \text{ ft}$$

24. **D.** To calculate 14% of 5,000, multiply 5,000 by 0.14.

$$5,000(0.14) = 700 \text{ votes}$$

25. **C.** To calculate the volume of a cylinder, use the formula $V = \pi r^2 h$. In this problem, the radius is 9 m. The height is given as 3.5 m. Substitute that information into the formula.

$$V = (3.14)(9^2)(3.5)$$
$$V = (3.14)(81)(3.5)$$
$$V = 890.19 \text{ or } 890.2 \text{ m}^3$$

26. **C.** Use the formula $C = 2\pi r$ or $C = \pi d$. You are given a diameter of 8,000 miles. Substitute that into the formula.

$$C = \pi(8,000)$$
$$C = 3.14(8,000)$$
$$C = 25,120 \text{ mi}$$

27. **B.** To calculate the answer to this question, add up the amount of the checks and subtract that from the current amount in the checking account.

$$\$4,355.19 - (\$1,204.90 + \$890.99)$$
$$= \$2,259.30$$

28. **B.** Multiply $585.00 by 0.75 to get $438.75.

29. **D.** Calculate the total cost of the items and then multiply by 20% to determine how much Danny will save.

$$\$750 + \$398.75 + \$149.98 + \$24.99$$
$$= \$1,323.72$$
$$\$1,323.72 \times 0.20$$
$$= \$264.74 \text{ savings from the sale}$$

30. **A.** In this problem, you know two of the three angles, one measuring 55° and the other measuring 90°. Together those two angles measure 145°. Since a triangle has a total of 180°, the third angle must measure 180° − 145° = 35°.

PART 3. WORD KNOWLEDGE

1. **C.** The word *exempt* means "to free someone from a rule or requirement." The words *empty*, *exit*, and *anoint* do not seem related to the definition. The only word that comes close to this is C, *excuse*.

2. **B.** To *recruit* means "to take on, sign up, or enlist members into an organization or club." The words *react*, *permit*, and *recur* do not relate to the definition. So B, *enlist*, is the correct answer.

3. **B.** *Unanimous* uses the prefix "uni-," or "one," as in *unity* or *unify*. The only word that relates to the concept of one or unity is *undisputed*, suggesting that everyone is in agreement. Choice B is the correct answer. The words *questioned*, *selective*, and *trivial* do not relate to the definition.

4. **C.** *Temperamental* most nearly means "volatile." Think of temperament or having a temper. Choice C is the correct answer.

5. **D.** *Economical* has to do with finances or the economy. The words *logical* and *comical* have no relation to finances, so they can be eliminated as possibilities. This leaves *costly* and *thrifty* as the two possible answers. Since being economical means to not waste money, time, or other resources, choice D, *thrifty*, is the correct answer.

6. **C.** The prefix *pro-* means "forward." *Motion* has to do with moving. Combining the two gives you "moving forward." The only word that relates to this is *elevation*, meaning "to move forward or higher up in the ranks." Choice C is the correct answer

7. **B.** The prefix *un-* means "not." So the word means "not predictable." If something cannot be predicted, it is not *consistent*, *dependable*, or *regular*, as these words are somewhat the opposite of the definition. So the correct answer has to be choice B, *random*.

8. **C.** To *deny* means "to refuse to accept as correct or to not grant or give something." *State*, *avoid*, and *open* do not relate to the definition. *Refuse*

is the closest word to that meaning, so choice C is the correct answer.

9. **D.** The prefix *ir-* means "not." *Irresponsible*, then, means "not responsible." *Hapless*, *giddy*, and *easy* do not relate to the definition. *Reckless* is the only word that reflects the correct meaning, so choice D is the correct answer.

10. **B.** *Persistent* means "refusing to give up." *Gentle*, *resigned*, and *manageable* do not relate to this idea. *Unrelenting*, choice B, is the correct answer.

11. **B.** *Dominant* means "being overpowering or prevailing" or "ruling by superior power." The words *joyous*, *understanding*, and *helpful* do not seem to match the definition. The word that most closely relates to the definition is *overriding*, so choice B is the correct answer.

12. **A.** *Conscientious* has to do with doing what is right and correct. By knowing that, you can eliminate the words *popular*, *friendly*, and *handsome*. The word *painstaking* is most closely related to the meaning of *conscientious*, so choice A is the correct answer.

13. **A.** A *diversion* is something that makes you turn aside or distracts your attention. The words *containment*, *pressure*, and *supply* do not match the definition. So, *distraction*, choice A, is the correct answer.

14. **A.** To *assume* means "to take for granted or to presume that something is true." So *assumption* has to do with taking something to be true or correct. The words *decision*, *rumor*, and *guess* do not match the definition. The word *belief* most closely relates to this idea. The correct answer is choice A.

15. **B.** To *profess* means "to proclaim, state, or declare." The words *work*, *disguise*, and *accuse* do not match the definition. The word that most nearly resembles this idea is *claim*, so choice B is the correct answer.

16. **B.** *Trivial* means "unimportant or insignificant." The words *sinister*, *sad*, and *demanding* do not match the definition. *Inconsequential*, choice B, is the correct answer.

17. **D.** *Optimistic* means "having a hopeful view of the world," so the words *grouchy*, *angry*, and *unusual* can be eliminated. *Positive*, choice D, is the correct answer.

18. **C.** To *vary* means "to change," so *variable* means *changeable*. Choice C is the correct answer.

19. **B.** A *clarification* is something that makes information clearer or more explicit, so the words *elimination* and *recognition* are clearly wrong. The word *expansion* could have to do with providing more information to clarify something, but *explanation*, choice B, is the best answer.

20. **D.** *Panoramic* means "providing a wide view." So *beautiful*, *unrealistic*, and *stylistic* are not related to the word. The best answer is choice D, *expansive*.

21. **C.** The prefix *com-* means "to bring together." *Passionate* has to do with strong, caring feelings. Combining those ideas suggests that *compassionate* has to do with bringing strong or deep feelings together. The words *unwieldy*, *modern*, and *restrictive* do not match the definition. The only word that deals with feelings is *concerned*. Choice C is the correct answer.

22. **B.** One event is *contingent* upon another if it will happen only if the second event occurs. The best synonym is choice B, *dependent*.

23. **C.** *Velocity* has to do with quickness or speediness. The words *variety*, *veracity*, and *overflow* do not match the definition. So *swiftness*, choice C, is the correct answer.

24. **B.** *Immaculate* means "without flaws or errors." The words *large* and *fancy* do not match the definition. A house that has been *cleaned* may be immaculate, but *spotless* is the closest word to "without flaws or errors." Choice B is the best answer.

25. **A.** *Visible* relates to "being able to see something" or "vision," so *visibility* must have to do with being able to see. *Test* and *dichotomy* do not match the definition. A view may be an *observation*, but that is not the best answer. *Clarity* is the only word that relates to the definition of "being able to see." Choice A is the best answer.

26. **A.** *Capsized* means "turned over." The words *accelerated*, *tilted*, and *docked* do not match the definition. Choice A, *overturned*, is the correct answer.

27. **D.** *Abdicated* means "gave up or surrendered." *Welcomed*, *cheered*, and *transitioned* are unrelated to this concept. *Abandoned* is most closely related to the word *abdicated*. Choice D is the correct answer.

28. **A.** *Eligible* means "fit to be chosen." The words *eager*, *adequate*, and *thankful* are unrelated to this concept. Choice A, *qualified*, fits most closely and is the best answer.

29. **A.** To be *buoyant* means "to be able to float." The words *happy*, *significant*, and *acrobatic* do not match the definition. Choice A, *floating*, is the correct answer. You can remember this by thinking of a buoy, which floats in the water.

30. **C.** *Teeming* means "swarming, abounding, or being full." *Empty* is the opposite of this concept. *Opaque* means "dense or thick," and *vigorous* means "energetic, strong, or active." These words don't relate to being full or swarming, but *crowded* does. Choice C is the correct answer.

31. **C.** *Bestow* means "to give," as in giving a gift or an award. The words *remove*, *have*, and *question* do not match the definition. *Give*, choice C, is the correct answer.

32. **A.** A river that *meanders* is one that winds and turns as it flows. The best synonym is choice A, *twist*.

33. **D.** *Infiltrate* means "to pass into" or "to seize control from within." The words *interfere* and *seek* are not related to the definition. *Enter* is similar in meaning, but the word *penetrate* is the closest to the stated meaning. Choice D is the correct answer

34. **B.** *Plummeted* means "fell." *Soared* has somewhat the opposite meaning. The words *increased* and *failed* are not related. Choice B, *plunged*, is the correct answer.

35. **B.** *Defiled* means "contaminated or dishonored." *Honored* is the opposite of this idea. The words *cleaned* and *respected* are not related. *Desecrated*, choice B, is the correct answer.

PART 4. PARAGRAPH COMPREHENSION

1. **A.** There is nothing in the passage indicating that people need to be convinced to exercise (choice B), so that is not a correct option. The same can be said for strenuous versus moderate exercise (choice D), as no such comparisons are made in the passage. It is true that exercise is good for preventing heart disease and for preventing stroke, but the passage does not indicate that exercise cures anything (choice C). The main thrust of the passage is that moderate exercise has many positive benefits. Choice A is the correct answer.

2. **B.** The passage does not indicate that people who exercise don't have weight problems, so choice D is clearly wrong. The passage also does not address the amount of exercise, so choice A cannot be correct. There is no mention of exercise eliminating any health problems, so choice C cannot be correct. That leaves choice B, which is implied fairly directly in the passage. Choice B is the correct answer.

3. **C.** The passage does not mention anything about fatty acids. Choice C is the correct answer.

4. **A.** In the third paragraph, the passage clearly states that 25 percent of adults do not exercise in their leisure time. Choice A is the correct answer.

5. **D.** The passage states that more than 50 percent of American adults do not get enough physical activity to provide health benefits. Choice D is the correct answer.

6. **B.** You need to read the passage carefully because it is somewhat complex. The passage indicates that the thought or perception or understanding at the time was that the West was geographically somewhat simple, that it was empty, and that it had only a single mountain range (choices A and C), but those ideas were proven incorrect. There is no information on discovering rivers (choice D). The passage states that what Lewis and Clark discovered was a far more complicated geographic landscape than they had originally thought. Choice B is the correct answer.

7. **A.** A labyrinth is a structure with winding passages; it is a maze. The only word that conveys that message is *jumble*.

8. **B.** The passage indicates that Major Stephen H. Long's Scientific Expedition (1819–1820) was the first military expedition to take along scientists and artists. The passage indicates that this set the pattern for increasingly scientific exploration of the West. Choice B is the correct answer.

9. **C.** There is no indication that Lewis and Clark's actions did anything to help the fur traders (choice A) or that the fur traders boosted the economy of the United States (choice B). The passage also does not give credit to the mapmakers for scouting out the route for the railroad. The only option that is true is choice C, that Lewis and Clark used information provided by native people.

10. **A.** The passage clearly indicates that the military had much to do with the scientific exploration of what is now the western part of the United States.

11. **C.** The word *dubious* means "skeptical or questionable." Choice C, *doubtful*, is the correct answer.

12. **A.** The passage explains that the Greeks admired Odysseus because of his quick thinking. Choice A is the correct answer.

13. **B.** Herakles may have been kind (choice A) and resourceful (choice C), but the passage does not

indicate that. It also does not indicate that he was a failure (choice D). The passage does indicate that Herakles was known for his strength and courage. So choice B, bravery, is the correct answer.

14. **C.** The passage indicates that it is not true that all Greek heroes were admired for the same reasons, so choice D cannot be correct. Although Odysseus was indeed a murderer (choice A) and a thief (choice B), the best answer is choice C, that the Greeks judged their heroes by their results.

15. **D.** The passage says that the Cyclops were shepherds. Choice D is the correct answer.

PART 5. MATHEMATICS KNOWLEDGE

1. **B.** Isolate the unknown on one side. Subtract 9 from each side

$$x = 13 - 9$$
$$x = 4$$

2. **A.** Perform operations in parentheses first, then exponents, and then the remaining operations.

$$9 + 4(3) =$$
$$9 + 12 = 21$$

3. **D.** Change the mixed numbers to improper fractions. Carry out the operation by multiplying the two numerators and the two denominators.

$$\frac{5}{4} \times \frac{5}{2} = \frac{1}{8}$$
$$\frac{25}{8} = 3\frac{1}{8}$$

4. **B.** $-8 \times -8 = 64$

5. **C.** Substitute 2 for x and 4 for y into the expression. Multiply out the exponents.

$$2(2)^3 (4)^2 = 2(8)(16)$$
$$16 \times 16 = 256$$

6. **D.** When adding such expressions, place all similar terms under each other. Then perform the necessary operations. For this particular problem, place all the x^2 under each other and add the numbers, giving $-2x^2$. Then go to the next terms (xy) and add those. Adding these terms gives you $5xy$. Next, add the y terms. Adding these terms results in $1y$ or y. Next, add the y^2 terms. Notice that adding these terms leaves you with zero y^2, so this term drops out of the final answer. The answer is $-2x^2 + 5xy + y$. In this instance, you could have worked from right to left to notice that the y^2 term drops out of the result and that D is the only answer with that term missing. Doing this would have saved you a little time, but not all problems like this have that type of clue.

7. **B.** Note the minus sign in the middle term and the plus sign in the third term. This should tell you that the factors will contain a minus sign in each term. In this problem, the fact that the two terms will have a minus sign should lead you to the correct answer, as only one answer has a minus sign in each term. However, not all questions of this type will have that kind of clue. Note that the first term in each factor will be y. Now you need to find two numbers that multiply to 48 and add to 16. Those numbers are 4 and 12. This gives the factors $(y - 4)(y - 12)$.

8. **A.** To solve such problems, move all the terms with an unknown to one side of the equal sign and the numbers to the other. In this problem, subtract 6 from both sides to move the numbers to the right side, giving you $12x = 8x + 10 - 6$. Next, work on the $8x$ term by subtracting $8x$ from both sides, giving $12x - 8x = 10 - 6$. Combine the terms, resulting in $4x = 4$. Solve for x: $x = 1$.

9. **C.** Set each equation equal to zero. Next, arrange one of the equations so that when it is subtracted from or added to the other equation, one of the terms becomes zero and drops out of the equation. Next, solve for the remaining unknown.

In this problem:

$$3y + 3x = 24$$
$$6y + 3x = 39$$

Set each equation to zero:

$$3y + 3x - 24 = 0$$
$$6y + 3x - 39 = 0$$

Note that each equation contains the term $3x$. If you subtract the second equation from the first, the x term will drop out, leaving

$$3y + 3x - 24 = 0$$
$$-6y - 3x + 39 = 0$$

Subtracting the terms results in $-3y + 15 = 0$. Solve for y:

$$-3y = -15; \ y = 5.$$

To solve for x, take that result and substitute it into one of the original equations.

$$3(5) + 3x - 24 = 0$$
$$15 + 3x - 24 = 0$$
$$3x = 24 - 15$$
$$3x = 9$$
$$x = 3$$

The answer to the problem is $x = 3$; $y = 5$

10. **B.** To divide numbers with exponents, subtract the exponents. In this problem, subtract the 4 from the 12, leaving x^8 as the correct answer.

11. **A.** Set the equation to equal zero:

$$y^2 + 7y + 10 = 0$$

Factor the equation.

$$(y + 5)(y + 2) = 0$$

Solve for y by setting each factor equal to 0:

$$y + 5 = 0; \ y = -5$$
$$y + 2 = 0; \ y = -2$$

The correct answer is $y = -5$, $y = -2$.

12. **C.** To multiply fractions, multiply the numerators and denominators. Simplify where possible.

$$\frac{6y}{11} \times \frac{2}{5x} = \frac{12y}{55x}$$

This cannot be simplified any further, so it is the final answer.

13. **B.** To divide fractions, invert the second term and multiply. Simplify where possible.

$$\frac{6x}{11} \div \frac{2}{5y} \quad \text{Invert and multiply.}$$

$$\frac{6x}{11} \times \frac{5y}{2}$$

$$\frac{3x}{11} \times \frac{5y}{1} \quad \text{Simplify.}$$

$$\frac{15xy}{11}$$

14. **A.** The cube root is the number that, when multiplied by itself three times, results in the answer. In this instance, $3 \times 3 \times 3 = 27$, so the cube root of $27 = 3$.

> To answer most questions of this type on the ASVAB, you just need to memorize some basic cubes. It would be wise to memorize all the cubes from 2 to 10, just to prepare for ASVAB test item possibilities.

15. **C.** Isolate the variable of interest on one side of the equation. In this problem, you must multiply each side by h, leaving g on the left side of the equation and h on the right side. Therefore, the answer is $g = ah$.

16. **D.** An equilateral triangle has three equal sides and angles, so D is the correct answer.

17. **C.** To find the perimeter of a quadrilateral, add the lengths of the four sides. In this instance, 9 meters + 9 meters + 3 meters + 3 meters = 24 meters, the correct answer.

18. **C.** The formula for the circumference of a circle is πd. The figure gives the radius, which is one-half the length of the diameter. To find the diameter, multiply the radius by 2. In this problem, the diameter is 8 meters. Multiplying 8 by π gives the correct answer of 25.12 meters.

19. **C.** An equilateral triangle has three equal angles. A triangle has a total of 180°, so each angle must be 60°.

20. **C.** A straight line or straight angle is 180°. So if $\angle 1$ is 34°, $\angle 2$ is the difference between 180° and 34°, or 146°.

21. **B.** $\angle C$ is a right angle, so you can employ the Pythagorean Theorem, $a^2 + b^2 = c^2$, to get your answer. $3^2 + 4^2 = c^2$. So $9 + 16 = 25$. $c^2 = 25$ and $c = 5$.

22. **D.** This is a right triangle, so 4 ft is the height of the triangle and 8 ft is the base. The formula for the area of a triangle is $\frac{1}{2}bh$. Substituting the information into the formula gives $\frac{1}{2}(8)(4)$, making the correct answer 16 ft^2.

23. **C.** The diameter is $2r$. $r = 80$ cm, so the diameter is 160 cm.

24. **A.** $\angle 2$ and $\angle 7$ are alternate exterior angles and therefore are equal in measure. So both $\angle 2$ and $\angle 7$ are 30°.

25. **B.** The formula for finding the volume of a cylinder is $V = \pi r^2 h$. In this problem, the cylinder has a height of 15 meters and a radius of 5 meters. Substituting that information into the formula gives $\pi (5)^2 (15)$, making the correct answer 1,177.50 m^2.

PART 6. ELECTRONICS INFORMATION

1. **B.** 1 hertz is defined as 1 cycle per second. A 60-hertz electric current oscillates at 60 cycles per second.

2. **C.** Alternating current travels much further through wire than direct current.

3. **B.** Without a circuit, electrons cannot travel, so there is no electric current.

4. **D.** Copper is a good conductor and inexpensive. Choice B, while true, is not the whole answer, as it also applies, for example, to gold.

5. **A.** Amperes measure the number of electrons moving through a conductor.

6. **B.** Ohms are a measure of resistance.

7. **D.** Using algebra, multiply both sides by ohms. You should get volts = amperes × ohms.

8. **C.** Volts = amperes × ohms = 20 × 12 = 240.

9. **B.** This circuit will have a dangerously high current and may burn up or cause a fire. It should trip a circuit breaker or blow a fuse before too much damage occurs.

10. **D.** The overall amount of power does not change. Power = amperes × volts. So if voltage rises, amperage must fall.

11. **D.** In a transformer, electricity is converted into magnetism and back into electricity. In an electromagnet, electricity is converted into magnetism. In an electric motor, electromagnets create a repulsive force that causes the rotor to spin.

12. **A.** Heat alone will not do the job, because the wires will start to corrode when you heat them. Solder does not adhere to corrosion.

13. **D.** Conduit (either flexible or rigid) provides a safe return path for electricity in many wiring systems.

14. **C.** Heaters use high current, which demands heavy wires.

15. **A.** Black is the most common color for hot wires, although red and blue are sometimes used.

16. **D.** Three-way switches are used in pairs to control one light from two locations.

17. **B.** Amperes × volts = power. Voltage does not change, so fewer amperes will carry less power.

18. **C.** Both the polarity and the voltage must be correct for a battery to work right.

19. **D.** Loose electrical connections can cause fire or equipment damage. Choice C, salt water, is wrong because it will cause corrosion, reducing conductivity across the connection.

20. **C.** A resistor reduces voltage.

PART 7. AUTO AND SHOP INFORMATION

1. **A.** The differential and driveshaft are the big moving parts at the rear of a rear-wheel-drive car that has the engine in front.

2. **C.** Either the spark plug is not working right or there is a problem with the incoming fuel–air mix.

3. **A.** Air pressure inside the tire pushes evenly against the entire inside of the tire, allowing it to hold its shape and to respond to unevenness in the road. Low tire pressure reduces efficiency, and engine power is converted into heat, which warms the tire.

4. **A.** Breaker points send quick jolts of electricity to the coil, which acts as a transformer to raise the voltage enough to spark a spark plug. The gap must be adjusted during a tune-up, and should be inspected whenever the engine runs rough.

5. **C.** Front-wheel-drive mechanisms are complicated because they both drive and steer.

6. **C.** Intake: The piston moves down, creating a partial vacuum that draws air (and sometimes fuel) into the cylinder. Compression: The piston moves up with both valves closed, compressing the air–fuel mixture. Power: The fuel explodes, creating heat and kinetic energy, pushing the piston down. Exhaust: The piston moves up, pushing burned gases from the cylinder through the exhaust valve. Choice A is tempting, but there is no "rest" cycle in a four-cycle engine. The cylinder is always working.

7. **A.** The fuel system must deliver enough fuel for combustion inside the engine. Choice D, the engine control module, is the second-best answer. Although it could cause the problem, the fuel system is the first thing to check.

8. **B.** Manual transmissions usually have one "direct-drive" gear, where the input speed is the same as the output speed. The drive shaft rotates along with the transmission's output shaft. Choice D is incorrect because the clutch moves at the same speed as the engine.

9. **C.** Because the pistons in the wheel cylinders are the only movable objects in the system, when you move the master cylinder, hydraulic fluid forces the pistons in the wheel cylinders to move.

10. **B.** The starter relay is a big switch located between the battery and the starter motor. Starter motors demand a huge current, which would require an expensive and clumsy cable from the battery to the ignition switch and the starter motor. Instead, the ignition switch activates the starter relay, which temporarily connects the battery to the starter motor.

11. **D.** After the power stroke, there is nothing left to burn inside the cylinder. The exhaust stroke clears the cylinder out through the exhaust valve, making room for a fresh fuel-air mixture. The engine will not run at all unless the exhaust valve opens.

12. **D.** The differential has reduction gears so that the engine can turn fast enough to make power. A 3-to-1 ratio means that 3 turns of the driveshaft produce 1 turn of the wheels. Why are the other answers incorrect? Choice A: The engine would run too slowly to make power. Choice B: Cars do not have a jack shaft. Choice C: The ring gear in a differential cannot be disengaged.

13. **B.** A blade that is as wide as the screw head will put maximum turning power on the screw.

14. **B.** Arc-joint pliers, commonly called Channel Locks, can adjust to grip a wide variety of objects.

15. **D.** Plywood splinters easily, and each small tooth takes a smaller "bite," reducing splintering. Why are the other answers incorrect? Choice A: Big teeth cause more splintering. Choice B: A plywood blade is fine for particle board. Choice C is dead wrong. In saw blades, sharper is better!

16. **C.** Choice D is tempting, but a claw hammer can drive some wood chisels without damage.

17. **A.** Bar clamps are also handy for holding parts for sawing or drilling, but not for the other three purposes listed.

18. **B.** Impact equals mass times velocity, so for any given hammer speed, a more massive hammer makes more impact. Why are the other answers incorrect? Choice A: Mass, not weight, determines the intensity of impact. Choice C: Size alone does not make the hammer more effective. Choice D: A slower-moving hammer will be less effective at driving the chisel.

19. **D.** Plywood is stronger than regular wood because the grain is crisscrossed, preventing splitting and reducing weakness at knots. Why are the other answers incorrect? Choice A: Plywood may be weatherproof, but subfloors don't face the weather. Choices B and C: Plywood may be lighter and thinner, but that's not as important as strength in a subfloor.

20. **D.** An oxyacetylene cutting torch is made for cutting metal. Why are the other answers incorrect? All the saws mentioned are designed to cut wood.

21. **A.** Tin snips act like big scissors to shear sheet metal.

22. **C.** A pipe wrench fits loosely on the pipe; turning activates the grabbing action. The other tools all must be tightened before use.

23. **A.** A sharp drill bit cuts faster, cooler, and with less pressure on the drill, but it makes the same amount of chips.

24. **D.** Cold water will cause the nut to contract and grab tighter on the bolt. A frozen (rusted) nut can be loosened with heat, which causes it to expand, by pounding, which creates vibration, or by penetrating oil, which softens rust.

25. **C.** A plumb line is vertical and a level line is horizontal, so the angle between them is square, or 90°.

PART 8. MECHANICAL COMPREHENSION

1. **D.** In this gear system, gear B is turning clockwise. That means that gears A and C are turning counterclockwise. Since gear C is turning counterclockwise, gears D and E are turning clockwise, the same direction as gear B.

2. **C.** For a class 1 lever, mechanical advantage is calculated as follows: MA = effort distance/load distance. In this case, MA = 18/6 = 3.

3. **C.** If the bolt has 20 threads per inch, each time the wrench makes one complete turn, the bolt will move 1/20 inch. So if the wrench makes 15 complete turns, the bolt will move $1/20 \times 15 = 15/20 = \frac{3}{4}$ inch.

4. **A.** The gear is rotating at a speed of 150 rpm. At a ratio of 5:1, the pinion makes 5 revolutions for each revolution of the gear. So the speed of the pinion is $150 \times 5 = 750$ rpm.

5. **D.** Total flow through the pipe must be the same everywhere, so water must flow faster in the 1-inch section. When a fluid moves faster, its pressure drops. Therefore, pressure is lower in the 1-inch pipe and rises as the water enters the 3-inch pipe.

6. **B.** To balance the lever, sum the moments of force around the fulcrum:

$$(5 \times 10 \text{ kg}) + (3 \times 20 \text{ kg}) = 5F$$
$$50 + 60 = 5F$$
$$110 = 5F$$
$$F = 22$$

7. **C.** Gear A, with 48 teeth, meshes with gear B, with 12 teeth. So for every complete rotation of gear A, gear B makes 48/12 = 4 complete rotations.

8. **D.** Of the choices, the only one that will make the job easier is to put the block on a wheelbarrow. The wheels will reduce friction and make it easier to move the block. Choice A, raising the angle of the plane, will decrease, not increase, the mechanical advantage. Choices B and C will have no effect.

9. **A.** Choice A is correct because the hydraulic system transfers movement between the cylinders. The cylinders are directly linked and contain an incompressible fluid.

10. **B.** The tank has an area of 18 square feet. Pressure = force/area. If the total weight of the water = 3,600 lb, weight per square foot = 3600/18 = 200 lb/ft^2.

11. **A.** Of the choices, only the teacup cannot bend without breaking. The spoon, nail, and automobile tire are all made of materials that can bend without breaking.

12. **C.** A knife blade is an example of a wedge.

13. **B.** The ratio of diameters determines the ratio of sheave speed. 6/2 = 3/1, so the driven sheave will rotate three times as fast as the drive sheave. 3 × 80 = 240.

14. **D.** The pulley with the greatest diameter is pulley D. Therefore, it is the one that is rotating the slowest.

15. **B.** The car jack is a lever. Of the choices, the only one that will make the job easier is choice B, getting a longer jack handle. This will increase the effort distance in the lever and thereby increase the mechanical advantage.

16. **C.** The floor of a fireplace must be made of a material that will not ignite or melt with the heat of the fire. Of the choices, the one that is best is stone.

17. **B.** Because the pulley has two supporting strands, it has a mechanical advantage of 2. Therefore, each pound of pull will produce 2 pounds of lift.

18. **A.** Air pressure is what makes a siphon function. The weight of the air on the surface of the water in the tank forces water through the hose and into the bucket.

19. **C.** To make two gears turn in the same direction, an idler gear is required between them. The idler gear turns in the opposite direction from the two gears on either side of it. If idler gear B is rotating clockwise, gears A and C are both rotating counterclockwise.

20. **D.** A wrench uses a wheel-and-axle effect to move a bolt or nut.

21. **B.** The post nearer to the load bears the greater part of the weight. In the diagram, post B is the one nearer to the weight of the box of books.

22. **C.** Water is flowing into the tank at the rate of 240 gallons per hour, or 240/60 = 4 gallons per minute. Water is flowing out at the rate of 1 gallon per minute, so the net gain per minute is 3 gallons. In 6 minutes, the tank will gain 6 × 3 = 18 gallons of water.

23. **B.** A point on the circumference must travel a longer distance than a point closer to the center.

24. **A.** Of the choices, the only one that will form the complete, totally enclosed box shown is pattern 1.

25. **A.** Ignore the full turns, because after any number of full turns, the shaft will return to the starting position. After a half turn, the shaft will be 180° degrees from the position shown, but the piston will be where it is shown.

PART 9. ASSEMBLING OBJECTS

1. **C.** The original figure has three objects. Choice A has four, so eliminate it. The most distinctive object is the one that looks like a rectangle with a tail, so look for it first. Choice D does not have that shape, and choice B has a rectangle without the tail part. Choice C is the correct answer.

2. **B.** The original figure has three objects, while choice D has four. None of the objects are very distinctive, so just choose one—perhaps the square. Choice C has a rectangle rather than a square. Now look for the two identical domes. Choice A shows semicircles rather than the domelike shapes in the original. Choice B is the correct answer.

3. **D.** The original figure has six objects, while choice B has five and choice C has four. Look for the object at the bottom right. It is not present in choice A. Choice D is the correct answer.

4. **B.** The original figure has three objects, while choice D has four. Look for the object at the bottom that is NOT a circle. Choice A does not have it. Look for the tiny circle. Choice C does not have it. Choice B is the correct answer.

5. **A.** The original figure has four objects. Look for the wavy figure. The wavy figure in choice D does not have flat ends as it does in the original. Look for the triangle. It is not present in choice B. Look for the oval. It is not present in choice C. Choice A is the correct answer.

6. **D.** The original figure has four objects, while choice B has five. None of the objects are distinctive, so just choose one to look for. Look for the two triangles. Choice A has only one triangle. Look for the two quadrilaterals. Choice C has a square instead of an irregular quadrilateral. Choice D is the correct answer.

7. **A.** There are four objects in the original and choice C has three, so you can eliminate C. Since all of the objects are rectangles, look for the largest one first. Choices B and D do not have it. Choice A is the correct answer.

8. **D.** The original figure has three objects. Choice C has two, so eliminate it. The most distinctive object is the one that looks like a flag. Choice B does not have it. Look for the triangle. Choice A has a right triangle instead of an isosceles triangle. Choice D is the correct answer.

9. **C.** The original figure has four objects, while choice A has five. Look for one object at a time, perhaps beginning with the rhombus. Only choice C appears to have the same rhombus. Choice C is the correct answer.

10. **B.** The original figure has three objects, while choice C has four and choice D has only two. Look for the L-shaped object at the left. Choice A does not have it. Choice B is the correct answer.

11. **A.** All the options have three objects, so look for a distinctive object such as the chord of a circle. Choice D does not have this object and choice B has two of them. Look for the two identical triangles. Choice C has only one. Choice A is the correct answer.

12. **D.** There are four objects in the original and choice B has five objects, so eliminate B. Look for the only triangle. Choices A and C each have two triangles. Choice D is the correct answer.

13. **C.** There are five objects in the original and six in choice D, so eliminate D. The hexagon is a distinctive object, so look for that. Choice B does not have it. Look for the rectangle. The rectangle in choice A is too short. Choice C is the correct answer.

14. **B.** There are four objects in the original and only three in choice A. Look for the long parallelogram. The long object in choice C is not a parallelogram. Look for the triangle. The triangle in choice D is too small. Choice B is the correct answer.

15. **A.** There are five objects in the original and only four in choice B. Eliminate B. Choice C has six objects, so eliminate it as well. Look for the large three-sided object at the upper right. Choice D does not have it. Choice A is the correct answer.

16. **C.** There are three objects in the original. Choice A has four and choice D has two, so eliminate both A and D. Look for the triangle. Choice B does not have it. Choice C is the correct answer.

17. **A.** There are four objects in the original and only three in choice D. Look for the object at the left of the box that looks like a long triangle but with a curved end. Choices B and C do not have this same shape. Choice A is the correct answer.

18. **C.** There are four objects in the original and three in choice A and five in choice B, so eliminate A and B. There is one small rectangle, so look for that in the remaining choices. Choice D does not have it. Choice C is the correct answer.

19. **D.** Point A is marked at the center of the flat side of the first object. Choices B and C do not have the first object joined at the center of the flat side. Point B is at the tip of the longest side of the second object. Choice A does not join the second object at that point. Choice D is the correct answer.

20. **B.** If you picture the slanted object standing on its flat base, then point A is marked at the shorter (right) top corner. Choice D has the point joined at the longer (left) corner. In choice C, the object is not the right shape at all. If you picture the semicircle lying on its flat side, point B is marked at the left corner. Choice A has the point at the right corner. Choice B is the correct answer.

21. **C.** Choice A has a rectangle instead of a wedge. If you picture the semicircle lying on its flat side, point A is marked at the right corner. Choice D has the point in the middle. The thin wedge has point B marked at the pointed tip. Choice B has the point marked at a rounded edge. Choice C is the correct answer.

22. **A.** If you picture the rounded object with the cutout at the bottom, point A is on the right corner of the cutout. Choice B has the point on the left side of the cutout. The triangle has point B in the center of the base. Choice C has the point on a corner of the triangle. Choice D has the point on the apex of the triangle. Choice A is the correct answer.

23. **D.** Point A is at the center of a ¾ circle. Choice B has point A at the edge of the ¾ circle. Choice C shows an entire circle. Point B is at the center of the second object. Choice A connects at the wrong point on the second object. Choice D is the correct answer.

24. **B.** Point A lies at the center of the rounded end of the curvy object. Choice A has this point in the wrong spot. If you picture the other object lying on the wider base, then Point B is at the lower right corner. Choice C has the point in the middle of the narrower base, and choice D has it at the lower left corner. Choice B is the correct answer.

25. **D.** Point A is in the indentation of the left figure. Only choice D connects at that point. Choice D is the correct answer.

Part 3

REVIEW OF ALL ASVAB TEST TOPICS

Chapter 7

General Science

YOUR GOALS FOR THIS CHAPTER:

- **Find out what science topics you need to know for the ASVAB.**

- **Review the basics of general science.**

- **Study and memorize important science terms and definitions.**

INTRODUCTION: ASVAB GENERAL SCIENCE QUESTIONS

The ASVAB General Science test measures your knowledge of the life sciences (biology, human physiology, ecology), the Earth and space sciences (geology, meteorology, oceanography, astronomy), and the physical sciences (physics and chemistry). You will not need to be a scientist to answer these questions. You will just need to know some very basic information in the major science areas. This section will help refresh your memory about the science information you will need to know to do well on the General Science test.

ASVAB General Science Test
Number of questions: • Paper-and-pencil ASVAB: 25   • CAT-ASVAB: 15   *Time limit:*   • Paper-and-pencil ASVAB: 11 minutes   • CAT-ASVAB: 10 minutes

The following pages offer a concise, systematic review of the general science topics you need to know in order to score well on the ASVAB. Make sure that you carefully review every topic covered in this section. Pay particular attention to scientific terms and their definitions. Also, look carefully at the diagrams and other illustrations. In many cases, they contain important information that you should know.

GENERAL SCIENCE REVIEW

Topics
Biology   Cells   Viruses   Classification of living things   Plants   Animals   Ecology, habitats, and biodiversity   **Human Physiology**   Cells, tissues, organs, and organ systems   Muscular system   Skeletal system   Integumentary system   Respiratory system   Digestive system   Circulatory system   Lymphatic system   Immune system   Excretory system   Nervous system   Endocrine system   Reproductive system   The senses   Genetics   Nutrition   **Chemistry**   Elements   Compounds   Mixtures   **Physical Sciences**   The laws of motion   Work, energy, and power

Waves
Heat
Magnetism

Geology
Earth's history
Earth's structure
Categories of rocks
Weathering and erosion
Plate tectonics and continental drift

The Atmosphere
Composition and structure
Winds and storms
Climate

Oceanography and Water
Tides
The hydrologic cycle

Astronomy
The solar system
Motion of the earth
Sun and stars

BIOLOGY

Cells

The *cell* is the basic unit of life. The average human body contains over 75 trillion cells, but many life forms exist as single cells that perform all the functions necessary for independent existence. There are two types of cells, one that has no membrane-bound structures inside the cell (*prokaryotic*), and one that does have a membrane-bound structure inside the cell (*eukaryotic*). The following table illustrates the difference.

	Prokaryotic	**Eukaryotic**
Cell structure	No internal membrane to contain structures	Internal membrane contains internal structures
Examples	Bacteria	Protists, fungi, plants, and animals

Cells are the basic building blocks of all organisms. Plants and animals are made up of one or more cells. Some organisms, such as *bacteria*, are made up of a single cell. Other organisms are far more complex, such as your pet dog or even you.

Cells are so small that you need magnification to see them well. Microscopes are generally used to observe and study cells. Using a powerful microscope, you would be able to see that cells are made up of parts that have special functions.

Cells are made up of several different parts, including an *outer membrane* (sometimes called the *plasma membrane* or *cell membrane*), *organelles*, and a large amount of mass called *cytoplasm*. The *cell membrane* is the dictator of the cell. It determines what goes into or out of the cell. Cytoplasm is a gelatin-like material that fills the cell. Cell functions are accomplished by structures called *organelles*. Organelles are specialized parts that move around the cell and perform functions that are necessary for life. Examples of organelles are the cell *nucleus*, *vacuoles*, and *mitochondria*.

The *nucleus* is a membrane that contains the cell's hereditary information and controls the cell's growth and reproduction. It is generally the most prominent organelle in the cell. The nucleus contains *chromosomes* that are made up of *DNA*. DNA determines the characteristics and traits of the organism, such as the color of your hair, the shape of a leaf, and so on.

Vacuoles are the storage containers of the cell. They may store waste until it is eliminated or food until it is needed. In plant cells, there are large vacuoles that hold water.

Energy for the cell is produced by *mitochondria* through a process called *respiration*. Respiration is a series of chemical reactions that combine food and oxygen to create energy and a waste by-product, carbon dioxide.

Plant cells have some additional components. They have a *cell wall* that gives the cell a firmer shape and support. The cell wall is made up of cellulose, which is not digestible by humans, but provides fiber for our good health. Plant cells also have organelles called *chloroplasts*; these contain *chlorophyll*, which uses the process of *photosynthesis* to make food for the plant cells. The chloroplasts are green; they are what give plants their green color. During photosynthesis, chloroplasts interact with light energy, combining carbon dioxide from the air with water to make food. With light, the green parts of plants produce a sugar called glucose ($C_6H_{12}O_6$) and oxygen (O_2) from carbon dioxide (CO_2) and water (H_2O).

$$\text{Carbon dioxide} + \text{water} \xrightarrow[\text{Chlorophyll}]{\text{Light}} \text{glucose} + \text{oxygen}$$

The basic difference between plant and animal cells is that plant cells have cell walls and chloroplasts, and animal cells do not.

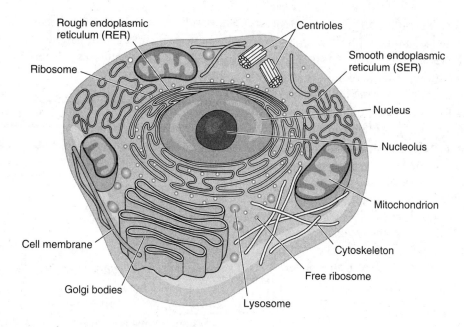

Rough endoplasmic reticulum (RER)

Ribosome

Centrioles

Smooth endoplasmic reticulum (SER)

Nucleus

Nucleolus

Mitochondrion

Cell membrane

Cytoskeleton

Free ribosome

Golgi bodies

Lysosome

Animal Cell

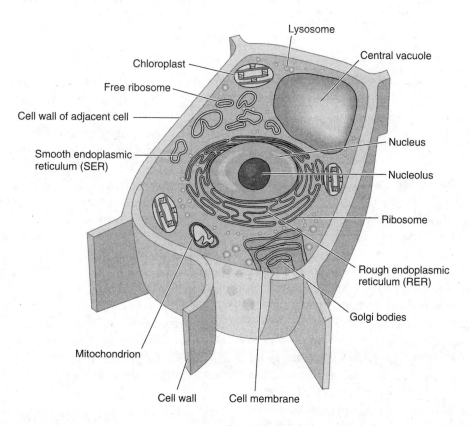

Lysosome

Central vacuole

Chloroplast

Free ribosome

Cell wall of adjacent cell

Smooth endoplasmic reticulum (SER)

Nucleus

Nucleolus

Ribosome

Rough endoplasmic reticulum (RER)

Golgi bodies

Mitochondrion

Cell wall

Cell membrane

Plant Cell

Cell and Species Reproduction Species of plants and animals cannot continue without reproduction. All cells have hereditary material that is passed on to the next generation. This hereditary material is called *DNA*, or deoxyribonucleic acid. DNA determines how one looks and how one functions. DNA is the instruction manual or blueprint for life.

The DNA molecule consists of two long strands that form a double helix. It spirals around like a twisted ladder. The DNA molecule has a sugar component, a phosphate component, and four different *bases—adenine* (A), *thymine* (T), *cytosine* (C), and *guanine* (G). Adenine is always paired with thymine, and cytosine with guanine. These pairs are held together to form the rungs of the ladder.

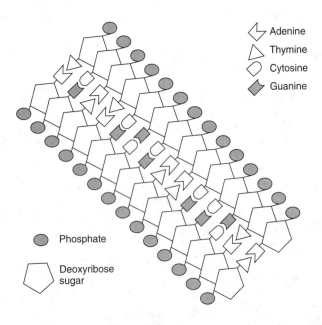

If a cell has a nucleus, the DNA is found in *chromosomes*. Chromosomes are duplicated before cell division takes place. The DNA unzips and separates. New bases are formed, and they link up in the proper order to form two new DNA helixes that are exactly alike. The instructions for the appearance and function of the new organism are contained in units called *genes*, which are parts of the DNA.

There are two reasons for cell division. One is to replace old or worn-out cells, and the other is for sexual reproduction. The process of *mitosis* takes place when cells are replacing themselves because they are old or worn out. The process called *meiosis* takes place with sex cell formation. In sex cell formation, cells with only 23 chromosomes are formed so that when an egg and a sperm join up, they have

the full set of 46 chromosomes and possess the genetic makeup of both parents.

The chart indicates the difference between the processes.

Mitosis	Meiosis
For cell division	For sex cell formation
Nucleus divides once	Nucleus divides twice
Number of cells formed = 2	Number of cells formed = 4
Chromosomes in each new cell = 46	Chromosomes in each new cell = 23

Organisms reproduce in different ways: *sexual reproduction* between a sperm and an egg or *asexual reproduction*.

Some plants and animals reproduce asexually. Some examples are bacteria, the hydra, and the eye of a potato, which can grow new potatoes.

Some plants reproduce sexually. Examples are flowers, which have male and female parts. When the sperm and egg join, they produce seeds, which, when planted, reproduce the species.

Viruses

Viruses are not alive in the strict sense of the word, but they reproduce and have a close relationship with all living organisms. Viruses are not plants, animals, or bacteria, although they may seem to be. A virus is basically a piece of hereditary material surrounded by a protein coating. Viruses do not have a nucleus or other organelles.

Viruses need a host cell in order to reproduce and function. They cannot generate or store energy, but take their energy from the host cell. They also take basic building materials from the host. All viruses contain nucleic acid, either DNA or RNA (but not both), and a protein coat, which encases the nucleic acid. Since viruses cannot penetrate plant cell walls, nearly all plant viruses are transmitted by insects or other organisms that feed on plants.

Viruses cause such conditions as the flu, the common cold, cold sores, measles, chicken pox, AIDS, and COVID-19.

Classification of Living Things

Scientists use an ordered taxonomy (classification system) to classify living things. The levels within this system are called, in descending order, *kingdom, phylum/division, class, order, family, genus,* and *species.*

All life is categorized into five kingdoms.

Kingdom	Description/Example
Monera	One-celled or a colony of cells, decomposers and parasites, move in water, and both producers and consumers. *Examples*: bacteria, blue-green algae
Protista	One-celled or multicelled, absorb food, move with flagella, both asexual and sexual reproduction, producers and consumers. *Examples*: plankton, algae, amoeba, protozoans
Fungi	One-celled or multicelled, decomposers, parasites, absorb food, asexual reproduction and budding, consumers. *Examples*: mushrooms, molds, mildew, yeast, fungi
Plantae	Multicelled, photosynthesis, mostly producers. *Examples*: angiosperms, gymnosperms, mosses, ferns
Animalia	Multicelled, parasites, prey, both asexual and sexual reproduction, consumers. *Examples*: sponges, worms, insects, starfish, mammals, fish, amphibians, reptiles, birds, gorillas, humans

Human beings can be categorized according to the taxonomy in this way:

Kingdom	Animalia
Phylum	Chordates
Class	Mammals
Order	Primates
Family	Hominidae
Genus	Homo
Species	Sapiens
Scientific Name	Homo sapiens

Within each kingdom, there are various *phyla* (the plural of *phylum*) for animals and divisions for plants.

How Organisms Obtain Energy There are many ways to classify living things. One way is to divide them into consumers and producers. Producers use an outside energy source, such as sunlight, to produce energy. Most producers have chlorophyll, and most, but not all, are plants. Consumers cannot make their own energy; to live, they need to eat other organisms. Consumers may eat only plants, only animals, or both. (See the classification of animals for more information.)

Plants

Botany is the scientific study of plants. Many people consider everything from bacteria to the giant sequoia trees to be plants. That would include algae, fungi, lichens, mosses, ferns, conifers, and flowering plants. Today, scientists believe that bacteria, algae, and fungi have their own distinct kingdoms. The kingdom *Plantae* includes 250,000 species of mosses, liverworts, ferns, flowers, bushes, vines, trees, and other plants. Aquatic (water) and terrestrial (land) plants are the basis of all food webs. They emit into the atmosphere the oxygen that is needed by animals for survival.

Categorizing Plants Plants are categorized according to the structure by which the plant absorbs water. Plants are either *vascular* or *nonvascular*. Vascular plants transport water from the roots to the stems and to the leaves by means of tubelike structures. Nonvascular plants absorb water only through their surfaces.

Another way to categorize plants is according to how they reproduce. Some plants reproduce by producing *seeds*. Others produce *spores*. Yet another way to categorize plants is by whether they produce flowers (*angiosperms*, such as roses or apple trees) or don't produce flowers (*gymnosperms*, such as pine trees). Seeds store food for the growing plant; the part of the seed that stores the food is called the *cotyledon*. Some seeds have one cotyledon and are called *monocots*. Others have two and are called *dicots*.

Plants can also be classified by their life cycle. *Annuals* go through their entire life cycle, from germination through seed production to death, in one growing season. Corn, zinnias, beans, marigolds, and mums all have to be planted each year. *Biennials* have a two-year growing cycle. In year one, the seed germinates, produces leaves and roots, and forms a compact stem for food storage. In year two, the plant forms an elongated stem, produces flowers and fruits, and then ends with seed production. After seed production, the plant dies. Examples are onions, parsley, hollyhocks, and carrots.

Perennials live for many years. Parts of the plant may die back during the winter, but the plant will grow back in the spring.

Plants can also be categorized by their behavior in winter. *Deciduous* plants, including shrubs and trees, lose their leaves in the winter. Maple and oak trees are examples. *Evergreen* plants keep their leaves or needles throughout the year, sometimes shedding only old leaves or needles that are more than two years old. Pine and spruce trees are examples of evergreen trees.

Major Parts of a Plant Plants are made up of a number of different parts, including roots, stems, leaves, flowers, and fruits.

Roots Roots have several functions: absorbing nutrients and water, anchoring the plant into the soil, holding up the stem and leaves, and storing food. They are usually below ground and do not have chloroplasts (see the section on plant cells).

There are basically two types of root systems, a taproot system and a fibrous root system. A taproot system has one fat or sturdy main root, with just a few branching roots. Carrots, radishes, and parsnips are good examples of a taproot. A fibrous root system has many branched roots. Examples of plants with fibrous roots are most grasses.

Roots have tiny root hairs. Roots can have a very large number of root hairs, increasing the capability of the root to absorb water and food.

Stems The stem is the main trunk of a plant. Leaves, flowers, and fruits get support from the stem. Stems also carry nutrients and water. Some stems store food. Stems are generally above ground and vertical. Some stems grow below ground (bulbs) or fasten themselves to the ground. The strawberry plant is an example of a plant with a stem that hugs the ground.

Nodes are places on the stem where buds form. Spaces between the nodes are called *internodes.*

One of the major functions of the stem is to move water, nutrients, and food through the plant. The system that does this is called the *vascular system* and is similar to the circulatory system in humans. *Phloem tubes* move food from the roots through the stem to the leaves. *Xylem tubes* move minerals and water. These tubes are surrounded by the main tissue in the stem, the *cambium.*

Leaves Leaves grow out from the stem. The leaves' major job is to make food for the plant. For the most part, leaves are flat, broad, and green. The flat surface maximizes their ability to absorb light and transform it into food. There is a protective layer on leaves called the *cuticle* that reduces the evaporation of water from the plant and helps to protect the plant from disease-causing organisms. Leaves have tiny openings called *stomata* that enable the plant to take in carbon dioxide and release oxygen into the atmosphere. *Guard cells* cover the stomata openings and regulate the exchange of water vapor, oxygen, and carbon dioxide into and out of the stoma.

Flowers The main job of flowers is sexual reproduction or seed production. Various parts of the flower are involved in sexual reproduction. The *pistil* is the female portion of the flower. It includes the *stigma*, the surface that captures and holds *pollen*, and the *style*, the area between the stigma and the *ovaries*. The *stamen* is the male portion of the flower. It includes the *filament* that holds the *anther*. The anther is where the pollen is formed and released.

Some plants have flowers with both male and female parts, and others make flowers with only male or only female parts. Corn is an example of a plant that has both male and female flowers on one plant. The tassel is the male flower, and the ear is the female flower. The holly is an example of a plant that has either a male or a female flower. Plants of both sexes need to be planted close together so that

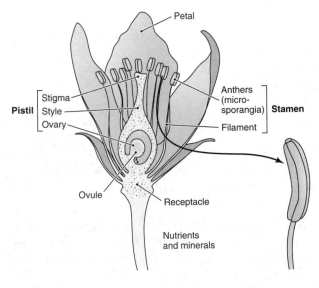

Parts of a Flower

the seeds (berries) can form on the female plant. Other parts of a flower are *sepals*, which enclose the various flower parts, and *petals*, the bright-colored parts like the petals of a rose or daisy.

Any edible part of a plant that is not a flower is considered to be a *vegetable*.

Fruits A fruit is a ripened ovary or group of ovaries containing the seeds. When the ovary is fertilized, the seeds develop and the ovary enlarges, forming the fruit. Examples of fruits are peanuts, sunflower seeds, barley, walnuts, tomatoes, grapes, oranges, apples, raspberries, cucumbers, squash, corn, eggplants, and strawberries.

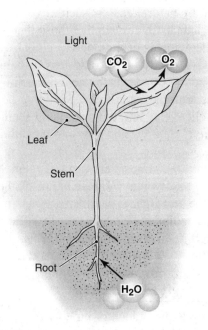

How Plants Make Food The process is called *photosynthesis*. Photosynthesis is a process that converts the energy in sunlight into chemical forms of energy that can be used by biological systems. Green plants are the only organisms that make their own food. Photosynthesis begins when light strikes a plant's leaves. Cells in the leaves, called *chloroplasts*, contain a green pigment called *chlorophyll*. Chlorophyll interacts with sunlight to split the water in the plant into its basic components.

Carbon dioxide enters the leaf through the stomata and combines with the stored energy in the chloroplasts through a chemical reaction to produce a simple sugar. This sugar is then transported through tubes in the leaf to the roots, stems, and fruits of the plants. Some of the sugar is used immediately by the plant for energy; some is stored as starch; and some is built into a more complex substance, like plant tissue or cellulose. Some of the sugar is stored for future use or for use by us when we eat the plants. During the process, oxygen is released into the atmosphere.

Plant Responses Plants respond either positively or negatively to various stimuli. These responses are called *tropisms*. For example, the roots of plants respond positively to gravity, while the stem responds negatively (moving against gravity). Response to gravity is called *gravitropism*. Also, plants respond to light, growing toward it. This is called *phototropism*. Plants also respond to touch *thigmotropism*.

Animals

Categorizing Animals Animals are generally categorized in terms of whether or not they have a *backbone*. Animals that have a backbone are called *vertebrates*. Animals that do not have a backbone are called *invertebrates*.

All invertebrates lack backbones, but their other physical characteristics are varied. About 95 percent of all animals are invertebrates. Some invertebrates, like worms, have soft bodies with no bones. Other invertebrates, like snails, have soft bodies, but have developed a hard shell for protection. These types of invertebrates are called *mollusks*. Mollusks can live in water or on land. Some examples are snails, clams, mussels, and squid.

Some other invertebrates have tough coatings made of chitin on the outside of their bodies, called *exoskeletons*. They also have jointed legs and a segmented body. This group is known as the *arthropods* and includes spiders (*arachnids*), centipedes and millipedes, and *insects*, such as beetles and butterflies. In addition, *crustaceans*, such as shrimp, lobsters, and crabs, are also arthropods. There are more insects—over 900,000 species—than any other kind of animal, vertebrate or invertebrate.

All vertebrates have a backbone. Despite that common characteristic, vertebrates vary widely in appearance. One way to categorize vertebrates is according to their diet. Animals that primarily eat plants are known as *herbivores*. Animals that feed mostly on meat are known as *carnivores*. Some animals, like humans, eat both plants and meat and are called *omnivores.*

Meat eaters have jaws and teeth that are designed for tearing and crushing. The *canine* teeth are sharp, and the *molars* have a flattened surface for grinding. The digestive system is adapted for handling the quick digestion of meat. Plant eaters usually have large incisors for cutting plants. Their large molars are designed for grinding tough plant fibers. Their digestive system has a long intestine adapted for the slow digestion of plant fibers.

Vertebrates can also be classified according to their body temperature. All vertebrates are either *cold-blooded* or *warm-blooded.* An animal is called cold-blooded if its body temperature follows or matches the external temperature around it. Fish, amphibians, and reptiles are examples of cold-blooded animals. Warm-blooded animals can control their body temperature. Their internal body temperature remains the same no matter what the temperature is outside. Only birds and mammals (like people, cats, cows, and the like) are warm-blooded. Birds rely on their feathers to help them adapt to temperature changes.

Mammals use their hair, skin, or fur to adapt to temperature changes. Another differentiating characteristic of mammals is that the female produces milk to feed her offspring. Whales, which are mammals that live in the sea, must deal with changing water temperatures. Wolves and bears that live in mountainous areas or in polar regions must adapt to cold temperatures. Warm-blooded animals try to avoid or find ways to cope with temperature extremes in order to survive.

Ecology, Habitats, and Biodiversity

The sum of all the places on Earth where life can exist is called the *biosphere.* An *ecosystem* is the collection of all the living creatures and nonliving features or conditions in a particular environment. For example, a rainforest has certain creatures living in it. The amount of rainfall, the temperature, the soil conditions, the air, and the contour of the land are some of the nonliving features of the ecosystem. The study of ecosystems—the interactions between and among these living creatures and nonliving features—is called *ecology.*

The many thousands of species of plants and animals that live on the planet contribute to *biodiversity.* Biodiversity is the variety of life forms that exist. Scientists believe that a rich biodiversity creates a condition in which plants and animals are more likely to recover from stresses caused by nature and people. Biodiversity tends to increase as one approaches the equator. Warmer weather tends to support biodiversity.

Life forms live in what is called a *habitat,* a geographic area with conditions that support the continued reproduction of the species. Some habitats may be small, such as the dark corners of an underground cave or a hot thermal pool where only a few creatures live. Other habitats may be large, such as the thousands of square miles of prairie land that support certain grasses.

There are many threats to biodiversity. One is the reduction or destruction of habitat. Habitats can be destroyed by natural causes, such as volcanic eruptions or changes in climate like those that occurred during the ice ages or now due to global warming. People have contributed to the loss of habitat and the destruction of species by actions such as building roads, cutting down forests, sending pesticides or other chemicals into the water, and sending pollutants from factories and cars into the atmosphere.

Global warming resulting from sending carbon dioxide into the atmosphere can raise temperatures and affect biodiversity. The thinning of the *ozone layer* in the atmosphere can increase the amount of ultraviolet radiation (UV) that reaches Earth's surface, resulting in harm to living organisms.

HUMAN PHYSIOLOGY

Physiology is the branch of biology that deals with the parts of the body, their functions, and the various bodily processes. Human physiology deals with the human body.

Cells, Tissues, Organs, and Organ Systems

In the human body, there are many different kinds of cells. Each kind is specialized for the primary function it performs. Examples are fat cells, skin cells, muscle cells, bone cells, and nerve cells. Groups of cells arrange themselves into *tissues,* and various tissues work together to form *organs,* such as the skin, liver, heart, gallbladder, and intestines. Organs work together to form *organ systems.* Organ systems include the muscular system, skeletal system, skin or integumentary system, respiratory system, digestive system, circulatory system, lymphatic system, immune system, excretory system, nervous system, endocrine system, and reproductive system.

Muscular System

The *muscular system* allows movement and locomotion. The muscular system helps you make body movements and supports the body in its activities. Muscles are involved in breathing, your heart beating, and the working of your digestive system.

Some actions are controlled by *voluntary* muscles, such as making a face, showing your biceps, or walking. Other muscle systems are *involuntary*, such as those that control breathing, your heart beating, and your digestive process.

Skeletal muscles help move the bones. They are attached to the bone by bands of tissue called *tendons*. Skeletal muscles work in pairs; when one muscle of the pair contracts, the other muscle relaxes. *Cardiac muscle* is found in the heart. *Smooth muscles* are found in some of your internal organs, such as your intestines and bladder.

Skeletal System

The skeletal system is a living system that provides shape and support to your body. It is built to protect your inner organs and to provide attachment points for muscles. The skeletal system provides a rigid framework for movement. It supports and protects the body and body parts, produces blood cells, and stores minerals. In lower animals, such as a grasshopper, the skeleton might be on the outside. This is called an *exoskeleton*.

Vertebrates have developed an internal mineralized *endoskeleton*. The endoskeleton is made up of *bone* and *cartilage*. Muscles are on the outside of the endoskeleton. Although our endoskeleton is mostly bone, some parts of our body are made of cartilage, such as the trachea, nose, and ears. The skeleton and muscles function together as the *musculoskeletal* system. Calcium and phosphorus are important components of bone; these elements make bone hard. *Osteoblasts* are bone-forming cells.

Places where your bones come together are called *joints*. Joints are held together by bands of tissue

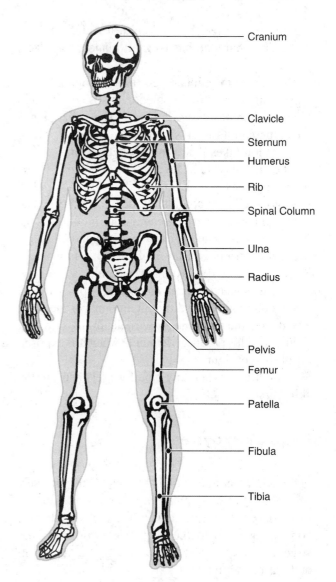

Skeletal System

Labels: Cranium, Clavicle, Sternum, Humerus, Rib, Spinal Column, Ulna, Radius, Pelvis, Femur, Patella, Fibula, Tibia

called *ligaments*. Bones move at the joints. There are three major types of joints: *ball and socket joints*, like the shoulder and hip; *pivot joints*, like the elbow; and *hinge joints*, like the knee. These joints, working with the muscles and tendons, allow the body to move in certain ways.

Integumentary System

Skin or *integument* is your outermost protective layer. The skin is the largest organ in your body. It protects you from losing water and from the invasion of foreign organisms and viruses. Special nerve cells in the skin help send information to the brain, helping you to feel hot and cold, texture, softness, and pain. Your skin has hair, oil glands, and sweat glands.

Skin helps to regulate your body temperature by expanding or constricting blood vessels and through the operation of the *sweat glands*. The sweat glands move perspiration or sweat onto the skin, where evaporation takes place and cools the skin.

The skin has three layers. The *epidermis* is the outer, thinner layer of skin. Underneath the epidermis, skin cells are continually being produced. The next layer is the *dermis*, containing blood vessels, nerves, muscles, and oil and sweat glands. Underneath the dermis is a layer of *fat*. This is where a lot of fat is stored as you gain weight.

Skin gets its color from *melanin*. People with different color skin have different amounts of melanin. The darker the skin, the more melanin the skin has. Skin also helps produce vitamin D when it is exposed to ultraviolet light.

Skin bruises when tiny blood vessels beneath the skin burst.

Respiratory System

The respiratory system takes in oxygen and moves out waste material of carbon dioxide. Our lungs perform this job by breathing in and out (gills do this in fish). This system includes the lungs, pathways connecting them to the outside environment, and structures in the chest involved with moving air in and out of the lungs.

The Respiratory System

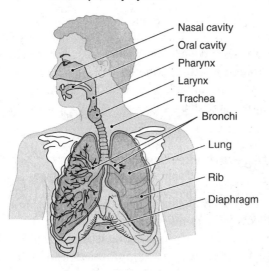

- Nasal cavity
- Oral cavity
- Pharynx
- Larynx
- Trachea
- Bronchi
- Lung
- Rib
- Diaphragm

When we breathe in, air enters the body through the nose, where it is warmed, filtered, and passed through the *nasal cavity*. Air passes the *pharynx* (which has the *epiglottis* that prevents food from entering the *trachea*). The upper part of the trachea contains the *larynx*. After passing the larynx, the air moves into the *bronchi*, which carry air in and out of the lungs.

The bronchi are lined with *epithelium* and mucus-producing cells. The bronchi branch into smaller and smaller tubes known as *bronchioles*. The bronchioles terminate in little sacks known as *alveoli*. Alveoli are surrounded by a network of thin-walled *capillaries*. The exchange of oxygen takes place there, between the alveoli and the capillaries. The capillaries take in the oxygen. The hemoglobin in the blood picks up the oxygen and carries it off to various parts of the body. At the same time, waste material is transferred to the alveoli and eventually out of our bodies during exhalation.

It is important to note that breathing is different from respiration. Breathing is the physical action of moving the *diaphragm* up and down, which allows air to enter our lungs. Respiration is how our bodies use the oxygen from the air we inhale and eliminate the carbon dioxide when we exhale.

Digestive System

Digestion is accomplished by *mechanical* and *chemical* means, breaking food into particles small enough to pass into the bloodstream. The digestive process begins as soon as you put food into your mouth. Mechanical breakdown begins in the mouth by chewing (teeth) and moving the food around with your tongue. The food interacts with your *saliva* to begin a chemical breakdown. This mixture of food and saliva is then pushed into the *pharynx* and *esophagus* when you swallow. The esophagus is a long tube with muscles that contract and move the food to the *stomach*. The stomach is a sack or bag. It breaks down the food by both mechanical and chemical means. The stomach mixes the food by a process called *peristalsis*. Peristalsis is merely waves of muscle contractions. The stomach releases *enzymes* and *hydrochloric acid* to break down the food even further. Eventually the food moves to your *small intestine.*

The small intestine is narrow in diameter but very long in length. The upper part of the small intestine is the *duodenum*. This is where most digestion takes place. *Bile* is introduced from your *liver*. Bile breaks up the fat particles. Next comes the *pancreas*, which introduces pancreatic fluid. This fluid promotes the chemical digestion of carbohydrates, proteins, and fats. The fluid also neutralizes the stomach acid and releases insulin. Absorption of the food takes place in the small intestine.

Food that is not digested continues to move through your system into the *large intestine* by means of peristalsis. The large intestine absorbs water, making the contents more solid. Eventually the *rectum* and *anus* control the release of the semi-solid waste called *feces*.

The Digestive System

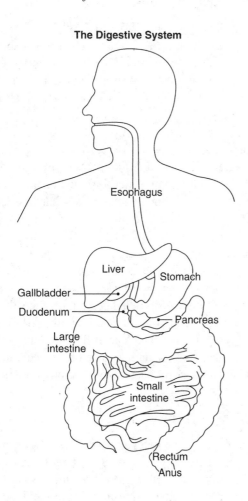

Circulatory System

The circulatory system is one of the transportation systems in your body. The main job of the circulatory system is to move oxygen, carbon dioxide, nutrients, waste products, immune components, and hormones through your body. The circulatory system is composed of vessels and muscles that control the flow of blood around the body. This process of blood flowing around the body is called *circulation*.

The main components of the circulatory system are the *heart*, *arteries*, *capillaries*, and *veins*.

There are three types of circulation, *coronary circulation*, *pulmonary circulation*, and *systemic circulation*. Coronary circulation is the circulation of blood within the heart itself by the coronary veins and arteries. If this circulation is blocked, it could result in a heart attack. Pulmonary circulation is the flow of blood from the heart to the lungs and back. Systemic circulation is the blood (with oxygen) moving through your body to your important organs.

The heart is a strong muscle that operates as the engine of the circulatory system. It has four compartments that are called *chambers*. The two upper chambers are called *atriums*, and the two lower chambers are called *ventricles*. When your heart beats, the two ventricles contract at the same time, followed by the two atria. Blood flows from an atrium to a ventricle and then from the ventricle to a blood vessel. The *pulmonary valves* keep the blood from flowing backward. There is a wall between the two atria and the two ventricles, keeping blood that has a lot of oxygen apart from blood that does not have as much oxygen.

Blood returns from the body to the right atrium of the heart. At this point, the blood is low on oxygen. When the right atrium contracts, it moves the blood to the right ventricle. The blood then moves via the *pulmonary artery* to the lungs, where it picks up oxygen and eliminates waste (carbon dioxide). The oxygenated blood leaves the lungs and comes into the heart through the left atrium. Blood leaves the heart from the left ventricle and goes into the biggest artery, called the *aorta*. Fresh blood from the aorta goes to various parts of your body,

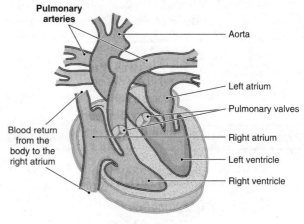

The Human Heart

including the brain, which needs a constant supply of oxygen. Blood returns to the heart through the veins, starting the cycle over.

Arteries are blood vessels that carry blood away from the heart. Veins carry blood to the heart. Arteries and veins are connected by thousands of miles of tiny vessels called capillaries.

Blood About five liters of blood flows through your body, delivering essential elements and removing harmful wastes. Blood transports oxygen from the lungs to body tissues and carbon dioxide from body tissues to the lungs. It also transports disease-fighting substances to the tissue and waste to the *kidneys*.

Blood contains *red blood cells* and *white blood cells*. These cells are responsible for nourishing and cleansing the body. Red blood cells are shaped like disks and contain *hemoglobin*, which carries oxygen and carbon dioxide. Unlike other cells, they do not have a nucleus. White blood cells fight bacteria, viruses, and other intruders in your body. The white blood cells are carried to the place where an intruder has invaded and go into that tissue. Their job is to destroy the bacteria or viruses.

Cell fragments called *platelets* are carried along with the red and white blood cells. Platelets plug holes in small blood vessels to stop bleeding. They help to clot blood.

Red and white blood cells and platelets are carried by a fluid called *plasma*. About 55 percent of the blood's volume comes from plasma. Nutrients, minerals, and oxygen are carried by the plasma to various parts of the body. Waste products are also carried by the plasma.

There are four different blood types: *A*, *B*, *AB*, and *O*. Types A, B, and AB have certain chemical tags called *antigens*. Type O has no antigens. Blood also has antibodies that destroy or neutralize substances that do not belong there. This prevents certain blood types from being mixed. The following chart indicates who can receive and give certain blood types.

Type	Can Receive	Can Donate to This Type
A	A, O	A, AB
B	B, O	B, AB
AB	All types	AB
O	O	All types

People with type O blood are called universal donors; those with type AB are called universal receivers.

Your blood also has Rh factors, making it even more unique. The Rh factor is an additional chemical tag. If the blood has the Rh factor, it is labeled Rh-positive. If the factor is not present, the blood is called Rh-negative. If the wrong blood is given to a person, the blood will try to destroy the other person's blood. That's why it is important to know your blood type.

Lymphatic System

The lymphatic system is composed of *lymph vessels*, *lymph nodes*, and certain organs. The system absorbs excess fluids from the body and returns them to the bloodstream. It also absorbs fat and transports it to the heart. The fluid contains *lymphocytes*, which are a type of white blood cell that tries to destroy disease-causing organisms.

The lymphatic system is a lot like the circulatory system. The fluid *lymph* passes through lymph nodes that remove any microorganisms and foreign materials. Lymph nodes generally occur in clusters in the neck, armpits, and groin. If you get sick with an infection, the lymphocytes fill the lymph nodes, and your lymph nodes may feel tender and swollen.

There are three organs that are part of the lymphatic system. These are the *tonsils*, the *thymus*, and the *spleen*. Your tonsils help keep out invaders that try to come in through your nose and mouth. The thymus makes lymphocytes. The spleen filters the blood and removes worn-out or damaged red blood cells. Cells in the spleen destroy bacteria and other invaders.

Immune System

The immune system defends our bodies from invading microorganisms and viruses called *pathogens*, as well as from cancerous cell growth. Immune-system components are grouped into *first-line defenses* and *second-line defenses*. First-line defenses include your skin and your respiratory, digestive, and circulatory systems.

Pathogens can't get through your skin unless it is cut or broken, but they can get though your mouth, nose, and eyes. The respiratory system uses *cilia*, little hairlike structures, and *mucus* to trap pathogens. When you cough or sneeze, you are expelling some mucus that contains pathogens. The digestive system uses saliva, enzymes, hydrochloric acid, and other substances to get rid of bacteria that can be harmful to you. The circulatory system uses white blood cells to surround and destroy foreign organisms and chemicals. Temperature destroys some organisms, so if your white blood cells cannot do the job fast enough, you might get a *fever*.

Second-line defenses are specific to the disease. Molecules that that are foreign to your body are called *antigens*. When your body determines that a foreign molecule has invaded, special lymphocytes, called *T cells*, attack. Special T cells stimulate other lymphocytes called *B cells* to form *antibodies*. Antibodies are made in response to a specific antigen.

With certain diseases, a lot of extra antibodies are formed so that when your disease is cured, a few antibodies hang around and stay on watch. If the pathogen enters your body again, these antibodies can reproduce very rapidly and eliminate the disease. That's why some diseases, like chicken pox, you get only once. This is an example of *active immunity*.

Passive immunity occurs when antibodies are acquired from another person such as a newborn receiving antibodies from the mother or antibodies being injected. A *vaccination* injects a type of antigen that gives you active immunity against the disease. It does this by stimulating the production of antibodies. Vaccinations are specific to one kind of virus or bacteria. For example, there is a new flu vaccine every year because the virus is different each year. Common vaccines include those for measles, diphtheria, tetanus, mumps, rubella, and whooping cough.

Diseases caused by bacteria include tetanus, tuberculosis, strep throat, and bacterial pneumonia. Viruses cause colds, influenza (flu), measles, polio, mumps, and smallpox. *Antibiotics* can cure some bacterial diseases, but not viral diseases.

Excretory System

The excretory system removes waste. It removes undigested material through the digestive system by way of the large intestine. It removes waste gases through the circulatory and respiratory systems. It removes salts through the skin when we sweat. It removes excess water and waste through the urinary system. The *urinary system* is responsible for maintaining the fluid levels in our bodies.

The *kidneys* play a major role in the excretory system. They are two bean-shaped organs that are responsible for filtering blood that contains waste from the cells. Once the blood has been purified by tiny filtering units called *nephrons*, it is returned to

the circulatory system. The leftover water from this process is called *urine*. It is collected in the *bladder* and then eliminated through the *urethra* during urination. Persons who have kidney disease need to have dialysis to remove the waste from their blood. Without waste removal, a person will die.

Nervous System

The nervous system coordinates and controls such actions as memory, learning, and conscious thought. The nervous system also maintains such functions as heartbeat, breathing, and control of involuntary muscle actions. It is the most complex and delicate of all our body systems.

The largest organ in the nervous system is the *brain*. The brain is a sort of control center, as it sends and receives messages through a network of nerves. It is made up of a hundred billion *neurons*, or brain cells. The brain has three major parts: the *cerebrum*, the *cerebellum*, and the *brain stem*.

The cerebrum is the largest part of the brain; it takes care of our thinking processes. The outer layer of the cerebrum is called the *cortex* and has a lot of ridges and grooves. More ridges and grooves allow more complex thinking to occur.

The cerebellum is the second-largest part of the brain. Its job is to coordinate our muscle movements and maintain normal muscle tone and posture. The cerebellum coordinates our balance while walking, riding a bike, and so on.

The brain stem is closest to the *spinal cord*. It has three parts: the *midbrain*, the *pons*, and the *medulla*. The midbrain and pons coordinate various parts of the brain so that it acts together. The medulla is involved in coordinating our heartbeat, breathing, blood pressure, and the reflex centers for vomiting, coughing, sneezing, swallowing, and hiccupping.

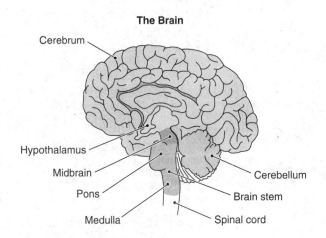

The Brain

Cerebrum

Hypothalamus

Midbrain

Pons

Medulla

Cerebellum

Brain stem

Spinal cord

Nerve Cells

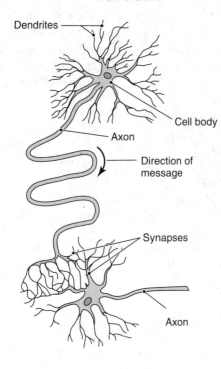

Dendrites

Cell body

Axon

Direction of message

Synapses

Axon

The *hypothalamus* regulates thirst, hunger, body temperature, water balance, and blood pressure, and links the nervous system to the endocrine system.

The spinal cord is a thick bundle of *nerves* running down the center of the spine. Along the spinal cord, smaller bunches of nerves branch out. From these bunches, still smaller bundles of nerves branch out again. Eventually they reach every part of the body. The spinal cord is protected by a column of vertebrae—part of the skeletal system.

The brain and the spinal cord make up the *central nervous system*. These are the nerves in your head and the nerves that come out from the spinal cord. The nerves outside the central nervous system are called the *peripheral nervous system*. The peripheral nervous system connects the central nervous system to the rest of the body. The peripheral nervous system has two parts: the *somatic system* and the *autonomic system*. The somatic system controls voluntary movements, like walking, running, and swiveling your hip. The autonomic system controls involuntary movements, such as heartbeat, breathing, digestion, and so on.

Messages that are transported through the nervous system are conducted by the *nerve cells*. Each microscopic nerve cell, or *neuron*, has a blob-shaped main part, the *cell body*, and thin, spiderlike *dendrites*. It has one lone fiber called the *axon*. The axon's branched ends have little bulbs that *almost* touch adjacent nerve cells. The spaces between the nerve cells are called *synapses*. Nerve signals travel in one direction only along the axon and jump across synapses to other nerve cells.

Gland	Hormone	What It Does
Pituitary gland	Growth hormone	Stimulates the growth of muscles and bones
	Follicle-stimulating hormone (FSH)	Stimulates the growth of reproductive organs, such as the ovaries and testes
	Thyroid-stimulating hormone (TSH)	Stimulates the thyroid gland, which controls metabolism
	Prolactin	Stimulates the secretion of milk
	Vasopressin	Helps the kidneys to absorb water
	Adrenocorticotropic hormone (ACTH)	Stimulates the adrenal cortex
Thyroid	Thyroxin	Regulates metabolism
Parathyroid	Parathyroid hormone	Increases the concentration of calcium in the blood
Adrenal cortex	Aldosterone	Helps the kidneys absorb water and sodium
Adrenal medulla	Epinephrine and norepinephrine	Get the body ready for strenuous activity by increasing the concentration of blood sugar
Pancreas	Insulin	Regulates the amount of sugar in the blood
	Glucagon	Increases the amount of sugar in the blood
Ovaries	Estrogen	Promotes female secondary sex characteristics
	Progesterone	Thickens endometrial lining
Testes	Testosterone	Promotes male secondary sex characteristics

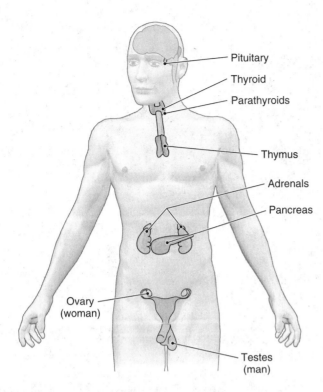

The Endocrine System

Endocrine System

Endocrine glands secrete hormones that regulate body metabolism, growth, and reproduction. These organs are not in contact with each other, but they communicate through chemical messages transported by the circulatory system. The preceding table lists the major glands in the endocrine system, the hormone(s) each one produces, and what the hormones do. On the ASVAB test, you probably will not need to know the specific hormones that are produced but just in general what each gland does.

Reproductive System

The purpose of the reproductive system is to continue the species for another generation. The organs of this system, called *gonads*, produce *gametes* that combine in the female system to produce the next generation. The male gonads are the *testes*, which produce *sperm* and male sex hormones. The female gonads are the *ovaries*, which produce *eggs* (ova) and female sex hormones. The sperm fertilizes the egg, and reproduction begins. Fertilization takes place when the sperm, using its *flagellum* (a sort of whiplike tail), moves through fluid to reach the egg. Each egg or sperm contains 23 *chromosomes*. These chromosomes carry DNA, which contributes to forming a new individual and determines the person's traits or characteristics.

Sex cells divide by a process called *meiosis*. Before this process begins, the nucleus divides twice, creating four cells, each with half the number of chromosomes of the original cell. So these cells have 23 chromosomes, half of what is needed. When a sperm and an egg combine, the fertilized egg contains the full complement of 46 chromosomes. These new cells keep dividing, growing, and developing over time, eventually creating a baby.

The Senses

The senses are sight, hearing, touch, taste, and smell. There are specialized organs for each sense.

Sight Let's suppose that you are looking at an object. Light waves from that object enter the eye first through the *cornea* (a transparent section of the eye) and then through the *pupil*, the opening in the *iris*. Light waves are brought to convergence first by the cornea, and then by the *lens*. The lens directs the light through the *vitreous humor* (a gelatinous substance) onto the *retina*. The retina has two types of cells, *rods* and *cones*. Rods detect light intensity, and cones respond to color. At this point, the light waves are changed to electrical signals and then sent along the optic nerve to the brain. The brain then interprets the signals as the picture of the object.

Some people have vision problems. Two typical problems are *nearsightedness* and *farsightedness*. Nearsighted people have trouble seeing distant objects. Farsighted people have trouble seeing things that are close up. Nearsighted people can have their vision corrected with a *concave* lens, and farsighted people can have their vision corrected with a *convex* lens.

Hearing Sound is vibrating air. Vibrating air creates sound waves. These waves can pass through solids, liquids, and gases to eventually reach your ear. When the sound waves reach your ear, they stimulate nerve cells that send signals to the brain. The brain interprets these signals as sound.

The ear is made up of several different parts, each contributing to the process of transforming sound waves into signals. The ear has three basic sections: the *outer ear*, the *middle ear*, and the *inner ear*.

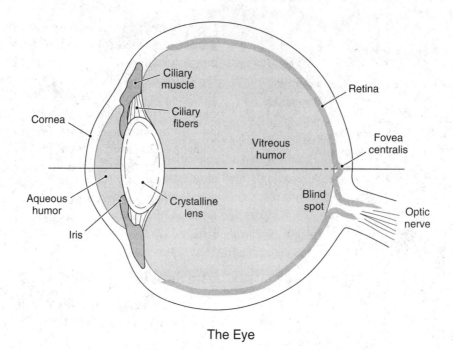

The Eye

The outer ear traps the sound waves and sends them down the ear canal to the middle ear, where the eardrum creates a vibration. This vibration then moves to some tiny specialized bones called the *hammer*, *anvil*, and *stirrup*. The stirrup bone rests against a membrane in the inner ear. The inner ear contains the *cochlea*, which is filled with fluid and is shaped like a snail. The cochlea vibrates when the stirrup vibrates, bending little hairs in the cochlea. The vibrating hairs send electrical signals to the brain, which in turn translates the signals into various sounds.

The semicircular canals in your inner ear are also responsible for maintaining your balance.

Smell Substances such as food, perfume, and gases give off molecules into the air. These molecules stimulate some nerve cells in your nose. These nerve cells are called *olfactory cells*. The olfactory cells are moist from mucus. Molecules are dissolved in the mucus, and, if there are a sufficient number of these molecules, a signal is sent to the brain. The brain then interprets these signals. If you have smelled this before, you may recognize the smell

The Ear

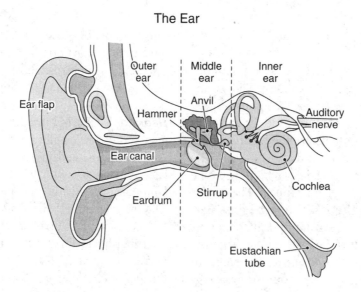

as a hamburger with onions, pumpkin pie, or your mother's favorite perfume.

Taste There are about 10,000 *taste buds* on your tongue. When you take food into your mouth, it begins to dissolve in liquid, your saliva. The mixture of saliva and food washes over your taste buds and stimulates the nerve fibers in the taste bud. The nerve fibers send a signal to your brain, and your brain translates the taste as sweet, sour, salty, bitter, or the taste of MSG (monosodium glutamate). The sense of smell and the sense of taste are related to each other.

Touch Your skin also has receptors allowing you to determine if something is hot, cold, rough, smooth, painful, hard, soft, and so on. The receptors send a signal to the brain, which in turn interprets the signals.

Genetics

All characteristics are inherited from genes of the mother and the father. Chromosomes come in pairs. The genes on a chromosome have information about the trait, but the information from the father and the mother is not necessarily exactly the same. Different forms of a gene are called *alleles*. For example, for the genes that determine hair color, the allele from your mother might be for blond hair and the allele from your father might be for black hair. Some alleles are *dominant* and some are *recessive*. The dominant allele will determine what traits actually become part of you.

Sometimes errors in mitosis or meiosis take place, altering a gene or chromosome. Some mutations are harmless, such as a four-leaf clover or a white tiger. Others can be harmful, setting the stage for certain diseases, conditions, or even death.

Determining the Probability of Certain Traits When describing genetic traits or alleles, a dominant allele is indicated with a capital letter (such as B). A recessive allele is indicated with a lowercase letter (such as b). In the genes, these alleles appear in pairs, such as BB, bb, or Bb. The sum of these genetic codes is the *genotype*, or genetic makeup of the organism. How the organism looks is called the *phenotype*. So if you have blond hair, that is your phenotype.

An organism that has two alleles that are the same, such as BB or bb, is called *homozygous*. An organism

that has one allele that is dominant and one that is recessive, such as Bb, is called *heterozygous*.

To determine the probability that an offspring will have a certain characteristic, you can use a device called a *Punnett square*.

Suppose a father has black hair with one dominant allele for black hair (B) and one recessive allele for blond hair (b). Then suppose a mother has blond hair with two recessive alleles (bb). The following Punnett square shows the probability that these parents will have a child with black or blond hair.

		Father	
		B	**b**
Mother	**b**	Bb	bb
	b	Bb	bb

There is a 2 in 4 (or 50 percent) probability that a child will have black hair (Bb) and a 2 in 4 (or 50 percent) probability that a child will have blond hair (bb).

Now suppose the father has two dominant alleles for black hair (BB).

		Father	
		B	**B**
Mother	**b**	Bb	Bb
	b	Bb	Bb

All offspring would have black hair (Bb).

Now suppose that the father and mother each have black hair with one dominant allele for black hair (B) and one recessive allele for blond hair (b).

		Father	
		B	**b**
Mother	**B**	BB	Bb
	b	Bb	bb

The probability of having a blond child (bb) would be 1 in 4, or 25 percent.

Determining the Sex of an Offspring The female carries two X chromosomes, and the male carries an X and a Y chromosome. The female egg contains an

Role of Some Important Vitamins

Vitamin	What It Does	Where to Get It
B	Helps in growth, use of carbohydrates, red blood cell production, and development of a healthy nervous system	Wheat germ, whole-grain cereals, poultry, eggs, fish, dairy products, green vegetables, pasta
A	Helps in growth, eyesight, and healthy skin	Fish and liver, green and yellow fruits and vegetables
E	Helps in the formation of cell membranes	Vegetable oils, whole grains, nuts and seeds, green vegetables
C	Helps in growth, good bones and teeth, and wound healing	Citrus fruits, tomatoes, vegetables, strawberries
D	Helps in absorption of calcium and phosphorus in bones and teeth	Milk, eggs, cheese
K	Helps with the clotting of blood and wound healing	Green vegetables, vegetable oils, pork, liver, egg yolks

X chromosome. A male sperm can contain either an X or a Y chromosome. Depending on which sperm fertilizes the egg, the result can be a female (XX) or a male (XY). The male parent determines the sex of the offspring.

Nutrition

You need to eat for energy and to maintain your biologic functions. Food provides you with the necessary nutrients to survive. There are six types of nutrients: *proteins*, *carbohydrates*, *fats*, *vitamins*, *minerals*, and *water*. Proteins, carbohydrates, fats, and vitamins are all *organic* nutrients. Minerals and water are *inorganic*. Organic means the substance contains carbon.

Proteins are used by the body for replacement and repair of body cells and for growth. They are made up of amino acids. Eggs, meat, and cheese are considered protein foods.

Carbohydrates are a good source of energy for your body. Three types of carbohydrates are sugar, fiber, and starch. Sugars are simple carbohydrates. Starches and fiber are complex carbohydrates. Foods like potatoes, pasta, beans, and breads are complex carbohydrates.

Fats provide energy and help absorb vitamins. Fats contain twice as much energy as carbohydrates. Excess energy from the foods you eat is stored as fat.

Vitamins are nutrients that are needed for certain bodily functions and for preventing some diseases. Water-soluble vitamins, such as B and C, need to be taken in every day. Fat-soluble vitamins are also stored in fat cells. Examples are vitamins A, D, E, and K.

Minerals can regulate the chemical reactions in your body. You need some level of about 14 different minerals. Most minerals are needed only in small (trace) amounts. Others, like calcium and phosphorous, are needed in larger amounts. Perhaps the most

Mineral	What It Does	Where You Get It
Calcium	Creates strong bones and teeth, good muscle and nerve activity	Dairy products such as milk and yogurt, green vegetables, soy
Phosphorus	Creates strong bones and teeth, regulates contraction of muscles	Cheese, meat, cereal
Potassium	Regulates water balance in cells, muscle contraction, nerve impulse conduction	Bananas, potatoes, oranges, nuts, meat
Sodium	Regulates fluid balance in tissues, nerve impulse conduction	Meat, milk, cheese, beets, carrots, salt, most prepared foods
Iron	Transports oxygen in red blood cells	Red meat, raisins, beans, spinach, eggs
Iodine	Controls thyroid activity, metabolic stimulation	Seafood, iodized salt

important contribution of minerals is the production and maintenance of our bones.

The preceding table lists the most important minerals, their benefits, and the foods that provide each one.

About two-thirds of your body is *water*. Your cells need water to carry out their work. You would die if you didn't have water for a few days. Some of your water comes from foods, but you must also drink water. Your body loses about four pints of water every day. That needs to be replenished. People who exercise need even more water.

CHEMISTRY

Everything is made up of *matter*. Matter is anything that takes up space and has energy. *Energy* causes change. Matter can exist as a *solid*, *liquid*, or *gas*—the three states of matter. An *atom* is the smallest building block of matter. Atoms are made of *neutrons* (with no electric charge), *protons* (with a positive electric charge), and *electrons* (with a negative electric charge). The center of an atom is the *nucleus*. The nucleus contains protons and neutrons. The nucleus of an atom is tiny in comparison to the atom itself. Electrons fly around the nucleus in orbits. Most of an atom is empty space. Energy holds the atom together. Atoms have the same number of electrons and protons, making them stable.

Elements

If a substance is made up entirely of the same kind of atom, it is called an *element*. An element cannot be broken down into a simpler form. The element carbon is made up of only carbon atoms, and the element sulfur is made up of only sulfur atoms.

Scientists have given all elements a one- or two-letter symbol. For example, oxygen is O, sulfur is S, calcium is Ca, and magnesium is Mg. All elements are arranged in a periodic table according to their similarities, providing important information about each element. You will not need to know the periodic table for the ASVAB test, but you may need to know that each block of the table has certain information.

All elements in the periodic table are ordered by their *atomic number*. The atomic number is the number of protons in the atom. In an atom with a neutral charge, the number of electrons equals the number of protons. The *atomic mass* number is equal to the number of protons plus the number of neutrons.

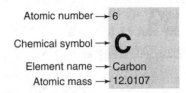

Atomic number → 6
Chemical symbol → **C**
Element name → Carbon
Atomic mass → 12.0107

Presence of Elements Most of the human body is made up of water, H_2O. Your cells contain about 65 to 90 percent water by weight. As much as 99 percent of the mass of the human body is made up of just six elements: oxygen, carbon, hydrogen, nitrogen, calcium, and phosphorus.

Elements in the Human Body	Percent
Oxygen	65
Carbon	18
Hydrogen	9.5
Nitrogen	3.2
Calcium	1.5
Phosphorus	1
Potassium	0.35
Sulfur	0.25
Sodium	0.15
Magnesium	0.05
Copper, zinc, selenium, molybdenum, fluorine, chlorine, iodine, manganese, cobalt, iron	0.70
Lithium, strontium, aluminum, silicon, lead, vanadium, arsenic, bromine	Trace amounts

By contrast, Earth's crust is made up of lesser amounts of oxygen, but a lot of silicon and aluminum.

Elements in Earth's Crust	Percent
Oxygen	47.2
Silicon	28.2
Aluminum	8.2
Iron	5.1
Calcium	3.7
Sodium	2.9
Potassium	2.6
Magnesium	2.1

Periodic Table of the Elements

Legend:

- Alkali Metals
- Alkaline Earth Metals
- Transition metals
- Lanthanide series
- Actinide series
- Other Metals
- Nonmetals
- Noble gases

C	Solid
Br	Liquid
H	Gas
Tc	Synthetic

New / Original group numbers

Group headings:

| 1 IA | 2 IIA | 3 IIIB | 4 IVB | 5 VB | 6 VIB | 7 VIIB | 8 VIIIB | 9 VIIIB | 10 VIIIB | 11 IB | 12 IIB | 13 IIIA | 14 IVA | 15 VA | 16 VIA | 17 VIIA | 18 VIIIA |

Period 1

- 1 **H** Hydrogen 1.00794
- 2 **He** Helium 4.002602

Period 2

- 3 **Li** Lithium 6.941
- 4 **Be** Beryllium 9.012182
- 5 **B** Boron 10.811
- 6 **C** Carbon 12.0107
- 7 **N** Nitrogen 14.00674
- 8 **O** Oxygen 15.9994
- 9 **F** Fluorine 18.9984032
- 10 **Ne** Neon 20.1797

Period 3

- 11 **Na** Sodium 22.989770
- 12 **Mg** Magnesium 24.3050
- 13 **Al** Aluminum 26.981538
- 14 **Si** Silicon 28.0855
- 15 **P** Phosphorus 30.973761
- 16 **S** Sulfur 32.066
- 17 **Cl** Chlorine 35.4527
- 18 **Ar** Argon 39.948

Period 4

- 19 **K** Potassium 39.0983
- 20 **Ca** Calcium 40.078
- 21 **Sc** Scandium 44.955910
- 22 **Ti** Titanium 47.867
- 23 **V** Vanadium 50.9415
- 24 **Cr** Chromium 51.9961
- 25 **Mn** Manganese 54.938049
- 26 **Fe** Iron 55.8457
- 27 **Co** Cobalt 58.933200
- 28 **Ni** Nickel 58.6934
- 29 **Cu** Copper 63.546
- 30 **Zn** Zinc 65.39
- 31 **Ga** Gallium 69.723
- 32 **Ge** Germanium 72.61
- 33 **As** Arsenic 74.92160
- 34 **Se** Selenium 78.96
- 35 **Br** Bromine 79.904
- 36 **Kr** Krypton 83.80

Period 5

- 37 **Rb** Rubidium 85.4678
- 38 **Sr** Strontium 87.62
- 39 **Y** Yttrium 88.90585
- 40 **Zr** Zirconium 91.224
- 41 **Nb** Niobium 92.90638
- 42 **Mo** Molybdenum 95.94
- 43 **Tc** Technetium (98)
- 44 **Ru** Ruthenium 101.07
- 45 **Rh** Rhodium 102.90550
- 46 **Pd** Palladium 106.42
- 47 **Ag** Silver 107.8682
- 48 **Cd** Cadmium 112.411
- 49 **In** Indium 114.818
- 50 **Sn** Tin 118.710
- 51 **Sb** Antimony 121.760
- 52 **Te** Tellurium 127.60
- 53 **I** Iodine 126.90447
- 54 **Xe** Xenon 131.29

Period 6

- 55 **Cs** Cesium 132.90545
- 56 **Ba** Barium 137.327
- 57 to 71
- 72 **Hf** Hafnium 178.49
- 73 **Ta** Tantalum 180.9479
- 74 **W** Tungsten 183.84
- 75 **Re** Rhenium 186.207
- 76 **Os** Osmium 190.23
- 77 **Ir** Iridium 192.217
- 78 **Pt** Platinum 195.078
- 79 **Au** Gold 196.96655
- 80 **Hg** Mercury 200.59
- 81 **Tl** Thallium 204.3833
- 82 **Pb** Lead 207.2
- 83 **Bi** Bismuth 208.98038
- 84 **Po** Polonium (209)
- 85 **At** Astatine (210)
- 86 **Rn** Radon (222)

Period 7

- 87 **Fr** Francium (223)
- 88 **Ra** Radium (226)
- 89 to 103
- 104 **Rf** Rutherfordium (261)
- 105 **Db** Dubnium (262)
- 106 **Sg** Seaborgium (266)
- 107 **Bh** Bohrium (264)
- 108 **Hs** Hassium (269)
- 109 **Mt** Meitnerium (268)
- 110 **Ds** Darmstadtium (271)
- 111 **Rg** Roentgenium (272)
- 112 **Uub** Ununbium (285)
- 113 **Uut** Ununtrium (284)
- 114 **Uuq** Ununquadium (289)
- 115 **Uup** Ununpentium (288)
- 116 **Uuh** Ununhexium (292)
- 117 **Uus** Ununseptium
- 118 **Uuo** Ununoctium

Lanthanide series (57–71)

- 57 **La** Lanthanum 138.9055
- 58 **Ce** Cerium 140.116
- 59 **Pr** Praseodymium 140.90765
- 60 **Nd** Neodymium 144.24
- 61 **Pm** Promethium (145)
- 62 **Sm** Samarium 150.36
- 63 **Eu** Europium 151.964
- 64 **Gd** Gadolinium 157.25
- 65 **Tb** Terbium 158.92534
- 66 **Dy** Dysprosium 162.50
- 67 **Ho** Holmium 164.93032
- 68 **Er** Erbium 167.26
- 69 **Tm** Thulium 168.93421
- 70 **Yb** Ytterbium 173.04
- 71 **Lu** Lutetium 174.967

Actinide series (89–103)

- 89 **Ac** Actinium (227)
- 90 **Th** Thorium 232.0381
- 91 **Pa** Protactinium 231.03588
- 92 **U** Uranium 238.0289
- 93 **Np** Neptunium (237)
- 94 **Pu** Plutonium (244)
- 95 **Am** Americium (243)
- 96 **Cm** Curium (247)
- 97 **Bk** Berkelium (247)
- 98 **Cf** Californium (251)
- 99 **Es** Einsteinium (252)
- 100 **Fm** Fermium (257)
- 101 **Md** Mendelevium (258)
- 102 **No** Nobelium (259)
- 103 **Lr** Lawrencium (262)

Atomic masses in parentheses are those of the most stable or common isotope.

Web Page Design Copyright © 1997 Michael Dayah. Last Modified: Fri May 06 2005 13:48:56

Note: The subgroup numbers 1-18 were adopted in 1984 by the International Union of Pure and Applied Chemistry. The names of elements 112-118 are the Latin equivalents of those numbers.

Compounds

In a *compound*, two or more elements combine chemically in a fixed proportion to create a new substance that has characteristics different from those of the combining elements. For example, water is a compound made up of two hydrogen atoms and one oxygen atom. It is commonly written as H_2O. This ratio holds whether you have a gram of water or an ocean full of water.

Common Compounds in the Human Body	Importance and Use
Water	Makes up of most of our body, including blood. Nearly all chemical reactions take place in water
Calcium phosphate	Strengthens bones
Hydrochloric acid	Breaks down food in the digestive system
Sodium bicarbonate	Helps with digestion
Various salts	Helps to send messages along the nerves

There are two types of compounds, *covalent compounds* and *ionic compounds*. A covalent compound is made up of *molecules* (in most cases). A molecule is a group of atoms that are held together by chemical bonds (mainly covalent bonds). *Chemical reactions* occur when the bonds are broken and new bonds form, making different molecules. *Ions* are atoms that are positively or negatively charged. If an ion has more electrons than protons, it is negatively charged. If it has more protons than electrons, it is positively charged. Ions that are positively charged are attracted to ions that are negatively charged. Often ions will come together to form neutral ionic compounds. Table salt is a good example of this, as it has a positive ion of sodium (Na^+) that is attracted to a negatively charged chlorine (Cl^-), forming NaCl.

Mixtures

A *mixture* is composed of two or more substances that do not combine with each other, but instead retain their original characteristics. Also, the substances in a mixture do not have to be in any fixed proportion. For example, if you poured flour and salt together into a bowl, you would create a mixture. Mixing a cup of sugar into a gallon of water would create a mixture called a *solution*, where the sugar dissolves into the water. Mixtures can be solids, liquids, or gases or any combination of those.

Another type of mixture is a *suspension*. A suspension is a mixture of a liquid or gas with another substance spread evenly through it. Common suspensions include salt in water, fine soot or dust in air, droplets of oil in air, and whole blood. Particles in a suspension precipitate to the bottom if the suspension is left standing.

PHYSICAL SCIENCE

The Laws of Motion

The English scientist Sir Isaac Newton (1642–1727) formulated three laws of motion.

Newton's First Law The first law reads as follows: *An object at rest tends to stay at rest and an object in motion tends to stay in motion with the same speed and in the same direction unless acted upon by an unbalanced force.* Basically, this means that an object will keep doing what it is doing unless something makes it change. The something that makes an object change its state is called a *force*. A force is a push or pull upon an object resulting from one object's interaction with another object. An example of this is a person pushing a swing or a person pulling a suitcase.

An object has *inertia* when it is moving. Inertia is merely the resistance to change. An example is the tendency of your body to keep moving forward when your car comes to a sudden stop. In a collision, a seatbelt can keep you in place when otherwise you might hurtle through the windshield because of inertia.

Friction is a force that results when the surface of one object touches the surface of another. Friction causes moving objects to slow down or stop. For example, if you shove a book across a table, it will soon slow to a stop because it is rubbing on the surface of the table. Friction between two objects causes *heat*.

Newton's Second Law According to Newton, *an object will accelerate only if there is an unbalanced force acting upon it.* The presence of an unbalanced force will accelerate an object, changing its speed, its direction, or both its speed and its direction. The second law states that the acceleration of an object is dependent upon two variables—the force acting upon the object and the mass of the object. The formula for acceleration is $a = \dfrac{F}{m}$.

Newton's Third Law Newton's third law reads: *For every action, there is an equal and opposite reaction.* When you sit in your chair, your body exerts a downward force on the chair and the chair exerts an upward force on your body. There are two forces resulting from this interaction—a force on the chair and a force on your body. When a bird flaps its wings downward, the air pushes it upward, allowing it to fly. A gun recoils when it is fired.

Describing Motion In physics, there are various ways to describe motion. *Speed* is how fast an object is moving. A fast-moving object has a high speed, while a slow-moving object has a low speed. An object with no movement at all has a zero speed. The average speed during the course of a motion can be calculated using the following equation:

$$\text{Speed} = \frac{\text{distance traveled}}{\text{time of travel}}$$

Velocity is defined as the rate and direction at which an object's position changes. Velocity includes both speed and direction, such as moving 55 miles/hour in a westerly direction or 6 meters/second upward.

Average velocity can be calculated using the equation:

$$\text{Velocity} = \frac{\text{change of position}}{\text{time}}$$

Acceleration is defined as "the rate at which an object changes its velocity." An object is accelerating if it is changing its velocity. If an object is not changing its velocity, then the object is not accelerating. A falling object accelerates as it falls. If you could measure the motion of a falling object, you would notice that the object has an average velocity of 5 m/s in the first second, 15 m/s in the second second, 25 m/s in the third second, 35 m/s in the fourth second, and so on. By pushing down the gas pedal, you can accelerate your car.

Work, Energy, and Power

Some of the most important concepts in physics are work, energy, and power.

Work Work results from a force acting upon an object, causing it to change position or move from one place to another. There are three key aspects to work: force, movement, and cause. In order for a force to qualify as having done work on an object, there must be a change of position caused by the force. Here are some common examples of work: a person carrying a box of books upstairs, a horse pulling a plow through the fields, a weightlifter lifting barbells, a person pushing a grocery cart down the aisle of a grocery store, a shot putter launching the shot, and an ice skater lifting his partner overhead. In each case, there is a force exerted upon an object that causes the object to be displaced. Work is measured in *joules*.

Energy In physics, *energy* is defined as the ability to do work. Energy can be either potential or kinetic.

Potential Energy An object can store energy as the result of its position. A can of soup on a shelf has potential energy. When it falls from the shelf onto the floor, it releases its energy. A roller-coaster car has potential energy when it is at the top of the track. It releases the energy when it plunges downward. An arrow in a drawn bow has potential energy resulting from its position. If the bow is not drawn, or pulled back, the arrow has no potential energy. The can of soup and the roller-coaster car have potential energy that is caused by *gravitation*—being pulled toward Earth. The second form of potential energy is *elastic* potential energy. Elastic potential energy is the energy stored when an object is stretched or compressed. For example the arrow in a drawn bow has elastic potential energy. Other examples include stretched rubber bands, bungee cords, trampolines, and springs. The more stretch or compression that is exerted on the object, the more potential energy it holds.

Kinetic Energy Kinetic energy is the energy of motion. Any object that has motion has kinetic energy. The direction does not matter. A moving car has kinetic energy, as does a moving ice skater or a soup can falling off the shelf. When the soup can falls off the shelf, its potential energy is changed to kinetic energy.

Forms of Energy Energy cannot be created or destroyed; it merely changes from one form of energy to another. Forms of energy include the following:

Chemical	Electrical
Gravitational	Heat
Light	Mechanical
Nuclear	Solar
Sound	Wind

Due to our depleting natural resources, such as coal and oil, scientists are developing new ways to harness other forms of energy to satisfy our energy and power needs. Some success has been created with transforming solar, wind, electrical energy into powering our cars, equipment, and gadgets.

Power Power is how much work is done over a given period of time. Work can be done very quickly or very slowly. For example, a rock climber scaling a sheer cliff may take a long time to raise their body just a few meters. But a hiker who selects an easier, less vertical path might raise their body the same few meters in a much shorter time. Even though the amount of work is the same, the hiker does the work in considerably less time than the rock climber. The hiker has a greater power than the rock climber.

The formula for power is

$$\text{Power} = \frac{\text{work}}{\text{time}}$$

The standard metric unit of power is the *watt*. Watts are used to measure the amount of energy consumed by an electric device.

Waves

Sound waves, light waves, radio waves, microwaves, water waves, earthquake waves, and waves on a piece of string are just a few of the kinds of waves we are likely to encounter every day. Waves transport energy. Waves have different parts. This section will focus on sound and light waves.

The *crest* of a wave is the point where there is the maximum amount of upward displacement from the rest or equilibrium position. The *trough* of a wave is the point where there is the maximum amount of downward displacement from the rest or equilibrium position. The *amplitude* of a wave refers to the height of the wave from its equilibrium point.

The *wavelength* of a wave is simply the length of one complete wave cycle. The wavelength can be measured as the distance from crest to crest or from trough to trough or at any two similar points on the wave. The *period* of a wave is the amount of time it takes to get from one point on the wave to the same point on the next wave.

The amount of energy carried by a wave is related to its amplitude (the distance from the wave's equilibrium position to the maximum height or depth). A high-energy wave is characterized by a high amplitude; a low-energy wave is characterized by a low amplitude.

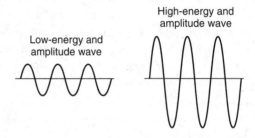

High-energy and amplitude wave

Low-energy and amplitude wave

The *frequency* of a wave is the number of complete cycles per second that the wave makes, from equilibrium to its crest, then back through equilibrium to its trough, and back again to equilibrium.

The diagram of the *electromagnetic spectrum* shows that waves have a large range of frequencies. The diagram on page 166 depicts the electromagnetic spectrum and its various regions. The longer-wavelength, lower-frequency regions are located on the far left of the spectrum; shorter-wavelength, higher-frequency regions are on the far right. The different wavelengths of waves determine what we see and how we communicate.

Sound Waves Sound is a wave that is created by vibrating objects and that moves through a medium from one location to another. Sound cannot travel through a *vacuum*. The medium is usually air, but it could be any material, like a liquid or solid. The

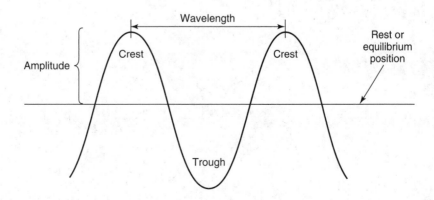

The Electromagnetic Spectrum

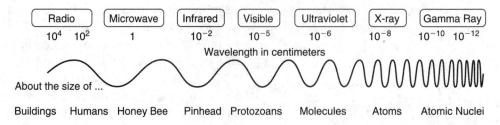

vibrating object could be the vocal chords of a person, the string of a guitar, or a tuning fork, or the vibrating diaphragm (see respiratory system earlier in this overview) of a television announcer.

The *loudness* of sound depends on the amplitude of the wave. The higher the amplitude, the louder the sound. Loudness is measured in *decibels*. The *pitch* of the sound depends on the frequency of the sound waves. A high-pitched sound is created by a high-frequency wave, and a low-pitched sound is created by a low-frequency wave.

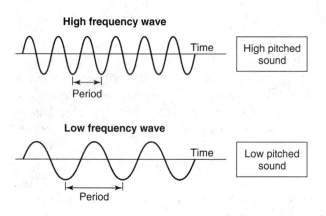

Sound travels faster in solids than in gases. Sound also travels faster in substances that have a higher temperature.

Sound waves are subject to a phenomenon called the *Doppler effect*. You may have heard the results of the Doppler effect without knowing it. Have you ever listened to an ambulance siren moving toward you and then detected a change in pitch when the siren started to move away from you? This is an example of the Doppler effect—a shift in the frequency of the waves coming from a moving object. As the object moves toward you, the pitch is higher; when it is moving away, the pitch is lower.

Light Waves *Visible light* occurs within a very tiny range of the electromagnetic spectrum. Each color in visible light has a distinct wavelength. This can be seen most clearly when you pass white light through a *prism* that separates the light into its "rainbow" of colors. Dispersion of visible light produces the colors red (R), orange (O), yellow (Y), green (G), blue (B), indigo (I), and violet (V). The red wavelengths of light are the longer wavelengths, and the violet wavelengths of light are the shorter wavelengths. The visible light spectrum is shown in the diagram below. To remember the order from long wavelengths to the shorter wavelengths, think of the letters *ROY G BIV*.

White is really not a color; rather, it is the combination of all the colors of the visible light spectrum. *Black* is not really a color, either. Black is

The Visible Spectrum

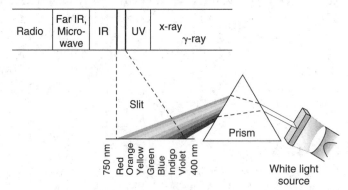

the absence of the wavelengths of the visible light spectrum or the absence of light or color. If you are in a room that is totally black/dark, it means that there are no wavelengths of visible light striking your eye.

Refraction Light *refracts*, or bends, when it passes through different media. Different substances bend light in different amounts. Every substance that can pass light has what is called a *refractive index*, the amount that the substance will bend light.

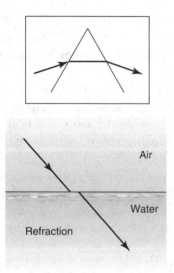

Air

Water

Refraction

Light is bent when it goes through *lenses*. Lenses can be *convex* or *concave*, and light behaves differently when it passes through each type of lens. Convex lenses are thicker in the middle than at the edges. Concave lenses are thicker at the edges than in the middle.

When light rays travel through a convex lens, the lens *converges* the light to a focal point. The more curved the lens, the closer the focal point is to the lens.

Refraction by a converging lens

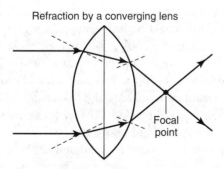

Focal point

A concave lens, by contrast, *diverges* the light so that the light rays will never meet.

Refraction by a diverging lens

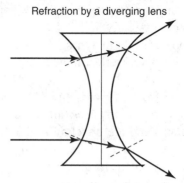

Reflection Light also has the property of reflection. A ray of light, perhaps coming from a flashlight or other light source, approaching a mirror is known as an *incident ray* (*I*). The ray of light that leaves the mirror is known as the *reflected ray* (*R*). At the point where the ray strikes the mirror, a perpendicular line can be drawn; this line is known as a *normal line* (*N*). The normal line divides the angle into two equal angles. The angle between the incident ray and the normal is known as the *angle of incidence*. The angle between the reflected ray and the normal is known as the *angle of reflection*.

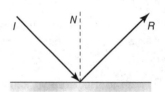

I   *N*   *R*

Like lenses, mirrors can be *concave* (converging) and *convex* (diverging).

A converging mirror takes light rays and converges them to a focal point. The focal point is the point in space at which light will meet after reflection. Flashlights and headlights on a car use concave mirrors to produce direct light beams in a narrow area.

A diverging mirror spreads the light after it is reflected so that the light beams will never intersect. For this reason, convex mirrors produce virtual images that appear to be located somewhere behind the mirror. Examples of these are the security mirrors you seen in stores so that clerks can see people in a large part of the store. They can also be found in the outside rearview mirrors in cars, which help you see a larger viewing area than a flat mirror. Convex mirrors make objects look farther away and smaller. That's why you have the warning in your outside rearview mirror that states that objects appear more distant than they really are.

Telescopes use both types of lenses to refract light and mirrors to reflect light.

Lasers Light from a *laser* is all of one wavelength, and all the waves are in *phase* (all crests and troughs are in tandem) and are traveling in the same direction. As a result, a laser beam of light does not spread out very much and directs its energy in a very small area. This characteristic allows lasers to be used as cutting tools in situations like eye surgery, for transferring digital music to CDs, and for reading bar codes in such places as grocery and department stores. Because laser light doesn't spread, lasers are used to carry information across long distances in space or through fiber-optic cable on land.

Blueshift and Redshift The *Doppler effect* is also a phenomenon of light. Scientists use this phenomenon to determine whether objects like stars and galaxies are moving toward Earth or away from it. If an object is moving toward Earth, the light from the object appears bluer (a blueshift). If an object is moving away from Earth, the light from the object appears redder (a redshift). Scientists have used this information to determine that the universe is expanding, creating a shift in wavelengths toward the red end of the spectrum

Heat

Heat is created by moving molecules. The faster the molecules move, the more heat is generated. Heat is basically *kinetic energy* (review the section on energy earlier in this chapter). Heat is measured by a *thermometer*.

Heat transfers or flows from warmer objects to colder objects when they are in contact. Transfer takes place in three ways: *radiation*, *conduction*, and *convection*.

Radiation The first method of heat transfer is *radiation*, which takes place via invisible waves through the air or even through a vacuum. The energy from the sun warms us through radiation, just as a fire in a fireplace warms our hands. A microwave oven heats food via radiation.

Conduction The second method of heat transfer is *conduction*, which refers to heat transferred between

atoms bumping into each other in a substance. What happens if you put a spoon in boiling water? Eventually the handle of the spoon gets too hot to hold. The heat from the water causes the molecules in the spoon to move and collide, creating heat that eventually is conducted up the spoon handle to your hand.

Some materials are good conductors of heat, and some are not. Metals are good conductors of heat, with gold, silver, and copper being the best conductors. Copper is used more because it is cheaper and more plentiful. Some materials are called *insulators* because they are not very good at conducting heat. Materials made of cotton or wool, for example, are not good conductors of heat. If you wrap yourself in a wool blanket, rather than conducting the heat away from your body, it traps the heat and you get warmer. Air is a good insulator, as are plastic, wood, rubber, and tile (think of the tiles on the space shuttle that protect the shuttle from the heat of friction created when the shuttle strikes the atmosphere during its return to Earth).

Think about heat conduction when you get into a car on a hot sunny day when the windows are closed. Vinyl or leather seats will feel very hot, as they are better conductors of heat then fabric seats.

Convection Heat also transfers via a process called *convection*. Convection generally occurs in gases or liquids. For example, as the temperature of a pan of water on the stove increases, the molecules start to move around more quickly. As they do this, the distance between the molecules increases and the liquid expands. The denser, cooler water then sinks to the bottom of the pan and forces the warmer liquid upward. A circulation of the colder and warmer liquid begins. On a much larger scale, this is the process that helps create currents in the oceans.

The same process creates currents in the atmosphere. When land is heated by the sun, it heats the air that is in contact with it. The warmer air expands and rises, and the cooler, heavier air sinks down. The rising air creates upward motion, which is used by birds and hang gliders to stay aloft. Sometimes these upward currents are called *thermals*. Thermals can reach many thousands of feet up and will give people in airplanes a bumpy ride.

Convection also creates land and sea breezes at a shoreline. Land heats faster than water. As the land heats up, it warms the air above it. The air expands and rises. It is then replaced by cooler air flowing in from over the water. This process creates a wind current that goes from the water to the land. We call this current a sea breeze. In the evening, the reverse happens. The land cools faster than the water, so the air current flows in the opposite direction. This creates what is called a land breeze—a breeze from the land to the sea.

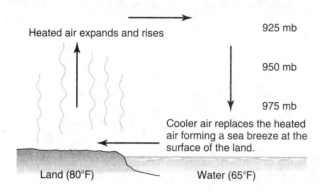

Heated air expands and rises

925 mb

950 mb

975 mb

Cooler air replaces the heated air forming a sea breeze at the surface of the land.

Land (80°F)   Water (65°F)

Magnetism

Magnets can be made by placing a certain material, such as iron or steel, in a strong magnetic field. Pieces of the steel or iron are generally arranged in a random order. When these pieces, or *domains*, are placed in a strong magnetic field, they line up in the direction of the field. When most of the domains are aligned, the material becomes a magnet.

Before magnetization

After magnetization

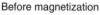

N   S

A bar magnet has three-dimensional lines of force around it on all sides. If the lines were visible, they would look like the following diagram.

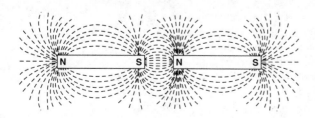

When opposite poles of two magnets are brought together, the lines of force join up and the magnets pull together.

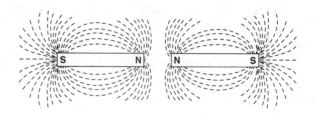

When like poles of two magnets are brought together, the magnets repel each other and the lines of force push away.

Earth as a Magnet Because of its iron-nickel core, Earth behaves like a giant magnet. It happens that Earth's magnetic poles are fairly close to the geographic poles. A compass has a needle that is attracted to the magnetic north pole and is repelled by the magnetic south pole.

Earth's magnetic field, called a *magnetosphere*, is produced by electric currents. Deep in Earth's core, there is convection as hot molten magma rises, cools, and sinks. The process repeats itself time and time again. Within these rising and falling masses of magma, the rotation of Earth creates organized patterns of circular electric currents, called eddies. The interior of Earth seems to act like a giant dynamo. Earth's magnetic field deflects most of the charged particles streaming from the sun.

Magnets are used to make electric motors and generators. Generators produce electricity, which supports our lifestyle in many ways.

Electricity and Magnetism The movement of electrons around the nucleus of an atom produces a *magnetic field*. The wire that contains an electric current is surrounded by a magnetic field. A wire with a current running through it that is wrapped around an iron core is called an *electromagnet*. Electromagnets are used in a variety of ways in our lives in everything from doorbells to magnetic high-speed trains.

GEOLOGY

Geology is the study of the solid Earth. It includes processes that have shaped Earth's surface, the ocean floors, and Earth's interior. It also involves the study of Earth's history from the past to the present.

Our planet has evolved since its creation about 4.6 billion years ago. Many processes have shaped Earth, some slow and others fast-acting. Slow processes include the formation of rocks, the chemical breakdown or weathering of rocks to form soil, the cementing of sand grains together to form rock, the metamorphosis of some kinds of rocks into other kinds of rocks through recrystallization, the uplifting of mountain ranges through tectonic activity, and the erosion of mountain ranges. Relatively faster processes include beach erosion during a storm, volcanic eruptions, landslides, dust storms, river flooding, and mudflows.

Earth's History

The chart below shows Earth's evolution over the past 4.6 billion years. On the ASVAB, you probably will not be asked about specific details on this chart, but you need to have a general idea of the history of Earth.

| **Phanerozoic Eon** (543 mya* to present) | **Cenozoic Era** (65 mya to today)

This era covers the last 10 percent of Earth's whole geologic history. After the disappearance of the dinosaurs, there were suddenly many empty places on Earth where other animals could live. Mammals diversified, and the Cenozoic Era became the "Age of Mammals." Whales took over the oceans; saber-tooth cats shared the land with elephants.

The global climate grew colder, and the last few million years saw the return of giant glaciers and ice caps to North America, Eurasia, and Antarctica. Ice ages occurred several times during this time period. There were mass extinctions during the Cenozoic Era, as there were during the Mesozoic and Paleozoic Eras.

Finally, humans appeared during the last 2 million years. In the last 10,000 years, humans spread across the planet. Humans changed the face of Earth with cities and farms, destroying some plants and animals and domesticating others. | **Quaternary** (1.8 mya to today)
Holocene (10,000 years to today)
Pleistocene (1.8 mya to 10,000 yrs)

Tertiary (65 to 1.8 mya)
Pliocene (5.3 to 1.8 mya)
Miocene (23.8 to 5.3 mya)
Oligocene (33.7 to 23.8 mya)
Eocene (54.8 to 33.7 mya)
Paleocene (65 to 54.8 mya) |
| | **Mesozoic Era** (248 to 65 mya)

The Mesozoic Era is divided into just three time periods: the Triassic, the Jurassic, and the Cretaceous. The supercontinents Gondwanaland and Laurasia collided to form a single super-super continent called Panagea. Plate tectonics continued, eventually causing Pangea to break up into the continents we know today.

Life was diversifying rapidly. The dominant animals on both land and sea were reptiles, the most famous of which are the dinosaurs. They are so prevalent in the Jurassic period that this is called "the Age of Reptiles." Birds and mammals also appeared during the Mesozoic | **Cretaceous** (144 to 65 mya)

Jurassic (206 to 144 mya)

Triassic (248 to 206 mya) |

Phanerozoic Eon (*Continued*) (543 mya* to present)	**Mesozoic Era** (*Continued*) (248 to 65 mya)	
	Era, as well as deciduous trees and flowering plants.	
	The climate during the Mesozoic Era was so warm that there were no ice caps. Plants grew profusely with all the warmth and moisture.	
	At the end of the Mesozoic Era, more than half of all the existing life-forms disappeared, including virtually all of the dinosaurs.	
	Paleozoic Era (543 to 248 mya)	**Permian** (290 to 248 mya)
		Carboniferous (354 to 290 mya) Pennsylvanian (323 to 290 mya) Mississippian (354 to 323 mya)
	Many different things happened during the Paleozoic Era. Earth's interior cooled down. Giant "hot-spot" type eruptions still occurred every hundred million years or so. Plate tectonics continued to push land masses across Earth's surface. At one particular time—the middle of the Silurian Period—most of the land was locked in two supercontinents called Gondwanaland (wandering over the southern hemisphere) and Laurasia (on the north side of the globe). Huge glaciers covered the interior of Gondwanaland, and Earth was experiencing one of its ice ages. Over the next hundred million years, Gondwanaland moved north over the equator and began to break up, and the climate warmed.	**Devonian** (417 to 354 mya) **Silurian** (443 to 417 mya) **Ordovician** (490 to 443 mya) **Cambrian** (543 to 490 mya) Tommotian (530 to 527 mya)
	The composition of the atmosphere continued to slowly change, mostly due to the increase in oxygen produced by photosynthetic algae floating on the ocean. By the Paleozoic Era, the composition of the air had reached something like what we breathe now: about 4/5 nitrogen, 1/5 oxygen, and small amounts of carbon dioxide, water vapor, and other gases. The air was capable of supporting large animals.	
	At the beginning of the Paleozoic Era, life existed only in or near the ocean. Trilobites, shellfish, corals, and sponges appeared, followed by the first fish. Land plants appeared near the end of the Ordovician Period.	
	The Paleozoic Era had several mass extinctions in which many life forms died out. Mass extinctions occurred at the end of the Ordovician and the Devonian Periods. During the Permian Period, about 95 percent of all life on Earth died! These mass changes were probably caused by volcanic eruptions and changes in the climate.	

(Continued)

Precambrian Time (4,500 to 543 mya)	Proterozoic Era (2,500 to 543 mya)	Neoproterozoic (900 to 543 mya) Vendian (650 to 543 mya)
This time period represents the very earliest rocks and fossils, and almost 90 percent of the entire history of Earth. It has been divided into three eras: the Hadean, the Archean, and the Proterozoic.		**Mesoproterozoic** (1,600 to 900 mya)
		Paleoproterozoic (2,500 to 1,600 mya)
This era began about 4.6 billion years ago with the formation of Earth from dust and gas orbiting the sun. Volcanoes were active, and asteroids kept pummeling the planet. The air was hot, thick, and steamy. It was not breathable by animals yet.	**Archaean Era** (3,800 to 2,500 mya)	
Later, cooling of Earth began and water in the atmosphere condensed to form the oceans. The air was now mostly nitrogen, like our air today. The lava had mostly cooled to form the ocean floor.		
Still later, blue-green algae came into existence. The surface of Earth was still very active. Continental collisions happened often. Life was still found only in the ocean. Eventually, multicelled creatures with no hard parts came into existence. Oxygen released by the algae floating in the oceans began to collect in the air. Earth at this time was also very cold, with huge, bluish glacial ice sheets visible across the planet.		
	Hadean Era (4,500 to 3,800 mya)	

*mya 5 million years ago

Earth's Structure

Earth is made up of three basic layers of rock: the crust, the mantle, and the core. The table on page 173 gives some characteristics of each of these layers.

Categories of Rocks

There are three categories of rock: *igneous*, *sedimentary*, and *metamorphic*. Igneous rocks are made of molten material that has solidified. Igneous rocks can be formed beneath Earth's surface (*intrusive*) or above Earth's surface (*extrusive*), as in the case of lava flows. When rock is subjected to weathering from chemical reactions, atmospheric conditions, or the results of gravity, the rock breaks into

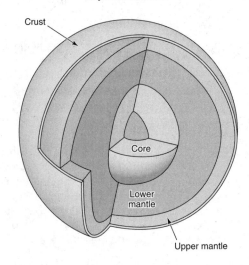

Layers of the Earth

Crust

Core

Lower mantle

Upper mantle

Characteristics of Earth's Layers

Layer	Characteristics	Composition
Crust	Crust is different on the continents and in the ocean; layers are very thin, only about 30 km; represents only 1 percent of Earth's volume; crust is lighter and floats on the mantle	Oceanic crust is dense rock called basalt; continental crust is made up of less dense andesite and granite Rich in the elements oxygen and silicon, with lesser amounts of aluminum, iron, magnesium, calcium, potassium, and sodium
Mantle	Represents about 85 percent of Earth by volume	Made of minerals rich in the elements iron, magnesium, silicon, and oxygen
Core	Two layers: a solid inner core and a liquid outer core; remains very hot after 4.5 billion years of cooling; represents about 15 percent of Earth's volume	Mostly iron and nickel

Common Elements Found in Earth's Rocks

Element	Chemical Symbol	Percent Weight in Earth's Crust
Oxygen	O	46.60
Silicon	Si	27.72
Aluminum	Al	8.13
Iron	Fe	5.00
Calcium	Ca	3.63
Sodium	Na	2.83
Potassium	K	2.59
Magnesium	Mg	2.09

smaller pieces, or sediments. These sediments can be compacted and cemented together to form solid rock called sedimentary rock. Both igneous and sedimentary rock can be placed under a great deal of heat and pressure. Under those conditions, the rock transforms into metamorphic rock. If the pressure and temperature are high enough, metamorphic rocks can melt and then solidify into igneous rock.

Igneous Rocks Igneous rocks are rocks that at one time were melted. They can form either underground or above ground. Melted rock underground is called *magma*, and it is sometimes trapped in pockets. As the melted magma cools, it forms masses of igneous rock. Igneous rocks form above ground when volcanoes erupt and magma reaches Earth's surface. When magma appears above Earth's surface, it is called *lava*. Typical igneous rocks are *pumice*, *obsidian*, and *granite*.

Pumice rocks are created when a volcano spews out lava that is filled with trapped gases. When the gases escape and the remaining materials harden, pumice is left. This rock is filled with air pockets

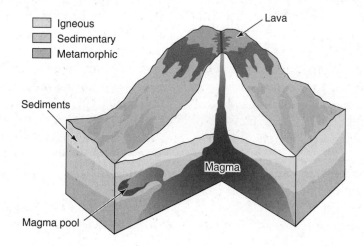

Igneous
Sedimentary
Metamorphic

Lava

Sediments

Magma

Magma pool

where the gases were once trapped. It is so light in weight that many pumice rocks will float in water. Obsidian rocks are created when lava is shot into the atmosphere and then cools so quickly that it forms a kind of glass. Slow cooling would give the lava time to form crystals like the ones in granite. Granite is an igneous rock formed by the slow cooling of magma beneath the earth's surface. The slower the magma cools, the larger the crystals.

Sedimentary Rocks Rock that is subjected to weathering from wind, water, gravity, and chemical reactions breaks down from larger pieces to smaller pieces. Gravity helps the pieces collect in low areas, such as valleys, ocean bottoms, lakes, and streams. Over thousands and even millions of years, these sediments form layer after layer of material, one on top of the other. As the layers get compressed under the weight of new layers of sediment, they eventually become rock. Typical sedimentary rocks include *sandstone*, *limestone*, *shale*, and *conglomerate*.

Sandstone rocks are made from small grains of the minerals quartz and feldspar. Sandstone typically forms in areas near ancient beaches and sand dunes. Limestone rocks are created over a long period of time in deep-water locations, such as the bottoms of oceans or deep lakes. The main mineral in limestone is calcite, which comes from the shells of sea animals and from minerals dissolved in the water. After the water evaporates, the calcite becomes sedimentary rock. Shale rock is formed from clays that are squeezed and compacted together by pressure. Conglomerates are made up of large sediments like sand and pebbles and are typically found in river beds, along a shoreline, or in valleys where pieces of rock of various sizes have gathered. The rock pieces are so large and varied in size that, in order to form rock, they are cemented together with dissolved minerals.

Metamorphic Rocks Metamorphic rocks are formed from igneous or sedimentary rocks that have been subjected to a tremendous amount of heat and pressure deep beneath Earth's surface. The pressure is so great that the rocks actually "morph" from one kind into another. By looking at some metamorphic rocks, you can actually see that the grains or crystals have been squeezed and flattened. Typical metamorphic rocks are *schist* and *gneiss*. Schist can be created from the igneous rock *basalt*, from the

sedimentary rock *shale*, or from the metamorphic rock *slate*. Slate itself is a metamorphic rock that was originally shale. Gneiss started out as the igneous rock called granite, but heat and pressure changed it. In some gneiss, you can see the flattened granite crystals.

Weathering and Erosion

Weathering and erosion are two processes by which nature works to flatten Earth's surface. *Erosion* involves the removal and transport of Earth's materials by natural agents, such as glaciers, wind, water, earthquakes, volcanoes, tornadoes, hurricanes, mud flows, and avalanches. *Weathering* is the breakup of rocks that are exposed to the atmosphere. Weathering includes both mechanical and chemical processes.

Causes Weathering and erosion take place as a result of various causes. The major ones are given here.

Freeze/Thaw Cycle Water collects in cracks in rocks. When the water freezes and expands, the crack grows larger, and eventually the rock breaks up into smaller pieces.

Wet/Dry Cycle This process occurs mostly in rocks that contain clay (i.e., shale and mudstone) when they are repeatedly wetted and dried. Clay expands when it is wet and contracts when it is dry. Eventually this process breaks up the rock into smaller parts.

Plants Plants can grow in the cracks in rocks. As their roots expand, the rocks can eventually break up. Mosses and lichens often grow on rocks and help to break the rock into smaller pieces through chemical means.

Sheet Jointing When rock layers are worn away, the pressure on the rocks below them is relieved. Those underlying rocks move up and expand. Joints or breaks in the rock occur. After some time, sheets of rock are loosened and fall away in a process called exfoliation. This typically occurs with domes of granite.

Wind Wind can cause larger rocks to break up into smaller particles. This happens in places where the winds are particularly strong and persistent. If winds are strong enough to carry particles like sand, larger rocks can be scoured.

Water Erosion takes place in water, such as streams where rocks are carried downstream, bumping into each other and making smaller particles. Strong rivers can carry very large boulders, which crash and smash, creating smaller pieces. Waves crashing on beaches cause erosion of the shoreline.

Chemical Processes When water is mixed with substances that form an acid, it can erode rock. Limestone is particularly vulnerable to this process, as seen in the formation of *caves* and *caverns*.

Glaciers Glaciers occur when the temperatures are so cold for so long that snow does not have a chance to completely melt. The snow accumulates and forms sheets of compressed snow or ice called glaciers. There are two categories of glaciers: *continental glaciers* and *mountain or alpine glaciers.*

Continental glaciers were prevalent during the Ice Ages, when massive sheets of ice covered much of the land. They spread from the poles down across the continents. At one time a large part of what is now the United States was covered by ice.

Mountain glaciers form high in the mountains and spread downward as a result of gravity.

Each type of glacier causes massive erosion through the scraping, plucking, scouring, and pushing of land material. Each produces distinctive land forms.

Continental glaciers act as sort of a conveyer belt, carrying rock and debris forward. When the glacier stops for a while and melts at its front, debris is dropped, forming a *moraine.* Continental glaciers have dammed up rivers to form lakes and have actually changed the direction of a river's flow. Alpine glaciers have distinct features as well. They scour mountains to form *cirques* (bowl-shaped depressions), block alpine rivers to form *hanging valleys,* and create the sharp peaks and characteristics shapes of *arêtes* and *horns.* Horns, like the Matterhorn in Switzerland, are pyramidal peaks that form when several cirques carve a mountain on three or more sides.

Plate Tectonics and Continental Drift

For many years, scientists discussed the theory of continental drift—the idea that the continents, as we know them now, were situated in different locations in the past and even were once combined into one large continent called Pangaea. This theory was based on fossil evidence, climatic evidence, the shapes of the continents and the way they seem to "fit together" like pieces of a gigantic jigsaw puzzle, and the similarity of geologic deposits found on different continents. But it was unclear how continental drift could take place.

Later, scientists developed the theory of *plate tectonics,* which provides the "how" for this worldwide geologic process. The theory is based on the assumption that Earth's crust is divided into several plates that move around and change over time. The places

Glaciers

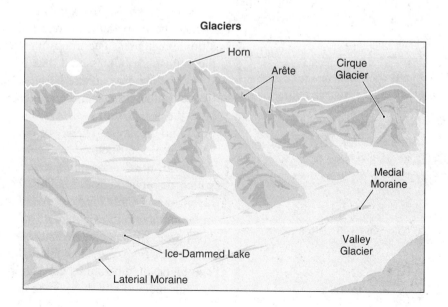

where these plates bump into each other or meet are locations of very intense volcanic and earthquake activity and mountain building. The force behind this movement is *sea-floor spreading*, a process in which molten rock beneath the seabed moves upward and forms underwater ridges, then spreads laterally away from the ridges and along the ocean floor. Sea-floor spreading creates new oceanic crust that slowly moves away from the mid-ocean ridges.

Sea-Floor Spreading

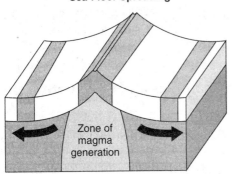

Further research indicated that sea-floor spreading is caused by very large convection currents within Earth's mantle. These currents cause the magma to rise up in the ocean floor, to spread outward, and to become *subducted* (sink back down), creating deep oceanic trenches. This process causes the plates on Earth's crust to move about Earth's surface. Where an oceanic plate *converges* with (bumps against) a continental plate, volcanic activity is prevalent and may create mountain ranges or volcanic islands like those in the Caribbean Sea. When two continental plates converge, mountains

Plate Tectonics

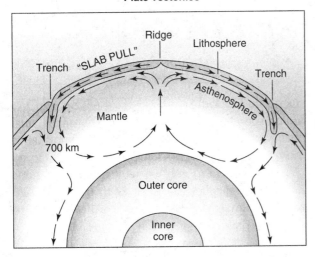

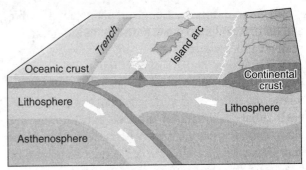

Oceanic–oceanic convergence

are formed on the land. (The Himalayas in Asia are one example.) The areas of convergence undergo tremendous stress, which is often relieved by earthquakes along fault lines.

THE ATMOSPHERE

Composition and Structure

Most of Earth's atmosphere is made up of nitrogen and oxygen.

Composition of Earth's Atmosphere

Gas	Formula	Abundance (percent by volume)
Nitrogen	N_2	78.0%
Oxygen	O_2	20.9%
Argon	Ar	<1%
Carbon dioxide	CO_2	<1%

The atmosphere is divided into various layers, determined by temperature. In the *troposphere*, the layer in which we live and in which our weather occurs, the temperature decreases dramatically as the altitude increases. The troposphere contains most of the atmosphere's mass and thus air pressure.

Above the troposphere is the *stratosphere*. Very little weather occurs in the stratosphere. Occasionally, the top portions of very large thunderstorms poke into this layer. The lower portion of the stratosphere is also influenced by the *polar jet stream* and the *subtropical jet stream*. Ozone gas molecules, which absorb ultraviolet sunlight, concentrate in this layer, causing this layer to experience a sharp rise in temperature.

The *mesosphere* contains the coldest temperatures, and the *thermosphere* has the highest temperatures. The highest temperatures come from the absorption of solar radiation by the oxygen molecules.

Layers of Earth's Atmosphere

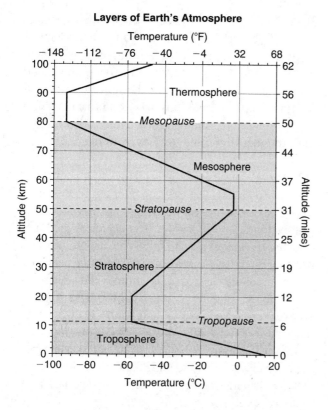

The line along which two air masses meet is called a front. Much of our severe weather takes place at or near a front. A *cold front* is the interface in the atmosphere where a cold, dry, stable air mass displaces a warm, moist, unstable subtropical air mass. *Cirrus* clouds are found before the front reaches an area and are followed by precipitation from cumulus and cumulonimbus clouds. A *warm front* is located at the place where an advancing warm, subtropical, moist air mass replaces a retreating cold, dry, polar air mass. An *occluded* front is produced when a fast-moving cold front catches and overtakes a slower-moving warm front.

Types of Cloud Clouds are classified according to their appearance and height. Clouds form when water vapor condenses onto dust particles floating in the air. Cloud formation also occurs when warm and cold air (or land) meet.

Winds and Storms

Wind is the movement of air in our atmosphere. Wind occurs because of pressure differences between one part of the atmosphere and another. The greater the pressure difference (called the *gradient*), the stronger the wind. Atmospheric pressure is measured by *barometers* and is shown on maps as *isobars*, or areas of equal pressure.

Air Masses Air masses are large bodies of air that have similar temperature and humidity. They can be cold air masses (*polar*) or warm air masses (*tropical*). If they are formed over water, they are called *maritime* (with much moisture), and if they are formed over land, they are called *continental* (dryer).

Types of Clouds

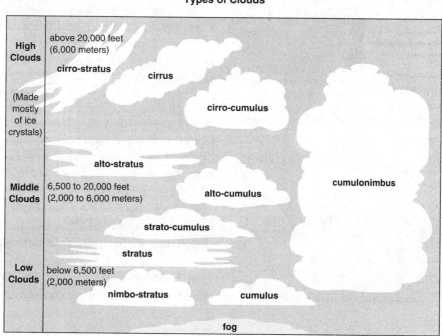

Wind Movements Wind does not necessarily travel in a straight line. In a high-pressure area (*anticyclone*), the wind flows out and around like water poured over an inverted bowl. In a low-pressure area (*cyclone*), the wind flows inward. Adding to the complexity, Earth's rotation creates another force called the *Coriolis force*. The Coriolis force deflects the air so that it moves to the right in the Northern Hemisphere and to the left in the Southern Hemisphere.

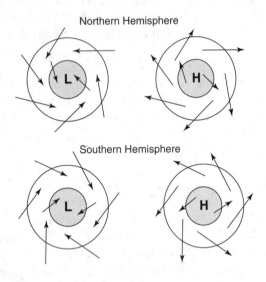

Effects of Mountains and Prevailing Winds When a prevailing wind meets up with an obstacle like a mountain range, the air rises and cools, forming clouds, and loses some moisture in the form of precipitation. This occurs on the windward side of the mountains. After the air passes over the mountain range, it descends and heats up. Thus, the leeward side of a mountain range is often dry and sometimes will become a desert.

Storms Two important kinds of storms are thunderstorms and hurricanes.

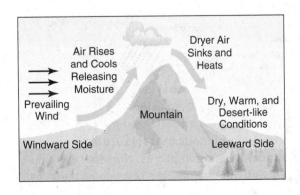

Thunderstorms These occur when moist, unstable air is lifted vertically. When lifting begins, the air starts to cool until the dew point is reached and condensation takes place, forming a cumulus cloud. For a thunderstorm to form, the uplifting needs to continue until a cumulonimbus cloud forms. Within these clouds are found strong winds, downdrafts, hail, thunder, lightning, and even tornadoes.

Hurricanes These are large cyclonic storms (areas of very low pressure) that develop over the oceans in the tropics. Other names for hurricanes are cyclones, tropical cyclones, and typhoons. The center of the hurricane has the lowest air pressure. The magnitude of the pressure gradient at Earth's surface from the outside to the inside of the hurricane determines the strength of the winds. Around the clear and calm eye are bands of thunderstorms, severe rain, and winds. Damage from hurricanes comes from strong winds, flooding, and storm surge. Storm surge is created when strong winds push ocean water upward and inland, much like a huge water wall. The water crushes and often washes away many of the obstacles in its way.

Climate

Climate is the general temperature and humidity of a region over time. For example, a climate can be tropical, desert, or polar. Climate is dependent upon several factors, including

- The latitude of the area
- What air masses generally influence the area
- Proximity of oceans and ocean currents
- Location of mountain barriers
- Direction of the prevailing winds
- Altitude

Over Earth's history, the climate has changed. Sometimes the planet was warmer and more tropical; other times it was colder and covered with ice. Today scientists have shown that Earth is warming, primarily because of atmospheric pollution caused by human activities.

Climate changes have been tracked through the chemical composition of rocks, the types of animals that lived in the past, and indicators such as tree rings.

The Greenhouse Effect The greenhouse effect is a phenomenon of Earth's atmosphere. It is similar to the heat buildup in your car on a sunny day. A huge percentage of the rays of visible light from the sun pass through Earth's atmosphere, warming the planet's surface. Part of the energy is re-radiated or bounced back toward space by long-wave infrared radiation and is absorbed by carbon dioxide molecules and water vapor in the atmosphere. This energy is reflected back to Earth's surface in the form of heat. Basically, all this is a good thing; without it, Earth's surface would be too cold to allow us to survive. Even the oceans would freeze if it were not for the greenhouse effect.

However, scientists worry that as we increase the amount of carbon dioxide in the atmosphere through industrial pollutants and other particles, the greenhouse effect can intensify. A significant and persistent warming of the planet can cause long-term climate changes. An increase in chlorofluorocarbons, nitrous oxide, and methane from human activity can aggravate this situation.

We are already seeing the effects of climate change and the warming of the earth with increased and more intense storms, intensified drought in some areas, and rising seas from melting glaciers.

OCEANOGRAPHY AND WATER

The oceans cover 71 percent of the planet. Around the edges of the continents is the *continental shelf*. Further out is the *continental slope*, then the *rise*, and then the *ocean floor*. The shape of the continental shelf often determines the magnitude of the tides along the coast.

Tides

An ocean tide refers to the cycle of rising and falling seawater caused by the gravitational pull of the moon and the sun. The moon, being closer to Earth, has the strongest effect on tides. Tides during full and new moons, when Earth, the moon, and the sun are in a nearly straight line (*spring tides*), are the highest. When the sun is more or less perpendicular to Earth and the moon, less dramatic tides are created, called *neap tides*. Neap tides take place during the first and last quarter of the moon.

The continental slope has the steepest angle. The ocean floor contains a thin layer of basaltic rock. The ocean floor also has many interesting features, such as volcanoes, fissures, trenches, and ridges.

Surface water moves around the oceans with currents that are generated by winds. The friction between the ocean and the wind causes the water to move in the direction of the wind. Some currents are temporary and others are permanent, such as the Gulf Stream on the east coast of the United States (bringing warm water northward) and the California Current along the west coast of the United States (bringing colder water southward).

The Greenhouse Effect

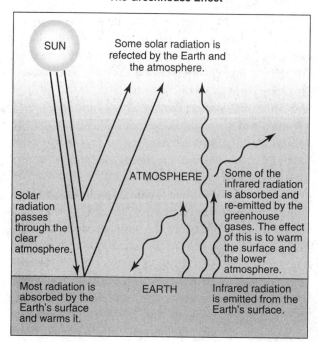

Undersea Features

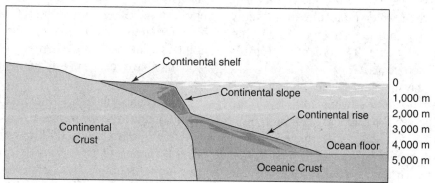

The Water Cycle
(The Hydrologic Cycle)

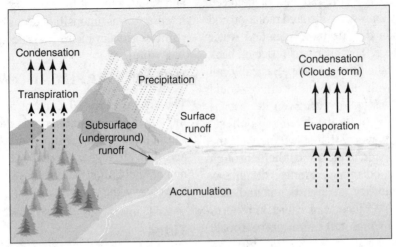

Water also moves around and up and down through the oceans as a result of temperature, density, salinity, and chemical differences between masses of water. Water that is denser (colder and/or saltier) sinks to the bottom, to be replaced by lighter water. This creates internal circulation or convection and mixing of the oceans.

The Hydrologic Cycle

The diagram above shows the cycle of water as it moves between Earth's surface and the atmosphere and back. Water is stored in reservoirs, such as the ocean, lakes or rivers, glaciers, and groundwater. From these reservoirs, the water *evaporates* into the atmosphere. When conditions permit, the water in the atmosphere *condenses* and returns to Earth's surface in the form of *precipitation* like rain or snow. Water can also flow downward over the land in the form of runoff from rivers or groundwater, eventually ending up back in the ocean, where it, too, evaporates, condenses, and precipitates back to Earth. The process of *sublimation* can occur when water molecules from snow or ice pass directly into the atmosphere without becoming a liquid. Some 97 percent of Earth's water is held in the oceans.

Water also reaches the atmosphere through a process called *transpiration*—water loss from the stomata of plants (see the relevant section in biology). It is difficult to tell the difference between evaporation and transpiration. As a result, the word *evapotranspiration* is used to describe the entire process.

ASTRONOMY
The Solar System

Earth is the third planet from the sun and, so far as we know, the only planet that currently supports life as we know it. Earth is part of the *solar system*, a group of planets that revolve around the sun. The planets in order from the nearest to the sun to the farthest are *Mercury, Venus, Earth, Mars, Jupiter, Saturn, Uranus,* and *Neptune.* For the ASVAB, it is good to memorize this order. Pluto used to be considered a planet, but in 2006, it was demoted because it was determined not to be a planet but rather a member of a category of dwarf planets called "plutoids." So now there are only eight planets recognized by scientists.

There is a gap between the orbits of Mars and Jupiter that is filled by tens of thousands of minor planets, known as *asteroids.* A few asteroids are several hundred kilometers in diameter, but most are just a few meters across. Scientists speculate that these asteroids are part of a planet that failed to form because of Jupiter's strong gravitational pull. Some asteroids have a very elliptical orbit, taking them outside the solar system for part of their orbit. Some asteroids have entered Earth's atmosphere. Generally they burn up before hitting the planet's surface, but a few have collided with the Earth. These are called *meteorites.* You may have seen some fly through the atmosphere as bright *meteors.*

Some planets have *satellites.* Earth has one satellite, called the *moon.* Jupiter has the most satellites or moons; 79 are currently known. Saturn also has several moons and rings made up of millions of tiny particles. Mars has conditions that are most similar to Earth's, but Mars is not habitable because

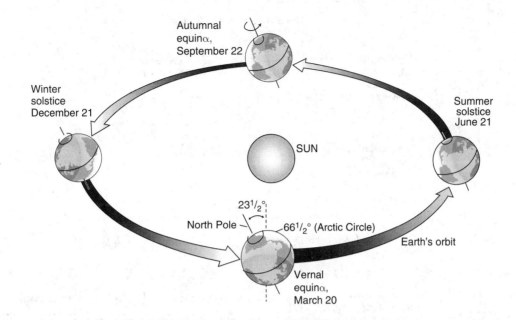

Winter
solstice
December 21

Autumnal
equinα,
September 22

SUN

Summer
solstice
June 21

North Pole

23½°

66½° (Arctic Circle)

Earth's orbit

Vernal
equinα,
March 20

of its cold temperature and lack of an atmosphere that would support humans. There is great interest in exploring Mars because it can be reached from Earth in about 6 1/2 months and may contain microbial life. Several countries have landed rovers on Mars to explore it composition and terrain. The United States even has a small helicopter device that can explore the planet at a larger range and from above. Periodic visitors to the inner solar system are *comets*, which are believed to be made up of ice and dust particles. Comets have their own orbits. Comets can be recognized by their bright core and tail and can usually be observed for several days.

Beyond Neptune, at the far reaches of the solar system, is the Kuiper Belt. This is a disc of millions of icy objects that were probably meant to be a planet but never made it.

Even further out is the Oort Cloud, which is a much more distant region of icy, comet-like bodies that surrounds the solar system, including the Kuiper Belt. Both the Oort Cloud and the Kuiper Belt are thought to be sources of comets.

Motion of the Earth

Seasons   The combination of Earth's revolution around the sun plus the tilt of Earth on its axis causes us to have *seasons*. Earth revolves around the sun in an elliptical fashion, as shown above. In January, Earth is closest to the sun, at a point called the *perihelion*. In July, it is farthest from the sun, at a point called the *aphelion*. On average, Earth is 93 million miles from the sun. The distance from sun to Earth is called an *astronomical*

unit. The distance light travels in one year is called a *light-year*.

When it is summer in the Northern Hemisphere, Earth's axis points toward the sun at an angle of 23.5°, allowing the sun's rays to hit the northern part of the planet in a more direct fashion and thus raise temperatures in that part of the globe. When it is winter in the Northern Hemisphere, the axis points away from the sun, causing the rays to come in at more of an angle and thus lower temperatures in that part of the globe. The opposite happens in the Southern Hemisphere.

Day and Night   Day and night are caused by Earth's rotation on its axis. One rotation occurs every 24 hours. Days are also longer in the summer and shorter in the winter because of the tilt of Earth's axis. At the *vernal* and *autumnal* equinoxes, days and nights are of equal length.

Eclipses   The diagram on the next page shows the lineup of the sun, Earth, and moon that results in a *lunar eclipse*. A *solar eclipse* occurs when the moon is in between the sun and Earth.

Sun and Stars

Our Sun   The *corona* is the outer layer of the sun's atmosphere. The corona extends for millions of miles, and its temperatures reach 1 million °C. The *chromosphere*, a reddish layer, is an area of rising temperatures. It appears red because hydrogen atoms are in an excited state and emit radiation near the red

The Structure of the Sun

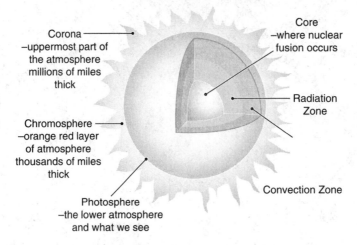

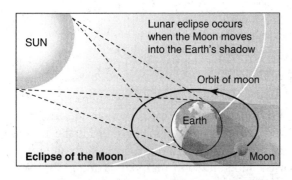

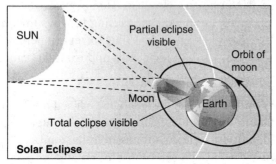

part of the visible spectrum. The chromosphere can be seen during solar eclipses. The next layer in is the *photosphere*, the part we can see, as it emits light in the visible range of wavelength. Next is the *radiative zone*, which emits radiation, and after that is the *convective zone*. The *core* is where nuclear fusion takes place. The temperature is about 15 million °C.

Our sun has *sunspots*, which are cool, dark patches on the surface. Individual sunspots last for one or two weeks, but they generally come in 11-year cycles. *Solar flares* are magnetic storms that appear to be an eruption of high-energy particles and gases. Solar flares are ejected thousands of miles from the surface of the sun. The energy from these flares can sometimes disrupt communications, and these flares are potentially hazardous to satellites.

Sometimes the sun's corona will emit a stream of ions called a *solar wind*. These ions can disrupt the working of artificial satellites, spacecraft, and radio communications.

Stars Stars are categorized by their composition, size, and temperature. Most stars, like our sun, are fueled by *nuclear fusion*, which converts hydrogen into helium. These stars are stable for most of their lives. When stars begin to die, they change into "red giants" or supergiants. Their hydrogen supply runs out, their core contracts, and eventually they explode into a *supernova*, later becoming a "white dwarf," a neutron star, or a black hole. Small stars like our sun will grow dim and will eventually become cold and dark "black dwarfs."

Galaxies Our solar system is part of a galaxy, a conglomeration of many stars. Our galaxy is called the Milky Way. The closest galaxy to ours is the Andromeda Galaxy. A galaxy is basically a huge group of stars and other celestial objects. Galaxies have between 100,000 and 3,000,000,000,000 (3 trillion) stars.

Chapter 8

Arithmetic Reasoning 1: The Basics

YOUR GOALS FOR THIS CHAPTER:

- Find out what arithmetic topics you need to know for the ASVAB.

- Review techniques and strategies for solving arithmetic problems.

- Study sample arithmetic problems and their step-by-step solutions.

- Measure your test readiness by taking an arithmetic quiz.

INTRODUCTION: ASVAB ARITHMETIC REASONING QUESTIONS

The ASVAB Arithmetic Reasoning test measures your ability to solve the kinds of arithmetic problems that you encounter every day at home or on the job. The questions on the test are word problems. That is, each one presents a real-life situation with a problem that must be solved using an arithmetic operation, such as addition, subtraction, multiplication, or addition. Arithmetic Reasoning problems also involve other arithmetic concepts, such as fractions, decimals, percents, exponents, and square roots.

It is important to do well on the Arithmetic Reasoning test because it is one of the four ASVAB tests that are used to calculate the AFQT—your military entrance score. That's why it pays to spend time reviewing topics in basic arithmetic and tackling plenty of sample ASVAB Arithmetic Reasoning questions.

The following pages offer a quick but important overview of the basic arithmetic you need to know to score well on the ASVAB. Make sure that you carefully review and test yourself on every topic covered in this section. Also make sure that you

---

ASVAB Arithmetic Reasoning Test

Number of Questions:

- Paper-and-pencil ASVAB: 30
- CAT-ASVAB: 15

Time Limit:

- Paper-and-pencil ASVAB: 36 minutes
- CAT-ASVAB: 55 minutes

Suggested Review Plan for Arithmetic

- Work through the arithmetic review at your own pace.
- Pay careful attention to the examples provided.
- After you complete the arithmetic review, take the arithmetic quiz.
- Score your quiz.
- Study any problems you answered incorrectly. Reread the corresponding review section, then try the problem again. Work until you understand how to solve the problem.

Remember that you probably won't get these exact problems on the ASVAB, but knowing how to solve them will help you do your best on the actual test.

learn how to use all of the problem-solving methods presented in the examples. There is much more to basic math, but if you master the core information presented here, you will be able to answer ASVAB Arithmetic Reasoning questions with relative ease. Note that your ASVAB arithmetic review should also include the next chapter of this book, which covers word problems.

ARITHMETIC REVIEW

Topics	
Place value	Decimals
Arithmetic operations	Percent
Positive and negative numbers	Finding a percent of a number
Multiplying and dividing by zero	Finding the percent increase or decrease
Factors	Exponents and square roots
Multiples	Scientific notation
Fractions	Mean, median, and mode
Mixed numbers	Graphs
Working with fractions and mixed numbers	Units of measure

Place Value

Each digit in a number occupies a particular place. In the number below, the digit 4 is in the hundreds and tens place.

32,768,445.123

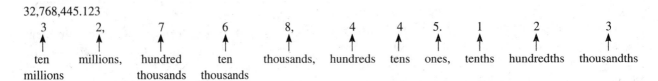

3	2,	7	6	8,	4	4	5.	1	2	3
ten millions	millions,	hundred thousands	ten thousands	thousands,	hundreds	tens	ones,	tenths	hundredths	thousandths

The *value* of a digit in a number is that number multiplied by its place. In the number above, 4 is in the tens place. That indicates a value of $4 \times 10 = 40$, or 4 tens.

Example

23,456,789.247

2 (the first 2) is in the ten millions place

3 is in the millions place

4 (the first 4) is in the hundred thousands place

5 is in the ten thousands place

6 is in the thousands place

7 (the first 7) is in the hundreds place

8 is in the tens place

9 is in the ones place

2 is in the tenths place

4 is in the hundredths place

7 is in the thousandths place

ARITHMETIC OPERATIONS

The basic operations of arithmetic are addition, subtraction, multiplication, and division.

Addition There are two ways to show addition:

$$234 \quad \text{or} \quad 234 + 123 =$$
$$\underline{+\ 123}$$

In either case, you need to add the two numbers to get the sum.

Subtraction There are also two ways to show subtraction:

$$234 \quad \text{or} \quad 234 - 123 =$$
$$\underline{-\ 123}$$

Subtraction of Whole Numbers with Renaming Sometimes you have to "rename" numbers in order to complete subtraction.

$$731$$
$$\underline{-465}$$

To subtract the 5 from the 1 in the ones place, "rename" the "31" in the top number as "2 tens and 11 ones." Write "2" above the 3 in the tens place and "11" above the 1 in the ones place. Subtract 5 from 11 to get 6.

$$
\begin{array}{r}
{}^{2\ 11}\\
7\cancel{3}\cancel{1}\\
-\ 465\\
\hline
6
\end{array}
$$

Next, you need to subtract the 6 from the (new) 2 in the tens place, so rename "72" as "6 hundreds and 12 tens." Write "6" above the 7 in the hundreds place and "12" above the 3 in the tens place. Subtract 6 from 12 to get 6. Then subtract 4 from 6 in the hundreds place to get 2, completing the subtraction.

$$
\begin{array}{r}
{}^{6\ 12\ 11}\\
7\cancel{3}\cancel{1}\\
-\ 465\\
\hline
266
\end{array}
$$

Multiplication Multiplication is a bit more complicated in that it can be shown in various ways.

$$10 \times 3 = 30$$
$$10(3) = 30$$
$$(10)3 = 30$$
$$10 \cdot 3 = 30$$
$$(10)(3) = 30$$

Division Division can also be shown in more than one way.

$$12 \div 3 = 4$$
$$12/3 = 4$$
$$\frac{12}{3} = 4$$

Symbols
= is equal to
≠ is not equal to
< is less than
≤ is less than or equal to
> is greater than
≥ is greater than or equal to
‖ is parallel to
≅ is approximately equal to
≈ is approximately
⊥ is perpendicular to
|−x| = x absolute value
± plus or minus
% percent
a:b ratio of a to b or $\frac{a}{b}$

Order of Operations If addition, subtraction, multiplication, and division are all in the same mathematical statement, the order of operations (what you do first, second, third, and so on) is

1. Perform operations shown within Parentheses
2. Attend to Exponents and Square Roots
3. Perform Multiplication or Division from left to right
4. Perform Addition or Subtraction from left to right

You can use the sentence "Please Excuse My Dear Aunt Sally" to help you remember the order of operations.

> **HINT** *Order is important!* You almost certainly will get an ASVAB math item that will require you to know order of operations rules.

Examples

$$
\begin{aligned}
2 + 3 \times 4 &= 2 + 12\\
&= 14
\end{aligned}
$$

$$
\begin{aligned}
3^2 + 4(5 - 2) &= 9 + 4(5 - 2)\\
&= 9 + 4(3)\\
&= 9 + 12\\
&= 21
\end{aligned}
$$

$$
\begin{aligned}
5(10 + 2) + 2 + 3 &= 5(12) + 2 + 3\\
&= 60 + 5\\
&= 65
\end{aligned}
$$

$$
\begin{aligned}
3^3(8 + 3) \times (15 \div 5) + (9 \times 2) - (298 + 300)\\
= 3^3(11) \times (3) + (18) - (598)\\
= 27(11) \times (3) + (18) - (598)\\
= 297 \times 3 + 18 - 598\\
= 891 + 18 - 598\\
= 909 - 598\\
= 311
\end{aligned}
$$

> **REMINDER:**
> **Order of Operations**
> 1. Simplify groupings inside parentheses, brackets and braces first. Work with the innermost pair, moving outward.
> 2. Simplify the exponents.
> 3. Do the multiplication and division in order from left to right.
> 4. Do the addition and subtraction in order from left to right.

Rounding Often you will be asked to round to the nearest 10, 100, or 1,000 or to the nearest tenth, hundredth, or thousandth. A good way to do this is to underline the number in the place to which you are rounding. Identify the number immediately to the right. If it is 5 or greater, round up by 1. If it is 4 or less, leave it as is. Change everything to the right of that to zeros.

Examples

Round 123,756 to the nearest thousand.
Underline the number in the thousands place.

123,756

Look at the number in the place immediately to the right.

7 is greater than 5, so round the underlined number to 4 and change the rest to zeros.

Result = 124,000

Round 123,456 to the nearest thousand.
Underline the number in the thousands place.

123,456

Look at the number in the place immediately to the right.

4 is less than 5, so keep the number in the thousands place as it is and change all numbers to the right to zeros.

Result = 123,000

Round 123.456 to the nearest hundredth.
Underline the number in the hundredths place.

123.456

6 is greater than 5, so change the 5 to 6 and make all numbers to the right zeros.

Result = 123.46

Positive and Negative Numbers

Positive numbers are numbers that are greater than 0, such as +6. Negative numbers are numbers that are less than 0, such as −4. (See number line at bottom of page.)

The *absolute value* of a number is the distance of the number from 0 on a number line. For example, the absolute value of +6 is 6. The absolute value of −4 is 4.

Operations with Positive and Negative Numbers

For the ASVAB, you'll need to know how to add, subtract, multiply, and divide with positive and negative numbers.

Adding Positive and Negative Numbers To *add numbers with the same sign*, add their absolute values. The sum will have the same sign as the numbers you added.

Example

Add: −4
$\underline{-3}$
−7

To add numbers with different signs, first find their absolute values. Subtract the lesser absolute value from the greater absolute value. Give the sum the same sign as the number with the greater absolute value.

Examples

Add: +4
$\underline{-3}$
+1

Add: +4
$\underline{-6}$
−2

> **HINT** *Know negative numbers!* Be sure to practice adding, subtracting, multiplying, and dividing using negative numbers.

Subtracting Positive and Negative Numbers Look again at the number line. (See number line at bottom of page.)

- When you subtract a positive number, you move *left* on the number line a distance equal to the absolute value of the number being subtracted.
- When you subtract a negative number, you move *right* on the number line a distance equal to the absolute value of the number being subtracted.

Examples

Subtract positive numbers (move *left* on the number line):

+7
$\underline{-(+12)}$
−5

−7
$\underline{-(+12)}$
−19

Subtract negative numbers (move *right* on the number line):

$$
\begin{array}{r}
+7 \\
-(-12) \\
\hline
+19
\end{array}
$$

$$
\begin{array}{r}
-7 \\
-(-12) \\
\hline
+5
\end{array}
$$

Note that when you subtract negative numbers, all you really need to do is change the sign of the number being subtracted from negative to positive, and then add.

Absolute Value A number regardless of its sign (+ or −) is called the *absolute value*. The absolute value of a number is shown as the number placed between two vertical parallel lines; for example, |−4| or |4|. In each of these cases, the absolute value of the number is 4. When using absolute values, just ignore the sign.

Examples

$|{-}x| = x$
$|{-}5 - 6| = |{-}11| = 11$
$8 + |{-}9| = 17$

Multiplying and Dividing Negative Numbers These operations are easy if you use just one trick. To multiply or divide negative numbers, first treat all the numbers as positive and perform the operation is normally.

- If there is an *odd number of negative signs*, the answer will be negative.
- If there is an *even number of negative signs*, the number will be positive.

Examples

$(-5)(+6)(+2) = -60$

This expression has 1 negative sign. Since 1 is an odd number, the result is negative.

$(-5)(-6)(+2) = +60$

This expression has 2 negative signs. Since 2 is an even number, the result is positive.

$\dfrac{-88}{-11} = +8$

This expression has 2 negative signs. Since 2 is an even number, the result is positive.

$\dfrac{-88}{+11} = -8$

This expression has 1 negative sign. Since 1 is an odd number, the result is negative.

Subtracting Numbers within Parentheses If there is a minus sign in front of parentheses, change the sign of all the numbers within the parentheses and then add.

Examples

$11 - (+3 - 5 + 2 - 8)$

changes to

$$
\begin{aligned}
11 + (-3 + 5 - 2 + 8) \\
= 11 + (-5 + 13) \\
= 11 + (+8) \\
= 19
\end{aligned}
$$

$$
\begin{aligned}
20 - (-3 + 5 - 2 + 8) \\
= 20 + (+3 - 5 + 2 - 8) \\
= 20 + (+5 - 13) \\
= 20 + -8 \\
= 12
\end{aligned}
$$

Multiplying and Dividing by Zero Remember the following rules:

- Any number multiplied by zero = zero.
- Zero divided by any number = zero.
- Dividing by zero is considered to be "undefined."

Examples

$45 \times 0 = 0$
$1{,}232{,}456 \times 0 = 0$
$10.45399 \times 0 = 0$
$0 \div 5 = 0$
$0 \div 500 = 0$
$7 \div 0$ is undefined
$\dfrac{32}{0}$ is undefined
$\dfrac{89}{0}$ is undefined

Factors

Factors are whole numbers that are multiplied together to create another number.

Examples

2 × 4 = 8 (2 and 4 are factors of 8.)
1 × 8 = 8 (1 and 8 are factors of 8.)
So 1, 2, 4, and 8 are all factors of 8.
1 × 32 = 32 (1 and 32 are factors of 32.)
2 × 16 = 32 (2 and 16 are factors of 32.)
8 × 4 = 32 (8 and 4 are factors of 32.)
So 1, 2, 4, 8, 16, and 32 are all factors of 32.

Types of Factors Knowing about factors will be very useful when solving equations.

Common Factors Common factors are factors that are shared by two or more numbers. As shown above, 8 and 32 share the common factors 1, 2, 4, and 8.

Multiples

Multiples of a number are found by multiplying the number by 1, 2, 3, 4, 5, 6, 7, 8, and so on. So multiples of 5 are 10, 15, 20, 25, 30, 35, 40, and so on.

Common Multiples Common multiples are multiples that are shared by two or more numbers.

Least Common Multiple The least common multiple is the smallest multiple that two numbers share. For example, 6, 12 and 18 are all common multiples of 2 and 3 because they are all evenly divisible by both 2 and 3. 6 is the least common multiple because it is the smallest number that 2 and 3 both divide evenly.

Fractions

Fractions have two numbers, a numerator and a denominator.

$$\frac{3}{6} = \frac{\text{numerator}}{\text{denominator}}$$

The denominator tells how many parts something is divided into. The numerator tells how many of those parts you have. The fraction as a whole tells the *proportion* of the parts you have to the parts there are in all.

So if a pie is divided into 8 pieces and you take 2 of them, the fraction telling the proportion of the

pie you have is $\frac{2}{8}$. The denominator (8) tells how many pieces of pie there are in all. The numerator (2) tells how many of those pieces you have.

$$\frac{2}{8} \text{ or a proportion of 2 to 8.}$$

Fractions can be positive or negative. A negative fraction is written like this:

$$-\frac{2}{8}$$

Types of Fractions For the ASVAB, you'll need to know about proper fractions, improper fractions, and mixed numbers.

Proper Fractions Fractions representing amounts smaller than 1 are called *proper fractions*. Fractions smaller than 1 have numerators that are smaller than the denominators.

Examples

$$\frac{1}{3} \qquad \frac{3}{7} \qquad \frac{5}{8} \qquad \frac{9}{11} \qquad \frac{6}{32} \qquad \frac{99}{100}$$

Improper Fractions *Improper fractions* are greater than 1. The numerator is larger than the denominator.

Examples

$$\frac{22}{3} \qquad \frac{3}{2} \qquad \frac{12}{5} \qquad \frac{17}{8} \qquad \frac{66}{3} \qquad \frac{11}{9}$$

Mixed Numbers

Expressions that include both whole numbers and fractions are called *mixed numbers*.

Examples

$$2\frac{7}{8} \qquad 3\frac{1}{2} \qquad 9\frac{2}{3} \qquad 27\frac{3}{4}$$

Renaming an Improper Fraction as a Mixed Number To rename an improper fraction as a mixed number, simply divide the numerator by the denominator.

Examples

$$\frac{22}{3} = 22 \div 3 = 7\frac{1}{3}$$

$$\frac{3}{2} = 3 \div 2 = 1\frac{1}{2}$$

$$\frac{12}{5} = 12 \div 5 = 2\frac{2}{5}$$

$$\frac{17}{8} = 17 \div 8 = 2\frac{1}{8}$$

HINT *You'll see these for sure!* On the ASVAB, you're sure to see items that require you to manipulate fractions and mixed numbers. Don't despair! Be sure to understand the procedures thoroughly and you should have no problem.

Renaming a Mixed Number as a Fraction To rename a mixed number as an improper fraction, simply multiply the whole number by the denominator and add the numerator. Put that number over the denominator.

Examples

$7\frac{1}{3}$

$3 \times 7 + 1 = 22$

Put 22 over the denominator $= \frac{22}{3}$

$2\frac{2}{5}$

$2 \times 5 + 2 = 12$

Put 12 over the denominator $= \frac{12}{5}$

Working with Fractions and Mixed Numbers

ASVAB problems are likely to require you to find equivalent fractions, reduce fractions to lowest terms, and add, subtract, multiply, and divide fractions and mixed numbers.

Finding Equivalent Fractions Two fractions are said to be *equivalent* (the same in value) if they use different numbers but represent the same proportion. For example, the following fractions are equivalent:

$$\frac{1}{2} \qquad \frac{4}{8} \qquad \frac{3}{6} \qquad \frac{5}{10}$$

To change a fraction into an equivalent fraction, multiply or divide the numerator and denominator by the same number.

Examples

$\begin{array}{l} 2 \times 3 = 6 \\ \overline{4 \times 3 = 12} \end{array}$ so $\frac{2}{4}$ is equivalent to $\frac{6}{12}$.

$\begin{array}{l} 9 \div 3 = 3 \\ \overline{12 \div 3 = 4} \end{array}$ so $\frac{9}{12}$ is equivalent to $\frac{1}{2}$.

Reducing Fractions to Lowest Terms Fractions are commonly shown in their lowest terms, that is, the smallest numbers that still represent the original proportion. When a fraction is not in its lowest terms, you can reduce it to its lowest terms by dividing the numerator and denominator by the largest number that will divide into both evenly.

Examples

$\frac{40}{80}$ can be divided evenly by 40, giving a fraction of $\frac{1}{2}$

$\frac{9}{15}$ can be divided evenly by 3, giving a fraction of $\frac{3}{5}$

$\frac{21}{28}$ can be divided evenly by 7, giving a fraction of $\frac{3}{4}$

Adding and Subtracting Fractions To add and subtract fractions with *like denominators*, simply add or subtract the numerators. The result is often shown in lowest terms.

Examples

$$\frac{3}{10} + \frac{1}{10} = \frac{4}{10} = \frac{2}{5} \qquad \frac{9}{10} - \frac{1}{10} = \frac{8}{10} = \frac{4}{5}$$

To add and subtract fractions with *unlike denominators*, you first need to find equivalent fractions that all have the same denominator. To do this, you need to find the *least common denominator (LCD)*, the least common multiple of the denominators of all the fractions. Use the LCD to create new fractions equivalent to the original ones. Then add the new numerators.

Examples

$$\frac{6}{8} + \frac{1}{2} =$$

The least common denominator is 8, so $\frac{6}{8}$ does not need to be changed. However, $\frac{1}{2}$ is changed to the equivalent fraction $\frac{4}{8}$.

$$\frac{6}{8} + \frac{4}{8} = \frac{10}{8} = 1\frac{2}{8} = 1\frac{1}{4} \text{ (reduced to lowest terms)}$$

$$\frac{1}{3} + \frac{1}{2} + \frac{7}{8} =$$

Find the least common denominator. The least common multiple of 8, 3, and 2 is 24, so 24 is the least common denominator. Use the LCD to create new fractions equivalent to the original ones:

$$\frac{1}{3}+\frac{1}{2}+\frac{7}{8}=\frac{8}{24}+\frac{12}{24}+\frac{21}{24}$$
$$=\frac{41}{24}$$
$$=1\frac{17}{24}$$

Use the same approach to add and subtract positive and negative fractions. Refer to the section earlier in this math review on adding and subtracting positive and negative numbers.

Adding and Subtracting Mixed Numbers Add and subtract mixed numbers by following the same rules previously outlined. Change the fractions to equivalent fractions using the least common denominator. Add or subtract the fractions. Add or subtract the whole numbers. If the fractions add up to more than a whole number, add that to the whole numbers.

Example (Addition)

$$3\frac{3}{4}$$
$$+2\frac{1}{2}$$

changes to

$$3\frac{3}{4}$$
$$+2\frac{2}{2}$$

$$\frac{3}{4}+\frac{2}{4}=\frac{5}{4}=1\frac{1}{4}$$
$$3 \text{ and } 2 = 5$$
$$5+1\frac{1}{4}=6\frac{1}{4}$$

Examples (Subtraction)

$$3\frac{3}{4}$$
$$-2\frac{1}{2}$$

changes to

$$3\frac{3}{4}$$
$$-2\frac{2}{4}$$

$$\frac{3}{4}-\frac{2}{4}=\frac{1}{4}$$
$$3 - 2 = 1$$
$$1+\frac{1}{4}=1\frac{1}{4}$$

$$3\frac{3}{4}$$
$$-5\frac{1}{2}$$

changes to

$$3\frac{3}{4}$$
$$-5\frac{2}{4}$$

Caution: Remember that subtraction means moving *left* on the number line. Start at $3\frac{3}{4}$ and move left $5\frac{2}{4}$ on the number line. The correct answer is $-1\frac{3}{4}$.

HINT *Subtracting mixed numbers: A special case.* Sometimes when you have mixed numbers, the fraction of the first mixed number is smaller than the fraction of the second mixed number. You need to rename before subtracting.

Example:

$$6\frac{1}{6}$$
$$-4\frac{3}{6}$$

Since you can't subtract $\frac{3}{6}$ from $\frac{1}{6}$ because $\frac{3}{6}$ is larger, you need to rename the whole number 6 as a 5 and rename the fraction as $\frac{7}{6}$. Now you can subtract:

$$5\frac{7}{6}$$
$$-4\frac{3}{6}$$

Subtract $\frac{3}{6}$ from $\frac{7}{6}$ to get $\frac{4}{6}$ and 4 from 5 to get 1. So the answer is $1\frac{4}{6}$. Since the fraction is not in lowest terms, change it to $\frac{2}{3}$, so the final answer is $1\frac{2}{3}$.

Multiplying Fractions To multiply fractions, just multiply the numerators and multiply the denominators. Reduce the resulting fraction to lowest terms.

Examples

$$\frac{5}{12}\times\frac{2}{3}=$$

Multiply the numerators: $5 \times 2 = 10$
Multiply the denominators: $12 \times 3 = 36$

Result: $\frac{10}{36}=\frac{5}{18}$

$$\frac{4}{5} \times \frac{7}{8} =$$

Multiply the numerators: $4 \times 7 = 28$
Multiply the denominators: $5 \times 8 = 40$

Result: $\dfrac{28}{40} = \dfrac{7}{10}$

HINT *Simplify, simplify.* Sometimes it is possible to simplify terms before performing any math operation. Take the same example that we have already completed:

$$\frac{4}{5} \times \frac{7}{8}$$

Find a number (or greatest common factor) that divides into one of the numerators and one of the denominators. In this example, the number 4 divides into 4 and into 8.

$$\frac{4}{5} \times \frac{7}{8} = \frac{{}^{1}\cancel{4}}{5} \times \frac{7}{\cancel{8}^{2}}$$

To finish the problem, multiply 1×7 and 5×2, making the final answer $\dfrac{7}{10}$.

Dividing Fractions To divide fractions, invert the second fraction and multiply. Reduce the result to lowest terms.

Examples

$$\frac{1}{2} \div \frac{1}{3} = \frac{1}{2} \times \frac{3}{1}$$

$$\frac{3}{2} = 1\frac{1}{2}$$

$$\frac{3}{4} \div -\left(\frac{1}{2}\right) = \frac{3}{4} \times -\left(\frac{2}{1}\right)$$

$$= -\left(\frac{6}{4}\right) = -1\frac{2}{4} = -1\frac{1}{2}$$

HINT *Dividing fractions.* Sometimes you might see division of fractions in this format:

$$\frac{\frac{1}{2}}{\frac{1}{3}}$$

Treat it the same as

$$\frac{1}{2} \div \frac{1}{3}$$

Multiplying Mixed Numbers To multiply mixed numbers, change each mixed number to a fraction and then multiply as usual. Reduce to lowest terms.

Examples

$$1\frac{1}{4} \times 2\frac{1}{2} = \frac{5}{4} \times \frac{5}{2}$$

$$= \frac{25}{8} = 3\frac{1}{8}$$

$$3\frac{2}{3} \times 9\frac{1}{2} = \frac{11}{3} \times \frac{19}{2}$$

$$= \frac{209}{6} = 34\frac{5}{6}$$

Dividing Mixed Numbers To divide mixed numbers, rename the mixed numbers as fractions and then follow the rule for dividing fractions: invert the second fraction and multiply.

Example

$$3\frac{1}{2} \div 1\frac{1}{4} = \frac{7}{2} \div \frac{5}{4}$$

$$= \frac{7}{2} \times \frac{4}{5}$$

$$= \frac{28}{10}$$

$$= 2\frac{8}{10}$$

Simplify the fraction.

$$2\frac{4}{5}$$

Decimals

A decimal is a number with one or more digits to the right of the decimal point. 0.862 and 3.12 are decimals. (Note that a zero is shown to the left of the decimal point when the decimal is between 0 and 1.)

Working with Decimals

On the ASVAB, you'll probably need to know how to add, subtract, multiply, and divide decimals. You'll also need to know how to change decimals to fractions.

Adding and Subtracting Decimals To add or subtract decimals, line up the decimal points one above the other. Then add or subtract as you would normally. Place a decimal point in the answer beneath the other decimal points.

Examples (Addition)

```
     14.50
    200.32
 +1,245.89
  1,460.71
```

```
     14.50
    200.047
     48.0075
   +10.6
```

Add zeros in the blank decimal places to make this problem easier to tackle.

```
    14.5000
   200.0470
    48.0075
   +10.6000
   273.1545
```

Examples (Subtraction)

```
   475.89
   −62.45
   413.44
```

```
        7 14
   475.8̸4̸      Since 9 is larger than 4, rename
   −62.69       84 as 7 tens and 14 ones.
   413.15
```

Multiplying Decimals To multiply decimals, follow the usual multiplication rules. Count the number of places to the right of the decimal point in each factor. Add the numbers of places. Put that many decimal places in the answer.

Examples

12.43	(two decimal places)
× 2.4	(one decimal place)
29.832	(total of three decimal places in the answer)
6.624	(three decimal places)
× 1.22	(two decimal places)
8.08128	(total of 5 decimal places)

Dividing Decimals To divide decimals, follow the usual division rules. If the divisor (the number you are dividing by) has decimals, move the decimal point to the right as many places as necessary to make the divisor a whole number. Then move the decimal point of the dividend (the number you are dividing) that same number of places to the right. (You may

have to add some zeros to the dividend to make this work.) Put the decimal point in the answer directly above the decimal point in the dividend.

Examples

$$12.76\overline{)58.696} = 1276\overline{)5{,}869.6}^{\,4.6}$$

Note that the decimal point is moved two places to the right in each term. The decimal point in the answer is directly above the decimal point in the dividend.

$$1.25\overline{)50} = 125\overline{)5{,}000}^{\,40}$$

In this example, the decimal point is moved two places to the right in 1.25 to make 125. The dividend 50 can also be expressed as 50.00 (adding zeros), and moving the decimal point the same number of places to the right makes 5,000. No decimal point needs to be shown in the answer because the answer is a whole number.

Changing Decimals to Fractions Read the decimal and then write the fraction. Reduce the fraction to its lowest terms.

Examples

$0.5 =$ five tenths or five over ten; $\dfrac{5}{10} = \dfrac{1}{2}$

$0.66 = 66$ hundredths or 66 over 100; $\dfrac{66}{100} = \dfrac{33}{50}$

$0.75 = 75$ hundredths or 75 over 100; $\dfrac{75}{100} = \dfrac{3}{4}$

$0.006 = 6$ thousandths or 6 over 1000; $\dfrac{6}{1000}$
$= \dfrac{3}{500}$

$0.0006 = 6$ ten thousandths or 6 over 10,000;
$\dfrac{6}{10{,}000} = \dfrac{3}{5{,}000}$

Changing Fractions to Decimals To change a fraction to a decimal, divide the numerator by the denominator.

Examples

$\dfrac{1}{2} = 1 \div 2 = 0.5$

$\dfrac{3}{4} = 3 \div 4 = 0.75$

$\dfrac{6}{1{,}000} = 6 \div 1{,}000 = 0.006$

$\frac{2}{3} = 2 \div 3 = 0.6\overline{6}$ (The bar over the final 6 indicates that this is a *repeating decimal*. That means that you could keep on dividing forever and always have the same remainder.)

Percent

Percent means "out of 100" or "per hundred." For example, "70%" is read as "70 percent," meaning 70 out of 100 equal parts. Percents are useful ways to show parts of a whole. They can also be easily changed into decimals or fractions.

Working with Percent

On the ASVAB, you'll probably need to know how to change percents to decimals and fractions, and how to solve problems involving percent.

Changing Percents to Decimals Percent means "per 100." So 70% means "70 per 100," which is 70/100 or 70 ÷ 100, which is 0.70. So to change percents to decimals, delete the percent sign and place the decimal point two places to the left. You may need to add zeros.

Examples

67% = 0.67
6% = 0.06 (A zero was added to the left of the 6.)
187% = 1.87
0.14% = 0.0014 (Two zeros were added to the left of the 14.)

Changing Decimals to Percents To change decimals to percents, merely move the decimal point two places to the right and add a percent sign. (You may need to add a zero on the right.)

Examples

0.67 = 67%
0.4 = 40% (A zero was added to the right of the 4.)
1.87 = 187%
28.886 = 2888.6%
0.0014 = 0.14%

Changing Percents to Fractions A percent is some number over (divided by) 100. So every percent is also a fraction with a denominator of 100. For example, $45\% = \frac{45}{100}$.

To change percents to fractions, remove the percent sign and write the number over 100. Reduce the fraction to lowest terms.

Examples

$50\% = \frac{50}{100} = \frac{5}{10} = \frac{1}{2}$

$25\% = \frac{25}{100} = \frac{1}{4}$

$30\% = \frac{30}{100} = \frac{3}{10}$

Changing Fractions to Percents To change fractions to percents, change the fraction to a decimal and then change the decimal to a percent.

Examples

$\frac{1}{2} = 0.5 = 50\%$

$\frac{1}{4} = 0.25 = 25\%$

$\frac{1}{10} = 0.10 = 10\%$

$\frac{6}{20} = \frac{3}{10} = 0.3 = 30\%$

$\frac{1}{3} = 0.33 = 33\%$

Time Savers: Fractions, Decimals, and Percents Memorizing some of these relationships might save you some calculation time.

Fraction(s)	= Decimal	= Percent (%)
$\frac{1}{100}$	0.01	1%
$\frac{1}{10}$	0.1	10%
$\frac{1}{5} = \frac{2}{10}$	0.2	20%
$\frac{3}{10}$	0.3	30%
$\frac{2}{5} = \frac{4}{10}$	0.4	40%
$\frac{1}{2} = \frac{5}{10}$	0.5	50%
$\frac{3}{5} = \frac{6}{10}$	0.6	60%
$\frac{4}{5} = \frac{8}{10}$	0.8	80%
$\frac{1}{4} = \frac{2}{8} = \frac{25}{100}$	0.25	25%
$\frac{3}{4} = \frac{75}{100}$	0.75	75%
$\frac{1}{3} = \frac{2}{6}$	$0.33\frac{1}{3}$	$33\frac{1}{3}\%$

Fraction(s)	= Decimal	= Percent (%)
$\frac{2}{3}$	$0.66\frac{2}{3}$	$66\frac{2}{3}\%$
$\frac{1}{8}$	0.125	12.5%
$\frac{3}{8}$	0.375	37.5%
$\frac{5}{8}$	0.625	62.5%
$\frac{1}{6}$	$0.16\frac{2}{3}$	$16\frac{2}{3}\%$
1	1.00	100%
1.5	1.50	150%

Finding a Percent of a Number   The ASVAB Math Knowledge test and Arithmetic Reasoning test frequently include problems that ask you to find a percent of a number. Problems are often worded like this: "What is 25% of 1,000?" There are two ways you can solve this kind of problem. You can start by changing the percent into a fraction, or you can change it into a decimal.

If you change 25% into a fraction, solve the problem like this:

$$\frac{25}{100} \times 1,000 = \frac{25,000}{100} = 250$$

If you change 25% into a decimal, solve the problem like this:

$$0.25 \times 1,000 = 250$$

You should use the approach that is easiest and best for you.

Percent problems are sometimes stated in another way. For example, a problem may ask, "25 is what percent of 200?" When you see a problem like this, make it into an equation:

$$25 = x(200) \text{ where } x \text{ is the percent}$$
$$x = \frac{25}{100} = \frac{1}{8} = 0.125 = 12.5\%$$

Examples

30 is what percent of 90?

$$30 = x(90)$$
$$\frac{30}{90} = \frac{1}{3} = 0.33 = 33\%$$

You can also solve the problem by setting up a proportion:

$$\text{Some unknown percent } (x\%) = \frac{20}{25}.$$

$x\%$ is really $\frac{x}{100}$. So the equation becomes

$$\frac{x}{100} = \frac{20}{25}$$

Reduce the fraction and solve:

$$\frac{x}{100} = \frac{4}{5}$$
$$5x = 400$$
$$x = 80$$

So 20 is 80 percent of 25.

Example

30 is what percent of 120?

$$\frac{x}{100} = \frac{30}{120}$$

$$\frac{x}{100} = \frac{3}{12}$$

$$12x = 300$$

$$x = \frac{300}{12} = 25$$

$$x = 25\%$$

Finding the Percent Increase or Decrease   You will likely also encounter these types of percent problems on the ASVAB. Here is an example: "What is the percent increase in Kim's salary if she gets a raise from $12,000 to $15,000 per year?" Set this up as an equation with the following structure

$$\frac{\text{Amount of change}}{\text{Original number}} = \text{percent of change}$$

Example (Increase)

For the problem above about Kim's salary, the amount of change is 3,000 because Kim's salary increased by that amount. The original number is 12,000. Plug those numbers into the formula:

$$\frac{3,000}{12,000} = \frac{3}{12} = 0.25 = 25\%$$

Kim's salary increased by 25%.

Example (Decrease)

A CD player has its price reduced from $250 to $200. What is the percent decrease? Use the same process as with the previous problem.

$$\frac{\text{Amount of change}}{\text{Original number}} = \text{percent of change}$$

The amount of change (decrease) is 50 (250 minus 200). The original number (original price) is $250. Plug these numbers into the formula:

$$\frac{50}{250} = \frac{5}{25} = \frac{1}{5} = 0.20$$

The CD price decreased by 20%.

Exponents and Square Roots

Arithmetic problems on the ASVAB may deal with exponents and square roots. One or two may even involve scientific notation, which is a way of writing large numbers using exponents.

Exponents An *exponent* is a number that tells how many times another number is multiplied by itself. In the expression 4^3, the 3 is an exponent. It means that 4 is multiplied by itself three times or $4 \times 4 \times 4$. So $4^3 = 64$. The expression 4^2 is read "4 to the second power" or "4 squared." The expression 4^3 is read "4 to the third power" or "4 cubed." The expression 4^4 is read "4 to the fourth power." In each of these cases, the exponent is called a "power" of 4. In the expression 5^2 ("5 to the second power" or "5 squared"), the exponent is a "power" of 5.

Examples

$3^5 = 3 \times 3 \times 3 \times 3 \times 3 = 243$
$6^2 = 6 \times 6 = 36$

Negative Exponents Exponents can also be negative. To interpret negative exponents, follow this pattern:

$$2^{-3} = \frac{1}{2^3} \quad \text{or} \quad \frac{1}{8}$$

Examples

$$3^{-2} = \frac{1}{3^2} = \frac{1}{9}$$

$$4^{-3} = \frac{1}{4^3} = \frac{1}{64}$$

Multiplying Numbers with Exponents To multiply numbers with exponents, multiply out each number and then perform the operation.

Examples

$3^4 \times 2^3 = (3 \times 3 \times 3 \times 3) \times (2 \times 2 \times 2)$
$= 81 \times 8$
$= 648$

$2^9 \times 12^2 =$
$(2 \times 2 \times 2 \times 2 \times 2 \times 2 \times 2 \times 2 \times 2) \times (12 \times 12)$
$= 512 \times 144$
$= 73,728$

Square Roots The *square* of a number is the number times the number. The square of 2 is $2 \times 2 = 4$. The *square root* of a number is the number whose square equals the original number. One square root of 4 (written $\sqrt{4}$) is 2, since $2 \times 2 = 4$. -2 is also a square root of 4, since $-2 \times -2 = 4$.

Examples

$\sqrt{1} = 1$ or -1
$\sqrt{4} = 2$ or -2
$\sqrt{9} = 3$ or -3
$\sqrt{16} = 4$ or -4
$\sqrt{25} = 5$ or -5
$\sqrt{36} = 6$ or -6
$\sqrt{49} = 7$ or -7
$\sqrt{64} = 8$ or -8
$\sqrt{81} = 9$ or -9
$\sqrt{100} = 10$ or -10
$\sqrt{121} = 11$ or -11
$\sqrt{144} = 12$ or -12
$\sqrt{169} = 13$ or -13
$\sqrt{196} = 14$ or -14
$\sqrt{225} = 15$ or -15

REMINDER
Watch the signs! Remember that a negative number multiplied by itself results in a positive number.

Scientific Notation

For convenience, very large or very small numbers are sometimes written in what is called *scientific notation*. In scientific notation, a number is written as the product of two factors. The first factor is a number greater than or equal to 1 but less than 10. The second factor is a power of 10. (Recall that in an expression such as 10^3, which is read "10 cubed" or "10 to the third power," the exponent 3 is a power of 10.)

For example, here is how to write the very large number 51,000,000 in scientific notation. This number can also be written 51,000,000.00. Move the decimal point to the left until you have a number between 1 and 10. If you move the decimal point 7 places to the left, you have the number 5.1000000 or 5.1. Since you moved the decimal point 7 places, you would have to multiply 5.1 by 10^7 (a power of 10) to recreate the original number. So in scientific notation, the original number can be written as 5.1×10^7.

For very small numbers, move the decimal point to the right until you have a number between 1 and 10. Then count the decimal places between the new position and the old position. This time, since you moved the decimal point to the right, the power of 10 is a negative. That is, the exponent is a negative number.

Examples (Large Numbers)

698,000,000,000 can be written 6.98×10^{11}
45,000,000 can be written 4.5×10^7

Examples (Small Numbers)

0.00000006 can be written 6×10^{-8}
0.0000016 can be written 1.6×10^{-6}
To change back to the original number from scientific notation, merely move the decimal point the number of places indicated by the "power of 10" number.

Example (Large Numbers)

2.35×10^5 becomes 235,000. The decimal point moved 5 places to the right.

6×10^8 becomes 600,000,000. The decimal point moved 8 places to the right.

Example (Small Numbers)

2.3×10^{-4} becomes 0.00023. The decimal point moved 4 places to the left.

7×10^{-12} becomes 0.000000000007. The decimal point moved 12 places to the left.

Multiplying in Scientific Notation To multiply numbers written in scientific notation, multiply the numbers and add the powers of 10.

Examples

$(2 \times 10^8) \times (3.1 \times 10^2) = 6.2 \times 10^{10}$

$(3.3 \times 10^{12}) \times (4 \times 10^{-3}) = 13.2 \times 10^9$

Here you are adding a 12 and a −3, giving you 9 for the power of 10. But remember the rule that the decimal point must be after a number that is between 1 and 9,

so you have to move the decimal place one more place to the left to put it after the 1, and you have to increase the power of 10 accordingly. So the final result of the multiplication is 1.32×10^{10}.

Dividing in Scientific Notation To divide in scientific notation, divide the numbers and subtract the powers of 10.

Examples

$$(9 \times 10^7) \div (3 \times 10^3) = 3 \times 10^4$$

Here you divide 9 by 3 (= 3), and you subtract the exponent 3 from the exponent 7 to get the new exponent 4.

$$(1.0 \times 10^4) \div (4 \times 10^{-3}) = 0.25 \times 10^7$$

Since the number in the first position must be between 1 and 9, move the decimal to the right one place and decrease the exponent accordingly, so the final result is 2.5×10^6.

Mean, Median, and Mode

The mean, the median, and the mode are mathematical measures that are used to understand and describe a given set of numbers.

Mean or Average The *mean* of a set of numbers is the average. Add the numbers and divide by the number of numbers.

Examples

What is the mean of the numbers 2, 4, 7, 4, and 5? The sum is 22. Since there are 5 numbers, divide 22 by 5 to get a mean or average of 4.4.

What is the mean of the numbers 3, 4, 2, 6, 7, 12, 56, and 104?
The sum is 194. There are 8 numbers, so divide 194 by 8 to get 24.25.

Median Order a set of numbers from least to greatest. If there is an odd number of numbers, the *median* is the number in the middle of that sequence of numbers. If there is an even number of numbers, the median is the mean or average of the two middle numbers.

Examples

What is the median of the following numbers?

14, 999, 75, 102, 456, 19, 10

Reorder the numbers from least to greatest: 10, 14, 19, 75, 102, 456, 999

The middle number is 75, so that is the median. What is the median of the following numbers?

15, 765, 65, 890, 12, 1, 10

Reorder the numbers from least to greatest: 1, 12, 15, 65, 765, 890

Since there is an even number of numbers, find the average of the middle two numbers.

$$15 + 65 = 80$$
$$80 \div 2 = 40$$
40 is the median

> **HINT** *When the median helps.* If there are extremes (a few very high or very low numbers) in a set of numbers, the median is a good way to describe the "central tendency" of the number set.

Mode The *mode* of a set of numbers is the number that appears most frequently in that set.

Example

What is the mode of the following set of numbers?

12, 14, 15, 15, 15, 17, 18

The number 15 appears most often in this set, so it is the mode of the set.

Graphs

Often it is helpful to represent numbers in a visual form called a *graph*. The most common types of graphs are circle graphs, bar graphs, and line graphs. You probably won't be asked to construct such graphs on the ASVAB, but you are likely to have to interpret one or more of them on the math and/or science sections of the test.

Types of Graphs Some of the most common types of graphs are circle graphs, bar graphs, and line graphs.

Circle Graph This kind of graph uses a circle divided into parts to show fractional or percentage relationships.

Example

In a survey at a local high school, students were asked to name their favorite lunch food. The results of that survey are shown in the circle graph below.

Favorite Lunch Food of High School Students

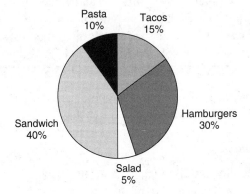

Questions such as the following are typical:

- Which food was most (or least) popular with the students? (Most: sandwich; least: salad)
- Which two food selections of the students make up 50% of the total? (Sandwich and pasta: 40% and 10% = 50%)
- What is the ratio of the students who selected sandwiches to the students who selected hamburgers? (4:3)
- If the total number of students surveyed was 2,500, how many chose salad as their favorite lunch? (125; 2,500 × 0.05 = 125)

Bar Graph This kind of graph uses bars to provide a visual comparison of different quantities.

Example

The following bar graph compares the number of different types of sandwiches sold on Monday at a certain sandwich shop.

Number of Sandwiches Sold on Monday

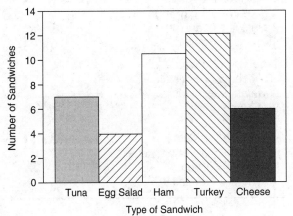

Questions such as the following are typical:

- Which kind of sandwich did the store sell the most of on Monday? (Turkey)
- How many more turkey sandwiches than egg salad sandwiches were sold on Monday? (8)
- How many tuna sandwiches and cheeses sandwiches were sold on Monday? (13)

Line Graph This kind of graph is generally used to show change over time.

Example

The following line graph shows the change in attendance at this year's football games.
Questions such as the following are typical.

- How many more people attended game 2 than game 1? (2,000)

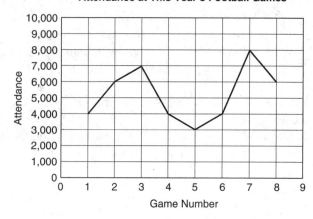

Attendance at This Year's Football Games

- Which game had the highest attendance? (Game 7)
- How many people attended the last game of the year? (6,000)

Units of Measure

If you don't know the following units of measure, you just need to memorize them.

Customary System

Length
12 inches (in.) = 1 foot (ft)
3 feet = 1 yard (yd)
36 inches = 1 yard
5,280 feet = 1 mile (mi)

Area
1 square foot (ft²) = 144 square inches (in.²)
9 square feet = 1 square yard (yd²)

Weight
16 ounces (oz) = 1 pound (lb)
2,000 pounds = 1 ton (t)

Liquid Volume
2 cups (c) = 1 pint (pt)
2 pints = 1 quart (qt)
4 quarts = 1 gallon (gal)

Time
60 seconds = 1 minute
60 minutes = 1 hour
24 hours = 1 day
7 days = 1 week
4 weeks = 1 month
12 months = 1 year
52 weeks = 1 year
10 years = 1 decade

Metric System

Length
100 centimeters (cm) = 1 meter (m)
1,000 meters = 1 kilometer (km)

Volume
1,000 milliliters (mL) = 1 liter (L)

Mass
1,000 grams (g) = 1 kilogram (kg)
1,000 kilograms = 1 metric ton

Converting Between Systems
1 meter ≈ 39.37 inches
1 kilometer ≈ 0.62 mile
1 centimeter ≈ 0.39 inch
1 kilogram ≈ 2.2 pounds
1 liter ≈ 1.057 quarts
1 gram ≈ 0.035 ounce

ARITHMETIC QUIZ

Circle the letter that represents the correct answer. Check your answers with those on page 204. If any of your answers were wrong, go back and study the relevant section of the math review.

1. Add:

   336 + 789

   A. 1,025
   B. 1,110
   C. 1,125
   D. 1,126

2. Subtract:

   475
   −397

   A. 83
   B. 82
   C. 78
   D. 68

3. Which of the following symbols means "is less than"?
   A. ≈
   B. ≠
   C. <
   D. ⊥

4. Which of the following means to multiply 3 times 6?
   A. (3)(6)
   B. 3 ÷ 6
   C. $\dfrac{3}{6}$
   D. 3 ± 6

5. 3 + 5(6 − 3) =
   A. 18
   B. 24
   C. 30
   D. 32

6. $4 + (19 − 4^2)^2 =$
   A. −240
   B. 13
   C. 85
   D. 240

7. $2^3 − (2 \times 3) + (−4) =$
   A. −2
   B. 7
   C. 17
   D. 19

8. 3(10 − 6)(3 − 7) =
   A. −48
   B. −43
   C. 18
   D. 48

9. Round to the nearest thousand:

   34,578

   A. 34,000
   B. 34,080
   C. 34,600
   D. 35,000

10. Round to the nearest tenth:

    12.8892

    A. 12.0
    B. 12.890
    C. 12.9
    D. 13.0

11. 4 + (−8) + 2 + (−6) =
    A. 10
    B. 4
    C. −8
    D. −20

12. Subtract:

    18
    − −2

    A. 20
    B. 16
    C. 12
    D. −16

13. Subtract:

$$-15$$
$$- \underline{-12}$$

    A. −27
    B. −3
    C. 3
    D. 27

14. $(-5)(-3)(-2) =$
    A. 30
    B. 13
    C. −10
    D. −30

15. $(14)(2)(0) =$
    A. 28
    B. 16
    C. 0
    D. −28

16. Reduce to lowest terms:

$$\frac{18}{36} =$$

    A. $\frac{1}{4}$

    B. $\frac{1}{2}$

    C. $\frac{2}{4}$

    D. $\frac{3}{6}$

17. Rename this fraction as a mixed number:

$$\frac{22}{7} =$$

    A. $2\frac{4}{7}$

    B. $2\frac{6}{7}$

    C. $3\frac{1}{7}$

    D. 4

18. Rename this mixed number as a fraction:

$$5\frac{3}{4} =$$

    A. $\frac{4}{15}$

    B. $\frac{15}{4}$

    C. $\frac{19}{4}$

    D. $\frac{23}{4}$

19. Add the following fractions:

$$\frac{1}{2} + \frac{3}{4} + \frac{3}{12} =$$

    A. $\frac{7}{24}$

    B. $\frac{7}{12}$

    C. $1\frac{1}{2}$

    D. $1\frac{3}{4}$

20. Subtract the following fractions:

$$\frac{3}{4} - \frac{1}{2} =$$

    A. $\frac{1}{4}$

    B. $\frac{1}{2}$

    C. $\frac{4}{4}$

    D. 1

21. Subtract the following fractions:

$$3\frac{1}{2}$$
$$-2\frac{7}{8}$$
_____

A. $-1\frac{7}{8}$

B. $\frac{5}{8}$

C. $\frac{7}{8}$

D. $1\frac{7}{8}$

22. Multiply the following fractions:

$$\frac{3}{8} \times \frac{9}{18} =$$

A. $\frac{27}{144}$

B. $\frac{3}{16}$

C. $\frac{1}{5}$

D. $\frac{78}{54}$

23. Divide the following fractions:

$$\frac{1}{9} \div \frac{2}{3} =$$

A. $\frac{2}{27}$

B. $\frac{1}{6}$

C. $\frac{3}{11}$

D. $\frac{2}{6}$

24. Multiply the following mixed numbers:

$$2\frac{3}{4} \times 4\frac{1}{3} =$$

A. $8\frac{1}{3}$

B. $8\frac{2}{3}$

C. $11\frac{11}{12}$

D. $12\frac{1}{6}$

25. Divide the following mixed numbers:

$$3\frac{1}{3} \div 4\frac{2}{9} =$$

A. $3\frac{1}{7}$

B. $\frac{45}{57}$

C. $\frac{16}{21}$

D. $\frac{15}{19}$

26. Change the following fraction to a decimal:

$$\frac{3}{4} =$$

A. 0.34
B. 0.50
C. 0.75
D. 0.80

27. Change the following decimal to a fraction:

0.008 =

A. $\frac{8}{50}$

B. $\frac{8}{100}$

C. $\frac{4}{100}$

D. $\frac{1}{125}$

28. Add the following decimals:

    $12.51 + 3.4 + 6.628 =$

    A. 21.52
    B. 22.528
    C. 22.538
    D. 223.258

29. Subtract the following decimals:

    $23.51 - 20.99 =$

    A. 3.62
    B. 3.49
    C. 2.52
    D. 2.42

30. Divide the following decimals.

    $40.896 \div 12.78 =$

    A. 5.6
    B. 4.4
    C. 3.6
    D. 3.2

31. Change the following percent to a decimal:

    9%

    A. 0.009
    B. 0.09
    C. 0.9
    D. 9.0

32. Change the following decimal to a percent:

    0.32

    A. 32%
    B. 3.2%
    C. 0.32%
    D. 0.032%

33. What is 75% of 1,250?
    A. 937.5
    B. 950
    C. 975
    D. 1000

34. 25 is what percent of 200?
    A. 10%
    B. 12%
    C. 12.5%
    D. 15%

35. Write the following number in scientific notation:

    $0.0002392 =$

    A. $2,392 \times 10^{-7}$
    B. 23.92×10^{-5}
    C. 23×10^{5}
    D. 2.392×10^{-4}

36. Multiply:

    $(2.3 \times 10^{5}) \times (7.1 \times 10^{-3}) =$

    A. 9.4×10^{-8}
    B. 9.4×10^{8}
    C. 16.33×10^{2}
    D. 16.33×10^{8}

37. What is the square root of 100?
    A. 0.05
    B. 5
    C. 10
    D. 75

38. What is the mean (average) of the following set of numbers?

    12 14 18 22 45 78

    A. 27.5
    B. 31.5
    C. 32.0
    D. 45.5

39. Last month, the librarian at a certain library kept track of all books checked out. The graph below shows the results by type of book. What percentage of all the books checked out last month were travel books?

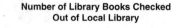
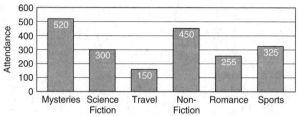

A. 15.3%

B. 15%

C. 8%

D. 7.5%

40. How many feet are in 2 miles?

A. 10,000

B. 10,560

C. 11,954

D. 16,439

41. How many square inches are in 3 square feet?

A. 144

B. 278

C. 432

D. 578

42. How many quarts are in 3 gallons?

A. 3

B. 9

C. 12

D. 16

43. How many meters are in 4 kilometers?

A. 40

B. 400

C. 4,000

D. 40,000

44. How many grams are in 2.5 kilograms?

A. 250

B. 2,500

C. 25,000

D. 250,000

45. 4 kilograms is about how many pounds?

A. 8.8

B. 9

C. 9.6

D. 10.6

ARITHMETIC QUIZ—ANSWERS

Number	Answer	Number	Answer
1.	C	24.	C
2.	C	25.	D
3.	C	26.	C
4.	A	27.	D
5.	A	28.	C
6.	B	29.	C
7.	A	30.	D
8.	A	31.	B
9.	D	32.	A
10.	C	33.	A
11.	C	34.	C
12.	A	35.	D
13.	B	36.	C
14.	D	37.	C
15.	C	38.	B
16.	B	39.	D
17.	C	40.	B
18.	D	41.	C
19.	C	42.	C
20.	A	43.	C
21.	B	44.	B
22.	B	45.	A
23.	B		

Chapter 9

Arithmetic Reasoning 2: Word Problems

YOUR GOALS FOR THIS CHAPTER:

- Find out what kinds of arithmetic word problems are on the ASVAB.

- Learn effective procedures for solving arithmetic word problems.

- Study sample word problems and their step-by-step solutions.

- Measure your test readiness by taking a word problem quiz.

INTRODUCTION: ASVAB ARITHMETIC WORD PROBLEMS

What is a word problem? Basically it is a description of a real-life situation that requires a mathematical solution. To solve a word problem, you need to translate the situation into mathematical terms and then calculate the answer.

The questions on the ASVAB Arithmetic Reasoning test are word problems. To do well, you need to learn good word problem–solving skills. The following pages will show you many different kinds of arithmetic word problems. For each kind, you'll learn how to translate the facts of the problem into mathematical terms. Then you'll see how to use those terms to set up an equation. Finally, you'll see how to solve that equation for the missing piece of information that you're looking for. To start your study program, look at the following word problem–solving tips.

How to Tackle Word Problems

Here are some helpful suggestions for dealing with word problems.

- Read the problem all the way through before you start making any calculations. That way, you will better understand what the problem is about and what you are supposed to find out.

- Be sure you understand what the problem is asking. Are you supposed to find a distance? How fast something is moving? When someone will arrive at point A? How much something costs? How many items you can buy if you have a certain amount of money? How much older Kim is than Kay? Draw a picture if that helps.

- List the information that the problem gives you. Note any units of measure (meters, pounds, feet per second, and the like).

- Look for key words in the problem that define relationships or indicate what mathematical operation you need to use. (See the list of key words that follows.)

- Figure out what unit of measure you must use to express your answer. Cubic meters? Centimeters per second? Dollars? Minutes? Number of items?

- Weed out unnecessary information that will not help you solve the problem.

- Use the information you have gathered to create an equation to help you solve the problem.

- Solve the equation.

Key Words

Many word problems contain key words that tell you what mathematical operation you need to use to solve the problem. It pays to know these key words, so be sure you study the following list.

These key words tell you to *add*.

Key Words	Example Phrases
Increased by	If the temperature is increased by
More than	If the bicycle costs $75 more than
Sum	If the sum of the paychecks is
Total	The total number of payments equals
Added	If a is added to b
Plus	If the interest plus the principal is
Combined	If the volume of the cube is combined with
In all	How many pounds in all
Successive	The cost of eight successive phone bills

These key words tell you to *subtract*.

Key Words	Example Phrases
Less than	The interest payment was less than
Difference	The difference between the time it takes to
Are left	How many pieces of pie are left if
Fewer than	If there are 15 fewer scholarships this year than
Minus	Some number minus 36
Reduced by	If the federal budget is reduced by

These key words tell you to *multiply*.

Key Words	Example Phrases
Times	There are 3 times as many red tiles as blue tiles
Product	The product of a and b is
Increased by a factor of	If the speed is increased by a factor of
Decreased by a factor of	If the temperature is decreased by a factor of
At	If you buy 12 cameras at $249 each
Per	If a ferry can carry 20 cars per trip, how many cars can it carry in six trips?
Total	If you spend $10 a week on movies for a total of 5 weeks
Twice	The house covers twice as many square feet as

These key words tell you to *divide*.

Key Words	Example Phrases
Quotient	What is the quotient if the numerator is 500?
Divided equally among	If 115 tickets are divided equally among 5 groups
Divided into equal groups	If the students were divided into 6 equal groups
Ratio of	If the ratio of oxygen to hydrogen is
Per	If a ferry can carry 24 cars per trip, how many trips will it take to carry 144 cars?
Percent	What percent of 100 is 30?
Half	If half the profits go to charity, then how much

These key words tell you to use an *equal sign*.

Key Words	Example Phrases
Is	If the total bill is $19.35
Sells for	If the car sells for $26,000
Gives	If multiplying a^2 and b^2 gives c^2

SETTING UP AN EQUATION AND SOLVING FOR AN UNKNOWN

Before you start tackling word problems, you need to know how to set up an equation and solve for an unknown. For each problem, you will be given certain pieces of information ("*what you know*"). You first need to translate this information into mathematical terms. Then you can use those terms to set up an equation. An *equation* is nothing more than a mathematical expression that indicates that one mathematical term is equal to another. The terms in an equation are shown on opposite sides of an equal sign. The missing piece of information that you are looking for ("*what you need to find*") is called the *unknown*. In your equation, you can represent an unknown by a letter such as *xf* or *a*.

In the pages that follow, you will see how to set up equations for many different kinds of word problems. The process of solving an equation for an unknown is pretty simple. Work through these examples until you get perfectly comfortable with the process.

Examples

Solve for *x*.

$$25x = 200$$

This equation is read, "25 times some unknown number *x* equals 200." Your task is to determine what number *x* is. Here is how to solve this equation.

Divide both sides of the equation by 25:

$$\frac{25x}{25} = \frac{200}{25}$$
$$x = \frac{200}{25}$$
$$x = 8$$

Solve for *x:*

$$\frac{x}{5} = \frac{3}{45}$$

To solve this type of problem, cross-multiply (45 times *x* and 3 times 5).

$$45x = 15$$
$$x = \frac{15}{45} = \frac{1}{3} = 0.33$$

Solve for *x:*

$$\frac{10}{x} = \frac{50}{350}$$

To solve this problem, cross-multiply (10 times 350 and 50 times *x*).

$$50x = 3,500$$
$$x = 70$$

To check your answer, merely substitute 70 back into the equation and see if it works. $\frac{10}{70} = \frac{50}{350}$ Reduce the fractions to $\frac{1}{7} = \frac{1}{7}$ or cross-multiply (10 times 350 and 70 times 50), making 3,500 = 3,500. Either way, you are correct.

TYPES OF ASVAB ARITHMETIC WORD PROBLEMS

In the pages that follow, you will learn about many different kinds of word problems that you are likely to see on the ASVAB Arithmetic Reasoning test. For each kind, you will see how to use the information you are given to set up an equation. Then you will see how to solve the equation for the unknown that is the answer to the problem. The following chart shows the different kinds of word problems discussed in this chapter.

Arithmetic Word Problem Types
Simple interest
Compound interest
Ratio and proportion
Motion
Percent
Percent change
Numbers and number relationships
Age
Measurement

Simple Interest

Interest is an amount paid for the use of money. *Interest rate* is the percent paid per year. *Principal* is the amount of money on which interest is paid. *Simple interest* is interest that is computed based only on the principal, the interest rate, and the time. To calculate simple interest, use this formula:

$$\text{Interest} = \text{principal} \times \text{rate} \times \text{time}$$
$$I = prt$$

Examples

Martina has $300 in a savings account that pays simple interest at a rate of 3% per year. How much interest will she earn on that $300 if she keeps it in the account for 5 years?

Procedure

What must you find? Amount of simple interest
What are the units? Dollars
What do you know? Rate = 3% per year, time = 5 years, principal = $300
Create an equation and solve.

$$I = (\$300)(0.03)(5)$$
$$I = \$45$$

Five years ago, Robin deposited $500 in a savings account that pays simple interest. She made no further deposits, and today the account is worth $750. What is the rate of interest?

Procedure

What must you find? Rate of simple interest
What are the units? Percent per year
What do you know? Interest = $250 ($750 − $500); principal = $500; time = 5 years

Create an equation and solve.

$$250 = (500)(x)(5)$$

$$250 = 2,500x$$

$$\frac{250}{2500} = x$$

$$x = \frac{250}{2,500} = 0.10 \text{ or } 10\%$$

Compound Interest

Compound interest is the interest paid on the principal and also on any interest that has already been paid. To calculate compound interest, you can use the formula $I = prt$, but you must calculate the interest for each time period and then combine them for a total.

Example

Ricardo bought a $1,000 savings bond that earns 5% interest compounded annually. How much interest will he earn in two years?

Procedure

What must you find? Amount of compound interest
What are the units? Dollars
What do you know? Principal = $1,000; rate = 5%; time = 2 years
Create an equation and solve.

To find the compound interest, calculate the amount earned in the first year. Add that amount to the principal; then calculate the interest earned in the second year. Total the amount of interest earned in the two years.

Year 1:

$$I = prt$$
$$I = (\$1,000)(0.05)(1)$$
$$I = \$50$$

New principal = $1,050

Year 2:

$$I = prt$$
$$I = (\$1,050)(0.05)(1)$$
$$I = \$52.50$$

So the total compound interest paid in two years is $50 + $52.50 = $102.50.

Ratio and Proportion

On the ASVAB, you will almost certainly encounter word problems that will require you to work with ratios and proportions.

Examples

Kim reads an average of 150 pages per week. At that rate, how many weeks will it take him to read 1,800 pages?

Procedure

What must you find? How long it will take to read 1,800 pages.
What are the units? Weeks
What do you know? 150 pages read each week; 1,800 pages to be read
Set up a proportion and solve.

$$\frac{\text{Number of pages}}{1 \text{ week}} = \frac{\text{number of pages}}{x \text{ weeks}}$$

Substitute values into the equation.

$$\frac{150 \text{ pages}}{1 \text{ week}} = \frac{1,800 \text{ pages}}{x \text{ weeks}}$$

Cross-multiply:

$$150x = 1,800$$
$$x = 12 \text{ weeks}$$

It takes 8 hours to fill a swimming pool that holds 3,500 gallons of water. At that rate, how many hours will it take to fill a pool that holds 8,750 gallons?

Procedure

What must you find? How long it will take to fill the 8,750-gallon pool
What are the units? Hours
What do you know? Number of hours for 3,500 gallons
Set up a proportion and solve.

$$\frac{\text{Number of gallons}}{\text{Number of hours}} = \frac{\text{number of gallons}}{\text{number of hours}}$$

$$\frac{3,500 \text{ gallons}}{8 \text{ hours}} = \frac{8,750 \text{ gallons}}{x}$$

Cross-multiply:

$$3,500x = (8)(8,750)$$
$$x = 20 \text{ hours}$$

An airplane travels the 1,700 miles from Phoenix to Nashville in 2.5 hours. Flying at the same speed, the plane could travel the 2,550 miles from Phoenix to Boston in how many hours?

Procedure

What must you find? Time it would take to fly
   2,550 miles
What are the units? Hours
What do you know? The plane traveled 1,700 miles
   in 2.5 hours
Set up a proportion and solve.

$$\frac{\text{Number of miles}}{\text{Number of hours}} = \frac{\text{number of miles}}{\text{number of hours}}$$

Substitute values.

$$\frac{1{,}700 \text{ miles}}{2.5 \text{ hours}} = \frac{2{,}550 \text{ miles}}{x}$$

Cross-multiply:

$$1{,}700x = (2.5)(2{,}550)$$
$$1{,}700x = 6{,}375$$
$$x = 3.75 \text{ hours}$$

Motion

Motion problems deal with how long it will take to get from point *a* to point *b* if you are traveling at a certain steady rate. To solve them, use this formula:

$$\text{Distance} = \text{rate} \times \text{time}$$
$$d = rt$$

Example

If a racing boat travels at a steady rate of 80 miles per hour, how many miles could it travel in 3.5 hours?

Procedure

What must you find? Distance traveled in 3.5 hours
What are the units? Miles
What do you know? Rate = 80 miles per hour;
   time = 3.5 hours

Create an equation and solve.

$$d = rt$$

Substitute values into the formula:

$$d = (80)(3.5)$$
$$d = 280 \text{ miles}$$

Percent

There are likely to be word problems involving percent on both the Arithmetic Reasoning and Math Knowledge tests of the ASVAB.

Examples

Lilly's bill at a restaurant is $22.00, and she wants to leave a 15% tip. How much money should her tip be?

Procedure

What must you find? Amount of tip
What are the units? Dollars and cents
What do you know? Total bill = $22.00; Percent
   of tip = 15

Create an equation and solve.

$$\text{Tip} = 15\% \times 22.00$$

Substitute and solve.

$$t = (0.15)(22.00)$$
$$t = \$3.30$$

Frederick earns $1,500 per month at his job, but 28% of that amount is deducted for taxes. What is his monthly take-home pay?

Procedure

What must you find? Monthly take-home pay
What are the units? Dollars and cents
What do you know? Monthly pay before
   taxes = $1,500; percent deducted = 28%

Create an equation and solve.

Take-home pay is 1,500 minus 28% × 1,500.

$$T = 1{,}500 - (1{,}500 \times 0.28)$$
$$T = 1{,}500 - (420)$$
$$T = \$1{,}080$$

40 is 80% of what number?

Procedure

What must you find? Number of which 40 is 80%
What are the units? Numbers

What do you know? 40 is 80% of some larger number

Create an equation and solve.

$$40 = 0.8x$$
$$0.8x = 40$$
$$x = \frac{40}{0.8}$$
$$x = 50$$

Percent Change

Some ASVAB word problems ask you to calculate the percent change from one number or amount to another.

Examples

Samantha now earns $300 per month working at a cosmetics store, but starting next month her monthly salary will be $375. Her new salary will be what percent increase over her current salary?

Procedure

What must you find? Percent change from current salary
What are the units? Percent
What do you know? Current pay = $300/month; pay after the raise = $375/month
Create an equation and solve.

$$\text{Percent change} = \frac{\text{amount of change}}{\text{starting point}}$$

Substitute values and solve:

$$\text{Percent change} = \frac{(375 - 300)}{300}$$

$$\text{Percent change} = \frac{75}{300}$$

$$\text{Percent change} = 0.25 \text{ or } 25\%$$

On his sixteenth birthday, Brad was 60 inches tall. On his seventeenth birthday, he was 65 inches tall. What was the percent increase in Brad's height during the year?

Procedure

What must you find? Percent change in height
What are the units? Percent
What do you know? Starting height = 60 inches; height after a year = 65 inches

Create an equation and solve.

$$\text{Percent change} = \frac{\text{amount of change}}{\text{starting point}}$$

Substitute values and solve.

$$\text{Percent change} = \frac{5 \text{ inch}}{60 \text{ inch}}$$

$$\text{Percent change} = 0.08 \text{ or } 8\%$$

At a certain store, every item is discounted by 15% off the original price. If Kevin buys a CD originally priced at $15.00 and a baseball cap originally priced at $11.50, how much money will he save?

Procedure

What must you find? Total amount saved
What are the units? Dollars and cents
What do you know? Percent change = 15%; original price for two items = $15.00 + $11.50 = $26.50
Create an equation and solve.

$$\text{Percent change} = \frac{\text{amount of change}}{\text{starting point}}$$

Substitute values and solve.

$$0.15 = \frac{x}{26.50}$$
$$x = 0.15(26.50)$$
$$x = \$3.98 \text{ (Note, this answer is rounded to the nearest tenth.)}$$

Numbers and Number Relationships

Pay attention to the key words in this type of word problem.

Examples

If the sum of two numbers is 45 and one number is 5 more than the other, what are the two numbers?

Procedure

What must you find? Value of each number
What are the units? Numbers
What do you know? Sum of two numbers is 45; one number is 5 more than the other
Create an equation and solve.
Let x be the smaller number.

$$x + (x + 5) = 45$$

Solve for x:

$$x + x + 5 = 45$$
$$2x = 40$$
$$x = 20$$

So the two numbers are 20 and $20 + 5 = 25$.

One number is twice the size of another, and the two numbers together total 150. What are the two numbers?

Procedure

What must you find? Value of each number
What are the units? Numbers
What do you know? One number is twice the size of the other, and the sum of the numbers is 150.
Create an equation and solve.
Let x be the smaller number.

$$x + 2x = 150$$
$$3x = 150$$
$$x = 50$$

So the smaller number is 50 and the larger number is 100.

Age

Some word problems ask you to calculate a person's age given certain facts.

Examples

Jessica is 26 years old. Two years ago she was twice as old as her brother Ned. How old is Ned now?

Procedure

What must you find? Ned's age now
What are the units? Years
What do you know? When Jessica was 24, she was 2 times as old as Ned

Create an equation and solve.

Prepare an equation that shows the relationship.
Let x = Ned's age two years ago.

$$2x = 24$$
$$x = 12$$

Ned's age two years later = $12 + 2 = 14$ years

Measurement

Some word problems will ask you to use what you know about units of measure to solve problems. If necessary, review the table of units of measure in Chapter 8, "Arithmetic Reasoning 1: The Basics."

Example

How many cups of milk are in 5 pints of milk?

Procedure

What must you find? The number of cups of milk in 5 pints
What are the units? Cups
What do you know? According to the chart in Chapter 8, there are 2 cups in 1 pint.
Create an equation and solve.
Let x = the number of cups in 5 pints.

$$5 \text{ pints} = x \text{ cups}$$
$$1 \text{ pint} = 2 \text{ cups}$$
$$5 \times 1 \text{ pint} = 5 \times 2 \text{ cups (Multiply both sides of}$$
$$\text{the equation by 5)}$$
$$5 \text{ pints} = 10 \text{ cups} = x \text{ cups}$$
$$x = 10$$

So there are 10 cups in 5 pints.

WORD PROBLEMS QUIZ

1. The U.S. Treasury Department spends 3.5 cents to manufacture every quarter. 3.5 cents is what percent of the value of the coin?
   A. 3%
   B. 7%
   C. 12%
   D. 14%

2. A person's recommended weight can be calculated using the following formula:

   $$w = \frac{11(h-40)}{2} \text{ where } w = \text{weight and}$$

   $$h = \text{height in inches.}$$

   If Stuart is 5 feet 10 inches tall, what is his recommended weight?
   A. 135 lb
   B. 155 lb
   C. 165 lb
   D. 167 lb

3. It takes 200 lb of sugar cane to produce 5 lb of refined sugar. How many pounds of sugar cane does it take to produce 200 lb of refined sugar?
   A. 2,500 lb
   B. 7,500 lb
   C. 7,700 lb
   D. 8,000 lb

4. Rudy is training for a race. On Monday he ran 2 km. On Tuesday he ran 15% farther than on Monday. If he increases his running distance by the same percentage each day, on which day will he first run more than 4 km?
   A. Wednesday
   B. Thursday
   C. Friday
   D. Saturday

5. Julie's salary is $600 per month, but after taxes and insurance are deducted, her monthly paycheck is $444. What percent of her monthly salary goes for taxes and insurance?
   A. 15%
   B. 26%
   C. 33%
   D. 34%

6. Keisha worked 75 hours during July at a pay rate of $12.50/hour. What were her total earnings for the month before taxes and insurance were deducted?
   A. $447.50
   B. $937.50
   C. $1,447.50
   D. $1,667.50

7. Debbie is walking to raise money for her favorite charity. She is being sponsored by Sally at 10 cents per mile, by Lucie at 12 cents per mile, and by Kelsey at 14 cents per mile. If she walks 8 miles, how much money will she raise?
   A. $2.88
   B. $3.12
   C. $5.25
   D. $10.75

8. Emilio has saved 150 quarters in a jar. Of that total, 25 were minted before 1970, 35 were minted between 1970 and 1980, and the remaining quarters were minted after 1980. If Mark picks a quarter out of the jar without looking, what is the probability that he will pick a quarter minted after 1980?
   A. 3:5
   B. 1:2
   C. 7:8
   D. 9:5

9. The bar graph shows the results of a survey asking workers in a company to tell what means of transportation they use to get to work. Based on the graph, what percent of all the workers surveyed said that they take the bus to work?

   A. 16%

   B. 20%

   C. 31%

   D. 35%

How Workers Get to Work

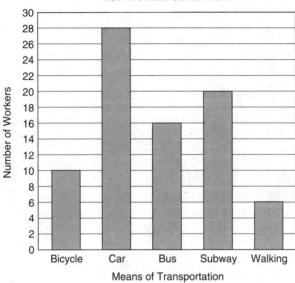

10. Javier was elected state senator with 60% of the vote. Which of the following represents that percent as a fraction in simplest form?

    A. $\frac{60}{100}$

    B. $\frac{6}{10}$

    C. $\frac{3}{5}$

    D. $\frac{3}{2}$

11. Samuel's current salary is $30,000. He has been promised a raise to $31,500 next year. His raise will be what percent increase over his current salary?

    A. 3%

    B. 3.5%

    C. 4%

    D. 5%

12. Phillip deposited $450 in an account that earns 5% interest, compounded yearly. If he makes no further deposits, how much will be in the account after 2 years?

    A. $472.50

    B. $491.14

    C. $496.13

    D. $512.24

13. If Kevin saves $4.75 per week, how many weeks will it take him to save $266?

    A. 27

    B. 34

    C. 51

    D. 56

14. If an airplane flies at an average speed of 320 miles per hour, how long will it take to fly 1,200 miles?

    A. 2.50 hours

    B. 3.25 hours

    C. 3.75 hours

    D. 4.5 hours

15. What is the difference between 2 meters and 180 centimeters?

    A. 2 cm

    B. 12 cm

    C. 18 cm

    D. 20 cm

ANSWERS TO WORD PROBLEMS QUIZ

Number	Answer
1.	D
2.	C
3.	D
4.	D
5.	B
6.	B
7.	A
8.	A
9.	B
10.	C
11.	D
12.	C
13.	D
14.	C
15.	D

Chapter 10

Word Knowledge

YOUR GOALS FOR THIS CHAPTER:

- **Find out what you can do to build a better vocabulary.**

- **Study common prefixes, suffixes, and word roots.**

- **Learn how to use context to figure out the meaning of an unfamiliar word.**

- **Make your own list of vocabulary words and their meanings.**

INTRODUCTION: ASVAB WORD KNOWLEDGE QUESTIONS

The ASVAB Word Knowledge test measures your ability to understand the meaning of words through synonyms. *Synonyms* are words that have the same or nearly the same meaning as other words. Your ability to recognize synonyms is an indicator of how well you comprehend what you read. Word Knowledge, along with Paragraph Comprehension, is part of the verbal ability portion of the AFQT.

Some Word Knowledge questions present a vocabulary word and ask you which of four answer choices the word "most nearly means." Other Word Knowledge questions present a vocabulary word in a sentence. You can use the meaning of the sentence to help you decide which of the four answer choices has the same meaning as the vocabulary word. The tested vocabulary words are not difficult scientific or technical terms. They are words that

you are likely to encounter in your ordinary reading or conversation but which may be unfamiliar to you. This chapter will teach you ways to figure out the meaning of unfamiliar words and improve your score on the Word Knowledge test.

If you take the CAT-ASVAB, you have only about half a minute to answer each Word Knowledge question. If you take the paper-and-pencil ASVAB, you'll have even less time, so you'll have to work fast if you want to get a good score. That's why it pays to spend time studying ways to build your vocabulary and tackling plenty of sample ASVAB Word Knowledge questions.

How Good Is Your Vocabulary Now?

Start your preparation for the Word Knowledge test by taking this short quiz. It will determine your level of strength in the area of vocabulary. Beside each word, write a word or phrase that helps to define the word. Correct definitions are given at the end of this chapter. Compare your definitions to the correct definitions. If your definition is the same or nearly the same as the correct definition, give yourself 1 point. If your definition is very different from the correct definition, give yourself 0 points.

18–20 points: pretty good work

15–17 points: not too bad

Below 15 points: you need some work

ASVAB Word Knowledge Test

Number of Questions:

- Paper-and-pencil ASVAB: 35
- CAT-ASVAB: 15

Time Limit:

- Paper-and-pencil ASVAB: 11 minutes
- CAT-ASVAB: 9 minutes

Vocabulary Quiz

Word	Definition	Score
abscond		
antagonist		
complacent		
consensus		
defensive		
didactic		
docile		
dormant		
exorbitant		
fidelity		
hyperbole		
intermittent		
neophyte		
parity		
pivotal		
reticent		
succinct		
tenacious		
unilateral		
venerate		
Total Points		

START BUILDING A BETTER VOCABULARY

If it's clear that you need to brush up on your vocabulary for the ASVAB, here's what you can do to start studying for this test.

Memorize (Not!)

It may help you to memorize some common prefixes, suffixes, and word roots. However, memorizing word lists will not be particularly helpful, especially if your study time is limited. You need to use study methods that are far more useful than memorizing words and definitions.

Read, Read, Read, and Read Some More

Read everything you can get your hands on, including school books, newspapers, magazines, and fiction and nonfiction books. The more you read, the more new words you'll come across that you can add to your vocabulary.

Develop a Word List

As you read and find words you don't know, start developing a word list. Use the Word List Chart provided at the end of this section. Write down each word you don't know. Take a guess at the meaning. Write down your guess. Then go to a dictionary, your thesaurus, or your word processing program to find synonyms (words that mean the same thing). Write the synonyms in the last column of the chart. You don't need to write down the formal dictionary definition because the synonyms are what the ASVAB asks for. Also use the Word List Chart for words that you hear people say, but that you do not know.

Often just writing down a word and some synonyms will help you retain the meaning. You may also want to use the Word List Chart to create flash cards that you can take with you anywhere to study when you have some free time. Using something like 3" × 5" cards, write the word on one side and some synonyms on the other. Later in this section, you will have a set of words that can be used for these flash cards or whatever other study system you decide to use.

Review, Review, Review

Once you have developed your word list, listed synonyms, and worked through the sample tests, you still need to keep reviewing. Review on the bus to work or school. Review during TV commercials. Review during your spare time. Review during study sessions. Review before you go to sleep at night. Review with your parents. Review with your friends. Review with a study group. Review in whatever way is most convenient and efficient for you. Be realistic. If reviewing with your friends ends up as a chat session or a party, then that's not helping you reach your goal of scoring high on the ASVAB. Stay disciplined. Stay focused. Stay on track.

USE WORD PARTS TO FIGURE OUT MEANING

Many words are made up of several parts: a *root* (the basic idea) and a *prefix* and/or a *suffix* (short letter combinations that change the meaning in some way). A prefix is attached to the root at the beginning of the word. A suffix is attached to the root at the end of the word. Not all words have prefixes and suffixes, but many do. On the ASVAB test, you will find many words that have roots with prefixes and/or suffixes. If you learn the most common word roots, prefixes, and suffixes, you can use what you know to pick apart unfamiliar words and figure out what they mean.

On the following pages you will find tables of common prefixes, common word roots, and common suffixes. Study these tables carefully. The more you know about these word parts, the better able you will be to figure out the meaning of unfamiliar words.

Study Some Examples

Now that you have studied these common word parts, you can start using what you know to figure out the meaning of unfamiliar words. Here are a few examples of picking words apart to find their meaning.

- *Antibiotic. Anti-* is a prefix that means "against," and *bio* is a root that means "life," so the word has something to do with "against life." An *antibiotic* is a substance that is designed to kill germs.
- *Inaudible. In-* is a prefix that means "not," *aud* has something to do with sound, and the suffix *-ible* mans "capable of," so the word means something like "not capable of being heard."

- *Audiovisual. Audio-* is a prefix that has something to do with hearing, and *visual* relates to seeing, so *audiovisual* involves both hearing and seeing.
- *Precede. Pre-* is a prefix that means "before," and *cede* is a root that means "to go," so the word means something like "going before."
- *Captain. Capit* is a root that means "head," so the word relates to the head of something (in this case, a ship's crew).
- *Capital. Capit* is a root that means "head," so the word has something to do with the head of something (in this case, the symbolic "head" of a country, the location of the government).
- *Chronometer. Chron-* is a prefix that has to do with time, and *meter* relates to measuring, so the word has to do with measuring time.
- *Pedicure. Pedi-* is a prefix that has to do with "foot," so this word has to do with something related to feet.
- *Visage. Vis* has something to do with seeing or looking, so the meaning of this word is "how a person looks."
- *Transform. Trans-* is a prefix that means "go across," so this word means "going across" or "changing form."
- *Microbiology. Micro-* is a prefix that means "small," *bio* is a root that has to do with life, and *-ology* is a suffix that means "the study of," so the word has to do with studying small living things.
- *Multitalented. Multi-* is a prefix that means "many," so this word has to do with many talents.
- *Thermometer. Therm-* has to do with heat, and *meter* means measuring, so this word has to do with measuring heat.
- *Geology. Geo* is a root that means "earth," and *-ology* is a suffix that means "the study of," so this word has to do with studying the earth.
- *Revolve. Re-* is a prefix that means "again," and *volve* means "to turn," so this word has something to do with turning again or turning repeatedly.
- *Mispronounce. Mis-* is a prefix that means "wrong" or "to do wrongly," so this word has to do with pronouncing something in a wrong way.
- *Vacate. Vac* means "empty," so this word has to do with making something empty or leaving an empty space.
- *Aerate. Aero* is a root that has to do with air, so this word has to do with putting or placing air in a particular place.

Look up each of these words in a dictionary to see the precise definition.

Common Prefixes

Prefix	Meaning	Examples
a-, an-	without	amoral, anaerobic
ante-	before	antedate, antechamber
anti-	against, opposing	antipollution, antipathy
auto-	self	autobiography, autopilot
bene-	good or well	beneficial, benediction
bi-	two	bicycle, bipolar
cen-	hundred, hundredth	century, centimeter
circum-	around	circumnavigate, circumvent
co-	together, with	coauthor, copilot
com-	bring together	complete, compile
contra-	against	contradict, contraband
counter-	opposite, against	counterclockwise, counterterrorism
de-	away from, down, undoing	deactivate, detract
dis-	the opposite, undoing	disagree, disarm
ex-	out of, away from	exhale, expropriate
hetero-	different	heterogeneous, heterodox
homo-	same	homogenized, homonym
hyper-	above, excessive	hyperactive, hypertension
il-/in-/ir-	not	illiquid, inaudible, irregular
inter-	between	intercontinental, interject
intra-	within	intramural, intranet
mal-	bad or ill	maladjusted, malevolent
micro-	small	microbiology, microscope
milli-	thousand, thousandth	millennium, millisecond
mis-	bad, wrong	misbehave, misunderstand
mono-	one	monosyllable, monorail
neo-	new	neoclassical, neophyte
non-	not	nonessential, nonconformist
paleo-	old, ancient	paleontology, paleobiology
pan-	all, all over	pandemic, panorama
poly-	many	polygon, polynomial
post-	after, behind	postpone, postnasal
pre-	before	preview, premeditate
pro-	forward, before, in favor of	promote, projection
re-	back, again	rewrite, retract
retro-	backward	retrofit, retroactive
semi-	half	semiannual, semifinal
sub-	under, below	submarine, subway
super-	above, over	supersede, supervise
sym-/syn-	together, at the same time	symbiosis, synchronize
tele-	from a distance	telecommute, telemetry
therm-	heat	thermal, thermometer
tor	twist	torsion, contort
trans-	across	transcontinental, transatlantic
un-	not	unannounced, unnoticed
uni-	one	unicycle, unify
vert, verse	change	revert, reverse

Common Word Roots

Root	Meaning	Examples
act	do	transact, activate
aero	air	aerobics, aerospace
ambu	walk	ambulate, ambulatory
anthrop	human	anthropology, philanthropy
annu, anni	year	annual, anniversary
aster, astr	star	astronomy, asterisk
audi	hear	audible, auditory
biblio	book	bibliography, bibliographic
bio	life	biosphere, biography
brev	short	brevity, abbreviate
capit	head	decapitate, capital
card, cord, cour	heart	cardiology, discord
carn	flesh	carnivorous, carnage
cede	go	recede, precede
chron	time	chronology, synchronize
cide	killing	suicide, homicide
cis	cut	precise, scissors
claim	shout, cry out	exclaim, proclaim
cogn	know	cognition, recognize
crat	rule	autocratic, democrat
culp	blame	culpable, exculpate
dem	people	democracy, demographics
dic, dict	speak	dictation, predict
dorm	sleep	dormant, dormitory
fer	carry	transfer, refer
fuge	flee	refuge, centrifugal
geo	earth	geography, geologic
gram	something written or recorded	telegram, cardiogram
graph	to write	graphic, calligraphy
jac, ject	throw	eject, trajectory
jur	law	jury, jurisprudence
labor	work	laboratory, collaborate
loc	place	location, collocate
luc	light	translucent, illuminate
manu	hand	manuscript, manufacture
meter, metr	measure	barometer, metric
morph	form	morphology, amorphous
mort	die	mortuary, mortal
omni	all	omnipresent, omnipotent
op, oper	work	cooperate, operator
path	feeling, suffering	empathy, pathetic
ped, pod	foot	pedal, podiatry
philo, phil	like, love	philanthropy, philosophy
phobe	fear	phobic, claustrophobia
phon	sound	phonograph, stereophonic
photo	light	photographic, telephoto
phys	nature, body	physique, physical
scrib	write	scribble, scribe
stro, stru	build	destroy, construction
ter	earth	extraterrestrial, territory
vac	empty	vacuous, evacuate
verb	word	verbose, proverb
vid, vis	see	video, television
vol, volv	turn, roll	revolve, evolution

Common Suffixes

Suffix	Meaning	Examples
-able, -ible	capable or worthy of	likable, possible
-ful	full of	healthful, joyful
-fy, -ify	to make or cause	purify, glorify
-ish	like, inclined to, somewhat	impish, devilish
-ism	act of, state of	capitalism, socialism
-ist	one who does	conformist, cyclist
-ize	make into	formalize, legalize
-logue, -log	discourse	dialogue, travelogue
-logy	discourse	paleontology, dermatology
-ment	state of being	entertainment, amazement
-oid	like, resembling	humanoid, trapezoid
-ty, -ity	state of being	purity, acidity

Start Using What You Know

Use the following activity to further expand and strengthen your vocabulary.

USE CONTEXT CLUES TO FIGURE OUT MEANING

One good way to figure out the meaning of an unfamiliar word is to use what are called *context clues*. The *context* of a word is the phrases and sentence or sentences that surround it. Often, even if you have never heard a word before, you can guess its meaning by looking at the context in which it appears. The context may give you clues to help you figure out the meaning.

Study Some Examples

Here are some examples of how to use context clues to figure out the meaning of an unfamiliar word.

- During the snowstorm, the government sent home all *nonessential* personnel. (*Nonessential* must mean "not important" or "not needed.")
- When giving a person driving directions, you must be very *explicit*, or else the person may make a wrong turn. (*Explicit* must mean "precise" or "exact.")
- Even though the club members disagreed about the project, their conversation was *amicable*. (*Amicable* must mean something that is the opposite of disagreeing.)
- The chairperson *convened* a panel of experts to discuss global warming. (*Convened* must mean something related to bringing together a group of people.)

- They *commemorated* the end of the war by holding a big parade. (*Commemorate* must mean something like "get together to remember.")
- Even though there was no sign of a problem, Kisha had a *premonition* that something would go wrong. (A *premonition* must be a suspicion or feeling about something that is going to happen.)
- Gene was only in fourth grade, but he was *precocious* in his school subjects because he studied with his older brothers. (*Precocious* must mean something related to being advanced or ahead of what is expected.)
- Scientists are worried that this new disease, which spreads rapidly and for which there is no vaccine, may cause a *pandemic*. (A *pandemic* must be a disease outbreak that is widespread and out of control.)
- Brad was usually stable and predictable, but to our surprise, his behavior in this instance was *unorthodox*. (*Unorthodox* must mean something like "different from stable and predictable.")
- Though most people in the audience were calm and quiet, Sally was bubbly and *loquacious*. (*Loquacious* must mean the opposite of calm and quiet.)
- Brian took *vicarious* pleasure in Joan's debut on stage, even though he was not able to attend the performance. (*Vicarious* must have something to do with experiencing pleasure without being there.)
- Even though the hurricane is expected to arrive on shore at any moment, the sea is still *placid*. (*Placid* must have something to do with not being stirred up with high surf and waves. It may mean "calm.")
- The weapons and stolen cash were *tangible* evidence that the defendant was guilty. (*Tangible* must mean "real" or "concrete.")

Start Using What You Know

Use the following activity to further expand and strengthen your vocabulary.

Word Meaning Activity: Prefixes, Roots, and Suffixes

The words below use common prefixes, roots, and suffixes. Read the word and try to guess its meaning. Write down your guess. Then look up the dictionary meaning and write it down. From the dictionary or a thesaurus, list at least one synonym.

Word	My Guess	Dictionary Meaning	Synonym(s)
achromatic			
ambidextrous			
antipathy			
apathetic			
atypical			
benefit			
biography			
cacophony			
chronicle			
chronological			
cognizant			
contiguous			
contradict			
contrarian			
decompress			
defile			
democracy			
demography			
deregulate			
devalue			
geography			
geopolitics			
hemisphere			
interject			
judicial			
metrics			
microwave			
patricide			
polychrome			
preamble			
prologue			
retrofire			
retrograde			
subterranean			
subversive			
sympathy			
terrestrial			
translate			
ungainly			

Word Meaning Activity: Guess the Meaning of Words in Context

In this activity, a sentence is given and a word is italicized. By reading the sentence, guess the meaning of the word. Then look up the dictionary meaning and list at least one synonym.

Sentence	My Guess	Dictionary Meaning	Synonym(s)
Even though Dexter did not believe he was at fault, he made the *magnanimous* gesture of apologizing.			
Valerie and Samuel were usually friendly and polite, but on the subject of politics, their discussions were *acrimonious*.			
After at first insisting on watching television, Edie finally *acquiesced* to her sister's wish to play a board game.			
During the storm, it was difficult to hold the umbrella because the wind was so *erratic*.			
Leon was mildly enthusiastic about football and baseball, but he cheered the hockey team with great *fervor*.			
Margaret usually spoke up in class discussions, but that day she was *reticent* about expressing her opinion.			
Generally, Mr. Smith considered the input of others before making a decision, but on the subject of overtime, he made a *unilateral* determination.			
The threat from the invading army *galvanized* the soldiers into preparing for battle.			

USE IT OR LOSE IT

Many experts say that the best way to learn a new word is to use the word in your speech and writing. When you learn a new word, use it in different sentences. Say the word out loud. Do this several times. If you don't use it, you are likely to forget it. Make up new sentences with the word. Find ways to use the word in your daily life. The following activities will help you practice these skills.

Answering ASVAB Word Knowledge Items

When you take the ASVAB Word Knowledge test, use the knowledge and skills you have learned in this chapter to figure out the meanings of unfamiliar words. Here are seven steps you can follow to attack Word Knowledge items:

1. When you read the item, see if there are any context clues.
2. Mentally guess at the meaning of the word.
3. Scan the possible answers.
4. Eliminate those that are clearly wrong.
5. Pick the word that is the closest match—the *best* answer.
6. If you can't make an initial determination, try picking the word apart. Look for the prefixes, roots, and suffixes.
7. Give it your best shot and don't be afraid of guessing, as there is no penalty for a wrong answer on the paper-and-pencil version.

WORD LIST CHART

Use the following chart to record unfamiliar words that you read or hear. Write down your guess at the meaning of the word. Then look up the word in a dictionary or thesaurus to identify synonyms.

Word Meaning Activity: Using New Words in a Sentence

In the activity below, you are given several words. If you don't know the meaning of the word, look it up in the dictionary. Then create a sentence using the word.

Word	Dictionary Definition	My Sentence
adapt		
attire		
automation		
calamitous		
cursory		
defile		
digress		
evade		
futile		
garner		
mundane		
overbearing		
pensive		
peripheral		
solvent		
thwart		
undermine		
wary		

Word Meaning Activity: Create Synonyms

For each word, list at least one synonym for it. A *synonym* is a word that has the same or nearly the same meaning. Use a dictionary or thesaurus if necessary.

Word	Synonyms
abate	
abhor	
acquit	
cajole	
disdain	
erroneous	
facade	
feasible	
ghastly	
haughty	
incognito	
incriminate	
legitimate	
leverage	
liaison	
meander	
opulent	
purge	
rebuke	
rescind	
sanction	
simulate	
stooped	
tangible	
truncate	
ultimate	
undermine	
vacillate	
vie	
viscous	

Word List Chart

Word	My Guess	Synonyms

Vocabulary Quiz: Correct Definitions

Word	Definition
abscond	escape quickly and secretly
antagonist	adversary; someone who opposes
complacent	contented or satisfied
consensus	general agreement; an opinion agreed upon by most or all
defensive	protecting; defending
didactic	instructive; moralizing
docile	willing to obey
dormant	quiet; sleeping
exorbitant	well beyond reasonable or normal
fidelity	faith; loyalty
hyperbole	a large exaggeration
intermittent	periodic; irregular
neophyte	a beginner or novice
parity	equal in status or value
pivotal	referring to a turning point; of critical importance
reticent	reserved; inclined to keep feelings to oneself
succinct	clear and brief
tenacious	holding on firmly; stubborn
unilateral	by one person or group only; one-sided
venerate	to honor or hold in deep reverence

Chapter 11

Paragraph Comprehension

YOUR GOALS FOR THIS CHAPTER:

- **Find out what you can do to build reading comprehension skills.**
- **Study samples of each type of Paragraph Comprehension question.**
- **Learn tips for answering each Paragraph Comprehension question type.**

INTRODUCTION: ASVAB PARAGRAPH COMPREHENSION QUESTIONS

The ASVAB subtest called Paragraph Comprehension is designed to find out how well you obtain information from written material. Basically, it measures your reading comprehension skills. Paragraph Comprehension is especially important because it is part of the AFQT, the primary score for entering the military. To get in and get good job and training opportunities, you need to score well on this test. Just think about it: During your basic and specialized training, you will need to understand written material provided to you by your instructors. You will need to read and understand manuals that relate to your job. The military, regardless of branch, needs you to have a good understanding of everything you read.

ASVAB Paragraph Comprehension Test

Number of questions:

- Paper-and-pencil ASVAB: 15
- CAT-ASVAB: 10

Time limit:

- Paper-and-pencil ASVAB: 13 minutes
- CAT-ASVAB: 27 minutes

Item format: paragraphs with multiple-choice questions

Subjects covered: history, health, science, social studies, sports, current events, culture

Each item on the Paragraph Comprehension test consists of a paragraph or paragraphs followed by multiple-choice questions. You will need to read the paragraphs in order to answer the questions. The paragraphs will cover a wide variety of topics. Note, however, that you won't need to know anything about each topic other than what is in the paragraph. The paragraph will contain everything you need to know to answer the questions.

If you do the math, you will see that on the paper-and-pencil ASVAB, you have only 52 seconds to answer each item, including reading the passage. That means that you will need a test-taking strategy that allows you to approach the test both quickly and accurately. This chapter gives you some practical advice on how to approach the test, what kinds of questions to expect, and how to prepare yourself to get your best score on the ASVAB Paragraph Comprehension test.

BUILD YOUR READING COMPREHENSION SKILLS

Start your preparation for the ASVAB Paragraph Comprehension test by learning what you can do to build your reading comprehension skills. Here are some suggestions.

Read

The best way to improve your reading comprehension skills is to *read, read, read*. The more you read and practice your reading comprehension skills, the better off you will be on the test. The following list suggests a variety of materials that you should be reading.

What to Read

- *Books on subjects you like.* Whatever topic interests you, there are books about it—and reading those books will help you. If you are a sports fan, read books about great teams and famous games. If you love science fiction, read this year's most popular science fiction stories. If you are a history buff, read about the famous people and events of the past that made our world what it is today. Read autobiographies, books on politics, health books, science books, books on bicycling, nutrition, ice skating, organizing your life, and more. The list is endless. You don't need to buy these books; use your local library. The library is also a great place to read because it is quiet and you won't get distracted.
- *School books.* If you are in school, devour those textbooks. Focus on paragraph headings, see how the information is organized, and highlight critical statements and facts. Reading the textbook will not only expand your knowledge base for the other ASVAB tests, but also improve your vocabulary and your ability to understand what you are reading.
- *Newspapers.* Daily newspapers, especially those from a large town or city, offer plenty of reading opportunity on a variety of subjects.
- *The Internet.* Be selective about what you read on the Internet. Look for more lengthy passages, such as articles from a newspaper or extracts from a book. Avoid Internet sites where information is short, choppy, and abbreviated. That kind of reading will probably not help your reading comprehension skills. Reading Internet material can be helpful if you select the right stuff.

Learn New Words

The better your vocabulary, the easier it will be to understand what you are reading. The previous chapter of this book explained how to improve your vocabulary for the ASVAB Word Knowledge test. Following those suggestions will help you on the Paragraph Comprehension test as well. (You may wish to go back to the earlier chapter and review.)

How to Learn Words

- *Develop a word list.* As you are reading, identify words that you don't know or don't know very well. Based on the sentence or paragraph that contains the word, try to guess its meaning. Look up the meaning in a dictionary to be sure that you are correct. The previous chapter of this book gave you a word list chart for recording these words and their definitions. Use it to help your vocabulary grow.
- *Use context clues.* The *context* of a word is the other words and sentences that surround it. Often you can determine the meaning of a word from its context. One way to determine the meaning is to see how the word is used in the sentence.
- *Use prefixes, suffixes, and roots.* These word parts can help you decode a word's meaning. The Word Knowledge chapter gives you a whole laundry list of common prefixes, suffixes, and roots.

LEARN HOW TO ANSWER EACH QUESTION TYPE

There are four different types of Paragraph Comprehension questions. Each type asks you something different about the paragraph you read. Here is a list of the question types, along with tips for answering each type.

Type 1: Words in Context Questions

Some Paragraph Comprehension questions will ask you to determine the meaning of an unfamiliar word in the paragraph by looking at the context in which that word appears.

Example

Space flight is a somewhat common experience these days. Nowadays the International Space Station generally is the home of several astronauts from the United States and Russia. Astronauts generally spend several months in a weightless condition. They sleep, eat, and work without the effects of gravity. One of the effects of lengthy space travel is the atrophy of major muscle groups and the stress of travel on the heart and lungs. When these space travelers return to Earth, it takes several days for them to adjust back to gravity and to walk normally.

In this instance the word <u>atrophy</u> means

A. strengthening.
B. defining.
C. weakening.
D. breaking.

If the correct meaning was "strengthening" (choice A), then it probably wouldn't take several days for the space travelers to adjust back to gravity. If choice D, "breaking," was the correct answer, then the space travelers would probably not adjust at all, as their muscles would be in very poor shape. Muscles that are "defining," choice B, would result from exercise performed against gravity, not in a weightless environment. Choice C, "weakening," is the correct answer, because when muscles do not have a chance to work out, they get weaker.

Type 2: Main Idea Questions

Another kind of Paragraph Comprehension question asks about the main idea of the paragraph. In this kind of question, you may be asked to identify the main idea, to choose the best title for the paragraph (which is simply another way of stating the main idea), to identify the theme of the paragraph, or to identify the author's purpose in writing the paragraph.

Example

Bologna, Italy, is a city with 26 miles of covered walkways dating from the 1200s. The atmosphere of this beautiful city and its residents envelop you like a warm hug. In the center piazza of the city are two leaning towers, forming the most notable landmarks. Around the corner is the famous Roxy coffee bar, a hangout for many of the young university students who are studying medicine and political science. The nearby open marketplace bustles with color and excitement. Listening closely, you can hear many languages spoken by the tourists who visit each year.

In the paragraph above, which of the following best states the main idea of the passage?

A. Bologna is an old city.
B. University students love Bologna.
C. Bologna is an interesting place to visit.
D. Bologna has two leaning towers.

Clearly, choice C is the best and correct answer. The main idea is that Bologna is an interesting and vibrant city that attracts visitors from many places.

The other answers may be true, but they do not sum up the gist of the entire passage.

Often in a paragraph there will be one sentence that sums up the main idea of the paragraph as a whole. The question may ask you to choose which sentence that is. A sentence that sums up the main idea of a paragraph is called the *topic sentence*. In the following example, the sentences are numbered. As you read, try to pick out the topic sentence.

Example

(1) It was 6 p.m. on a cloudy, frigid day in the forest. (2) It had been sleeting the entire afternoon. (3) Simone was worried because her son hadn't returned home, as he usually did by this time. (4) Todd was always punctual, but tonight was different. (5) Todd had a basketball game after school, and a friend was supposed to drive him home. (6) Simone waited by the window in hopes of seeing Todd's smiling face as he came up the stairs.

Which of the following sentences best reflects the main idea of the paragraph?

A. 1
B. 3
C. 4
D. 5

The best and correct answer is choice B. In most paragraphs, the topic sentence is the first sentence. But in this case, as you might have noticed, it is not. Of the six sentences, sentence 3 best conveys the meaning of the entire paragraph. It tells you that for some reason, Todd was supposed to be home and wasn't, and that Simone was worried. Sentence 3 is the topic sentence. Sentences 1 and 2 just give an indication of the setting. Sentences 4 and 5 tell us that Todd was usually on time and that this day he had a basketball game. These sentences don't reflect the basic idea of the paragraph: that Simone was worried and why she was worried. Sentence 6 conveys no sense that there was any kind of problem or reason for worry, only that Simone was looking out the window hoping to see Todd.

"Best Title" Questions Another kind of main idea question will ask you to choose the best title for the paragraph. The best title is the one that best expresses the main idea of the paragraph as a whole. Here is an example.

Example

Visiting New York City and taking in a Broadway musical has always been a dream of Kevin's, but he knew a visit to the big city would be very expensive. He wanted to plan a short vacation to the city over a holiday weekend. Could he visit the city on a limited budget? Searching the Internet, he found opportunities for reduced-price hotels, half-price Broadway play tickets, and a special on train transportation. He was ecstatic as he started to pack his bags.

> For the paragraph above, which of the following is the best title?
> A. New York—An Expensive City
> B. Kevin's Budget
> C. Finding Cheap Tickets
> D. Kevin Goes to the Big City

Of the options, choice D is the best and correct answer. You don't know about Kevin's actual budget, so choice B is not correct. You do know that New York City is expensive to visit, but that is not the major point of the paragraph. Finding cheap tickets is part of the process, but it is not by any means the major thrust of the paragraph. Kevin's dream and all its parts is the most important theme of the paragraph.

Note that there are a number of other possible good titles for this paragraph. A few include *Kevin Goes to the Big City*, *New York City on a Budget*, and *Kevin's Dream Comes True*. Each of these titles conveys the main idea of the paragraph and would make a possible correct answer choice.

Author's Purpose Questions Another kind of main idea question will ask you to identify the author's purpose in writing the paragraph. Authors write for a variety of purposes: to describe, to raise issues or concerns, to move readers to action, to persuade readers to think in a particular way, to frighten, to give directions, to describe steps or procedures, to compare and contrast, or to entertain. Here is an example of this kind of question.

Example

Global warming is an increasingly serious environmental problem. It is caused by "greenhouse gases," which are created by things we do to the environment every day. But there are many little things people can do to reduce greenhouse gas emissions. You can do your part by carpooling to save gasoline. Four people can ride to school or work in a carpool instead of each person taking a car and driving alone. Save electricity by turning off the lights, the television, and the computer when you are through with them. Save energy by taking the bus or by riding your bicycle to school or to run errands. Walk to where you want to go. Recycle your cans, bottles, plastic bags, and newspapers. If you care about the future of this planet, help protect the environment! Get with the program!

> In the paragraph above, what is the author's purpose?
> A. To entertain readers
> B. To describe global warming
> C. To offer directions
> D. To move readers to action

The best answer is choice D. The purpose of the paragraph is to move you to act to protect the environment. The paragraph makes no attempt to entertain readers (choice A), nor does it really describe what global warming is (choice B). It does give suggestions for fixing global warming, but these are not really directions, so choice C is also incorrect.

Author's Attitude or Tone Questions Another kind of main idea question will ask you to identify the author's attitude or tone. In other words, you will need to determine how the author feels toward the subject of the paragraph. Is the author angry? Discouraged? Excited? Happy? For clues to how the author feels, look at the words he or she has chosen to use. Here is an example.

Example

Shooting a cat with a BB gun or anything else is animal cruelty and is illegal. The recent incident in our neighborhood should be reported to the Society for the Prevention of Cruelty to Animals, the local humane society, or the police. We must as a community band together to find the perpetrators, prosecute them, and get the person or persons into some serious counseling program. It's important for all of us to be watchful and to speak up about this horrific behavior. These incidents *must* be stopped before these individuals cause even more serious harm.

> In the above paragraph, which of the following best describes the author's tone?

A. Happy about the situation
B. Biased in favor of cats
C. Angry about the situation
D. Depressed about the situation

The correct answer is choice C, "angry about the situation." The author's anger is apparent in words and phrases like "horrific," "we must as a community band together," "prosecute them," and "these incidents *must* be stopped," There is nothing in the paragraph to support any of the other choices.

Type 3: Specific Details Questions

The third kind of ASVAB Paragraph Comprehension question will ask you to pick out specific details in the paragraph that you have read. In these paragraphs, a lot of information is often provided. You have to find the specific item or detail that the question asks about. For this kind of question, it is very helpful to read the question before you read the paragraph. Here is an example.

Example

Dental assistants perform a variety of patient care, office, and laboratory duties. They work chairside as dentists examine and treat patients. They make patients as comfortable as possible in the dental chair, prepare them for treatment, and obtain their dental records. Assistants hand instruments and materials to dentists and keep patients' mouths dry and clear by using suction or other devices. Assistants also sterilize and disinfect instruments and equipment, prepare trays of instruments for dental procedures, and instruct patients on postoperative and general oral health care.

In the above paragraph, where is the dental assistant when the doctor is examining the patient?
A. Right next to the dentist and the patient
B. In the laboratory
C. Finding dental records
D. Sterilizing instruments

The correct answer is choice A. The detail you are looking for is found in the sentence, "They work chairside as dentists examine and treat patients." So the correct answer is "right next to the dentist and the patient." At other times, a dental assistant may be in the laboratory (choice B), finding dental records (choice C), or sterilizing instruments (choice D), but when the dentist is working with the patient, the assistant is right by the dentist's side.

Sequence of Steps Questions One type of specific details question will ask you about the sequence (order) of steps in a process. The process may be something in nature, or it may be preparing a food item, operating a machine, making a repair, or something similar. To answer this kind of question correctly, you need to pay close attention to the order in which things happen or in which things are done. Reading the question before reading the passage is a very good strategy with this type of item. Here is an example.

Example

Making a Brownie

Preheat the oven to 350° and lightly spray the pan with cooking oil. In a saucepan, combine the butter and chopped chocolate. Set this over a low heat until melted. Stir the mixture and set aside to cool. In a mixing bowl, mix together the flour, cocoa, baking powder, salt, and cinnamon. Measure the sugar into a large bowl and mix in the cooled butter-chocolate mixture. Add the eggs, vanilla, and water. Mix very well. Add the dry ingredients and mix until blended. Spoon the batter into the pan and bake for 23 to 25 minutes. Cool in the pan and cut into squares.

When should you add the eggs, vanilla, and water?
A. After melting the chocolate
B. Before cooling the melted butter and chocolate
C. After spooning the batter into the pan
D. Before adding the dry ingredients

The correct answer is choice D. There is a lot of detail in this paragraph, but the time for adding the eggs, vanilla, and water is before combining the melted butter and chocolate with the dry ingredients.

Type 4: Interpretation Questions

The fourth type of ASVAB Paragraph Comprehension question will ask you to interpret something that you read. Often an author will suggest or hint at a certain idea, but will not state it directly. It is up to you to figure out the author's meaning by "reading between the lines" and drawing your own conclusion. When you do this, you analyze the author's words, you think about what they mean, and you put your ideas together to create something new and original. This process is called *making an inference*. Here is an example of a question that asks you to make an inference.

Example
_____

The dinosaurs became extinct at the end of the Cretaceous Period. Reasons for this event are still undetermined. Some scientists attribute it to a cataclysmic occurrence, such as a meteor that struck the Earth, kicking up vast quantities of dust. Another possibility is the great increase in volcanic activity that is known to have taken place at the end of the Cretaceous period. Either cause could have filled the atmosphere with enough dust and soot to block out the sunlight, producing a dramatic climate change. Recent discoveries indicate that in many places on several continents, there is a layer of iridium in geologic strata associated with the Cretaceous period. Iridium is an element associated with lava flows.

An inference test item might look like this:

According to this passage, the dinosaurs became extinct because of which of the following conditions?
A. Disappearance of vegetation
B. Radiation from the sun
C. Climate changes
D. Volcanic activity

The best answer is choice C, climate changes. The paragraph does not come to a conclusion about which of the two events caused the extinction of the dinosaurs, but both seem to point to the fact that climate changes were the eventual cause of their disappearance. This conclusion is not stated in the passage. You needed to infer this from reading the passage.

Answering ASVAB Paragraph Comprehension Questions

Here are a few basic tips for answering ASVAB Paragraph Comprehension questions.

1. *Read the questions before reading the paragraph.* This is an absolute must—no exceptions. You need to do this so that you can focus on the pertinent parts of the paragraph and ignore the remainder. Remember that you have a very limited amount of time and you need to get to the right answer as quickly as possible.

2. *Read the paragraph next.* When you read the paragraph, focus on the answers to the questions and ignore all the extraneous information that might be in the paragraph. There will be a lot of fluff in the passage that has nothing to do with the questions you are being asked. Don't try to fully understand all the information that is given. Your job is to answer the question, not to be an expert on the subject at hand.

3. In your mind, *try to answer the question* as you read each paragraph. Try to guess the answers in your own mind. Your answer will probably be similar to one of the answer choices.

4. *Now look at the answer choices* to see which one matches the answer you reached in your mind. Pick the choice that is closest.

5. *Guess if you must.* If you cannot decide which answer choice is correct, try to eliminate choices that are clearly wrong. Then guess, even if you cannot eliminate more than one or two choices. The more choices you eliminate, the better your chance of guessing correctly. There is no penalty for wrong answers on the paper-and-pencil version, so be sure to mark an answer for every question, even if you have to guess.

Chapter 12

Mathematics Knowledge 1: Algebra and Probability

YOUR GOALS FOR THIS CHAPTER:

- Find out what algebra and probability topics you need to know for the ASVAB.

- Review techniques and strategies for solving algebra and probability problems.

- Study sample algebra and probability problems and their step-by-step solutions.

- Measure your test readiness by taking an algebra and probability quiz.

INTRODUCTION: ASVAB MATHEMATICS KNOWLEDGE QUESTIONS

The ASVAB Mathematics Knowledge test measures your ability to solve problems using concepts taught in high school math courses. These concepts include various topics in algebra, probability, and geometry. You'll need to know about solving equations, setting up ratios and proportions, graphing on a coordinate plane, determining the probability of a given event, identifying plane and solid geometric figures, and calculating perimeter, area, and volume.

It is important that you do well on the Mathematics Knowledge test because it is one of the four ASVAB tests that are used to calculate the AFQT—your military entrance score. That's why it pays to spend time reviewing topics in algebra, probability, and geometry and tackling plenty of sample ASVAB Mathematics Knowledge questions.

The following pages offer a quick but important overview of the basic algebra and probability that you need to know if you are to score well on the ASVAB. Make sure that you carefully review and test yourself on every topic covered in this section. Also make sure that you learn how to use all

---

ASVAB Mathematics Knowledge Test

Number of Questions:

- Paper-and-pencil ASVAB: 25
- CAT-ASVAB: 15

Time Limit:

- Paper-and-pencil ASVAB: 24 minutes
- CAT-ASVAB: 23 minutes

Suggested Review Plan for Algebra and Probability

- Work through the algebra and probability review at your own pace.
- Pay careful attention to the examples provided.
- After you complete the review, take the algebra and probability quiz.
- Score your quiz.
- Study any problems you answered incorrectly. Reread the corresponding review section, then try the problem again. Work until you understand how to solve the problem.

Remember that you probably won't get these exact problems on the ASVAB, but knowing how to solve them will help you do your best on the actual test.

---

of the problem-solving methods presented in the examples. Chapter 13 will provide a similar review of basic concepts and problem-solving methods in geometry. If you master the core information presented in these two chapters, you will be able to answer ASVAB Mathematics Knowledge questions with relative ease.

ALGEBRA AND PROBABILITY REVIEW

Topics	
The language of algebra	Monomials, binomials, and polynomials
Evaluating expressions	Solving quadratic equations
Solving equations for one unknown	Algebraic fractions
Inequalities	Graphing on a number line
Ratios and proportions	
Solving equations for two unknowns	Graphing on a coordinate plane
	Probability

The Language of Algebra

Algebra uses arithmetic functions and processes, but some of the numbers are replaced by letters. The letters merely represent numbers that are currently unknown or that can change in value according to circumstances. In algebra, a letter representing a number that can change in value is called a *variable*.

- An expression such as $6x$ means "6 times some number, currently unknown" or "some number, currently unknown, times 6."
- An expression such as $x + 7$ means "some number, currently unknown, plus 7."
- An expression such as $x - 12$ means "some number, currently unknown, less 12."
- An expression such as $\dfrac{x}{5}$ means "some number, currently unknown, divided by 5," or "the ratio of some number and 5."

Very often verbal expressions in word problems need to be translated into algebraic expressions before they can be solved. Here are some examples of verbal expressions and their algebraic counterparts.

Verbal Expression	Algebraic Expression
Some number plus 7	$x + 7$
Some number subtracted from 8	$8 - x$
8 subtracted from some number	$x - 8$
The product of some number and 12	$12x$
The product of 5 and the sum of x and y	$5(x + y)$
Some number divided by 4	$\dfrac{x}{4}$
The ratio of 6 and some number	$\dfrac{6}{x}$
9 times some number plus the sum of 5 and y	$9x + (5 + y)$
12 less the sum of 3 and some number	$12 - (3 + x)$

Evaluating Expressions

To evaluate an algebraic expression, substitute the given value for the unknown and then perform the arithmetic as indicated.

Examples

Evaluate $a + b + c$ if $a = 2$, $b = 4$, and $c = 3$.

Substitute each value for the corresponding letter and then do the addition as indicated.

$2 + 4 + 3 = 9$

Evaluate $2x^2 + 4y + 5$ if $x = 2$ and $y = 3$.

$$2(2)^2 + 4(3) + 5 = 2(4) + 12 + 5$$
$$= 8 + 12 + 5$$
$$= 25$$

Evaluate $\dfrac{a + b}{4} + \dfrac{a}{b + c}$ if $a = 2$, $b = 6$, and $c = 10$.

$$\frac{2 + 6}{4} + \frac{2}{6 + 10} = \frac{8}{4} + \frac{2}{16}$$
$$= 2 + \frac{2}{16}$$
$$= 2\frac{1}{8}$$

HINT *Don't forget:* Order of operations is critical!
- Simplify anything within the parentheses.
- Apply the powers or exponents.
- Multiply and divide in order from left to right.
- Add and subtract in order from left to right.

Solving Equations for One Unknown

An *equation* is a mathematical statement that contains an equal (=) sign. When an equation contains a letter standing for an unknown number, you can use the equation to find the value of that unknown. This is called *solving the equation for the unknown.*

Think of an equation as a balanced scale. Everything to the right of the = sign has to balance with everything on the left side of the = sign.

Because an equation is balanced, it will stay in balance if you do the same thing to the numbers on both sides of the = sign. For example, the equation $10 = 10$ will stay balanced if you add 3 to both sides. The new equation will be $13 = 13$. Similarly, the equation $x + y = x + y$ will stay balanced if you subtract 10 from both sides. The new equation will be $x + y - 10 = x + y - 10$.

Similarly, the equation $x + y = a + b$ will stay balanced if you subtract 8 from both sides. The new equation will be $x + y - 8 = a + b - 8$.

To solve an equation and find the value of an unknown, you need to get the *unknown* on one side of the equation and *all the other terms* on the other side of the equation. Consider this example.

Solve: $y - 4 = 20$

Add 4 to both sides:

$$y - 4 + 4 = 20 + 4$$
$$y = 20 + 4$$
$$y = 24$$

> **HINT** *Another way to think about it.* Here is a simple way to think about using addition or subtraction to solve an equation: Just move the number you are adding or subtracting from one side of the = sign to the other and change its sign (either − to + or + to −). So in the equation shown, move the − 4 to the other side of the = sign and make it + 4. This makes the equation $y = 20 + 4$ or 24.

An equation will also stay balanced if you multiply or divide both sides by the same number. So you can also use these operations to solve equations.

Example (Division)

Solve: $3x = 18$

You want to get x all alone on the left side of the equation, so divide $3x$ by 3. Since $3/3 = 1$, $3x/3 = x$. To maintain the balance, divide the right side of the equation by 3 as well: $\frac{18}{3} = 6$. So $x = 6$.

> **HINT** *Another way to think about it.* Here is another way to think about using division to solve an equation: In a problem such as $3x = 18$, instead of trying to divide both sides by 3, simply move the 3 across the = sign and make it the denominator of the other side of the equation:
>
> $$3x = 18$$
> $$x = \frac{18}{3}$$
> $$x = 6$$

Example (Multiplication)

$$\frac{x}{3} = 12$$

Multiply both sides by 3.

$$3\frac{x}{3} = 3(12)$$
$$x = 3(12)$$
$$x = 36$$

> **HINT** *Another way to think about it.* Here is another way to think about using multiplication to solve an equation: In a problem such as $\frac{x}{2} = 20$, move the denominator 2 across the = sign and make it the multiplier on the other side.
>
> $$\frac{x}{2} = 20$$
> $$x = 2(20)$$
> $$x = 40$$

Examples

Solve for z:

$$16 + z = 24$$

Subtract 16 from both sides.

$$z = 24 - 16 = 8$$

Solve for x:

$$\frac{x}{5} - 4 = 2$$

Add 4 to both sides.

$$\frac{x}{5} = 2 + 4$$

$$\frac{x}{5} = 6$$

Multiply each side by 5 to isolate x on the left side.

$$x = 5(6) = 30$$

Solve for a:

$$\frac{2}{3}a - 5 = 9$$

Add 5 to both sides.

$$\frac{2}{3}a = 9 + 5$$

$$\frac{2}{3}a = 14$$

Multiply each side by $\frac{3}{2}$.

$$a = 14\left(\frac{3}{2}\right)$$

$$a = \frac{42}{2} = 21$$

HINT   *Divide out common factors.* Another way to do this final step is to divide out common factors.

$$\frac{\overset{7}{\cancel{14}}}{1} \times \frac{3}{\underset{1}{\cancel{2}}} = 21$$

Solve for y:

$$7y = 3y - 12$$
$$7y - 3y = -12$$
$$4y = -12$$
$$y = -\frac{12}{4}$$
$$y = -3$$

Inequalities

Unlike equations, inequalities are statements that show that certain relationships between selected variables and numbers are *not* equal. Instead of using the equal sign, you use the "greater than" sign (>) or the "less than" sign (<). At times you may see signs for "greater than or equal to" (≥) or "less than or equal to" (≤).

Examples

$x > 13$ means that the value of x is greater than 13.

$y < 45$ means that the value of y is less than 45.

$x - y < 33$ means that when y is subtracted from x, the result is less than 33.

$\frac{x}{5} \geq z$ means that when x is divided by 5, the result is greater than or equal to the value of z.

Solving Inequalities   If you work problems with inequalities, you can treat them much like equations. If you multiply or divide both sides by a negative number, you must reverse the sign.

Examples

Solve for x:

$$3x + 5 > 8$$
$$3x + 5 - 5 > 8 - 5 \quad \text{Subtract 5 from both sides.}$$
$$3x > 3 \quad \text{Divide each side by 3.}$$
$$x > 1$$

Solve for x:

$$5 - 2x > 9$$
$$5 - 5 - 2x > 9 - 5 \quad \text{Subtract 5 from both sides.}$$
$$-2x > 4 \quad \text{Divide by } -2 \text{ (Be sure to reverse the sign, since you are dividing by a negative number.)}$$
$$x < -2$$

Ratios and Proportions

A *ratio* is a comparison of one number to another. A ratio can be represented by a fraction.

Example

On a certain road, there are 6 cars for every 4 trucks. The ratio is $\frac{6}{4}$ or $\frac{3}{2}$.

A *proportion* is an equation stating that two ratios are equivalent. Ratios are equivalent if they can be represented by equivalent fractions. A proportion may be written

$$\frac{a}{b} = \frac{c}{d}$$

where a/b and c/d are equivalent fractions. This proportion can be read "a is to b as c is to d."

Like any other equation, a proportion can be solved for an unknown by isolating that unknown on one side of the equation. In this case, to solve for a, multiply both sides by b:

$$\frac{(b)a}{b} = \frac{(b)c}{d}$$

$$a = \frac{bc}{d}$$

Examples

Solve for p:

$$\frac{c}{p} = \frac{h}{j}$$

Multiply both sides by p.

$$(p)\frac{c}{p} = (p)\frac{h}{j}$$

$$c = \frac{ph}{j}$$

Solve for a:

$$\frac{a}{4} = \frac{3}{6}$$

$$(4)\frac{a}{4} = (4)\frac{3}{6}$$

$$a = \frac{12}{6}$$

$$a = 2$$

HINT *Another way to solve it.* Another way to solve proportion problems is to find the cross products.

Solve for a:

$$\frac{a}{4} = \frac{3}{6}$$

Cross-multiply:

$$a \times 6 = 4 \times 3$$

$$6a = 12$$

$$a = 2$$

Solve for k:

$$\frac{2}{5} = \frac{8}{k}$$

$$2k = 40$$

$$k = 20$$

Solving Equations for Two Unknowns

An equation may have two unknowns. An example is $3a + 3b = 9$. If you are given two equations with the same unknowns, you can solve for each unknown. Here is how this process works:

Solve for a and b:

$$3a + 4b = 9$$
$$2a + 2b = 6$$

Follow these steps:

Step 1. Multiply one or both equations by a number that makes the number in front of one of the unknowns the same in both equations.

Multiply the second equation by 2 to make $4b$ in each equation.

$$2(2a + 2b = 6)$$
$$4a + 4b = 12$$

Step 2. Add or subtract the two equations to eliminate one unknown. Then solve for the remaining unknown.

$$3a + 4b = 9$$
$$\underline{-(4a + 4b = 12)}$$
$$-a = -3$$
$$a = 3$$

Step 3. Insert the value for the unknown that you have found into one of the two equations. Then solve for the other unknown.

$$3a + 4b = 9$$
$$3(3) + 4b = 9$$
$$9 + 4b = 9$$
$$4b = 0$$
$$b = 0$$

Answer: $a = 3$, $b = 0$

Examples

Solve for j and k: $j + k = 7$
$$j - k = 3$$

You can skip Step 1 of the solution process because the number in front of each of the unknowns is understood to be 1.

Step 2. Add the equations.

$$j + k = 7$$
$$\underline{j - k = 3}$$
$$2j = 10$$
$$j = 5$$

Step 3. Substitute the solution for j into one of the equations and solve for k.

$$5 + k = 7$$
$$k = 7 - 5$$
$$k = 2$$

Solve for c and d: $3c + 4d = 2$
$$6c + 6d = 0$$

Step 1. Multiply the first equation by 2.

$$2(3c + 4d = 2)$$
$$6c + 8d = 4$$

Step 2. Subtract the two equations and solve for one unknown.

$$6c + 8d = 4$$
$$\underline{-(6c + 6d = 0)}$$
$$2d = 4$$
$$d = 2$$

Step 3. Substitute the solution for *d* into one of the equations and solve for *c*.

$$3c + 4(2) = 2$$
$$3c + 8 = 2$$
$$3c = -6$$
$$c = -2$$

Answer: $d = 2$, $c = -2$

Monomials, Binomials, and Polynomials

You can guess by the prefix *mono-* that a monomial has something to do with "one." A *monomial* is a mathematical expression consisting of only one term. Examples include $12x$, $3a^2$, and $9abc$.

A *binomial* (the prefix *bi-* means "two") has exactly two terms: $12z + j$.

A *polynomial*, as indicated by the prefix *poly-*, meaning "many," has two or more terms. Examples include $x + y$, $x + y + z$, and $y^2 - 2z + 12$.

Adding and Subtracting Monomials　If the variables are the same, just add or subtract the numbers.

Examples

Add:

$$\begin{array}{r} 9y \\ 11y \\ \hline 20y \end{array}$$

Subtract:

$$\begin{array}{r} 30b \\ -15b \\ \hline 15b \end{array}$$

Add:

$$\begin{array}{r} 12a^2bc \\ 4a^2bc \\ \hline 16a^2bc \end{array}$$

Subtract:

$$\begin{array}{r} 12a^2bc \\ -4a^2bc \\ \hline 8a^2bc \end{array}$$

Multiplying Monomials　When multiplying monomials, multiply any numbers, then multiply unknowns.

Add any exponents. Keep in mind that in a term like x or $2x$, the x is understood to have the exponent 1 even though the 1 is not shown.

Examples

$$(2k)(k) = 2k^2$$
$$(3x)(2y) = 6xy$$
$$(k^2)(k^3) = k^5$$
$$(j^3k)(j^2k^3) = j^5k^4$$
$$-5(b^4c)(-3b^3c^5) = 15b^7c^6$$

Dividing Monomials　To divide monomials, divide the numbers and subtract any exponents (the exponent of the divisor from the exponent of the number being divided).

Examples

$$\frac{2g^5}{6g^3} = \frac{1}{3}g^2$$
$$3\frac{g^5}{g^3} = 3g^2$$
$$\frac{12a^6b^2}{3a^3b} = 4a^3b$$
$$\frac{m^5}{m^8} = m^{-3} = \frac{1}{m^3}$$
$$\frac{-5(ab)(ab^2)}{ab} =$$

This example can be handled in two ways. One way is to simplify the numerator:

$$\frac{-5a^2b^3}{ab} = -5ab^2$$

Another way is to divide out similar terms:

$$\frac{-5(\cancel{ab})(ab^2)}{\cancel{ab}} = -5ab^2$$

Either way will give you the correct answer.

Adding and Subtracting Polynomials　Arrange the expressions in columns with like terms in the same column. Add or subtract like terms.

Examples (Addition)

Add:

$$\begin{array}{r} j^2 + jk + k^2 \\ 3j^2 + 4jk - 2k^2 \\ \hline 4j^2 + 5jk - k^2 \end{array}$$

Add:

$$3p^3 + 2pq + p^2 + r^3$$
$$2p^3 - 6pq + 2p^2 + 3r^3$$
$$\underline{-p^3 + pq + 3p^2 - 6r^3}$$
$$4p^3 - 3pq + 6p^2 - 2r^3$$

Examples (Subtraction)

Remember that when you subtract negative terms, you change the sign and add.

Subtract:

$$j^2 + jk + k^2$$
$$\underline{-(3j^2 + 4jk - 2k^2)}$$
$$-2j^2 - 3jk + 3k^2$$

Subtract:

$$9pq^3 + 3ab - 12m^2$$
$$\underline{-(3pq^3 + 2ab - 15m^2)}$$
$$6pq^3 + ab + 3m^2$$

Multiplying Polynomials

To multiply polynomials, multiply each term in the first polynomial by each term in the second polynomial. The process is just like regular multiplication. For example, if you multiply 43 times 12, the problem looks like this:

$$43$$
$$\underline{\times 12}$$
$$86 \quad \text{Multiply } 2 \times 3 \text{ and } 2 \times 4 \text{ to make } 86$$
$$\underline{43} \quad \text{Then multiply } 1 \times 3 \text{ and } 1 \times 4 \text{ to make } 43.$$
$$516 \quad \text{Add the results.}$$

Using polynomials, the process is just the same.

Examples

Multiply:

$$2g - 2h$$
$$\underline{\times 3g + h}$$
$$2gh - 2h^2 \qquad \text{Multiply } h \times 2g \text{ and } h \times -2h$$

Multiply $h \times 2g$ and $h \times -2h$ to make $2gh - 2h^2$.
Multiply $3g \times 2g$ and $3g \times -2h$ to make $+6g^2 - 6gh$.

$$6g^2 - 6gh$$
$$\overline{6g^2 - 4gh - 2h^2} \qquad \text{Add.}$$

Multiply:

$$3b + a$$
$$\underline{2b - 2a}$$
$$-6ab - 2a^2 \qquad \text{Multiply } -2a \times 3b \text{ and } -2a \times$$

Multiply $-2a \times 3b$ and $-2a \times a$ to make $-6ab - 2a^2$.
Multiply $2b \times 3b$ and $2b \times a$ to make $6b^2 + 2ab$.

$$6b^2 + 2ab$$
$$\overline{6b^2 - 4ab - 2a^2} \qquad \text{Add.}$$

Dividing a Polynomial by a Monomial

Just divide the monomial into each term of the polynomial.

Examples

Divide:

$$\frac{8b^2 + 2b}{2b} = \frac{8b^2}{2b} + \frac{2b}{2b}$$
$$= 4b + 1$$

Divide:

$$\frac{9c^3 + 6c^2 + 3c}{3c} = \frac{9c^2}{3c} + \frac{6c^2}{3c} + \frac{3c}{3c}$$
$$= 3c^2 + 2c + 1$$

Dividing a Polynomial by a Polynomial

To divide a polynomial by another polynomial, first make sure the terms in each polynomial are in descending order (i.e., cube → square → first power).

For example, $6c + 3c^2 + 9$ should be written $3c^2 + 6c + 9$.

$10 + 2c + 5c^2$ should be written $5c^2 + 2c + 10$.

Then use long division to solve the problem.

Example

Divide $(a^2 + 18a + 45)$ by $(a + 3)$

$$a + 3 \overline{)a^2 + 18a + 45}$$

$$\begin{array}{r} a + 15 \\ a + 3 \overline{)a^2 + 18a + 45} \\ \underline{a^2 + 3a} \\ + 15a + 45 \\ \underline{+ 15a + 45} \\ 0 \end{array}$$

Multiply a times $a + 3$ and subtract the result, leaving $15a$. Bring down the 45 and continue by dividing a into $15a$. Subtracting $15a + 45$ from $15a + 45$ gives zero.

Check your work by multiplying $(a + 3)(a + 15)$ to get $a^2 + 18a + 45$.

> **HINT** *A Faster Way?* If you run into a problem like this one on the ASVAB, it might be quicker to multiply the divisor (in this case, $a + 3$) by each answer choice to see which choice produces the dividend (in this case, $a^2 + 18a + 45$).

Factoring a Polynomial   A *factor* is a number that is multiplied to get a product. *Factoring* a mathematical expression is the process of finding out which numbers, when multiplied together, produce the expression.

To factor a polynomial, follow these two steps:

- Find the largest common monomial in the polynomial. This is the first factor.
- Divide the polynomial by that monomial. The result will be the second factor.

Examples

Factor the following expression:

$$5y^2 + 3y$$

y is the largest common monomial.

$$\frac{(5y^2 + 3y)}{y} = 5y + 3$$

y and $5y + 3$ are the two factors

Factor the following expression:

$$8x^3 + 2x^2$$

$2x^2$ is the largest common monomial.

$$\frac{(8x^2 + 2x^2)}{2x^2} = 4x + 1$$

$2x^2$ and $4x + 1$ are the two factors.

To check your work, multiply the two factors. The result should be the original expression.

$$2x^2 \times (4x + 1) = 8x^3 + 2x^2$$

Special Case: Factoring the Difference between Two Squares

Examples

Factor the following expression:

$$y^2 - 100$$

In this expression, each term is a perfect square; that is, each one has a real-number square root. The square root of y^2 is y, and the square root of 100 is 10.

When an expression is the difference between two squares, its factors are the sum of the squares $(y + 10)$

and the difference of the squares $(y - 10)$. Multiplying the plus sign and the minus sign in the factors gives the minus sign in the original expression.

> **HINT** *Remember the rules for multiplying positive and negative terms.* Don't forget that the product of a positive term (+) and a negative term (−) is a negative term (−).

Factoring Polynomials in the Form $ax^2 + bx + c$, Where a, b, and c Are Numbers   Remember that you want to find two factors that when multiplied together produce the original expression.

Examples

Factor the expression:

$$x^2 + 5x + 6$$

First, you know that x times x will give x^2, so it is likely that each factor is going to start with x.

$$(x\quad)(x\quad)$$

Now you need to find two factors of 6 that when added together give the middle term of 5. Some options are 1 and 6 and 2 and 3. 2 and 3 add to 5, so add those numbers to your factors. Now you have

$$(x\quad2)(x\quad3)$$

Finally, deal with the sign. Since the original expression is all positive, both signs in the factors must be positive. So the two factors must be

$$(x + 2)(x + 3)$$

Check your work by multiplying the two factors to see if you come up with the original expression.

$$(x + 2)(x + 3) = x^2 + 5x + 6$$

Factor the expression:

$$6x^2 + 8x - 8$$

$6x^2$ can be factored into either $(6x)(x)$ or $(2x)(3x)$. Using the latter, the first terms in our factors are as follows:

$$(2x\quad)(3x\quad)$$

Now let's consider the −8. Factors of 8 can be $(8)(1)$ or $(2)(4)$. Let's try 2 and 4, so our factors are now:

$$(2x\quad2)(3x\quad4)$$

Now for the signs. In order to get a minus 8 in the original expression, one of the numbers must be a negative and the other a positive. Let's try making the 4 negative, making the factors:

$$(2x + 2)(3x - 4) = 6x^2 - 2x - 8$$

That's close, but the original expression was $6x^2 + 8x - 8$, not $6x^2 - 2x - 8$. What if we switched the numbers 2 and 4?

$$(2x + 4)(3x - 2) = 6x^2 + 8x - 8$$

Now the factors give the original expression when multiplied together, so this is the correct answer.

> **HINT** *It helps to practice.* Don't worry about factoring. Factoring is a little bit of intuition and a lot of practice. Once you do a lot of these kinds of problems, factoring will become almost second nature.

Solving Quadratic Equations

A quadratic equation is one that is written in the form $ax^2 + bx + c$, where a, b, and c are numbers.

To solve an equation in this form for x, set the expression equal to zero. Note that x will have more than one value.

Follow these steps:

- Put all the terms of the expression on one side of the = sign and set it equal to zero.
- Factor the equation.
- Set each factor equal to zero.
- Solve the equations.

Examples

Solve:

$$x^2 + 7x = -10$$
$$x^2 + 7x + 10 = 0 \qquad \text{Add 10 to both sides}$$

in order to get all the terms on one side of the = sign. Now the expression is equal to 0.

$(x + 5)(x + 2) = 0$	Factor the equation.
$x + 5 = 0 \quad x + 2 = 0$	Set each factor equal to zero.
$x = -5 \quad\quad x = -2$	Solve the equations.

Check your answer by substituting each value back into the original equation.

$$x^2 + 7x = -10$$
$$(-5)^2 + 7(-5) = -10$$
$$25 - 35 = -10$$
$$(-2)^2 + 7(-2) = -10$$
$$4 - 14 = -10$$

Solve for y:

$$y^2 - 4y = 12$$
$$y^2 - 4y - 12 = 0 \qquad \text{Move all terms to one}$$

side and set the equation equal to zero.

$(y - 6)(y + 2) = 0$	Factor.
$y - 6 = 0 \quad y + 2 = 0$	Set each factor equal to zero.
$y = 6 \quad\quad y = -2$	Solve the equations.

Algebraic Fractions

An algebraic fraction is a fraction containing one or more unknowns.

Reducing Algebraic Fractions to Lowest Terms To reduce an algebraic fraction to lowest terms, factor the numerator and the denominator. Cancel out or divide common factors.

Example

Reduce to lowest terms:

$$\frac{3y - 3}{4y - 4} = \frac{3(y - 1)}{4(y - 1)} \qquad \text{Factor each expression.}$$

$$= \frac{3(y - 1)}{4(y - 1)} \qquad \text{Divide out common terms.}$$

$$= \frac{3}{4} \qquad \text{Reduce the fraction to lowest term}$$

> **HINT** *Division reminder:* Don't divide out common terms from a term that includes an addition or subtraction sign.
>
> For example, in a term such as $\frac{x + 2}{2}$, you cannot divide out the 2s.
>
> **Wrong:** $\frac{x + 2}{2}$

Example

Reduce to lowest terms:

$$\frac{k^2 + 2k + 1}{3k + 3} = \frac{(k + 1)(k + 1)}{3(k + 1)} \qquad \text{Factor each expression.}$$

$$= \frac{(k + 1)(k + 1)}{3(k + 1)} \qquad \text{Divide out common terms.}$$

$$= \frac{k + 1}{3}$$

Adding or Subtracting Algebraic Fractions with a Common Denominator

To add or subtract algebraic functions that have a common denominator, combine the numerators and keep the result over the denominator. Reduce to lowest terms.

Examples

$$\frac{3y+9y}{m}=\frac{12y}{m}$$

$$\frac{9y-8}{z}-\frac{6y-6}{z}=\frac{9y-8-6y-(-6)}{z}$$

$$=\frac{9y-6y-8+6}{z}$$

$$=\frac{3y-2}{z}$$

Adding or Subtracting Algebraic Fractions with Different Denominators

To add or subtract algebraic fractions that have different denominators, examine the denominators and find the least common denominator. Then change each fraction to the equivalent fraction with that least common denominator. Combine the numerators as shown in the previous section. Reduce the result to lowest terms.

Examples

$$\frac{7}{a}+\frac{12}{b}=\frac{7b}{ab}+\frac{12a}{ab}$$

$$=\frac{(12a+7b)}{ab}$$

Note that ab is the least common denominator.

In this example, $12x$ is the least common denominator, as $4x$ and $6x$ both divide into it.

$$\frac{g+2}{4x}+\frac{g-5}{6x}=\frac{3}{3}\times\frac{g+2}{4x}+\frac{2}{2}\times\frac{g-5}{6x}$$

$$=\frac{3g+6}{12x}+\frac{2g-10}{12x}$$

$$=\frac{3g+6+2g-10}{12x}$$

$$=\frac{5g-4}{12x}$$

Note that you have to multiply in order to make each term contain the least common denominator.

Multiplying Algebraic Fractions

When multiplying algebraic fractions, factor any numerator and denominator polynomials. Divide out common terms where possible. Multiply the remaining terms in the numerator and denominator together. Be sure that the result is in lowest terms.

Examples

Multiply the following fractions:

$$\frac{y^2}{2j}\times\frac{2j}{3y}=\frac{2y^2j}{6yj}$$

$$=\frac{2\,y^2\,j}{6\,yj}=\frac{1y}{3}=\frac{y}{3}$$

Divide out common terms, then multiply. Reduce to lowest terms.

Multiply the following fractions:

$$\frac{6}{x+1}\times\frac{3x+3}{6}=\frac{6}{x+1}\times\frac{3(x+1)}{6}\quad\text{Factor where possible.}$$

$$=\frac{6}{x+1}\times\frac{3\,(x+1)}{6}\quad\text{Divide out common terms and multiply.}$$

$$=3$$

Dividing Algebraic Fractions

To divide algebraic fractions, follow the same process used to divide regular fractions: invert one fraction and multiply.

Examples

Divide the following fractions:

$$\frac{3k^2}{5}\div\frac{2k}{z}=\frac{3k^2}{5}\times\frac{z}{2k}\quad\text{Invert the second fraction and multiply.}$$

$$=\frac{3k^2}{5}\times\frac{z}{2k}\quad\text{Simply where possible.}$$

$$=\frac{3kz}{10}$$

Divide the following fractions:

$$\frac{8m^3}{15}\div\frac{6m^2}{3}=\frac{8m^3}{15}\times\frac{3}{6m^2}\quad\text{Invert and multiply.}$$

$$=\frac{8m^3}{15}\times\frac{3}{6m^2}\quad\text{Simply where possible.}$$

$$=\frac{4m}{5}\times\frac{1}{3}=\frac{4m}{15}$$

Graphing on a Number Line

You can represent a number as a point on a number line, as shown in the following examples. Representing a number on a number line is called *graphing*.

Note that whole numbers on the line are equally spaced. Note too that in these examples, both positive and negative numbers are represented.

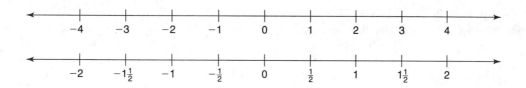

On the number line, positive numbers are shown to the right of zero. Negative numbers are shown to the left of zero. The positive number +3 is three

units to the right of zero. The negative number −2 is two units to the left of zero.

Graphing on a Coordinate Plane

A coordinate plane is based on an *x* axis (horizontal number line) and a *y* axis (vertical number line). The axes intersect at their zero points. This point of intersection is called the *origin*. Every point on the plane has both an *x* coordinate and a *y* coordinate. The *x*

coordinate tells the number of units to the right of the origin (for positive numbers) or to the left of the origin (for negative numbers). The *y* coordinate tells the number of units above the origin (for positive numbers) or below the origin (for negative numbers).

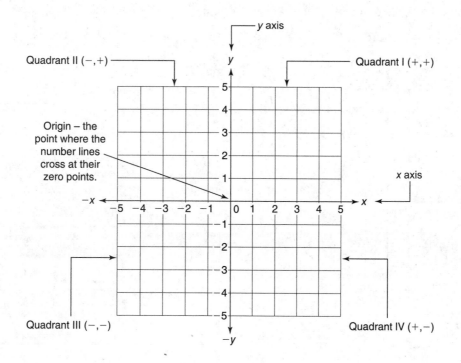

The coordinates of each point are often shown in what is called an *ordered pair* of numbers. An ordered pair looks like this: (2, 3). In every ordered pair, the first number is the *x* coordinate, and the second number is the *y* coordinate. So the ordered pair (2, 3) identifies a point with an *x* coordinate of 2 and a *y* coordinate of 3. The point is located at the intersection of the vertical line that is 2 units to the right of the origin ($x = +2$) and the horizontal line that is 3 units above the origin ($y = +3$). The point (2, −3) is located at the intersection of the vertical line that is 2 units to the right of the origin ($x = +2$) and the horizontal line that is 3 units below the origin ($y = −3$). The origin is identified by the ordered pair (0, 0).

The *x* and *y* axes separate the graph into four parts called quadrants.

- Points in Quadrant I have positive numbers for both the *x* and the *y* coordinates.
- Points in Quadrant II have a negative number for the *x* coordinate but a positive number for the *y* coordinate.
- Points in Quadrant III have negative numbers for both the *x* and the *y* coordinates.
- Points in Quadrant IV have a positive number for the *x* coordinate but a negative number for the *y* coordinate.

Examples

The graph below shows the following points:

(2,3), (−3,2), (−4,−4), and (0,−2)

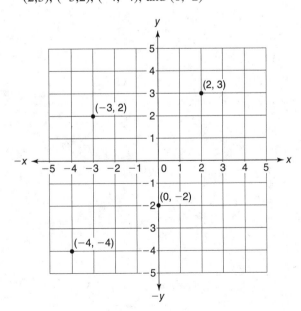

The graph below shows the following points:

A (4,−2), B (−1,1), C (3,3), and D (−4,−3).

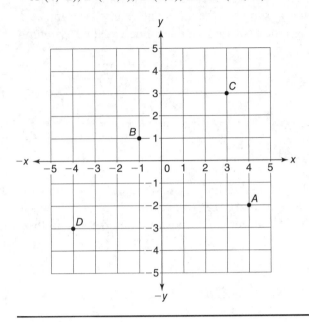

Graphing Equations on the Coordinate Plane An equation with two variables *x* and *y* can be graphed on a coordinate plane. Start by plugging in values for either *x* or *y*. Then solve the equation to find the value of the other variable. The *x* and y values make ordered pairs that you can plot on the graph.

Examples

Graph the equation $x + y = 4$

Solving for *x*, the equation becomes $x = 4 − y$

If $y = 1$, then $x = 3$.
If $y = 2$, then $x = 2$.
If $y = 3$, then $x = 1$.
If $y = 4$, then $x = 0$.

HINT *Use a function machine to generate ordered pairs!* Plug in numbers for *x* to get *y*.
If $x = $ _____, then $y = $ _____.

Function Machine

x	y
3	1
2	2
1	3
0	4

Plot the ordered pairs (3,1), (2,2), (1,3), and (0,4) on a graph. If you connect the points, you will see that the result is a straight line.

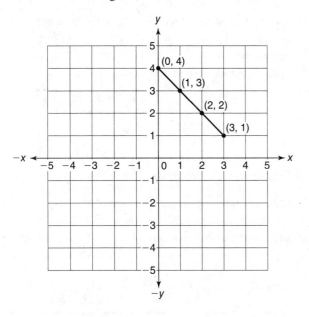

Graph the equation $y - x^2 = 2$

$y = 2 + x^2$

If $x = 0$, then $y = 2$.
If $x = 1$, then $y = 3$.
If $x = 2$, then $y = 6$.
If $x = 3$, then $y = 11$.
If $x = 4$, then $y = 18$.

Graph these points and connect the points with a line. Note that when you connect the points, you get a curved line.

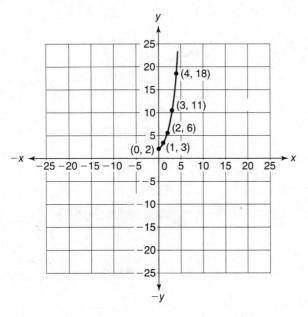

Probability

When every event in a set of possible events has an equal chance of occurring, probability is the chance that a particular event (or "outcome") will occur. Probability is represented by the formula

$$\text{Probability} = \frac{\text{number of positive outcomes}}{\text{number of possible outcomes}}$$

Let's say you have a spinner with an arrow that spins around a circle that is divided into six equal parts. The parts are labeled from 1 to 6. When you spin the arrow, what is the probability that it will land on the part labeled 4? Following the formula:

$$\text{Probability} = \frac{\text{number of positive outcomes}}{\text{number of possible outcomes}}$$

$$\text{Probability} = \frac{1}{6} \text{ or 1 in 6}$$

Let's take that same spinner. What is the probability that the arrow will land on the number 2 or the number 3 when it is spun? Using the formula:

$$\text{Probability} = \frac{\text{number of positive outcomes}}{\text{number of possible outcomes}}$$

$$\text{Probability} = \frac{2}{6} = \frac{1}{3} \text{ or 1 in 3}$$

Examples

Solve: The National Fruit Growers' Association is conducting a random survey asking people to tell their favorite fruit. The chart shows the results so far.

Favorite Fruit	Number of People
Apples	236
Peaches	389
Oranges	250
Pears	125

What is the probability that the next randomly selected person will say that pears are their favorite fruit?

To solve probability problems, follow the word problem solution procedure outlined in Chapter 9.

Procedure

What must you find? The probability that a certain event will occur

What are the units? Fractions, decimals, or percents

What do you know? The number of people selecting each fruit as their favorite.

Create an equation and solve.

$$\text{Probability} = \frac{\text{number of positive outcomes}}{\text{number of possible outcomes}}$$

Substitute values and solve.

Number of positive outcomes = 125 people who named pears as their favorite fruit

Number of possible outcomes = all people surveyed = 236 + 389 + 250 + 125 = 1,000

$$\text{Probability} = \frac{125}{1,000} = \frac{1}{8}$$

$$= 1 : 8 = 0.125 = 12.5\%$$

Solve: A box is filled with 25 orange balls, 50 green balls, and 75 red balls. If Wendell reaches into the box and picks a ball without looking, what is the probability that he will pick a orange or a green ball?

Procedure

What must you find? The probability that either of two events will occur

What are the units? Fractions, decimals, or percents

What do you know? How many of each kind of ball are in the box

Create an equation and solve.

$$\text{Probability} = \frac{\text{number of positive outcomes}}{\text{number of possible outcomes}}$$

Substitute values and solve.

Number of positive outcomes = number of orange balls + number of green balls = 25 + 50 = 75

Number of possible outcomes = 25 + 50 + 75 = 150

$$\text{Probability} = \frac{75}{150} = \frac{1}{2} = 1 : 2 = 0.5 = 50\%.$$

ALGEBRA AND PROBABILITY QUIZ

Circle the letter that represents the best or correct answer. Check your answers with those on page 249. If any of your answers were wrong, go back and study the relevant section of the algebra review.

1. The phrase "some number plus 20" is the same as
   A. $j + 20$
   B. $j - 20$
   C. $\dfrac{j}{20}$
   D. $j \times 20$

2. Evaluate: $a + 2(b) - c$ if $a = 5$, $b = 3$, and $c = 2$
   A. 7
   B. 9
   C. 10
   D. 11

3. Solve for z: $32 + z = 12$
   A. -20
   B. -12
   C. 12
   D. 20

4. The proportion "a is to b is as 3 is to 6" can be represented as
   A. $a + b = 3 + 6$
   B. $a \times b = 3 \times 6 + 45$
   C. $a(3) + b(6)$
   D. $\dfrac{a}{b} = \dfrac{3}{6}$

5. Solve for each variable.

   $y + g = 12$ and $2y + 3g = 16$

   A. $g = 15$, $y = -3$
   B. $g = 5$, $y = 7$
   C. $g = 1$, $y = 11$
   D. $g = -8$, $y = 20$

6. Add:

   $45g$
   $\underline{-90g}$

   A. $-45g$
   B. $45g$
   C. $45g^2$
   D. $135g$

7. Subtract:

   100
   $\underline{-130}$

   A. -230
   B. -30
   C. 30
   D. 230

8. Multiply:

   $(g^3)(g^{12})$

   A. g^{-3}
   B. g^{15}
   C. g^{-15}
   D. g^{36}

9. Add:

   $g^2 + gh + h^2 + 3g^2 + 4gh + 3h^2 =$

   A. $2g^2 + 3gh + 2h^2$
   B. $2g^2 + 5gh + 4h^2$
   C. $4g^2 + 5gh + 4h^2$
   D. $4g^2 + 3gh + 2h^2$

10. Multiply:

    $3h - 2m$
    $\underline{\times\ 2h + m}$

    A. $5h^2 + 3m^2$
    B. $6h^2 - hm - 2m^2$
    C. $6h^2 + 5hm + 2m^2$
    D. $6h^2 + 4hm + 2m^2$

11. Factor: $x^2 + 4x - 12$
    A. $(2x^2 + 1 + 6)(-x^2 + 3x - 2)$
    B. $(x + 6)(x - 2)$
    C. $(x + 4)(x + 3)$
    D. $(x^2 + 2)(x^{-2} + 4)$

12. Reduce to lowest terms:

    $\dfrac{6x - 6}{8x - 8}$

    A. $x - 2$

    B. $\dfrac{3x}{4x}$

    C. $\dfrac{6x}{8x}$

    D. $\dfrac{3}{4}$

13. Multiply:

    $\dfrac{j^2}{2j} \times \dfrac{2j}{3g}$

    A. $\dfrac{3j^2}{6j}$

    B. $\dfrac{2j^3}{6jg}$

    C. $\dfrac{2j^2}{6jg}$

    D. $\dfrac{j^2}{3g}$

14. Divide:

    $\dfrac{8g^3}{15} \div \dfrac{6g^2}{3}$

    A. $\dfrac{48g^5}{45}$

    B. $\dfrac{2g^5}{12}$

    C. $\dfrac{4g}{15}$

    D. $\dfrac{4g^3}{15g^2}$

15. Which equation is represented by the following graph?

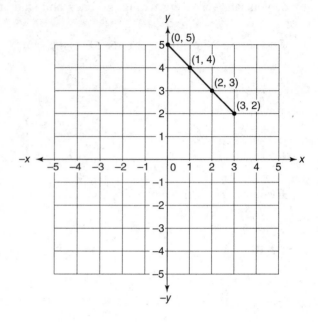

    A. $2x + y = 0$
    B. $x^2 + 1 = 0$
    C. $x + y = 5$
    D. $2x + y = 7$

16. Using a number cube with sides labeled 1 through 6, what is the probability of throwing a 3 or a 4?
    A. 2 in 4
    B. 1 in 3
    C. 2 in 8
    D. 1 in 9

ALGEBRA QUIZ ANSWERS

Number	Answer
1.	A
2.	B
3.	A
4.	D
5.	D
6.	A
7.	B
8.	B
9.	C
10.	B
11.	B
12.	D
13.	D
14.	C
15.	C
16.	B

Chapter 13

Mathematics Knowledge 2: Geometry

YOUR GOALS FOR THIS CHAPTER:

- Find out what geometry topics you need to know for the ASVAB.

- Review techniques and strategies for solving geometry problems.

- Study sample geometry problems and their step-by-step solutions.

- Measure your test readiness by taking a Geometry Quiz.

INTRODUCTION: GEOMETRY ON THE ASVAB MATHEMATICS KNOWLEDGE TEST

Along with problems in algebra and probability, the ASVAB Mathematics Knowledge test also includes problems in geometry. To do well on this portion of the test, you'll need to know the basic geometry concepts taught in high school math courses. Topics tested include classifying angles, identifying different kinds of triangles and parallelograms, calculating perimeter and area, finding the circumference and area of circles, identifying different kinds of solid figures, and solving geometry word problems.

It is important that you do well on the Mathematics Knowledge test because it is one of the four ASVAB tests that are used to calculate the AFQT—your military entrance score. That's why it pays to spend time reviewing topics in algebra, probability, and geometry and tackling plenty of sample ASVAB Mathematics Knowledge questions.

The following pages offer a quick but important overview of the basic geometry you need to know to score well on the ASVAB. Make sure that you carefully review and test yourself on every topic covered in this section. Also make sure that you learn how

to use all the problem-solving methods presented in the examples. Chapter 12 provides a similar review of basic concepts and problem-solving methods in algebra and probability. If you master the core information presented in these two chapters, you will be able to answer ASVAB Mathematics Knowledge questions with relative ease.

Suggested Review Plan for Geometry

- Work through the geometry review at your own pace.
- Pay careful attention to the examples provided.
- After you complete the review, take the geometry quiz.
- Score your quiz.
- Study any problems you answered incorrectly. Reread the corresponding review section, then try the problem again. Work until you understand how to solve the problem.

Remember that you probably won't get these exact problems on the ASVAB, but knowing how to solve them will help you do your best on the actual test.

GEOMETRY REVIEW

Topics
Points, lines, and angles
Triangles
Quadrilaterals
Circles
Perimeter and area
Three-dimensional (solid) figures
Geometry word problems

Points, Lines, and Angles

To work with geometry, you need to understand points, lines, and angles.

- A *point* is an exact location in space. It is represented by a dot and a capital letter.

- A *line* is a set of points that form a straight path extending in either direction without end. A line that includes points *B* and *D* is represented as follows: $\overleftrightarrow{BD}$.

- A *ray* is a part of a line that has one endpoint and continues without end in the opposite direction. A ray that ends at point *A* and includes point *B* is represented as follows: $\overrightarrow{AB}$.

- A *line segment* is a part of a ray or a line that connects two points. A line connecting points *A* and *B* is represented as follows: $\overline{AB}$.

An *angle* is a figure formed by two rays that have the same endpoint. That endpoint is called the *vertex* (plural: *vertices*) of the angle. An example is shown below.

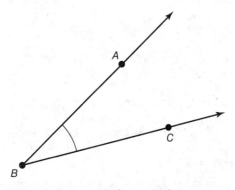

In this example, rays $\overrightarrow{BA}$ and $\overrightarrow{BC}$ have the same endpoint, which is point *B*. So point *B* is the vertex of the angle. The two line segments $\overline{BA}$ and $\overline{BC}$ are called the *sides* of the angle. The symbol ∠ is used to indicate an angle.

An angle is labeled or identified in several different ways:

- By the vertex: ∠*B*
- By the letters of the three points that form it: ∠*ABC* or ∠*CBA*. (The vertex is always the middle of the three letters.)

The measure of the size of an angle is expressed in *degrees* (°).

Classifying Angles   There are three types of angles that you should know for the ASVAB test. They are right angles, acute angles, and obtuse angles.

Right Angles   A right angle measures exactly 90°. Right angles are found in squares, rectangles, and certain triangles. ∠*ABC* is a right angle.

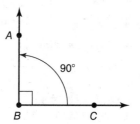

Examples

The angles below are both right angles.

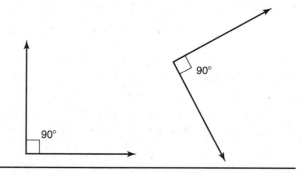

> **HINT**   *Look for the "box."* A little "box" tucked into the corner of an angle always means that the angle is a right angle.

Acute Angles   An angle that measures less than 90° is called an acute angle. ∠*STU* is an acute angle.

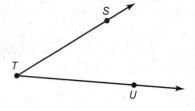

Examples

The angles below are all acute angles.

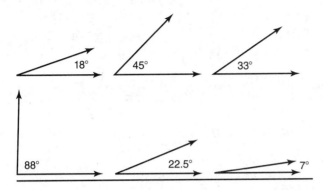

Obtuse Angles   An angle with a measure that is greater than 90° but less than 180° is called an obtuse angle. ∠*MNO* is an obtuse angle.

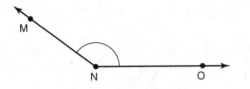

Examples

The angles below are all obtuse angles.

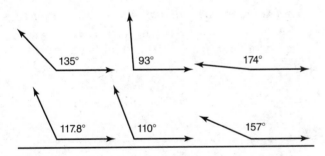

Straight Angles   A straight angle is one that measures exactly 180°. This kind of angle forms a straight line. ∠*EFG* is a straight angle.

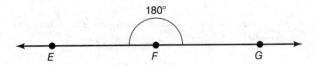

Classifying Pairs of Lines

Intersecting Lines   Intersecting lines are lines that meet or cross each other.

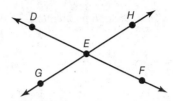

Line *DF* intersects line *GH* at point *E*.

Parallel Lines   Parallel lines are lines in a plane that never intersect.

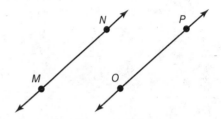

Line *MN* is parallel to line *OP*. In symbols, *MN* ∥ *OP*.

Perpendicular Lines   Perpendicular lines intersect to form right angles.

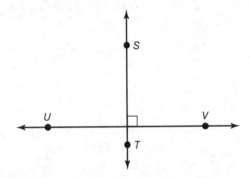

Line *ST* is perpendicular to line *UV*. In symbols, *ST* ⊥ *UV*.

Classifying Pairs of Angles

Adjacent Angles   Adjacent angles have the same vertex and share one side. ∠*ABC* and ∠*CBD* are adjacent angles.

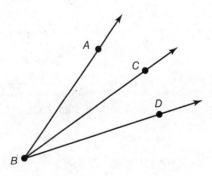

Complementary Angles   Two adjacent angles whose measures total 90° are called complementary angles.

∠*MNO* and ∠*ONP* are complementary. Their measures total exactly 90°.

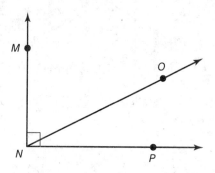

> **HINT** Figures on the ASVAB are not necessarily drawn exactly to scale.

Example

These two angles are complementary. Together they measure 90°.

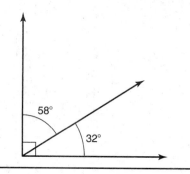

Supplementary Angles Two adjacent angles whose measures total 180° are called supplementary angles. Together they make a straight line. ∠*KHG* and ∠*GHJ* are supplementary because together they add to 180° or a straight line.

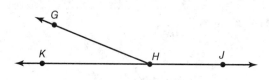

Example

The two angles below are supplementary. Together they measure 180° and form a straight line.

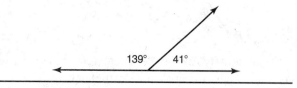

Vertical Angles Two angles formed by intersecting lines are called vertical angles if they are not adjacent. In the figure below, ∠*AED* and ∠*BEC* are vertical angles. ∠*AEB* and ∠*DEC* are also vertical angles. Vertical angles are often said to be "opposite" to each other, as shown in the figure.

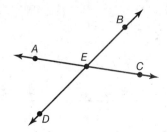

Vertical angles are *congruent*. That is, their measures are the same. ∠*AED* = ∠*BEC* and ∠*AEB* = ∠*DEC*.

Example

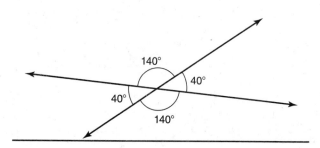

Identifying Congruent (Equal) Angles In the figure below, lines *AC* and *DF* are parallel. They are intersected by a third line *GH*. This third line is called a *transversal*.

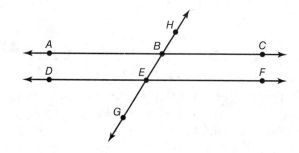

This intersection creates eight angles. There are four pairs of vertical congruent angles:

∠*ABH* = ∠*EBC*

∠*ABE* = ∠*HBC*

∠*DEB* = ∠*GEF*

∠*DEG* = ∠*BEF*

Alternate Interior Angles In addition, four of these angles make two pairs of alternate interior angles. These are angles that are on opposite sides of the transversal, are between the two parallel lines, and are not adjacent. When parallel lines are intersected by a transversal, alternate interior angles are congruent. The two pairs are

$$\angle ABE = \angle BEF$$

$$\angle DEB = \angle EBC$$

Alternate Exterior Angles Four of the angles also make two pairs of alternate exterior angles. These are angles that are on opposite sides of the transversal, are outside the two parallel lines, and are not adjacent. When parallel lines are intersected by a transversal, alternate exterior angles are congruent. The two pairs are

$$\angle ABH = \angle GEF$$

$$\angle HBC = \angle DEG$$

Corresponding Angles Eight of the angles also make four pairs of corresponding angles. These are angles that are in corresponding positions. When parallel lines are intersected by a transversal, corresponding angles are congruent. The four pairs are

$$\angle ABH = \angle DEB$$

$$\angle HBC = \angle BEF$$

$$\angle ABE = \angle DEG$$

$$\angle EBC = \angle GEF$$

> **HINT** *Angles count!* Pay attention to these angle relationships! They are almost certain to appear in some form on the ASVAB.

Solving Angle Problems On the ASVAB, you will most likely be asked to use what you know about angles and angle relationships to solve problems. You may be asked to tell which angles in a figure are congruent. Or you may be given the measure of one angle and asked for the measure of an adjacent angle or some related angle in a figure.

Examples

In the following diagram, parallel lines *MO* and *RT* are intersected by transversal *WV*.

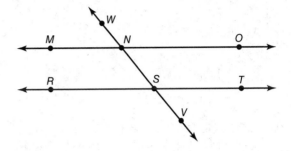

Which angle is congruent to $\angle MNW$?

A. $\angle MNS$
B. $\angle WNO$
C. $\angle RSV$
D. $\angle VST$

Of the choices, the only one that is congruent to $\angle MNW$ is $\angle VST$ because they are alternate exterior angles.

Which angle is congruent to $\angle MNS$?

A. $\angle RSV$
B. $\angle SNO$
C. $\angle VST$
D. $\angle MNW$

Of the choices, the only one that is congruent to $\angle MNS$ is $\angle RSV$ because they are corresponding angles.

If $\angle RSN$ measures 50°, what is the measure of $\angle RSV$?

A. 90°
B. 110°
C. 130°
D. 150°

$\angle RSN$ and $\angle RSV$ are supplementary angles. That is, together they form a straight line and their measures add up to 180°. So if $\angle RSN$ measures 50°, then $\angle RSV$ measures $180 - 50 = 130°$.

Triangles

A *polygon* is a closed figure that can be drawn without lifting the pencil. It is made up of line segments (sides) that do not cross. A *triangle* is a polygon with three sides. Every triangle has three angles that total 180°.

HINT *Look for tick marks and arcs.* When sides of a polygon are congruent (equal), they may be marked with an equal number of tick marks. When angles are congruent, they may be marked with an equal number of arcs.

Example

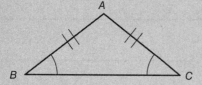

The tick marks indicate that sides *AB* and *AC* have the same length. The arcs indicate that ∠*ABC* and ∠*ACB* are congruent.

If sides have different numbers of tick marks, they are not congruent. If angles have different numbers of arcs, they are not congruent.

Identifying the Longest Side of a Triangle The longest side of a triangle is always opposite the largest angle. So, if a triangle has angles of 45°, 55°, and 80°, the side opposite the 80° angle would be the longest.

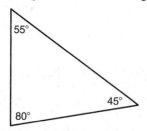

Types of Triangles There are four main types of triangles. They are equilateral, isosceles, scalene, and right. Each has special characteristics that you should know.

Equilateral Triangle This kind of triangle has three congruent (equal) sides and three congruent (equal) angles. In an equilateral triangle, each angle measures 60°.

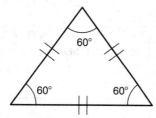

Isosceles Triangle This type of triangle has at least two congruent sides, and the angles opposite the congruent sides are also congruent. In the isosceles triangle shown below, sides *AB* and *BC* are congruent. ∠*BAC* and ∠*BCA* are also congruent. In an isosceles triangle, if you know the measure of any one angle, you can calculate the measures of the other two.

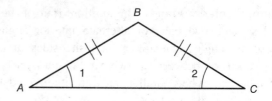

Examples

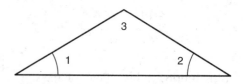

In this isosceles triangle, if ∠1 measures 30°, what is the measure of ∠3?

Since ∠1 and ∠2 are congruent, ∠2 must also measure 30°. Together, ∠1 and ∠2 add up to 60°. Since the sum of all three angles in any triangle is 180°, ∠3 must be 180 − 60 = 120°.

If ∠3 measures 100°, what are the measures of ∠1 and ∠2?

Since the sum of all three angles in any triangle is 180°, the sum of the measures of ∠1 and ∠2 must be 180 − 100 = 80°. Since angles 1 and 2 are congruent, each one must measure 80° ÷ 2 = 40°.

Scalene Triangle This kind of triangle has no equal sides or angles.

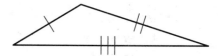

Right Triangle This kind of triangle has one angle that measures 90°. This angle is the *right angle*. It is identified in the figure by the little "box." Since the sum of all three angles in any triangle is 180°, the sum of the two remaining angles in a right triangle is 180 − 90 = 90°.

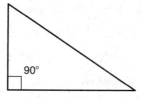

In a right triangle, there is a special relationship among the lengths of the three sides. This relationship is described by the *Pythagorean Theorem*.

In the following right triangle, ∠*C* is the right angle. The side opposite the right angle is called the *hypotenuse* (*c*). It is always the longest side. The other two sides (*a* and *b*) are called *legs*.

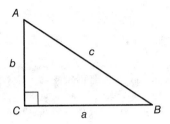

According to the Pythagorean Theorem, in any right triangle, the sum of the squares of the legs equals the square of the hypotenuse. In symbols,

$$a^2 + b^2 = c^2$$

So if you know the lengths of any two sides of a right triangle, you can calculate the length of the third side.

Examples

If *a* = 2 and *b* = 3, what is *c*?
Following the Pythagorean Theorem,

$$2^2 + 3^2 = c^2$$
$$4 + 9 = c^2$$
$$13 = c^2$$
$$\sqrt{13} = c$$

If *a* = 3 and *b* = 4, what is *c*?

$$3^2 + 4^2 = c^2$$
$$9 + 16 = c^2$$
$$25 = c^2$$
$$5 = c$$

If *a* = 6 and *c* = 10, what is *b*?

$$6^2 + b^2 = 10^2$$
$$36 + b^2 = 100$$
$$b^2 = 100 - 36$$
$$b^2 = 64$$
$$b = \sqrt{64}$$
$$b = 8$$

Base and Height of a Triangle Any side of a triangle can be called the *base*. The *height* is the length of a line segment that connects a base to the vertex opposite that base and is perpendicular to it.

Look at the following triangle. Dashed line *CD* is the height. Line *CD* is perpendicular to the base

AB. Where line *CD* meets base *AB*, it creates two right angles, $\angle CDA$ and $\angle CDB$.

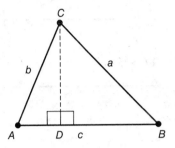

> **HINT** Sometimes you will see the word *perpendicular* represented by the symbol $\perp$. So in the picture above, $\overline{CD} \perp \overline{AB}$.

Median of a Triangle A *median* of a triangle is a line drawn from any vertex to the middle of the opposite side. This line splits the opposite side into two equal lengths.

Example

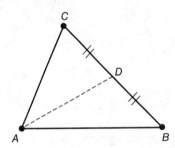

Dashed line *AD* is a median of triangle *ABC*. It splits side *BC* into two equal lengths, $\overline{CD}$ and $\overline{BC}$.

Quadrilaterals

A quadrilateral is a polygon with four sides and four angles. The sum of the four angles is always 360°.

Types of Quadrilaterals There are several different kinds of quadrilaterals. Each type is classified according to the relationships among its sides and angles. The square, rectangle, parallelogram, and rhombus are all types of quadrilaterals.

Parallelogram A parallelogram is a quadrilateral with both pairs of opposite sides parallel and congruent. The opposite angles are also congruent. Around the edge of the parallelogram, each pair of consecutive angles is supplementary; that is, their sum is 180°. Diagonal lines drawn from opposite vertices *bisect* each other (divide each other exactly in half), but the diagonals themselves are not equal in length.

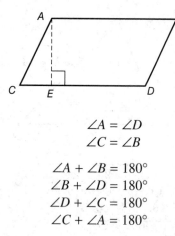

$$\angle A = \angle D$$
$$\angle C = \angle B$$

$$\angle A + \angle B = 180°$$
$$\angle B + \angle D = 180°$$
$$\angle D + \angle C = 180°$$
$$\angle C + \angle A = 180°$$

Like triangles, quadrilaterals have bases and height. Any side of this parallelogram can be a base. Dashed line $\overline{AE}$ is a height of this parallelogram. The *height* is a line originating at a vertex and drawn perpendicular to the opposite base. The height forms two right angles where it meets the base.

Rhombus   A rhombus is a parallelogram with four congruent sides. Opposite angles are also congruent.

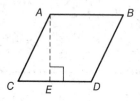

> **HINT** Watch out for ASVAB questions that use parallelograms, rhomboids, or trapezoids but involve the Pythagorean Theorem. In the trapezoid, for example, you might be asked to calculate the hypotenuse by knowing the length of $\overline{CE}$ and $\overline{AE}$.

Rectangle   A rectangle is a parallelogram with four right angles. Diagonal lines drawn from opposite vertices of a rectangle bisect each other (divide each other exactly in half) and are equal in length.

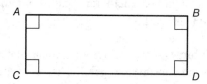

Square   A square is a rectangle with four congruent sides. Diagonal lines drawn from opposite vertices of a rectangle bisect each other (divide each other exactly in half) and are equal in length.

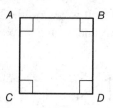

> **HINT**   In any square like the one shown, if you draw diagonals from point A to point D and from point C to point B, the lines are equal in length and bisect each other.

Trapezoid   A trapezoid is a quadrilateral with only one pair of parallel sides. Like other quadrilaterals, it has bases and height. In the example below, dashed line CE is the height. Sides AB and CD are parallel, but sides AC and BD are not parallel.

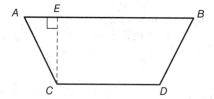

Circles

A circle is a closed figure with all points the same distance from a *center*. A circle with its center at point A is called circle A.

Parts of a Circle   A *chord* is a line segment that has endpoints on a circle. A *diameter* is a chord that passes through the center of a circle. A *radius* is a line segment that connects the center of a circle and a point on the circle. Its length equals half the length of the diameter. In the figure below, A is the center of the circle. EF is a chord. BC is a diameter of the circle. AC and AB are each a radius of the circle.

An *arc* is two points on a circle and the part of the circle between the two points. In the figure below, CD is an arc of the circle. A *central angle* is an angle whose vertex is the center of a circle. In the figure below, $\angle CAD$ is a central angle. Its measure is $50°$. The sum of the measures of the central angles in a circle is $360°$.

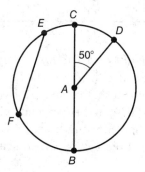

Perimeter and Area

The *perimeter* is the distance around a closed two-dimensional figure. The *area* is the amount of surface a two-dimensional figure covers. Area is measured in square units such as square inches (in^2) or square centimeters (cm^2). A square inch is the area of a square with sides 1 inch long.

Finding the Perimeter of a Polygon   To find the perimeter of a polygon, just add the length of each side to find the total.

Example

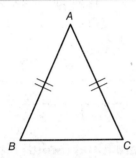

$\triangle ABC$ is an isosceles triangle. If side *AB* has a length of 25 cm and side *BC* has a length of 15 cm, what is the perimeter?

Since $\triangle ABC$ is an isosceles triangle, side AB = side AC. So if side AB has a length of 25 cm, side AC also has a length of 25 cm. Thus, the perimeter is 25 + 25 + 15 = 65 cm.

Finding the Area of a Polygon   There are special formulas you can use to calculate the areas of various types of polygons. You will want to memorize these, as you will almost certainly be asked a question about area on the ASVAB.

Area of a Triangle   The area (*A*) of a triangle is one-half the base (*b*) multiplied by the height (*h*), or

$$A = \frac{1}{2}bh$$

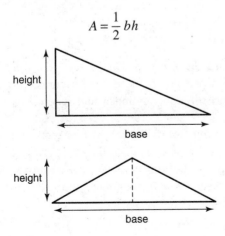

> **HINT**   *Watch the squares!* Anytime you see units squared, you are dealing with area. Examples: square inches (in^2), square centimeters (cm^2), square yards (yd^2), and so on.

Example

If a triangle has a height that measures 30 cm and a base that measures 50 cm, what is its area?

$b = 50$ cm

$h = 30$ cm

$A = \dfrac{1}{2}bh$

$A = \dfrac{1}{2}(50)(30)$

$A = \dfrac{1}{2}(1,500)$

$A = 750$ cm^2

Area of a Square or Rectangle   The area (*A*) of a square or rectangle is its length (*l*) multiplied by its width (*w*), or $A = lw$

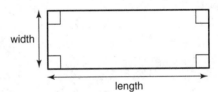

Example

If this rectangle is 10 miles long and 6 miles wide, what is its area?

$A = lw$

$A = (10)(6)$

$A = 60$ square miles (60 mi^2)

Area of a Parallelogram (Including a Rhombus)   The area (*A*) of a parallelogram is its base (*b*) multiplied by its height (*h*), or $A = bh$.

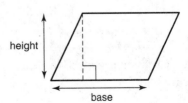

Example

If a parallelogram has a base of 6 meters and a height of 4 meters, what is the area?

$A = bh$
$A = (6)(4)$
$A = 24$ square meters (24 m^2)

Area of a Trapezoid   The area (A) of a trapezoid is one-half the sum of the two bases (b_1 and b_2) multiplied by the height (h), or $A = \frac{1}{2}(b_1 + b_2)h$.

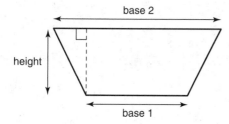

Example

If a trapezoid has one base of 30 meters and another base of 60 meters, and its height is 20 meters, what is the area?

$A = \frac{1}{2}(b_1 + b_2)h$

$A = \frac{1}{2}(30 + 60)20$

$A = \frac{1}{2}(90)20$

$A = \frac{1}{2}(1,800)$

$A = 900$ square meters (900 m^2)

Finding the Circumference and Area of a Circle   The *circumference* of a circle is the distance around the circle. The circumference (C) divided by the diameter (d) always equals the number π (pi). Pi is an infinite decimal, meaning that its decimal digits go on forever. When you use it to solve problems, you can approximate as π 3.14 or 22/7.

To find the circumference of a circle, use the formula $C = \pi d$.

Example

If a circle has a radius of 3 inches, what is the circumference?

Since the diameter is twice the radius ($2r$), the diameter is 6 inches.

$C = \pi d$
$C = 3.14(d)$
$C = 3.14(6)$
$C = 18.84$ in.

To find the area of a circle, multiply π times the square of the radius: $A = \pi r^2$

Example

If a circle has a radius of 4 centimeters, what is its area?

$A = \pi r^2$
$A = 3.14(4)^2$
$A = 3.14(16)$
$A = 50.24$ cm^2

Three-Dimensional (Solid) Figures

A figure is *two-dimensional* if all the points on the figure are in the same plane. A square and a triangle are two-dimensional figures. A figure is *three-dimensional* (solid) if some points of the figure are in a different plane from other points in the figure.

On solid figures, the flat surfaces are called *faces*. *Edges* are line segments where two faces meet. A point where three or more edges intersect is called a *vertex*.

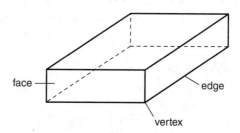

Types of Solid Figures   On the ASVAB, you may see problems related to these solid figures: rectangular solid (prism), cube, cylinder, and sphere.

Rectangular Solid (Prism)   On a rectangular solid (also called a prism), all of the faces are rectangular. The top and bottom faces are called *bases*. All opposite faces on a rectangular solid are parallel and congruent.

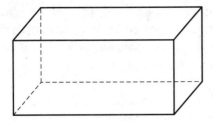

Cube A cube is a rectangular solid on which every face is a square.

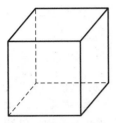

Cylinder A cylinder is a solid figure with two parallel congruent circular bases and a curved surface connecting the boundaries of the two faces.

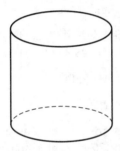

Sphere A sphere is a solid figure that is the set of all points that are the same distance from a given point, called the center. The distance from the center is the radius (r) of the sphere.

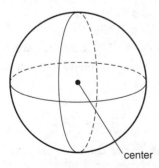

Finding the Volume of Solid Figures Volume is the amount of space within a three-dimensional figure. Volume is measured in cubic units, such as cubic inches (in^3) or cubic centimeters (cm^3). A cubic inch is the volume of a cube with edges 1 inch long.

Volume of a Rectangular Solid To find the volume (V) of a rectangular solid, multiply the length (l) times the width (w) times the height (h). The formula is $V = lwh$.

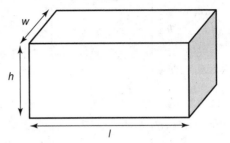

Example

If a rectangular solid has a length of 3 yards, a height of 1.5 yards, and a width of 1.5 yards, what is its volume?

$V = lwh$
$V = (3)(1.5)(1.5)$
$V = 6.75$ cubic yards (6.75 yd^3)

Volume of a Cube On a cube, the length, width, and height are all the same: Each one equals 1 side (s). To find the volume (V) of a cube, multiply the length × width × height. This is the same as multiplying side × side × side. The formula is $V = s \times s \times s = s^3$.

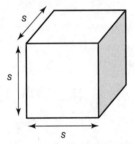

Example

If each side of a cube measures 9 feet, what is its volume?

$V = s^3$
$V = (9)^3$
$V = 729$ cubic feet (729 ft^3)

Volume of a Cylinder To find the volume (V) of a cylinder, first find the area of the circular base by using the formula $A = \pi r^2$. Then multiply the result times the height (*h*) of the cylinder. The formula is $V = (\pi r^2)h$.

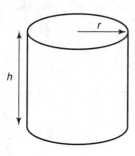

Example

If a cylinder has a height of 7 meters and a radius of 2 meters, what is its volume?

$V = (\pi r^2)h$
$V = 3.14(2)^2(7)$
$V = 3.14(4)(7)$
$V = 87.92$ cubic meters (87.92 m³)

Volume of a Sphere   To find the volume of a sphere, multiply 4/3 times π times the radius cubed. The formula is $V = \frac{4}{3}\pi r^3$.

Example

If the radius of a sphere measures 12 inches, what is the volume?

$V = \frac{4}{3}\pi r^3$

$V = \frac{4}{3}\pi(12)^3$

$V = \frac{4}{3}(3.14)(1728)$

$V = \frac{4}{3}(5,425.92)$

$V = 7,234.56$ cubic inches (7,234.56 in.³)

Geometry Word Problems

Word problems on the ASVAB may deal with geometry concepts such as perimeter, area, and volume. To solve these problems, you may also need to use information about different units of measure. To review units of measure, see Chapter 8.

Just as with other kinds of word problems, you can solve geometry problems by following a specific procedure. In the examples that follow, pay special attention to the procedure outlined in each solution. Follow this same procedure whenever you need to solve this kind of word problem.

Examples

Sally is buying wood to make a rectangular picture frame measuring 11 in. × 14 in. The wood costs 25 cents per inch. How much will Sally have to pay for the wood?

Procedure

What must you find? Cost of the wood for the frame
What are the units? Dollars and cents
What do you know? Cost of the wood per inch; shape of the frame, measure of the frame
Create an equation and solve.

$$\text{Length of wood needed for rectangular frame} = 2l + 2w$$

Substitute values and solve:

Length of wood = 2(14) + 2(11)
Length of wood = 28 + 22 = 50 in.

If each inch costs 0.25, then

0.25 × 50 = $12.50

Sergei is planting rosebushes in a rectangular garden measuring 12 ft × 20 ft. Each rosebush needs 8 ft² of space. How many rosebushes can Sergei plant in the garden?

Procedure

What must you find? Number of rosebushes that can be planted in the garden
What are the units? Numbers
What do you know? Shape of the garden, garden length and width, amount of area needed for each rosebush
Create an equation and solve.

$A = lw$

Substitute values and solve.

$A = 12 \times 20 = 240$ ft²

Each rosebush needs 8 ft²

240 ÷ 8 = 30

Sergei can plant 30 rosebushes in the garden.

Shapes and Formulas: Summary

Shapes	Formulas

Triangle
Area = 1/2 of the base × the height
$A = 1/2\ bh$
Perimeter = $a + b + c$

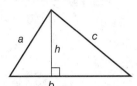

Square
Area = length × width
$A = lw$
Perimeter = side + side + side + side
$P = 4s$

Rectangle
Area = length × width
$A = lw$
Perimeter = 2 × length + 2 × width
$P = 2l + 2w$

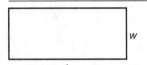

Parallelogram and Rhombus
Area = base × height
$A = bh$
Perimeter = 2 × length + 2 × width
$P = 2l + 2w$

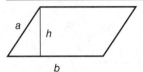

Trapezoid
Area = the sum of the two bases divided by 2 × height

$$A = \left(\frac{b_1 + b_2}{2}\right)h$$

Perimeter = $a + b_1 + b_2 + c$

$$P = a + b_1 + b_2 + c$$

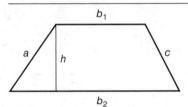

Circle
The distance around the circle is its circumference (C). The length of a line segment passing through the center with endpoints on the circle is the diameter (d). The length of a line segment connecting the center to a point on the circle is the radius (r). The diameter is twice the length of the radius ($d = 2r$).
$C = \pi d = 2\pi r$
$A = \pi r^2$
$\pi = 3.14$ or $22/7$

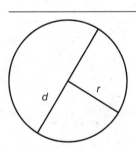

Cube
Volume = side × side × side
$V = s^3$

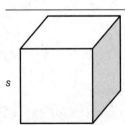

(Continued)

Shapes and Formulas: Summary (*Continued*)

Shapes	Formulas

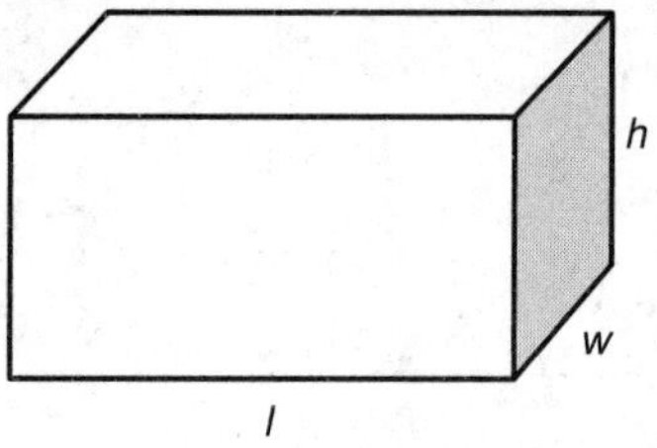

Rectangular Solid
Volume = length × width × height
$V = lwh$

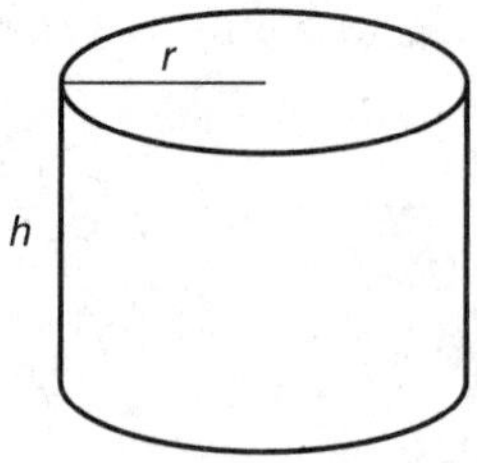

Cylinder
Volume = πr^2 × height
$V = \pi r^2 h$

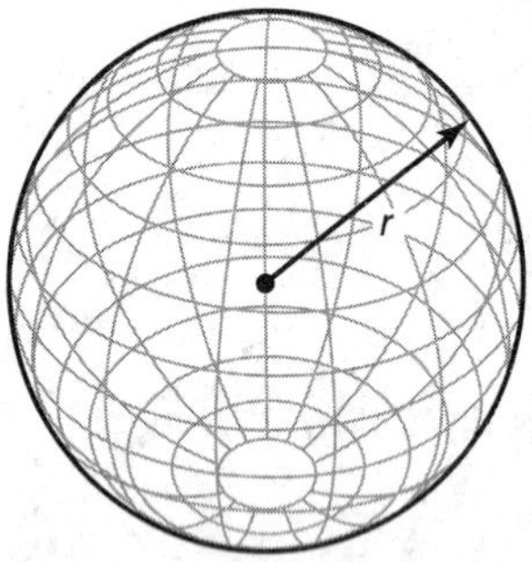

Sphere
Volume = $4/3\pi r^3$
$V = 4/3\pi r^3$

GEOMETRY QUIZ

1. The endpoint shared by two rays that form an angle is called a
   A. line segment.
   B. degree.
   C. vertex.
   D. straight line.

2. A right angle is an angle that measures
   A. exactly 90°.
   B. greater than 90°.
   C. less than 90°.
   D. 45°.

3. An acute angle is an angle that measures
   A. exactly 90°.
   B. greater than 90°.
   C. less than 90°.
   D. exactly 180°.

4. An obtuse angle is an angle that measures
   A. exactly 90°.
   B. between 90° and 180°.
   C. exactly 180°.
   D. greater than 180°.

5. In the figure shown, if angle 1 measures 65°, what is the measure of angle 2?

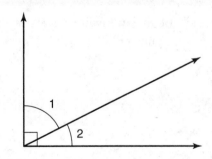

   A. 120°
   B. 90°
   C. 25°
   D. 10°

6. In the figure shown, if angle 1 measures 35°, what is the measure of angle 2?

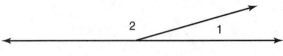

   A. 145°
   B. 95°
   C. 65°
   D. 45°

7. In the figure shown, if angle 2 measures 45°, what is the measure of angle 4?

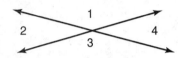

   A. 135°
   B. 75°
   C. 45°
   D. 35°

8. In the figure shown, lines *AB* and *CD* are parallel and angle 1 measures 120°. What is the measure of angle 2?

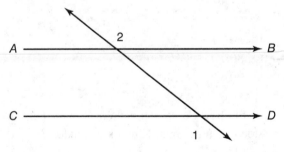

   A. 60°
   B. 90°
   C. 120°
   D. 180°

9. A certain triangle has exactly two equal sides, and the angles opposite those sides are also equal. This triangle is a(n)
   A. isosceles triangle.
   B. scalene triangle.
   C. equilateral triangle.
   D. right triangle.

10. In the equilateral triangle shown, what is the measure of each angle?

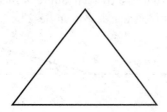

A. 60°
B. 90°
C. 120°
D. 180°

11. In the triangle shown, ∠ACB measures 55° and ∠CAB measures 65°. What is the measure of ∠CBA?

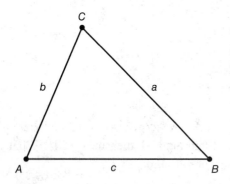

A. 50°
B. 60°
C. 80°
D. 90°

12. In the triangle shown, side *a* is 5 in. long and side *b* is 6 in. long. How long is side *c*?

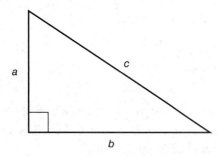

A. 7.81 in.
B. 8.51 in.
C. 9.11 in.
D. 10.71 in.

13. In the rectangle shown, side *a* measures 13 cm and side *b* measures 36 cm. What is the perimeter of the rectangle?

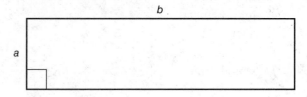

A. 49 cm
B. 98 cm
C. 469 cm
D. 512 cm

14. In the triangle shown, side *a* is 4 ft long, side *b* is 6 ft long, and side *c* is 8 ft long. What is the area of the triangle?

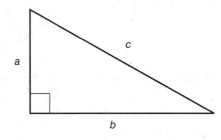

A. 12 ft²
B. 18 ft²
C. 24 ft²
D. 36 ft²

15. In which of the following circles is the line segment a diameter?

A.

B.

C.

D.

16. The circle shown has a diameter of 12 m. What is the area of the circle?

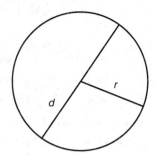

A. 452.16 m²
B. 360 m²
C. 144 m²
D. 113.04 m²

17. In the trapezoid shown, $b_1 = 4$ ft, $b_2 = 8$ ft, and $h = 5$ ft. What is the area of the trapezoid?

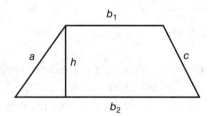

A. 16 ft²
B. 30 ft²
C. 90 ft²
D. 160 ft²

18. Figure *WXYZ* is a parallelogram. Which of the following is NOT necessarily true?

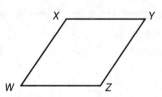

A. Side *WX* is parallel to side *ZY*.
B. Side *XY* is parallel to side *WZ*.
C. ∠*W* has the same measure as ∠*Y*.
D. Side *WX* is the same length as side *XY*.

19. Which of the following figures contains line segments that are perpendicular?

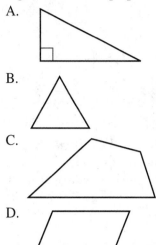

A.

B.

C.

D.

20. What is the ratio of the length of a side of an equilateral triangle to the triangle's perimeter?
A. 1:1
B. 1:2
C. 1:3
D. 3:1

21. Jules needs to purchase a ladder that just reaches the top of a 16-foot building. If the bottom of the ladder will be placed 12 feet from the base of the building, how long should Jules's ladder be?
A. 18 ft
B. 20 ft
C. 25 ft
D. 30 ft

22. Kai must buy canvas to make a sail for his sailboat. The sail will have the shape of a right triangle. It will be as tall as the mast (the vertical pole) and as wide at the bottom as the boom (the horizontal pole). If the mast measures 18 meters and the boom measures 9 meters, about how much canvas does Kai need to buy?
A. 27 m²
B. 61 m²
C. 81 m²
D. 103 m²

23. Renee wants to carpet her living room. If the room measures 15 feet × 18 feet, how many square yards of carpet must she purchase?
    A. 20 yd²
    B. 30 yd²
    C. 32 yd²
    D. 35 yd²

24. Saul wants to wallpaper one wall in his dining room. The wall is 11 ft high and 14 ft long. It has a window that measures 3 ft by 4 ft. About how many square feet of wallpaper must Saul buy?
    A. 25 ft²
    B. 32 ft²
    C. 112 ft²
    D. 142 ft²

25. In the following figure, line *AB* intersects two parallel lines. If angle 1 measures 35°, what is the measure of angle 4?

    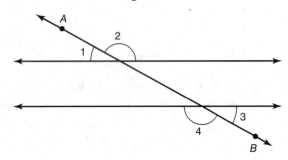

    A. 35°
    B. 55°
    C. 90°
    D. 145°

26. The Lakeville skating rink is a perfect circle. If the diameter of the rink is 50 ft, which of the following best represents its area? (Use π = 3.14)
    A. 78.5 ft²
    B. 157 ft²
    C. 1,962 ft²
    D. 7,850 ft²

27. A tomato paste can has a diameter of 8 cm and a height of 20 cm. Which of the following tells about how much tomato paste it will hold?
    A. 4,019 cm³
    B. 1,005 cm³
    C. 201 cm³
    D. 50 cm³

28. In the figure shown, what is the area of the rectangle if the radius of each circle is 6 cm?

    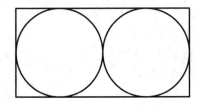

    A. 72 cm²
    B. 112 cm²
    C. 288 cm²
    D. 310 cm²

29. A square garden is doubled in length and 4 feet is added to its width. Which of the following expressions represents the new area of the garden?
    A. $2x(x + 4)$
    B. $4x^2$
    C. $4x + 4$
    D. $(2x)(4x)$

30. A rectangular garden is 40 yards long and 15 yards wide. Darryl runs once around the edge of the garden. How far does Darryl run?
    A. 55 yards
    B. 80 yards
    C. 100 yards
    D. 110 yards

GEOMETRY QUIZ ANSWERS

Number	Answer
1.	C
2.	A
3.	C
4.	B
5.	C
6.	A
7.	C
8.	C
9.	A
10.	A
11.	B
12.	A
13.	B
14.	A
15.	A
16.	D
17.	B
18.	D
19.	A
20.	C
21.	B
22.	C
23.	B
24.	D
25.	D
26.	C
27.	B
28.	C
29.	A
30.	D

Chapter 14

Electronics Information

YOUR GOALS FOR THIS CHAPTER

- **Find out what Electronics Information topics are covered on the ASVAB.**

- **Learn basic facts about electricity, electric circuits, and electrical devices and systems.**

- **Study examples of ASVAB Electronics Information questions and learn ways to solve them.**

- **Take an Electronics Information quiz.**

INTRODUCTION: ASVAB ELECTRONICS INFORMATION QUESTIONS

The electronics information questions that appear on the ASVAB measure how much you understand about electricity, electric circuits, and electrical and electronic devices and systems.

The questions may ask you to identify a particular device on a circuit diagram, explain how to measure voltage or current, or identify particular types of circuits. If you have tinkered with electricity or electronics at home or in school, you may be familiar with some of the topics covered here.

Whichever ASVAB version you take, you'll have only about half a minute to answer each electronics information question, so you'll have to work fast if

you want to get a good score. That's why it pays to spend time studying the test topics and tackling plenty of sample ASVAB Electronics Information questions.

The topic review that follows will help prepare you to answer ASVAB Electronics Information questions. At the end of the chapter there is a short quiz with questions modeled on those on the actual test. Read carefully through the review materials in this chapter, then use the quiz to find out how well you have mastered this subject area. Go back and reread the review materials for any quiz items you miss.

ELECTRICITY

Basics

Let's start by getting acquainted with some basic concepts in electricity. To understand electricity, you need to know the following:

- *Electricity* is a form of energy that can travel invisibly through conductors. It can be used in so many ways that we could call it the most versatile form of energy. Electricity is carried by moving charged particles, especially by electrons. Electrons are tiny negative charges that orbit the nucleus of an atom.
- A *conductor* is a material that allows an easy flow of electrons. Silver, copper, and aluminum are all good conductors.
- An *insulator* is a material that resists the flow of electrons. Rubber, plastic, and ceramic are good insulators.

ASVAB Electronics Information Test

Number of Questions:
- Paper-and-pencil ASVAB: 20
- CAT-ASVAB: 15

Time Limit:
- Paper-and-pencil ASVAB: 9 minutes
- CAT-ASVAB: 10 minutes

Topics covered:
- Electric current
- Electric circuits
- Electrical devices and systems

- A *circuit* is a loop of conductor that takes electricity from its source to the load (the place where it does some work) and back to the source.
- A *load* is anything in the circuit, such as a heater, a light, or a motor, that uses power.
- *Direct current* (DC) is a steady-flowing type of electricity, produced by batteries and used in flashlights, boom boxes, and computers.
- *Alternating current* (AC) is a type of current that changes direction many times per second. AC is used in home wiring, mainly because it can be transported long distances over transmission wires.
- *Electronics* is a branch of science that deals with complicated uses of electricity, such as in radios, televisions, and computers.

Technical Terms

To understand electricity, you also need to know a few technical terms. Study the following list.

- *Electric current* is the amount of electrons flowing through a conducting material.
- *Electric power* is the amount of power consumed by an electrical device.
- *Voltage* is a force that affects the rate at which electricity flows through a conductor. It is sometimes called electrical pressure. The higher the voltage, the more likely electricity is to "leak" across an insulator or an air gap. That's one reason higher voltages are more dangerous. *Voltage drop* tells how much electrical pressure is used in a part of the circuit.

- *Frequency* is the number of complete alternations—from one direction to the other and then back again—that alternating current makes per second. Each complete alternation is called a cycle.
- *Resistance* is the opposition of a material to the flow of electricity through it. All circuits must have a resistance. If they don't, they are called *short circuits*, and wires can overheat.

Units of Measure and Measuring Devices

Different aspects of electricity are measured using different units of measure. Special measuring devices are used. The following table shows the different units and devices.

Ohm's Law

Ohm's law describes the relationship among electrical pressure (voltage), current strength (amperage), and resistance (ohms) in any circuit:

$$\text{Amperes} = \text{volts/ohms}$$
$$A = V/\Omega$$

If you know two of these three quantities, you can always calculate the third.

$$\text{Amperes} = \text{volts/ohms}$$
$$\text{Volts} = \text{amperes} \times \text{ohms}$$
$$\text{Ohms} = \text{volts/amperes}$$

What Is Measured	Unit of Measure	Unit Definition	Symbol	Measuring Device
Electric power (*amount of energy consumed*)	Watt	Watts = volts × amperes 1 *kilowatt* = 1,000 watts 1 *kilowatt hour* = 1,000 watts flowing for 1 hour.	W	Electric meter
Electric current (*strength*)	Ampere	1 ampere = the flow of 1 coulomb of electricity past a given point per second (1 coulomb = 6.25×10^{18} electrons)	A	Ammeter
Voltage (*electrical pressure*)	Volt	Volts = watts/amperes	V	Voltmeter
Resistance	Ohm	1 ohm = 1 volt per ampere	Ω	Ohmmeter
Frequency	Hertz	Number of cycles per second (in North America, AC is delivered at 60 Hz)	Hz	

Here's an easy way to remember how to use Ohm's law to find the third quantity if you know two already. On the circle (below), place your finger over the quantity you want to find. Look at the remaining two quantities to see how to calculate the third quantity.

$$\frac{\text{Volts}}{\text{Amps} \times \text{Ohms}} = \frac{V}{A \times \Omega}$$

Ohm's Law Circle.

Memorize the equations on the circle; you'll need them on the ASVAB. Let's look at a couple of examples of how to use the Ohm's law circle.

Examples

Find the amperes if a 120-volt current runs through 6 ohms.

Amperes = volts/ohms
Amperes = 120/6 = 20 amperes

Or cover amperes in the circle. Notice that what remains is volts divided by ohms (the same result you would get by remembering the three formulas above).

Amperes = 120/6 = 20 amperes

In a 6-volt circuit with 24 amperes flowing, what is the resistance?

Ohms = volts/amperes
Ohms = 6/24 = 1/4 ohms

The Law of Electric Power

The amount of power consumed by an electrical or electronic device can be calculated using the following formula:

Watts = volts × amperes

You can use this equation to find any one of the three factors as long as you know the other two. Here are two examples.

Examples

How much power is consumed by a lamp that draws 10 amperes of current at 120 volts?

Watts = volts × amperes
Watts = 120 × 10 = 1,200 watts

A clothes dryer is rated at 2,400 watts. At 120 volts, how much current does it draw?

Amperes = watts/volts
Amperes = 2400/120 = 20 amperes

ELECTRICITY AND MAGNETISM

Electricity and magnetism are tightly connected: It's easy to change from an electric current to a magnetic field and back again. This close relationship explains electromagnets, transformers, motors, and generators. Let's start with electromagnets.

A current passing through a conductor creates a magnetic field around it. In most electromagnets, the conductor (wire) is wrapped around an iron core.

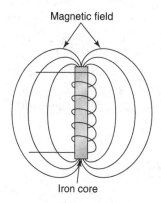

An Electromagnet with Magnetic Field.

A transformer is like two electromagnets placed next to each other. If the transformer has more turns of wire on the output side, it is a *step-up* transformer, and the output voltage will be greater than the input voltage. If there are more turns on the input side, it's a *step-down* transformer, and the output voltage is smaller than the input voltage.

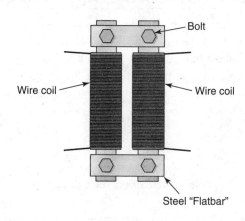

Transformer.

Motors and Generators

Electromagnetism also explains motors and generators. In fact, motors and generators are really the same machines, operating backward.

- A *generator* changes rotating (kinetic) energy into electric energy.

- A *motor* changes electric energy into kinetic energy.

Here's how a motor works. Each magnet has two poles: north and south. Opposite poles attract, and like poles repel: North attracts south, but repels north. A motor has two magnets: a *rotor* that spins inside a *stator*, a fancy name for a stationary magnet. One of these magnets, usually the rotor, is an electromagnet. It is wired so that the magnetic field changes twice per rotation. When the rotor starts rotating, the rotor and the stator repel each other, forcing the rotor to start turning. At just about the point where the magnets would stop repelling, the rotor changes polarity, and it again repels the stator. This change in polarity is what drives the motor. We call them electric motors, but motors are all about magnetism.

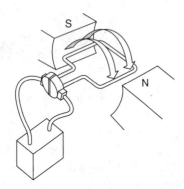

Electric Motor.

ELECTRICAL DEVICES AND SYMBOLS

Electrical devices are the guts of many electrical and electronic systems. You'll need a basic acquaintance with these devices. You should also know the symbols used for them in electrical diagrams.

A *capacitor* is a device that can briefly store electricity.

This Is the Symbol for a Capacitor.

A *resistor* creates resistance to the flow of electrons. If a circuit does not have any resistance, it's called a short circuit. Excess current will cause such a circuit to heat up, which may cause a fire.

This Is the Symbol for a Resistor.

A *transformer* changes the voltage and amperage of a current. Transformers work by changing electricity into magnetism, then back into electricity. As you can see from the diagram, transformers have two coils. One gets current (usually AC) from a source; the other supplies the output. The first coil creates a magnetic field, which rises and falls as the current alternates. The second coil is inside a changing electric field, and any wire in a changing magnetic field will pick up current from that magnetism.

This Is the Symbol for a Transformer.

A *battery* stores electricity as chemical energy that can be readily converted into electric current. Batteries may be wet (like the lead-acid storage batteries used in cars) or dry (like the nickel-cadmium or metal hydride batteries used in flashlights, computers, and the like). Batteries always make direct current. Depending on their chemistry, some batteries can be recharged.

The electrochemical reactions between the cathode and the electrode result in free electrons, which can travel through a circuit to make an electric current.

This Is the Symbol for a Battery.

This Is the Symbol for Alternating Current.

SERIES, PARALLEL, AND SERIES-PARALLEL CIRCUITS

Electric current exists only when electrons can flow through a circuit. There are two basic types of circuits, and a third type that blends the two.

Let's start with the basics: the series and parallel circuits.

In a *series circuit*, all moving electrons pass through every part of the circuit, including all the loads and switches. Current (the quantity of electrons) is the same at all points of the circuit, but voltage drops as the current goes through each device. The total voltage of the loads must equal the voltage of the circuit: In a series circuit supplied by a 12-volt car battery, a single light bulb must have a 12-volt drop. If the circuit has two lights, their combined voltage drop is 12 volts.

When a series circuit is used in a string of Christmas tree bulbs, the whole string goes dark if any bulb burns out. Thus, although they are simple, series circuits are less common than the next basic type, the parallel circuit.

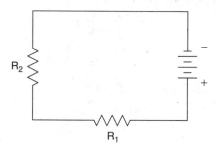

Series Circuit.

In a *parallel circuit*, the loads are placed between the two supply wires, so that they all get the same voltage. A second advantage of the parallel circuit is this: Current can flow through any of the loads, even if one is switched off. (In a series circuit, a switch controls the current in the entire circuit.)

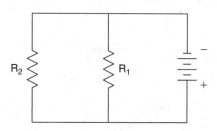

Parallel Circuit.

Remember this rule: Current is the same at all points in a series circuit. Voltage is the same at all points in a parallel circuit.

To find the *total resistance* in a series circuit, add the resistance of each load. For example, in a series circuit with one 12-Ω and one 8-Ω resistor, total resistance = 12 + 8 = 20 Ω.

It's more complicated to calculate total resistance in a parallel circuit since you must add the inverse of the resistances. What is the total resistance in a parallel circuit with one 12-Ω and one 8-Ω resistor?

$$1/R_{total} = 1/12 + 1/8 = 2/24 + 3/24 = 5/24 = 1/R_{total}$$

Solve for R_{total}:

Multiply both sides by R_{total}: $5 \times R_{total}/24 = 1$

Multiply both sides by 24: $5 \times R_{total} = 24$

Divide both sides by 5: $R_{total} = 24/5$ ohms

Note that R_{total}, 24/5, is simply the inverse of 5/24, the fraction equaling $1/R_{total}$.

The third type of circuit is the *series-parallel circuit*, which combines features of series and parallel circuits. The series-parallel circuit is a hybrid with many advantages of each. You'll find series-parallel circuits throughout your house. The branch circuits that bring electric power to the lights and outlets are series-parallel: All the current goes through a fuse or circuit breaker, but then it is distributed in parallel. Why is this circuit needed? Because voltage must be the same for all outlets and lights, and because the circuit must work whether any particular light is switched on or not.

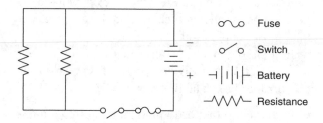

Series-Parallel Circuit.

SEMICONDUCTORS

So far, we've talked about insulators and conductors. But there is an important category of materials between these two categories.

A semiconductor can act as a conductor or as an insulator. Silicon, the main semiconductor, is the basis for computer memory and logic boards. Chemicals called *dopants* are applied to the silicon to determine whether it will act as an insulator or as a conductor. They do this by making electrons available or not available to flow. (When electrons can flow, a material becomes a conductor.)

The basis for computer applications is a group of components, particularly transistors and diodes.

Transistors are devices that can switch a current, regulate its flow, or amplify a current, all based on the presence of a smaller current. Millions of tiny transistors are built on small pieces of semiconductor, which are the basis of computer logic and memory.

This Is the Symbol for a Transistor.

Diodes are devices that allow a current to flow in one direction only. In addition to electronics, diodes are also used in devices called *rectifiers*, which convert AC into DC.

 Diode

This Is the Symbol for a Diode.

PRACTICAL ELECTRICITY

The ASVAB will also ask you about more practical matters, including simple electric circuits. You'll have an advantage here if you've ever worked on your home wiring (which is usually much easier than most people think).

In a simple electric circuit like the one in your home, electricity is distributed at a fuse box or circuit breaker box. The box has two functions:

- Breaking up the load in the building into a number of circuits
- Preventing excess current from flowing into the circuits

Many simple circuits have two separate conductors: hot and grounded. The hot conductor is usually black, but it can be red or another color. The grounded (sometimes called neutral) conductor is white. Together, the black and white wires are called the *supply* wires, because they form the circuit that electric current needs to travel.

To work safely on any electric circuit, you need to shut off the electricity, and check that it is off. The fuses and circuit breakers can shut off the circuit. Before starting work, use an electrical tester to test whether the hot wires are energized.

Bigger (heavier) wires can carry more current. Home wiring systems are usually rated for 15 or 20 amperes. Fifteen-ampere circuits require 14-gauge wires. Larger 12-gauge wires are needed for 20-ampere circuits.

Grounding Systems

Electricity always "wants" to complete a circuit, to get back to where it came from. The wiring in a building always includes two options for returning current to the circuit breaker or fuse box. The white (grounded) wire is the normal return path for electricity. The bare (equipment grounding) wire is the alternative path, available if something goes wrong with the grounded wire. In many systems, the alternative return path is provided by a steel sheath on the wires, called a *conduit*.

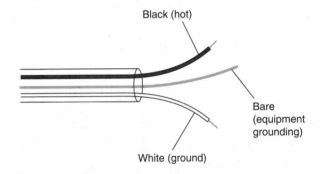

Black (hot)

Bare (equipment grounding)

White (ground)

House Wiring.

Because the two supply wires are different (black brings the current, and white drains it away), modern electric plugs are *polarized*. This means that they can fit into the socket in only one way, because one lug on the plug is bigger than the other.

Polarized Plug.

Switches and Rheostats

Switches turn electric circuits on and off. One single-pole switch controls a circuit. Two *three-way* switches can work together to control a circuit from either of two locations, such as at the top and bottom of a stairway. A rheostat continuously controls the voltage, and is used to dim lights. In home wiring, switches always control the hot side of the circuit—the one that supplies the current.

Electrical Connections

Electrical connections must be secure and tight. Otherwise, electric arcs will develop, causing heat and ruining the connection. One of the best connections comes from solder, a metal that is melted over a connection. Solder joints must have flux, a chemical that prevents oxidation from happening when the wires get heated. Solder can be used to join wires to other wires or wires to terminals.

Circuit Breakers and Fuses

Because electric circuits get hot when something goes wrong, circuit breakers and fuses are used in most circuits. Both have the same job: breaking the circuit (cutting off the power) when something goes wrong. A circuit breaker usually has an electromagnet that opens a circuit when the current gets too high. A fuse has an element that melts when the current is too high. Fuses are easier to make, but circuit breakers can just be reset after they trip, whereas a fuse must be replaced. If a fuse continually blows, do not replace it with a larger-capacity fuse; this invites fire. Find the problem and fix it.

Coaxial Cables

Coaxial means "having the same axis." Coaxial cables, usually used in cable TV systems, have an inner and an outer conductor.

ELECTRONICS INFORMATION QUIZ

Take the following quiz to check your understanding of ASVAB Electronics Information topics. There is no time limit. Answers and explanations are given immediately after the quiz.

1. Alternating current is so called because
   A. it is an alternative to direct current.
   B. it alters the magnetic field.
   C. resistance alternates many times per second.
   D. the current changes direction many times per second.

2. An electric current is
   A. a form of energy.
   B. a flow of electrons.
   C. the ability to do work.
   D. both A and B.

3. A battery
   A. always produces direct current.
   B. may produce alternating current.
   C. converts electric energy into chemical energy.
   D. must have a liquid electrolyte.

4. In a series circuit, total voltage drop equals
   A. the drop produced by the biggest resistance in the circuit.
   B. the sum of the individual voltage drops.
   C. one-half the total individual voltage drops.
   D. the inverse of the largest individual voltage drop.

5. A volt is a unit of electric(al)
   A. current.
   B. resistance.
   C. pressure.
   D. capacitance.

6. If a 120-volt current is running through a circuit with a resistance of 6 Ω, what is the current strength in amperes?
   A. 15 A
   B. 20 A
   C. 30 A
   D. 60 A

7. The transformer depends on the relationship between electricity and
   A. voltage.
   B. amperage.
   C. chemical energy.
   D. magnetism.

8. In an electric motor, magnets
   A. repel each other.
   B. attract each other.
   C. cancel each other.
   D. amplify each other.

9. A circuit without resistance is called
   A. an open circuit.
   B. a short circuit.
   C. a closed circuit.
   D. a hot circuit.

10. A transformer changes the _____ in a circuit.
    A. resistance
    B. current
    C. voltage
    D. both B and C

11. In a series circuit, a switch controls
    A. all devices on the circuit.
    B. only devices with a positive charge.
    C. only devices with a negative charge.
    D. voltage, but not current.

12. Which is the same at all points in a parallel circuit?
    A. Voltage
    B. Resistance
    C. Current
    D. Ohms

13. A switch in a series-parallel circuit controls
    A. all devices.
    B. some or all devices.
    C. only electronic devices.
    D. only nonelectronic devices.

14. The purpose of dopant in a semiconductor is to make it act as a(n)
    A. insulator.
    B. resistor.
    C. conductor.
    D. A or C.

15. In a parallel circuit, what is the total resistance if the individual resistances are 3 Ω and 12 Ω?
    A. 5/12 Ω
    B. 12/5 Ω
    C. 12 Ω
    D. 5 Ω

ANSWERS AND EXPLANATIONS

1. **D.** The polarity, or direction of electron flow, alternates in AC. Choice B is something that happens with alternating current, but this is not the reason for the name.

2. **D.** Electricity is a form of energy, and a current is defined as a flow of electrons.

3. **A.** Batteries always produce direct current (DC).

4. **B.** Total voltage drop is the sum of the individual voltage drops.

5. **C.** Electrical pressure is what "pushes" current through a circuit.

6. **B.** Amperes = volts/ohms = 120/6 = 20.

7. **D.** When electricity goes through a coil of wire in a transformer, it creates a magnetic field. When this field interacts with another coil of wire, it induces an electric current.

8. **A.** Electric motors are based on magnetic repulsion.

9. **B.** A short circuit has no resistance.

10. **D.** A transformer changes both current and voltage.

11. **A.** A series circuit has only one pathway, so when a switch is open, the whole circuit shuts down.

12. **A.** Voltage is the same at all points of a parallel circuit.

13. **B.** A switch in a series-parallel circuit can control some or all devices, depending on the way it's wired.

14. **D.** Dopants control the electrical properties of a semiconductor.

15. **B.** $1/R_{total} = 1/R_1 + 1/R_2 + 1/R_3 = 1/3 + 1/12 = 5/12$. Solve for R_{total} to get 12/5 ohms.

Auto Information

INTRODUCTION: ASVAB AUTO INFORMATION QUESTIONS

The automobile questions that appear on the ASVAB measure how much you understand about automobile components and systems, and how much you know about maintaining and repairing them. The questions may ask you to describe the function of a particular part, to tell what might be causing a given problem, or to explain how to repair a given malfunction. You won't necessarily be asked to explain *why* auto parts function as they do, but you will be asked *what to do* and *how to do it* when a part needs maintenance or repair. If you own a car and maintain it yourself, you may already be familiar with many of the topics covered on the test. You may also have learned about cars by watching or helping family members or friends maintain an automobile, or by working in a garage yourself.

ASVAB Auto Information Questions

Number of Questions:
- Paper-and-pencil ASVAB: 10–12 (as part of the Auto and Shop Information test)
- CAT-ASVAB: 10 (as a separate test)

Time Limit:
- CAT-ASVAB: 7 minutes

On the paper-and-pencil version of the ASVAB, automobile questions are one part of the Auto and Shop Information test. On the CAT-ASVAB, they form a separate test of their own.

Whichever ASVAB version you take, you'll have only about a minute to answer each automobile question, so you'll have to work fast if you want to get a good score. That's why it pays to spend time studying the test topics and tackling plenty of sample ASVAB automobile questions.

The topic review that follows will help prepare you to answer ASVAB automobile questions. It describes each of the major systems of today's automobiles and reviews the functions of all the most important automobile parts. At the end of the chapter, there is a short quiz with questions modeled on those on the actual test. Read carefully through the review materials in this chapter, then use the quiz to find out how well you have mastered this subject area. Go back and reread the review materials for any quiz items you miss.

AUTOMOBILE ENGINES

The engine is the heart of an automobile. Cars use an internal-combustion engine, meaning that the fuel is burned inside the engine. (Steam engines are "external-combustion engines" that burn fuel outside the engine; steam is piped to a turbine that creates the rotary motion.) All car engines, including diesel engines, use the Otto cycle, named for Nicholas Otto, the German who invented the four-stroke gasoline engine in the 1870s.

Here is an overview of what happens inside an Otto-cycle engine: A mix of fuel and air is brought inside a closed space, called a cylinder. The mix is compressed and then explodes. The explosion moves a piston, which rotates the crankshaft. The crankshaft is connected through the drive train to the driving wheels, which move the car. Waste heat from the explosions is removed by the cooling system.

Cylinders are located in the large cast-iron engine block. Cylinders are laid out in a straight line or a V shape. Straight-line engines usually have four cylinders. For a six- or eight-cylinder engine, the V design (called a V-6 or V-8) saves space.

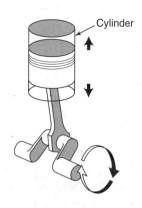

Piston, Connecting Rod, and Crankshaft.

Cylinder and Piston

The *cylinder* is the heart of the internal combustion engine, since it's where combustion takes place. The cylinder is a finely machined chamber that holds a *piston* as it slides up and down. Thin rings called *piston rings* seal the gap between the cylinder and the piston, containing the explosions and increasing efficiency.

- If the piston rings wear, oil can enter the cylinder. Burning oil makes blue smoke and cuts power output. When an engine starts to burn oil, a major repair called an engine overhaul is needed.
- Changing the oil regularly is the best way to prevent excess wear to piston rings.

Cylinder Head

The *cylinder head* is a complex metal casting that closes the top of the cylinders. The head is bolted to the engine block. A *head gasket* separates the head and the block. Like all gaskets, the head gasket creates a seal between two rigid objects that would leak if there were not something compressible between them.

Each cylinder needs at least one intake valve and one exhaust valve. These valves close off a port that allows intake gases to enter or exhaust gases to leave. Many engines increase their power output by using two intake valves and/or two exhaust valves. The cylinder head also has passages for coolant and holes for the bolts that hold it to the engine block.

When you bolt a cylinder head to the engine block, both the order of tightening and the torque (tightening force) are important. Tightening bolts in the correct order prevents the head from warping. Tightening to the right torque ensures that the head is tight enough to seal the gasket evenly. For American cars, torque is measured in foot-pounds. For other cars, it is measured in newton-meters.

The cylinder head also has threaded holes for the *spark plug*. These electrical devices create a spark when they get a high-voltage jolt of electricity from the ignition system. Spark plugs are screwed into the cylinder head and should be replaced periodically.

Crankshaft

Pistons move in a straight line, but the engine produces rotating motion. The *connecting rods* and *crankshaft* change linear motion into rotary motion. Think of the knee of a bicyclist. It moves up and down in a straight line, just like a piston. The knee is connected to the pedal by the lower leg, which acts like a connecting rod. The pedals and cranks act like a crankshaft to convert linear motion to rotary motion.

Connecting rods are attached to the crankshaft by the *main bearings*. The crankshaft itself rotates on *journal bearings* attached to the engine block. The crankshaft is housed inside an *oil pan*, and the bearings also get lubrication from oil tubes or channels in the block.

How the Four-Cycle Engine Works

Understanding an Otto-cycle engine starts with firing order. Memorize this order: intake, compression, power, exhaust.

- *Intake.* The piston moves down, creating a partial vacuum in the cylinder. The fuel–air mixture enters the cylinder through the open intake valve. The exhaust valve is closed.

- *Compression.* Both valves are closed. The piston moves up, compressing the fuel–air mixture to about 10 times atmospheric pressure.
- *Power.* The spark plug fires, starting an explosion inside the cylinder. The resulting high pressure pushes the piston down.
- *Exhaust.* The piston moves up again, with the exhaust valve open and the intake valve closed. The piston pushes burned exhaust gases into the exhaust manifold and out of the engine.

fuel and air. Valves must open and close precisely and quickly, several thousand times a minute when an engine is running at full throttle. Valves, particularly the exhaust valve, are in the hottest part of the engine. They cannot be cooled by water but must conduct away their heat by contact with the valve seat in the cylinder head.

Helix-shaped *valve springs* hold valves against the *valve seat*, a polished, sloping surface that closely fits the outside edge of the valve.

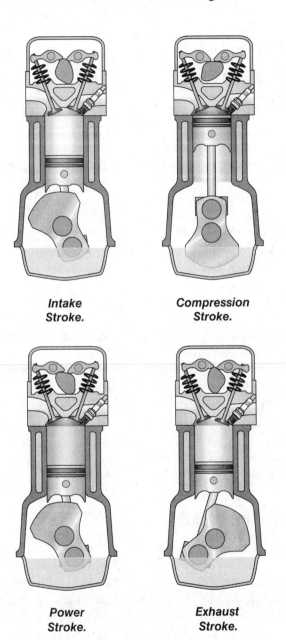

Intake Stroke.

Compression Stroke.

Power Stroke.

Exhaust Stroke.

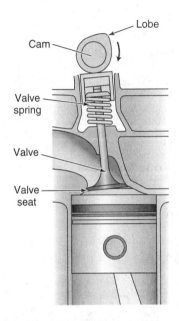

Overhead Camshaft and Valve.

In overhead-camshaft engines, the top of the valve rides against the *camshaft*. Once every two revolutions of the crankshaft, the *camshaft lobe* pushes against the valve stem, opening the valve.

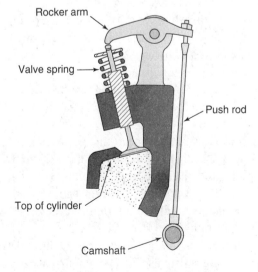

Conventional Camshaft and Valve.

Valves and Valve Train

Valves play a critical part in the Otto engine because they admit fresh fuel and air and discharge burned

In conventional-camshaft engines, the camshaft pushes against a *push rod*, which pushes the *rocker arm*. The rocker arm pivots on the *rocker-arm shaft*, so the ends move in opposite directions. When the push rod raises one end of the rocker arm, the other end pushes down on the valve, so it opens.

The camshaft is driven by a *timing chain* from a sprocket on the crankshaft. The chain keeps the camshaft perfectly in sync with the crankshaft. A broken timing chain is one of the most serious of all problems that can affect an Otto-cycle engine.

Firing Order

To make the engine run more smoothly, nearby cylinders do not ignite in sequence. Instead, the firing is spread around the engine. A typical four-cylinder engine might fire in the order 1-3-2-4. Firing order gets a bit more complicated in V-type engines, but the general rule is the same.

Lubrication System

Engines contain hundreds of metal parts that rotate or slide against metal. Pistons and piston rings slide against the cylinder. The crankshaft rotates inside journal bearings. Camshafts, push rods, rocker arms, and valves must also be protected against friction with a thin film of lubricant—engine oil.

Engines have a complicated set of tubes and internal passages that bring oil to the contact points. A gear-driven *oil pump* pushes oil through these passages. Oil also splashes onto the cylinder walls, lubricating the piston rings and making compression more effective.

- Oil gets dirty and wears out, so it must be replaced periodically. Oil filters clean the oil, but they must also be replaced.
- Oil and filter manufacturers estimate their product lifetime in miles and/or months. Replace the oil or filter when you reach the first of these milestones.
- To change oil, warm up the engine, place a pan under the oil plug in the crankcase, remove the plug, and drain the oil. Replace the plug and refill the oil through the filler cap on top.
- Recycle used oil; do not dump it down the drain or in a field.
- New cars should not need a top-up of oil between oil changes. But as cars age, engines wear and oil

consumption increases, so it makes sense to check the oil periodically. Much of this wear occurs at the piston rings and cylinders.

- When a car starts "burning oil," blue smoke indicates wear of the rings and/or cylinders.
- If the oil pressure light comes on while you are driving, pull over as soon as possible and check for trouble. If you are lucky, the oil level may simply be low, and adding oil should take care of the problem. Otherwise, to prevent severe engine damage, get the car towed to a shop. Sudden loss of oil pressure can also result from oil-pump failure or other serious engine problems.

Viscosity Oil, like most fluids, gets thicker as the temperature drops. You can see this with molasses. In the refrigerator, it is almost solid. But if you heat molasses on the stove, it starts to flow like water. The viscosity, or thickness, of oil is measured by S.A.E. numbers, which range from a very light S.A.E. 5 to a molasses-like S.A.E. 90. Typically, for summer driving, oil is rated at S.A.E. 30 to 40.

Auto engines start out cold and warm up as they operate. Thick, cold oil creates a lot of resistance when you try to start a cold engine. But if the oil is too thin at operating temperature, it won't separate the metal parts. To resolve this dilemma, lubrication engineers created "multiweight" oil, which is rated with two S.A.E. numbers, for low/high-temperature viscosity. Thus S.A.E. 10W-40 flows more easily at low temperatures (where it's S.A.E. 10) and is thicker when it warms up (where it's S.A.E. 40).

Transmissions and differentials also need lubrication, usually from heavier oil, such as S.A.E. 80 or 90. (These transmission oils or transmission greases are different from automatic transmission fluid, which is used only in automatic transmissions.)

Cooling System

Internal-combustion engines develop a great deal of heat from all those explosions. Only about 30 percent of the energy in gasoline is converted into energy to drive the car. The rest becomes waste heat. Unless the engine can get rid of this heat quickly, it will overheat. Waste heat is removed and delivered to the atmosphere by coolant. At first, water was used as the coolant. But water freezes in winter, and the expansion when it does so can crack the engine block. Chemicals called *antifreeze* are added to prevent freezing. Dissolving most chemicals in water usually lowers the freezing point. Water can also rust—corrode—the iron in an engine. Coolants also contain a chemical that halts rust, called, logically, *rust inhibitor*. Because rust inhibitor eventually breaks down, coolant must be replaced every few years.

Coolant flows through hollow passages in the engine block and cylinder head. These *water jackets* must bring enough coolant to every part of the engine block and cylinder head. From the engine block, coolant flows to the radiator, which has many small tubes covered by fins. The radiator is located on the front of the engine, where it can get plenty of fresh air.

A *radiator fan* pulls air through the radiator, removing heat from the tubes. The radiator is connected to the engine block through a series of hoses, which can wear out or leak.

- Checking hose and fan belt condition is good preventive maintenance, although modern engines are packed so tightly that the hoses and belts may be hard to see.
- Don't fill the radiator through the cap. Instead, add coolant to the translucent reservoir that's attached by hoses to the radiator. The reservoir has "low" and "full" markings; just keep coolant between these marks.
- Repair shops have testers to determine the freezing point of coolant. If you always add coolant (50–50 antifreeze and water), your coolant should stay liquid down to about −30°F. But if you top up the radiator with water, get the freezing point checked in the fall; you may have to replace the coolant to protect the engine.

A *water pump*, often located on the front of the crankshaft, circulates coolant through the engine. If the water pump fails for even a few minutes, overheating and severe engine damage can result. To increase heat removal, the coolant is pressurized. Pressurized liquids stay liquid at higher temperatures, and hotter liquids can move more heat.

Although we've talked about the need to cool an engine, engines operate best after they have warmed up. To help an engine reach operating temperature, a *thermostat* prevents the circulation of cold coolant. Thermostats are usually located on the top of the engine, inside a housing that connects the engine block to the top radiator hose. Inside the thermostat, a metal valve opens when the coolant reaches operating temperature.

In newer cars, the fan may also help with engine warm-up. In older cars, the fan was driven from a pulley on the crankshaft by a *fan belt*. These fans rotated when the engine ran. Now, electrically driven fans start running only when the engine is warm.

- Cars have either a warning light to indicate engine overheating or a gauge that measures engine temperature. Another sign of overheating is the smell of antifreeze; overheating raises engine pressure, and eventually coolant blows out through the radiator cap, releasing a cloud of sweet-smelling, greasy steam.
- A stuck thermostat is probably the largest cause of engine overheating. If you don't drive until you correct the problem, it's usually an inexpensive one as well. Problems with fan belts, radiators, coolant level, and sensors can also cause overheating.

Engine Troubleshooting

Engines are getting more complex all the time, but it helps to know a few troubleshooting hints to score big on the ASVAB:

- A "ping" or "knocking" sound on acceleration usually means that you need a higher-octane gasoline. The noise indicates that the fuel-air mixture is igniting too soon inside the cylinder.
- A squealing noise that increases with engine speed indicates a loose or worn fan belt.
- An engine that runs extremely rough may have a failed spark plug or some other problem in the ignition system.

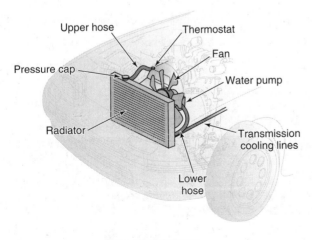

Cooling System.

- General sluggishness, roughness, or poor fuel mileage all indicate the need for a tune-up.
- Loud clanking sounds may indicate major engine problems that will only get worse if you ignore them.

DRIVE TRAIN

The *drive train* gets power from the engine to the wheels. The drive train includes the transmission, driveshaft, differential, and axles on the driving wheels, which may be in front, in back, or both. In cars with front-wheel drive, the transmission and differential are combined in a "transaxle."

The drive train must

- Provide different gears so that the engine can always work at an efficient rpm (revolutions per minute), no matter what the driving speed.
- Allow the car to move backward (in reverse gear).
- Allow the engine to run when the car is not moving.
- Drive two or four wheels.
- Allow the car to turn without tire slippage.

Manual Transmission

In a manual transmission, you change gears with a *clutch* and *gearshift*. Manuals, which once had only two or three speeds, today generally have five forward speeds plus reverse. In general, transmissions have an input shaft, a layshaft, and an output shaft. The input shaft is connected to the clutch, and the output shaft is connected to the driveshaft and eventually the driving wheels.

When a transmission "changes gears," it changes the ratio of input speed to output speed. In first gear, roughly four revolutions of the input shaft turn the output shaft once. First gear is used to accelerate from a stop, and the engine must turn fast while the wheels turn slowly. First gear allows fast acceleration because it multiplies torque at slow driving speed.

When you shift gears in a manual transmission, the gears do not engage or disengage; all the gears are always engaged. Instead, the gears are shifted by a small collar attached to the output shaft. *Dog teeth* on the side of this collar catch holes in the side of the gear, connecting the gear to the output shaft. When the collar is disengaged, the gears spin freely on the output shaft.

The gearshift moves the collar, and synchronizers between the collar and the gear allow the dog teeth to engage the gears. If you shift too fast, these synchronizers don't have time to engage, and the gears clash.

To change direction for reverse gear, there is an *idler gear* between the layshaft and the output shaft. In forward gears, the input and output shafts rotate in opposite directions. But the idler gear causes the output shaft to turn in the same direction as the input shaft.

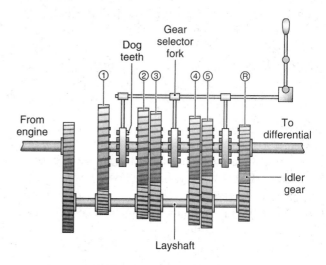

Five-Speed Manual Transmission.

Clutch The clutch disconnects the engine from the transmission, so that you can shift gears. The clutch also allows you to idle at a traffic light without shifting into neutral. A clutch usually has three plates, which are controlled by the clutch pedal.

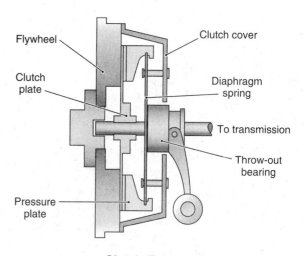

Clutch, Engaged.

The clutch is *engaged* when the clutch pedal is up and *disengaged* when the pedal is down.

When the clutch is engaged, springs push the pressure plate against the clutch disk, pressing the clutch disk against the flywheel. Because of friction between the pressure plate and the flywheel, the input shaft and flywheel rotate together. In this position, neither the throwout bearing nor the clutch plates wear.

When the clutch is disengaged, a cable or hydraulic piston moves the release fork, pressing the throwout bearing against the diaphragm spring, moving the pressure plate away from the clutch disk, and disconnecting the flywheel from the input shaft. The throwout bearing does get wear in this position. The pressure plate wears when the clutch is partly engaged, mainly when starting from a standstill in first gear.

Automatic Transmission

Automatic transmissions change gears automatically to suit driving conditions. Modern automatics often have four forward gears, plus reverse, park, and neutral. The transmission itself uses a complex set of gears to do the shifting.

Instead of the manual clutch just described, automatics link the engine to the transmission with a hydraulic mechanism called a *torque converter*. The torque converter has two blades that look much like propellers. The drive blade, attached to the engine output shaft, churns up hydraulic fluid the way a propeller moves water. The driven blade is caught by the stream of hydraulic fluid and turns like a pinwheel in a breeze.

When the engine is idling at a stoplight, little hydraulic fluid is pumped, little energy is transmitted to the driven blade, and just a bit of pressure on the brake holds the car still. When the engine speeds up, the drive blade pumps more fluid, and more energy is transferred to the driven blade. At highway speeds, almost all of the energy is transmitted to the driven blade. Because of inefficiencies, automatics waste more gasoline than manual transmissions, but they are "hands-free" and much more popular.

Differential

When a car turns a corner, the outside wheels must drive further than the inside wheels. If both driving wheels were locked to the axle, the tires would be

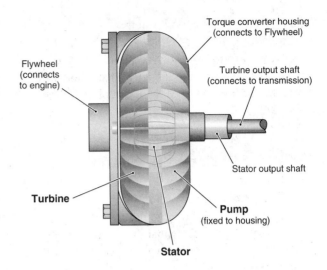

Torque Converter.

forced to slip against the pavement. The differential solves this problem.

The differential has three jobs:

- *Allow different (differential) movement of the two axles* so that the car can turn a corner.
- *Change drive directions.* In rear-wheel-drive cars, the differential connects to a driveshaft that runs from front to back and powers the axles, which run from side to side.
- *Increase power.* The differential usually reduces the drive speed, so that three to four input revolutions create one turn of the axles.

Normally, if one wheel is slipping, the differential will direct all power to it, allowing the other wheel to sit still without doing any work. This causes problems when traction is poor. "Limited-slip" or "positraction" differentials limit this slip.

Front-Wheel or Rear-Wheel Drive

Traditionally, cars steered with the front wheels and drove with the rear wheels. Driving with the front wheels is more complicated, but it offers many advantages:

- The engine weight is on the driving wheels, increasing traction.
- The steering wheels also drive, pulling the car to follow the steering wheels on ice, snow, or sand.
- There is no driveshaft, saving interior room.

The disadvantage is complexity. Stacking the transmission and differential together in a "transaxle" makes front-wheel-drive systems harder to design and repair.

Cars with four-wheel drive, of course, have both a conventional driveshaft and a transaxle, and a front-back differential. They have the most complicated drive trains.

ELECTRICAL SYSTEM

The first cars had barely any electrical system. A hand crank started them, and gas or oil lamps lit the road for those foolhardy enough to drive after dark. The only electrical mechanism was the magneto that fed electric current to the spark plugs. Let's start our examination of the electrical system by looking at the ignition system.

Ignition System

The key parts of the ignition system are the breaker points, the coil, the distributor cap, and the distributor rotor. The breaker points ride against a lobed shaft inside the distributor. The shaft completes one revolution for every two strokes of the crankshaft (intake, compression, power, exhaust).

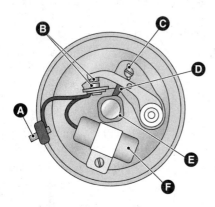

- Ⓐ Connection to coil-$\frac{1}{2}$ volt
- Ⓑ Breaker points
- Ⓒ Adjusting screw
- Ⓓ Cam follower
- Ⓔ Distributor Camp
- Ⓕ Condenser

Breaker Points.

When the breaker points close, they complete a circuit, and a pulse of 12-volt current goes through the primary winding of the ignition coil, creating a

brief magnetic field. This changing magnetic field induces a current in any nearby wire. The voltage of this induced (output) current depends on the ratio of the primary and secondary windings. In an ignition coil, the primary winding has few loops, and the secondary winding has many.

The ignition coil is a direct-current transformer. The induced current, which creates the spark at the spark plugs, is at 10,000 volts or higher. This high-voltage output goes via heavy, high-voltage cable to the center of the distributor cap.

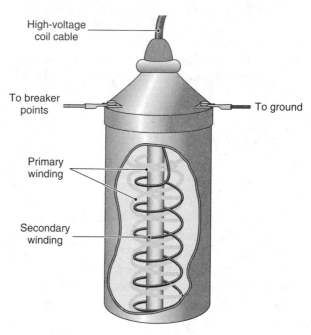

Ignition Coil.

The distributor cap and rotor direct the high-voltage current to the spark plugs. The rotor is on top of the distributor shaft. The rotor receives the high-voltage current from the coil wire and directs it to the spark plug wires connected to the distributor cap. Because the shaft rotation is synchronized with the crankshaft rotation, the spark occurs at the right time.

All the important parts of the electrical system—the breaker points, rotor, distributor cap, and high-voltage wires—can wear out. That's why tune-ups are needed every few thousand miles. Distributor ignition has largely been replaced by electronic ignition, which we'll get to shortly.

Spark Plug The spark plug receives the high-voltage spark current from the distributor and creates an electric spark that sets off the explosion in the cylinder. Spark plugs operate in hot conditions, and they must be replaced occasionally.

- If a spark plug is coated with a greasy black substance, the cylinder is leaking oil, probably because of worn piston rings. Oil burns incompletely in the cylinders, leaving this black deposit.
- Spark plugs must have the correct gap, measured in thousandths of an inch. Feeler gauges are used to set the correct gap.
- When spark plug electrodes get thin, the plug should be replaced.

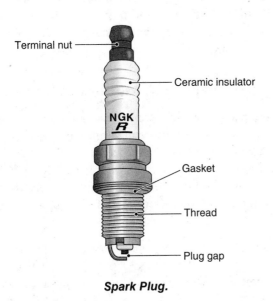

Spark Plug.

Diesel engines need no spark plug because the fuel–air mixture ignites when it is compressed. Diesels don't need a coil or distributor. They are somewhat more fuel-efficient, but also more polluting, than gasoline engines. Diesels require fuel injectors and special fuel but otherwise are quite similar to gasoline engines.

Electronic Ignition

Electronic ("solid-state") *ignitions* have been introduced over the last 20 years to eliminate the many weak points of distributor ignition. Instead of one centrally located coil, there is a coil at each spark plug, and instead of a mechanical distributor, an electronic unit directs a low-voltage current to those coils. The system allows precise spark timing, which increases power and gasoline mileage while reducing pollution.

Battery

Cars use a 12-volt battery to store electric energy for starting. These lead-acid batteries are compact and affordable electricity storage devices that last for several years. Lead-acid batteries can quickly deliver the large current needed by the starter motor. They are easily recharged, because the electrochemical reaction that makes electricity is reversible, allowing electricity to be stored as chemical energy.

Inside a lead-acid battery, a chemical reaction between sulfuric acid (the *electrolyte*) and lead plates (the *electrodes*) creates extra electrons at the negative pole. When you connect the starter motor to the negative and positive electrodes, a current flows, discharging the battery and turning the starter motor. Each battery cell makes about 2 volts, so a 12-volt battery has six cells. During charging, a 12-volt battery requires about a current of about 14 volts.

- Lead-acid batteries eventually wear out. However, the biggest problem with a lead-acid battery is often the simplest. Corrosion on the battery terminals or battery cables can break the circuit, causing a fully charged battery to appear "dead."
- Corrosion-preventing chemicals can avoid problems blamed on "dead batteries." These chemicals often appear as a red spray on the terminal.
- Older lead-acid batteries had a tendency to run dry, but modern batteries really need no checking. They do wear out after a few years, however. Repair shops have battery testers that can measure how much life is left in a battery.

Starter Motor

The battery is connected to a large direct-current motor called the *starter motor*. When the starter spins, its gear is inserted against teeth on the flywheel, cranking the engine until (we hope) it starts. When you release the ignition key, the starter motor disengages from the flywheel and the motor stops.

If you've ever tried to start a car that was already running, you've heard a loud grinding noise. This racket comes from the starter-motor gear clashing against the spinning flywheel.

Starter motors require a very large current, which requires a large cable. Instead of running a large cable from the battery to the ignition switch and then to the starter, cars use a *relay*. This electromagnetic switch closes a circuit between the battery and the starter motor. Because they save weight and money, these *solenoid relays* are also used in other circuits.

Alternator

Batteries don't create electricity, they only store it, and they must be recharged while the engine runs. On early cars, a generator made electricity to run the lights and recharge the battery. These generators made direct-current (DC) electricity, the kind used in a car, and were most effective at higher engine rpms. Nowadays, the *alternator* handles this essential task. Alternators make alternating-current (AC) electricity, which must be "rectified" to DC for use in a car. Alternators, however, make more power at lower rpms.

An alternator is basically an electric motor working in reverse. Instead of converting electricity *into* rotary motion, it makes electricity *from* rotary motion. The rotation comes from the engine, via a fan belt.

The principle of an alternator is similar to the principle of an ignition coil: Electricity is induced in a wire that moves in a magnetic field. In a coil, although the wires don't move, the magnetic field does.

To make an electric current, you can rotate a magnet inside a coil or rotate a coil inside a magnet. In an alternator, a magnet rotates inside a stationary coil, called a *stator*. The magnet is called the *rotor* because it rotates.

As the name implies, alternators create AC. To make the DC that the car needs, the AC goes through a device called a *diode*. Diodes, housed inside the alternator, are a common source of trouble. The rest of the alternator is more reliable, because it has three separate circuits for creating AC:

- If an alternator circuit fails, you may not notice it until you place a heavy demand on the electrical system.
- Alternators can fail suddenly if a belt breaks.
- If the alternator goes out, the battery must supply all electricity. By cutting off all unnecessary electrical devices, you may be able to get home or to a repair shop before the engine dies.

Engine Control Unit

Auto engines are now run by a small computer called the *engine control unit*, or ECU. The ECU is the car's brain, and while it is not serviceable, technicians can diagnose trouble by connecting the ECU to a computer.

ECUs get information from sensors that detect

- The mass of air going into the engine
- Engine speed
- The position of the throttle (showing how much power the driver wants)
- The amount of oxygen in the exhaust (showing whether the fuel mix is right)
- Coolant temperature (cold engines need a different mix from warm ones)
- Pressure in the intake manifold (showing how hard the engine is working)
- Alternator voltage (whether the engine needs to speed up to make more current)

Electrical System Troubleshooting

Electrical problems were once about the most common source of driving complaints. Fortunately, all of the big problems have been improved over the years through better materials, better design, and the elimination of problematic parts, especially the distributor and coil. Here are a few troubleshooting points for electrical systems:

- Loose and corroded connections are a major source of intermittent problems. Corrosion is an insulator, remove it with fine sandpaper to ensure a good contact.
- Many times, the best diagnosis for electrical problems comes from hooking a car up to a diagnostic computer.

FUEL SYSTEM

The fuel system contains the fuel tank, the fuel filter, and the carburetor or fuel injector. To increase reliability and reduce pollution, fuel systems have changed radically over the years.

The fuel tank, usually located at the rear of the car, stores gasoline. A charcoal filter in the tank absorbs gasoline fumes, reducing pollution. A fuel pump delivers gas through a different fuel filter to the fuel injectors.

A throttle plate controls how much air enters the engine. In older cars, the plate was connected directly to the gas pedal. The plate controls how fast the engine is running and how much power it puts out.

Carburetor

Carburetors are another "old-style" component that has largely been replaced. Carburetors combine gasoline with air in a *venturi*, where a rapid stream of air flows past a small fuel port. The partial vacuum in the fast-moving air draws fuel into the airstream, where it vaporizes and mixes with air.

The exact ratio of fuel to air is critical to performance. A mixture with too much fuel (a "rich" mixture) will waste gas and increase pollution. A mixture with too little fuel (a "lean" mixture) will burn too hot and be short on power. Carburetors always struggled to create the perfect mix for constantly varying driving conditions. Now, fuel injectors have solved that problem through a combination of electronic control and quick response.

Fuel Injector

Fuel injectors are electronically controlled valves that squirt fuel into the cylinder. The valve is closed until the injector receives an electric current and an electromagnet opens the valve. Fuel sprays into the cylinder until the valve closes.

The first fuel injectors were called *throttle-body* injectors. One injector sprayed fuel into the intake manifold. Newer, *multiport* injectors spray fuel directly upstream of the intake valve. The injectors, one per cylinder, get fuel from the *fuel rail*, a supply pipe filled with high-pressure liquid fuel.

The key advantages of injectors over carburetors are accuracy and reliability. By precisely timing the moment and quantity of spray, injectors allow engine designers to improve fuel mileage and performance while cutting pollution. And since injectors have few moving parts, they are more reliable than carburetors.

Fuel systems have improved greatly over the decades. Most problems stem from dirty or stale gasoline. While problems are rare, they still arise:

- Water or dirt in the fuel can clog lines or filters.
- Water can freeze in the fuel line.
- Vapor lock, a condition caused by overheated fuel lines, is no longer a problem with modern vehicles.

EXHAUST SYSTEM

After combustion, burned gases enter the *exhaust system*. The *exhaust manifold* connects the cylinders to the muffler, tailpipe, and parts of the pollution control equipment. Exhaust systems operate at high temperatures, and must protect the car and its occupants from heat.

The *muffler* is a chamber with baffles that deaden the noise of the explosions inside the engine. It's something that you take for granted until you hear a car with a hole in the muffler—and then you realize how loud an internal-combustion engine can be!

POLLUTION CONTROL

When gasoline burns cleanly, it produces mainly carbon dioxide and water vapor. While carbon dioxide is a cause of global warming, it is not usually considered a pollutant. However, gasoline engines make several types of pollutants, especially when combustion is incomplete. One goal of the ECU is to cut pollution by allowing complete combustion.

Engines can produce these kinds of pollution:

- *Carbon monoxide*, a product of partial combustion, is a colorless, odorless, but poisonous gas.
- *Partially burned hydrocarbons*, also known as particulate matter, include various pollutants. In large quantities, small pieces of hydrocarbon make black soot that you can see.
- *Volatile organic compounds* are another group of pollutants made by partial combustion. Some cause cancer, others irritate lung tissue, and many are a source of smog.
- Air is 79 percent nitrogen, and several *nitrogen oxides* are created in the high temperatures and pressures inside an engine. Sunlight can convert nitrogen oxides into *ozone*, a molecule with three oxygen atoms that is a key part of smog. Ozone damages the lungs. Nitrogen oxides also cause acid rain, which damages forests and surface waters.

Positive Crankcase Ventilation

The bottom of an internal-combustion engine is full of a smoky, polluted gas made from the burning of oil and gasoline. To prevent crankcase gases from polluting the atmosphere, the *positive crankcase ventilation* system pipes this gas to the intake manifold, where the gas is burned in the cylinders. These simple, effective controls were the first pollution controls on gasoline engines.

Catalytic Converter

Catalysts—rare metals like platinum, rhodium, and palladium—cause other chemicals to react without being consumed themselves. Catalysts are embedded in heat-resistant ceramic elements. Pollutants in the exhaust gases are converted to simpler, less toxic compounds in the catalytic converter. From the outside, catalytic converters look like mufflers.

A *three-way* catalytic converter deals with carbon monoxide, volatile organic compounds, and nitrogen oxides. Often, two separate catalysts are needed to perform these chemical reactions:

A *reduction catalyst* reduces nitrogen oxides to oxygen and nitrogen.

An *oxidation catalyst* oxidizes carbon monoxide into carbon dioxide and volatile organic compounds into carbon dioxide and water.

SUSPENSION

The job of the suspension is to hold the wheels to the road and make the ride safe and comfortable. *Axles* connect the car to the wheels. Driving axles move the wheels and must rotate themselves. In rear-wheel-drive cars, the driving axles connect the differential to the wheel. In front-wheel-drive cars, the axles are part of the transaxle. These are quite complicated.

Steering Gear

Steering gear connects the steering wheel to the front wheels. In *rack-and-pinion* steering, a small *pinion* gear at the end of the steering shaft turns against a flattened gear called a *rack*. In other systems, a steering gearbox converts rotary motion of the steering wheel to a linear motion that steers the wheels.

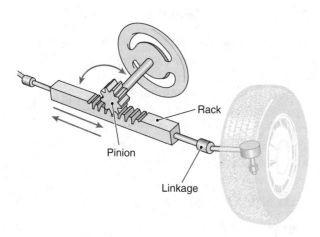

Rack

Pinion

Linkage

Rack-and-Pinion Steering Gear.

Modern steering equipment can be quite complex, especially with front-wheel or all-wheel drive.

Steering may be power-assisted, where the steering wheel does only part of the work, or manual, in which case the driver provides all the effort. Generally, larger cars have power steering, while in smaller cars, the driver's force is sufficient. Power steering can supply a varying level of assistance. A type of hydraulic fluid called *power steering fluid* is used in the hydraulic system in power steering.

Suspension

Autos use many types of suspension, but the two key categories are *leaf spring* or *coil spring*. Springs are made of a kind of steel that is tough and resilient.

Shock absorbers ("shocks") prevent axles from bouncing back, which makes for an uncomfortable ride. Springs resist movement when the car is moving downward, and shocks resist the rising motion. Ideally, after hitting a bump, the axle will move up toward the frame, then return to its original position.

When shocks wear out, a car tends to wander on the road because the springs aren't limited by the shocks.

Bearings

Bearings are mechanical gadgets that allow a part to rotate with almost no friction. Cars have bearings wherever rotating parts are found: in the steering wheel, engine, transmission, drive shaft, differential, and axles. Wheel bearings hold a wheel to a stationary (not driving) axle. Axle bearings hold driving axles. The individual bearings may be shaped like balls (*ball bearings*) or like tiny cylinders (*roller bearings*). Many individual bearings are housed together in an assembly that is also called a bearing.

Wheels and Tires

As we reach the end of the suspension, we meet two familiar objects: the *wheel* and the *tire*. Wheels are usually made of steel, and are bolted to the axle (for a driving wheel) or the brake apparatus (for a nondriving wheel). Most tires are called *radial-ply*, meaning that their plies (the invisible reinforcing belts that hold the tire together) run radially—in a line starting at the center of the wheel. Radials offer better life and performance than older *bias-ply* tires.

Tires are made of artificial rubber, inflated to a pressure of 30 or 40 pounds per square inch. Air pressure holds the tire's shape, and also permits it to absorb road shock.

- Proper inflation is important for critical gas mileage, long tire life, and safe handling. The suggested pressures are listed on a tire sidewall or inside the glove compartment. Normally, hot air expands. But air trapped inside a tire cannot expand, so its pressure rises instead. Suggested tire pressures apply only to a cold tire. Do not release air from a hot tire if it seems overpressurized. Instead, wait until the tire is cold to check the pressure.
- Tires may gradually lose pressure without having a major problem. But if you have to pump up a tire frequently, it's got a problem that needs attention.

BRAKE SYSTEM

Mechanically, brakes are the opposite of engines. Engines convert heat that originates in fuel into motion; brakes use friction to convert motion into heat. Brake systems have two key parts: the hydraulics and the braking mechanism at the wheel that does the actual stopping. The hydraulics transfer motion from the driver's foot to the brake mechanism.

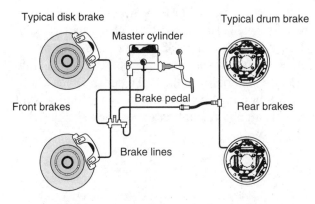

A Dual Braking System Has Two Master Cylinders and Two Sets of Tubes to the Individual Brakes.

Hydraulic systems are based on the fact that hydraulic fluid cannot be compressed. In the simplest hydraulic mechanism, two cylinders are connected by a tube. When a piston moves in the first cylinder, the piston in the second cylinder must also move. Why? Because the fluid is not compressible and must always occupy the same volume.

Air is easily compressed. If air enters the hydraulic system, the brake pedal will get "spongy." The cure is to "bleed" out the air through bleeder valves located at each brake.

In a car, the brake pedal is connected to the *master cylinder*, and four tubes run to the *slave* or *wheel cylinders*. These are located in the brake mechanism at the wheels. Cars now have dual master cylinders and dual hydraulic lines. If one part of the system fails, the other half should still work.

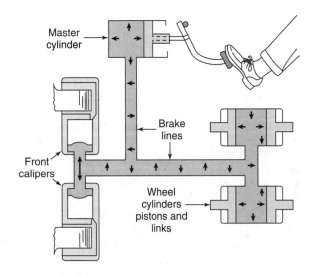

In This Simple Hydraulic System, Moving the Piston in the Master Cylinder Causes the Piston to Move in the Slave (Wheel) Cylinder.

Drum Brakes

Drum brakes operate by expanding two *brake shoes* inside a large cylinder called a *drum*. The drum rotates with the wheel, but the shoes do not rotate. A *wheel cylinder* operates the brake by moving the shoes toward the drum. Friction between the shoes and the drum creates the stopping power.

- When the shoes wear, a self-adjusting mechanism moves them closer to the drum. Wheel cylinders have rubber pistons that must be replaced when they wear out.
- Brake shoes and drums must also be replaced occasionally. If you let the shoes wear too far, they will damage the drums, forcing a more expensive repair.

Disk Brakes

A disk brake has a rotor, or disk, that revolves with the wheel and a pair of stationary pads that create

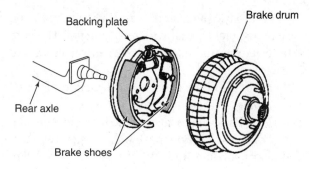

Typical Drum Brake.

friction when they press against the rotor. Disk brakes are a big improvement over drum brakes because they offer shorter stopping distances and better control.

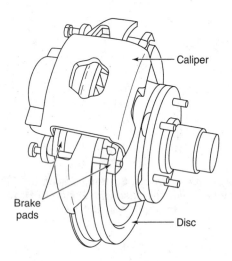

Typical Disk Brake.

The horseshoe-shaped *caliper* holds the pads. Hydraulic force squeezes the pads against the disk.

- Over time, pads wear out. Many disk brakes make a warning noise when the pads are wearing out. You usually will hear the noise when your foot is not on the brake pedal.
- Eventually, the pads wear concentric grooves in the disk. Before putting on new pads, the disks must be machined flat so that the new pads press against a flat surface. After being machined several times, the disks become too thin, risking a heat buildup and reduced stopping power ("fading"). For safety's sake, these disks must be replaced.

Parking Brakes

Parking brakes need a separate mechanical system to operate the brakes; they cannot use hydraulics. Usually, cables from the parking-brake lever to a drum brake operate the parking brake. Disk brakes require a more complicated parking brake. They may have a miniature drum brake built inside them. In other cases, a corkscrew mechanism pushes the brake pads against the rotor.

Antilock Brakes

When tires skid, they lose braking power, and steering becomes difficult or impossible. To improve safety during sudden stops, many cars have an antilock braking system (ABS). An ABS uses sensors to detect wheel rotation. If a wheel starts to skid during braking, the ABS reduces pressure to that brake, then starts pumping the brake. Pumping may produce a vibration in the brake pedal. Other wheels get full braking pressure until they too start to skid. ABS is a good feature on ice or snow.

AUTO INFORMATION QUIZ

Take the following quiz to check your understanding of ASVAB Auto Information topics. Circle each answer that you choose. There is no time limit. Answers and explanations are given immediately after the quiz.

1. If the shock absorbers are defective,
    A. the springs will wear out too soon.
    B. the front wheels will go out of alignment.
    C. the brakes will lock up too soon.
    D. the wheels will rebound excessively after the car hits a bump.

2. Some cylinder heads are made of aluminum because
    A. it is light and inexpensive.
    B. it is heavy and inexpensive.
    C. cast iron warps too easily.
    D. it is light and conducts heat well.

3. Disk brakes are sometimes used only on the front wheels because
    A. disk brakes are easier to steer.
    B. the front wheels provide most of the stopping force.
    C. disk brakes are cheaper to make than drum brakes.
    D. disk brakes are less likely to seize up than drum brakes.

4. In a 12-volt system, when the electric current is 12 volts, the battery is
    A. charging.
    B. discharging.
    C. neither charging nor discharging.
    D. I cannot tell. There is not enough information to answer the question.

5. In a hydraulic system, a 2-inch-diameter master cylinder drives a 1-inch-diameter wheel cylinder. If we move the piston 3 inches into the master cylinder,
    A. the force will be multiplied at the wheel cylinder.
    B. the force will be reduced at the wheel cylinder.
    C. the force will be the same at the wheel cylinder.
    D. nothing will happen. This hookup will not work because the slave cylinder needs to have a larger diameter than the master cylinder.

6. A higher viscosity rating is applied to oil that is
    A. thicker.
    B. thinner.
    C. less likely to freeze in winter.
    D. more likely to overheat.

7. If a car has a fuel injector, it will not have
    A. a regulator.
    B. a carburetor.
    C. an Otto-cycle engine.
    D. a fuel filter.

8. When the brake shoes wear down too much, you are likely to see _____ in the drums.
    A. glazing
    B. scoring (gouges)
    C. shot spots
    D. rust

9. When you see blue smoke coming from the tailpipe, you should check the
    A. piston rings.
    B. intake valves.
    C. positive crankcase ventilation system.
    D. fuel system.

10. If a car suddenly stops responding to the starter and the headlights do not light, what would be the most likely problem?
    A. The battery terminals or cables are corroded.
    B. The battery has gone dead.
    C. The starter switch is bad.
    D. The starter motor has died.

11. If a manual transmission is in neutral,
    A. the wheels cannot turn.
    B. the engine can run without moving the car.
    C. the differential is locked up.
    D. the transmission cannot be shifted.

12. When you bolt a cylinder head into place, you should
    A. align the cylinders with the pistons.
    B. align the carburetor.
    C. check the main bearings.
    D. observe proper tightening torque.

13. If the oil light comes on while you are driving,
    A. drive home and check for problems.
    B. replace the oil within 1,000 miles.
    C. do not worry unless you see other evidence of trouble; the light is probably defective.
    D. as soon as it's safe, pull over to find the problem.

14. Which components are not found in a car with electronic ignition?
    A. Coil wire and distributor
    B. Coil and alternator
    C. Distributor and ECU
    D. ABS and ECU

15. If it's hard to shift gears in a car with manual transmission,
    A. add transmission oil.
    B. check the free play on the clutch.
    C. don't worry, the oil is probably cold.
    D. try starting in gear.

ANSWERS AND EXPLANATIONS

1. **D.** Without shock absorbers, the axle and wheel would continue to bounce (think of a weight bouncing up and down on a long spring). That would cause an unstable, unpleasant ride. The other answers are partly correct, but D is the most accurate answer.

2. **D.** Why are the other answers incorrect? A, aluminum is relatively expensive; B, aluminum is light; C, cast iron does not warp easily (it is commonly used in engine blocks).

3. **B.** Vehicles lurch forward while braking, and the engines are in front, so up to two-thirds of the stopping force occurs on the front wheels. Because disk brakes are more effective than drum brakes, a cost-saving compromise is to put disks in front and drums in the rear.

4. **B.** Despite the name, a 12-volt battery is actually discharging at 14 volts. To charge the battery, the alternator must supply a higher voltage.

5. **B.** When a larger cylinder drives a smaller one, the smaller one moves further, but with less force.

6. **A.** Viscosity measures resistance to flow. Oils with higher viscosity are thicker, and flow more slowly.

7. **B.** Fuel injectors replace carburetors. (Internal-combustion engines do use the Otto cycle, named for the German who invented the gasoline engine.)

8. **B.** Worn brake shoes can score, or gouge, the brake drums, which then must be machined to produce good stopping power.

9. **A.** Worn piston rings allow oil to enter the cylinder, causing blue smoke.

10. **A.** Oxidized metal (corrosion) is often an insulator. So corroded battery terminals or cables can cause a sudden loss of current to the starter and lights. Why are the other answers incorrect? B is possible, but batteries seldom go completely dead this quickly. C, a bad starter switch, and D, a dead starter motor, would not affect the headlights.

11. **B.** The purpose of neutral gear in any transmission is to allow the engine to run while the car stands still.

12. **D.** Proper tightening torque prevents warping of the cylinder head.

13. **D.** Low oil pressure can quickly destroy an engine; pull over and determine the cause of the problem as soon as it's safe to do so.

14. **A.** Electronic ignition replaces the coil wire, which feeds high-voltage current from the coil to the distributor. It also replaces the distributor, which routes high-voltage current to the spark plugs. Why are the other answers incorrect? The car would still have an alternator, ECU (engine control unit), and ABS (antilock braking system).

15. **B.** If the clutch does not have the right amount of free travel at the top of the stroke, it may not be disengaging completely, and shifting may become more difficult.

Shop Information

INTRODUCTION: ASVAB SHOP INFORMATION QUESTIONS

The shop information questions that appear on the ASVAB measure how much you understand about shop tools, practices, materials, and procedures. The questions may ask you to identify a particular shop tool, to tell how it is used, or to choose the correct procedure when working with wood, metal, or construction materials. If you took shop courses in high school, or if you have done any woodworking or metalworking on your own, you may already be familiar with many of the topics covered on the test. You may also have learned shop and construction tools and procedures on the job.

ASVAB Shop Information Questions

Number of Questions:

- Paper-and-pencil ASVAB: 13–15 (as part of the Auto and Shop Information test)
- CAT-ASVAB: 10 (as a separate test)

Time Limit:

- CAT-ASVAB: 6 minutes

Topics Covered:

- Tools
- Wood and metal shop practices
- Construction materials and procedures

On the paper-and-pencil version of the ASVAB, shop questions are one part of the Auto and Shop Information test. On the CAT-ASVAB, they form a separate test of their own.

Whichever ASVAB version you take, you'll have only about half a minute to answer each shop information question, so you'll have to work fast if you want to get a good score. That's why it pays to spend time studying the test topics and tackling plenty of sample ASVAB shop information questions.

The topic review that follows will help prepare you to answer ASVAB shop information questions. It lists and describes the different tools you should know, with an emphasis on traditional hand tools. It also covers common fasteners and materials. There is also important information about tool safety. At the end of the chapter, there is a short quiz with questions modeled on those on the actual test. Read carefully through the review materials in this chapter; then use the quiz to find out how well you have mastered this subject area. Go back and reread the review materials for any quiz items you miss.

MEASUREMENT AND LAYOUT

New projects and many repairs start with measurement and layout. This is where you amass the proper tools and materials, and mark out the cutting and drilling.

Measuring Tools

Carpenters usually use a *tape measure*, marked in 1/16-inch increments. The tape retracts automatically into the case but locks in place when it must be extended for a while. The tape has a hook on the end that moves slightly for accurate inside and outside measurements. Retractable tape measures range from 6 feet to 25 feet long.

When greater accuracy is needed, machinists use a rigid *steel rule*. These rules are often marked in 1/32- or 1/64-inch increments and are often 1 foot long. A metric steel rule would usually be marked in 1-millimeter increments.

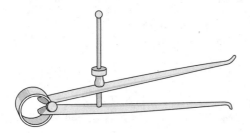

Inside Caliper.

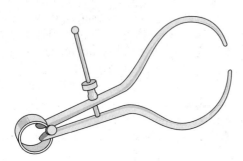

Outside Caliper.

Vernier Caliper.

When even greater accuracy is needed, machinists use *calipers* or *micrometers*. Some calipers are simply two legs that can transfer a measurement to a steel rule. These calipers can take inside or outside measurements, depending on the shape of the legs. Calipers can also have straight legs.

The Vernier caliper is even more accurate, thanks to the clever Vernier system. First identify the Vernier scale and the main scale. Now look at the "0" on the Vernier scale (see Figure below). This is just past the third mark on the main scale, indicating 3 millimeters (mm) on this metric caliper. So the measurement is a bit more than 3 mm, but how much more? Notice that line 3 on the Vernier scale lines up with a line on the main scale. That means that you should add 0.3 mm to the measurement, making a total of 3.3 mm. (Only one mark on the Vernier scale will line up with the main scale, and it doesn't matter which line it lines up with. Remember, read the number *on the Vernier scale* to get the right-hand digit in the measurement.)

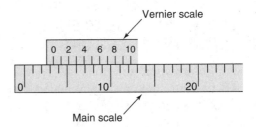

Close-up of Vernier Scale.

Micrometer.

Micrometers are even more accurate than Vernier calipers, but they are usually designed to read only in a certain range, say, up to 1 inch or 1 inch to 2 inches. Unlike a Vernier caliper, where you slide the adjuster, you turn a screw on a micrometer.

Layout often calls for square (90°) lines. A *carpenter's square* is used to draw these lines: When you hold one leg against the edge of a board, the second makes a square line across the board. A smaller version is called the *try square*.

Carpenter's Square.

You may also see a *sliding bevel*, which has a metal leg fastened to a wooden block. By loosening the adjustment screw, you can set the tool to mark almost any angle. Sliding bevels can be used to transfer angles from place to place.

Sliding Bevel.

The easiest way to tell if something is level (horizontal) or vertical (plumb) is with a *level*, sometimes called a spirit level. Levels use glass or plastic tubes that are curved or slightly swollen in the middle. When the bubble in the liquid (spirit) is centered, the level is horizontal or vertical.

Level.

CUTTING AND SHAPING

After the layout step is done, it's time to cut and shape the materials. We'll take up woodworking tools first, then metalworking tools.

Woodworking Tools

Sharp hand saws are the most basic way to cut wood. Saws cut a *kerf* that is wider than the blade itself;

the kerf allows the saw to move freely through the cut. *Crosscut saws* are designed to cut at 90° to the grain, while *ripsaws* cut parallel to the grain. Ripsaws have larger teeth. *Backsaws* have a rigid steel backing that improves accuracy; they are used in miter boxes that guide them for 45° or 90° cuts.

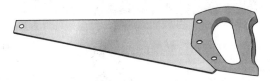

Hand Saw.

Keyhole saws are made to cut complicated profiles. An electric version is called the *jigsaw*. A *coping saw* has a thin blade held in a P-shaped handle. The saw is used to cut molding.

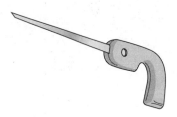

Keyhole Saw.

An electric *circular saw*, usually with a 7-1/4-inch-diameter blade, is much faster for cutting wood, especially for ripsawing, and for sawing plywood or other panels. These saws are dangerous; read the instruction manual carefully.

Circular Saw.

Wood chisels, sold in widths from 1/4 inch to 1-1/2 inches, cut wood when they are struck with a hammer or mallet.

Wood Chisel.

A *hand plane* removes thin strips of wood and is used to shape, smooth, or reduce the size of boards. It's especially useful for removing saw marks from the edge of a board. The "jack" plane is a general-purpose type of hand plane.

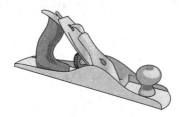

Jack Plane.

Metalworking Tools

A *hacksaw* has a replaceable metal blade with small teeth and is used for cutting iron, steel, and other, softer metals. Choose a blade with finer teeth for thinner metal, and one with larger teeth for thicker metal. The hacksaw should cut on the forward stroke.

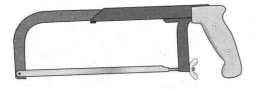

Hacksaw.

Tin snips cut steel, copper, or aluminum sheet metal, using a shearing action. Some snips have replaceable blades; others can be sharpened. Special snips are designed to cut curves.

Tin Snips.

A *right-angle grinder* can polish metal before painting, or otherwise smooth or shape metal. Grind toward the edge of the wheel; do not hold it flat to the surface of the metal. The same tool will also drive a wire brush for removing rust.

A *pipe cutter*—used for copper, not steel, pipe—has a sharp cutting wheel. Gradually tighten the handle as you rotate the tool around the pipe.

Pipe Cutter.

Taps and *dies* cut or restore threads in metal. A die cuts threads on a rod; a tap cuts threads in a hole drilled in a plate. Either tool can be used to restore mangled threads. Both taps and dies cut only one diameter and pitch (number of threads per inch). To select a die, you must know the outside diameter (O.D.) of the pipe.

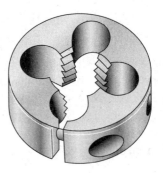

Thread-Cutting Die.

Thread-Cutting Tap.

DRILLS AND DRILLING TOOLS

Electric *drills* have become a centerpiece of all workshops, for making holes, driving screws, and other purposes. Drills are sized by the largest diameter of bit that will fit in the chuck (the rotating clamp that holds the bit). You sometimes see 1/4-inch drills, but 3/8-inch drills are an all-around tool for the home workshop. Larger drills handle bits of 1/2- or 3/4-inch diameter. However, the shank (the part that gets grabbed in the chuck) can be smaller than the tip of the drill bit, so it's possible to drill 1-inch holes in wood with a 3/8-inch drill.

Iron and steel are much harder to drill than wood, and thus call for a larger, more powerful drill. Often, it's best to drill a small "pilot" hole in metal. This is because metal-cutting drill bits have a blind spot near the center where they don't drill very well. The pilot hole makes a clear space for the blind spot on the larger drill. While drilling metal, it often helps to oil the bit for cooling; excess heat can destroy the heat treatment that makes a bit hard enough to cut metal.

The chuck may be tightened with a chuck key, as shown. Newer, self-tightening chucks do not use a key. Instead, the parts tighten when they are turned against each other.

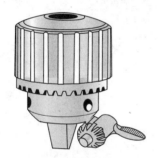

Drill Chuck with Chuck Key.

3/8-Inch Drill.

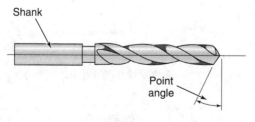

Drill Bit.

The helix on a twist drill brings chips up out of the hole. Twist drills will start accurately in wood,

but they wander across the surface in metal. A *center punch* makes a dimple in metal to locate the bit as it starts to drill. Use a hammer to hit the center punch.

Center Punch.

Auger bits are made only for wood, which they cut much faster than twist drills. Auger bits were originally driven by a brace and bit, but they can also be used in electric drills.

Auger Bit.

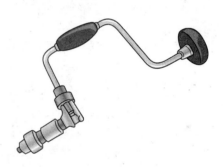

Brace and Bit.

A *countersink* is a conical depression in a surface that allows a flat-head screw to sit flush (flat) to the surface. A special drill bit, also called a countersink, makes the countersink.

Countersink.

A *hole saw* makes large-diameter holes in wood and some metals. The type of hole saw shown screws onto a mandrel, allowing one mandrel to handle several size saws. Hole saws are more economical than big drills for drilling wood, but they do not work in hard metal.

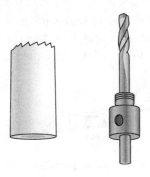

Hole Saw.

POUNDING TOOLS

Many of the oldest tools are designed to hit something. A *claw* (carpenter's) *hammer* pounds nails with the face and pulls them with the claw. A straight claw is better for longer nails and is also handier for doing demolition. Some hammer faces have a checkered pattern, called a *waffle head*, to increase the grip on the nails. The standard size of claw hammer is 16 oz. Hammers 24 oz. in size are used for large nails.

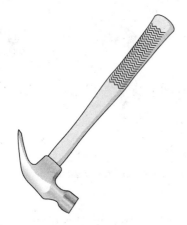

Claw Hammer.

Metal workers use a *ball-peen hammer*. One face is flat, like a claw hammer, the other face has a ball peen, used for shaping metal and riveting. Ball-peen hammers may weigh up to 3 lb.

Ball-Peen Hammer.

Sledgehammers are used for heavy purposes. Some people can swing the hand sledge shown here with one hand. True sledgehammers have a 32-inch handle and require two hands.

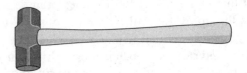

Hand Sledge.

A *rubber mallet* may be used to adjust parts without damage, or to drive wooden-handled chisels. Wooden mallets are also used for striking chisels. Many modern chisels, however, have steel shanks and can safely be struck with a hammer.

Rubber Mallet.

TOOLS FOR TURNING OR GRABBING

Shop work, including maintenance and repair, requires a good assortment of tools that turn or grab. An *adjustable* (Crescent) *wrench* can be adjusted to hold various sizes of hexagonal or square bolts. Typical overall lengths range from 6 to 12 inches.

Adjustable Wrench.

A *combination wrench* combines two basic types of wrench: the box-end and the open-end wrench. Each end of a combination wrench fits the same size bolt. Because most bolt heads have six faces (called *hexagonal bolts* or hex bolts) the box end has 6 or 12 facets. With 12 facets, you can turn the wrench 1/12 of a turn, which is handy in tight quarters. That's why they are offset 15° at the end.

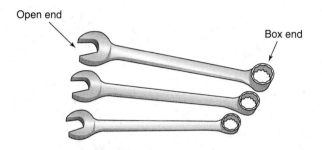

Combination Wrench.

A *socket wrench* holds a bolt or nut from above. The deep socket, shown, allows you to grab a nut even if some threads are sticking out above it. Socket wrenches connect to the sockets with a square drive; possible drive dimensions include 1/4, 3/8, 1/2, or 3/4 inch. Larger drives are available for heavier-duty use.

Deep Socket.

A *ratchet wrench* is one way to drive a socket. It will grab when it swings in one direction and slip in the other, for convenient tightening or loosening.

Ratchet Wrench.

A *torque wrench* fits a socket and drives bolts to a specified tightness. You can buy them marked in American or metric units.

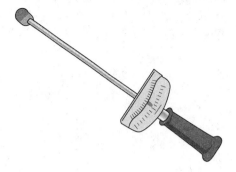

Torque Wrench.

An *Allen wrench* fits screws with a hexagonal recess in the head. Allen wrenches are sold in inches and millimeters.

Allen Wrench.

Arc-joint pliers (often called Channel Locks) are used to grab various sizes of material. To adjust the jaws, open them wide and engage a different set of arcs. Arc-joint pliers are not good for grabbing bolts or nuts, as they will scar the metal.

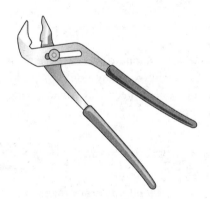

Arc-Joint Pliers.

Locking pliers (often called Vise Grips) have a lever system that gives a very strong, locking grab. They are one of the handiest tools in the box, but they can damage bolts and nuts.

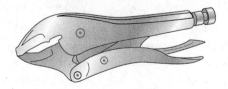

Locking Pliers.

A *needle-nose* (longnose) *pliers* gets at small parts and is especially handy for electric wiring.

Needle-Nose Pliers.

A *bar clamp,* among other types of clamp, can hold parts in position while you work or hold joints while the glue sets.

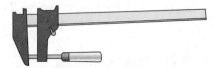

Bar Clamp.

A *pipe wrench* has steel teeth that hold steel pipe. They are sold in lengths ranging from 12 to 48 inches. Flip the wrench over to turn the pipe in the other direction.

Pipe Wrench.

Flat-bladed (standard) *screwdrivers* are a major part of any toolbox. They come in many sizes. A long shank will protect your hands when you are pressing hard.

Standard Screwdriver.

A *Phillips screwdriver* drives Phillips screws—the ones with a cross-shaped head. They also are sold in various sizes.

Phillips Screwdriver.

FASTENERS AND FASTENING TECHNIQUES

Keeping things together requires a variety of different types of fasteners. We'll start with threaded fasteners and nails and then move to fasteners that use molten metal.

Threaded Fasteners

Some threaded fasteners—*screws* and *bolts*—are designed to cut threads in the material. These wood and sheet-metal screws come in many sizes and materials, with any number of head styles. Sheet-metal screws cut threads in sheet-metal.

Round-Head Wood Screw.

Wood screws are sized by length and diameter. Length is measured in inches; diameter by a numbering system (#6, #8, #10, and so on). Larger numbers indicate a larger diameter.

Sheet-Metal Screw.

A separate category of threaded fasteners uses what are called *machine threads.* These screws and bolts must be screwed into a nut with the same diameter and number of threads per inch (also called pitch). Most machine threads are right-hand, meaning that they tighten when you turn the top to the right—clockwise. Left-hand bolts tighten in the opposite direction. Machine screws often use lock washers to prevent the nut from loosening. The lock washer is compressed under the nut and holds the nut tight.

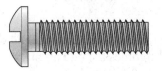

Machine Screw.

A *hex bolt* is a machine screw. It can range in diameter from 1/4 inch on up, and in length from 1/2 inch on up.

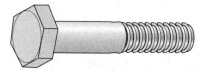

Hex Bolt.

A *carriage bolt* is a machine screw with a round head and a square shank. The shank fits a square slot, so the bolt does not turn while being tightened. The large head substitutes for a flat washer, distributing the bolt's force across a wider area.

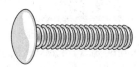

Carriage Bolt.

Nails

Nails are an ancient form of fastener, but they have gotten a lot more complicated over the past 20 years. Let's start with some terminology:

- *Brad*: A small, thin nail with a small head, used for picture frames and other light fastening.
- *Tack*: A small nail with a big head, used to attach carpet and upholstery.
- *Finishing nail*: A sturdy, small-diameter nail with a small head, used to attach trim and molding.
- *Common nail*: A big, large-headed nail used for rough construction.
- *Sinker*: A smaller-diameter version of the common nail that causes less splitting.
- *Spike*: A heavy, large nail for fastening timbers.
- *Ring-shank*: A nail with rings that improve grip.
- *Spiral*: A nail with a spiral on the shank, used to increase grip, for example, on flooring nails.

Nails are sized by length and by pennies ("d"). A 4d nail is 1-1/2 inches long, while a 16d nail is 3-1/2 inches long. Nails are sometimes coated with zinc (*galvanized*) to resist rust. These days, nails are often driven with pneumatic devices called *nail guns*.

When selecting a nail, choose a large head to hold soft materials and a small head for an exposed location. In construction, structural nails must grab about 1-1/2

inches. A larger diameter not only gets a stronger grip but is also more likely to split the wood. Finally, for strength, a ring- or spiral-shank nail is a good choice.

Welding, Soldering, and Brazing

Welding, soldering, and brazing are fastening methods that use molten metal. *Welding* melts the metal that is being fastened. *Soldering* and *brazing* melt a separate metal, called *solder* or *brazing rod*, to make the attachment. Soldering is used for plumbing and for electrical or electronic equipment. Solder was traditionally a mixture of tin and lead, but lead is toxic, so in modern solder, tin is now mixed with other metals, such as antimony.

To solder, clean the parts, apply flux, heat, and then touch the solder to the heated parts (do not just melt it with the iron and drip it over the joint). *Flux* is needed to prevent oxidation during heating and to allow the solder to flow into the joint. A good solder joint makes a tight seal in copper pipe and a durable connection in electronics.

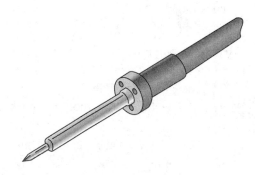

Soldering Iron.

Brazing is similar to soldering except that the joining material is bronze, and the metal must be heated hotter. This is usually done with an oxyacetylene torch, which burns acetylene gas and oxygen. The same torch can be used to weld steel. An oxyacetylene cutting torch heats the steel and then floods it with oxygen, causing the steel to rapidly oxidize, or burn.

An arc welder heats metal with an electric current and is widely used for welding. The current, the composition of the welding rod, and the gas surrounding the weld can all be controlled to get the desired results. Arc welders such as TIG (tungsten inert gas) are highly versatile machines that can weld a wide variety of metals, including steel, tungsten, and aluminum.

SHOP INFORMATION QUIZ

Take the following quiz to check your understanding of ASVAB Shop Information topics. Circle each answer that you choose. There is no time limit. Answers and explanations are given immediately after the quiz.

1. The reason for a long shaft on a screwdriver is to
   A. give leverage for opening paint cans.
   B. protect your fingers when you're pushing hard on a screw.
   C. allow plenty of room to grab the shaft with a locking pliers.
   D. hang easily on the tool board.

2. What is the purpose of a saw kerf?
   A. To make room for the saw blade in the cut
   B. To increase tension on the saw blade, to make a straight cut
   C. To reduce space between the teeth, so that the saw cuts faster
   D. To hold the blade straight so that the saw will cut straight

3. What is the purpose of a 15° offset on an open-end wrench?
   A. To protect your knuckles
   B. To "torque down" the bolt to the correct tightness
   C. To allow the wrench to grab the nut when working in tight quarters
   D. To allow a hex wrench to fit a square bolt head

4. An electric drill used to drive wood screws should be
   A. reversible.
   B. variable speed.
   C. a plug-in type.
   D. variable speed and reversible.

5. Tin snips are a type of
   A. nipper.
   B. wrench.
   C. shear.
   D. torque wrench.

6. A common nail has a _____ than a finishing nail.
   A. smaller head and smaller diameter
   B. smaller head and larger diameter
   C. larger head and smaller diameter
   D. larger head and a larger diameter

7. In a wood screw labeled #8 × 1-1/4", "#8" indicates that the
   A. screw is #8 galvanized steel.
   B. screw has threads per inch.
   C. screw threads are at an 8° angle.
   D. the screw is #8 diameter.

8. If a steel pipe has an inside diameter (I.D.) of 3/4 inch,
   A. you have enough information to select a thread-cutting die.
   B. the outside diameter must be larger than 3/4 inch.
   C. the pipe is legal for carrying natural gas, but not potable water.
   D. the inside diameter must be smaller than 3/4 inch.

9. If a bolt has a left-hand thread,
   A. it requires a right-hand-thread nut.
   B. a clockwise turn will tighten the nut.
   C. the switch on a ratchet wrench should be pushed to the left side.
   D. a counterclockwise turn will tighten the nut.

10. The purpose of a lock washer is to
    A. prevent vibration from loosening the joint.
    B. hold the flat washer in position.
    C. prevent anyone from taking the nut off.
    D. keep the screw threads clean.

11. An arc welder
    A. creates heat with an electric arc.
    B. is used to make curved welds.
    C. can be used only on stainless steel.
    D. can also be used to solder.

12. The best tool for making accurate layout measurements is
    A. a try square.
    B. a steel rule.
    C. a tape measure.
    D. a spirit level.

13. A machine screw
    A. will fit any nut with the right diameter.
    B. can also be used to cut threads in sheet metal.
    C. cannot cut threads in wood.
    D. is used only in autos and related machinery.

14. In a Vernier caliper, the Vernier scale is used to
    A. find the rough measurement.
    B. find the most accurate digit in the measurement.
    C. find the fractional measurement.
    D. convert fractions to decimals.

15. What is the purpose of a center punch?
    A. To find the center of a dowel
    B. To locate the starting point for a circular chisel (gouge)
    C. To start a drill hole in wood
    D. To start a drill hole in metal

ANSWERS AND EXPLANATIONS

1. **B.** With a long shaft, your hands are a long way from the screw, protecting you from injury if something slips.

2. **A.** The kerf, or overall width of a cut, must be wider than the blade, or else the blade will jam in the cut.

3. **C.** In tight quarters, the 15° allows you to turn the wrench over after moving the bolt slightly to get another grab on the bolt.

4. **D.** A variable-speed, reversible electric drill is best for driving and removing wood screws.

5. **C.** Tin snips cut with a shearing action. A is incorrect; the cutting blades of nippers meet instead of sliding past each other. B and D are incorrect; tin snips are not made to turn nuts or bolts.

6. **D.** A common nail has a larger head and a larger diameter than a finishing nail.

7. **D.** The diameter of wood screws is sized by these numbers; larger numbers indicate larger diameter.

8. **B.** If the inside diameter is 3/4 inch, the outside diameter (O.D.) must be larger than 3/4 inch. Why are the other answers incorrect? A, to choose a thread-cutting die, you also must know the number of threads per inch (the pitch). C, in most places, steel pipe is acceptable for carrying potable (drinking) water. D, 3/4 inch is the inside diameter.

9. **D.** A left-hand thread works opposite to the usual thread, so a counterclockwise turn will tighten, not loosen, the nut. Why are the other answers incorrect? A, bolts with left-hand threads require left-hand nuts. B, a clockwise turn will loosen the nut. C, it depends on the ratchet wrench.

10. **A.** Lock washers spring out against the nut, to prevent unwanted loosening.

11. **A.** An arc welder creates heat with an electric arc.

12. **B.** A steel rule makes the most accurate measurements.

13. **C.** Machine screws do not cut threads in wood. Why are the other answers incorrect? A, the pitch and diameter must be the same. B, a machine screw cannot cut threads in sheet metal. D, machine screws can be used in building and other applications.

14. **B.** The Vernier scale provides the most accurate digit in the measurement.

15. **D.** Center punches make a dimple that locates the drill bit as you start to drill.

Chapter 17

Mechanical Comprehension

YOUR GOALS FOR THIS CHAPTER
• Find out what topics are covered on the ASVAB Mechanical Comprehension test.
• Learn basic facts about mechanical devices and mechanical motion.
• Study examples of Mechanical Comprehension questions and learn ways to solve them.
• Take a Mechanical Comprehension quiz.

INTRODUCTION: ASVAB MECHANICAL COMPREHENSION QUESTIONS

The ASVAB Mechanical Comprehension test is all about the basic materials and mechanical devices that you see around you every day. The questions deal with things like levers, pipes, water wheels, gears, pulleys, and the like. This is not a test of high school physics, so don't worry if you can't explain things in terms of Newton's laws or use words like *inertia* and *elastic rebound*. You won't have to know *why* machines function as they do. You won't have to know a lot of complicated scientific terms. But you will have to know *what* simple machines and materials are and *how* they operate. Most questions on the Mechanical Comprehension test can be answered using simple common sense, but you may need to do a few calculations. Many questions include a picture of some kind of simple machine, and you can use the picture to help you answer the question. Spinning gears, heat transfer from one material to another, and barrels rolling up inclined planes are all standard situations that you'll find on the Mechanical Comprehension test.

Whichever ASVAB version you take, you'll have only about a minute to answer each Mechanical Comprehension question, so you'll have to work fast if you want to get a good score. That's why it pays to spend time studying the test topics and tackling plenty of sample Mechanical Comprehension questions.

ASVAB Mechanical Comprehension Test
Number of Questions:
• Paper-and-pencil ASVAB: 25
• CAT-ASVAB: 15
Time Limit:
• Paper-and-pencil ASVAB: 19 minutes
• CAT-ASVAB: 22 minutes

MECHANICAL COMPREHENSION QUESTION TOPICS

The topics covered on the Mechanical Comprehension test will probably be familiar to you if you have ever operated machines or taken them apart to repair them. (See the box on the next page.)

This chapter will help you prepare for the Mechanical Comprehension test. It starts by listing the important simple machines and explaining how they operate. Then it covers other important topics, such as properties of materials, mechanical motion, and fluid dynamics. At the end of the chapter, there is a short quiz with questions modeled on those on the actual test. Read carefully through the review materials in this chapter, then use the quiz to find out how well you have mastered this subject area. Go back and reread the review materials for any quiz question you miss.

Mechanical Comprehension Question Topics
Principles of Mechanical Devices
• Mechanical advantage
• Simple machines
• Compound machines
• Structural support
• Properties of materials
Mechanical Motion
• Systems of pulleys
• Systems of gears
• Rotating wheels and disks
• Cams and cam followers
• Cranks and pistons
Fluid Dynamics
• Air pressure
• Water pressure
• Filling and emptying tanks

SOME BASIC CONCEPTS: WORK, ENERGY, AND FORCE

To do well on Mechanical Comprehension questions, you'll need to understand just a few basic concepts. For starters, you should know that in mechanics, *work* refers to a specific force applied over a specific distance. For example, your arm does work when it uses force to pick up a book. A lever does work when it uses force to lift a heavy object.

The ability to do work is called *energy*. Energy comes in several forms:

- *Kinetic energy*: Energy in a moving object.
- *Potential energy*: Energy that can be released under certain conditions. For example, potential energy is stored in objects when they are lifted off the ground. It is released when the objects fall.
- *Chemical energy*: Energy stored in chemicals, such as in a flashlight battery. Chemical energy is potential until it is released in a chemical reaction.
- *Electric energy*: Energy in moving electrons in an electric current.
- *Nuclear energy*: Energy released by reactions in the nucleus of an atom.
- *Solar energy*: Energy in the heat and light from the sun.

On the ASVAB, you may be asked to tell which kind of energy is present in a given situation. For example, when a child's swing reaches its highest point and pauses momentarily before swinging back down, the swing has only potential energy and no kinetic energy. When it swings through its lowest point, the opposite is true: The swing has no potential energy (because it cannot fall any further) and all kinetic energy. As it swings back up the other side, the kinetic energy is converted back into potential energy.

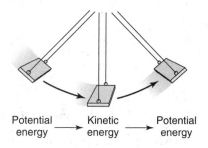

Potential energy → Kinetic energy → Potential energy

Similarly, a battery stores electric energy as chemical energy. When the flashlight is turned on, electric energy passes through the filament in the flashlight bulb, where it is converted into heat and light.

Forces are powers that push or pull objects. A force has a *magnitude* (strength) that you can measure, and it has a *direction*. Some forces are obvious—when a bat hits a baseball, you can even hear the force being applied. Other forces, like gravity and air pressure, are much less obvious, but they are still real.

One kind of force is *gravity*. Gravity is an attractive force between objects. All objects create a gravitational attraction for each other. On Earth, gravity causes objects to fall toward the center of the Earth. Falling objects accelerate (fall faster) as they fall. (*Acceleration* is defined as the change in velocity, or speed in a particular direction.) If you discount air resistance, all objects fall at the same rate—a fact proved centuries ago by the Italian scientist Galileo. In real life, however, air resistance often disguises the fact that objects fall at the same rate.

Air resistance is a kind of friction. *Friction* is a force that results from the interaction between two surfaces that are touching each other. Friction acts as a resistance to the movement of an object. For example, friction makes it harder to push a heavy crate up a ramp. Friction also helps to keep the crate from sliding back down the ramp. Without friction, a car would slide all over the place because the tires could not get a grip on the pavement. Tables would slide all over the room, and billiard balls would never stop rolling! In machinery, friction can often be a problem. It can prevent machines from running smoothly and efficiently. That's why lubricants are used to reduce friction.

Two other forces are *compression* and *tension*. Compression is a force that pushes materials together. Tension is a force that pulls materials apart. Air pressure and water pressure are forms of compression. The force exerted by the cable in a pulley is a kind of tension. In a bridge, some parts are in compression and others in tension. The weight of the structure squeezes (compresses) the top and puts tension on the bottom. Steel reinforcing is strong in tension, so it is placed at the bottom of a bridge, where its tensile strength can support the load.

PRINCIPLES OF MECHANICAL DEVICES

Machines are devices that multiply force or motion. Some machines are simple devices that involve only a single force. A lever is an example. Other machines involve combinations of devices working together. A bicycle is an example. The essential thing about all machines is that in order to make them multiply your force, you must exert that force over a longer distance. You'll see how this works in the following section, which describes the main simple machines one by one. First, however, you need to learn how to calculate how much a machine multiplies your force.

Mechanical Advantage

The amount your force is multiplied by a machine is called the *mechanical advantage*, or MA. There are two ways to calculate MA.

1. Divide the output force (called the *load* or sometimes the *resistance*) by the input force (called the effort): Load/effort = MA.

Example

With a lever, you use a 50-lb force (the effort) to lift a 200-lb weight (the load). What is the mechanical advantage of the lever? 200/50 = 4. MA = 4.

2. Divide the length of the effort (called the *effort distance*) by how far the load moves (called the *load distance*): Effort distance/load distance = MA.

Example

With a pulley, you use 5 feet of rope (the effort distance) to lift a load 1 foot (the load distance). What is the mechanical advantage of the pulley? 5/1 = 5. MA = 5

Simple Machines

The simple machines are a group of very common, basic devices that have all been in use for a very long time. They are called *simple* because each one is used to multiply just one single force. The simple machines include the lever, the pulley, the inclined plane, the gear, the wedge, the wheel and axle, and the screw. You can count on the ASVAB to test your knowledge of simple machines.

Levers The first kind of simple machine is the lever, a device that helps you apply force to lift a heavy object. To understand levers, you'll need to know the following terms:

- *Fulcrum*: The stationary element that holds the lever but still allows it to rotate.
- *Load*: The object to be lifted or squeezed.
- *Load arm (load distance)*: The part of the lever from load to fulcrum.
- *Effort*: The force applied to lift or squeeze.
- *Effort arm (effort distance)*: The part of the lever from force to fulcrum.

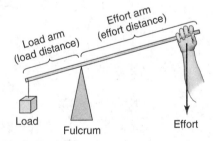

There are three classes of levers. Let's examine them one at a time and see how to calculate MA for each.

Class 1 Lever In a class 1 lever, the fulcrum is between the load and the effort. If the fulcrum is closer to the load than to the effort (as it usually is), the lever has a mechanical advantage.

Mechanical advantage
= effort distance/load distance = load/effort

Class 1 lever

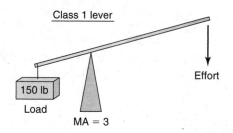

MA = 3

Example

The figure shows a class 1 lever. What force (effort) is needed to lift the load? Since you know that MA = 3, use this formula to find the effort.

$$MA = load/effort$$
$$3 = 150 \text{ lb}/effort$$
$$3 \times effort = 150 \text{ lb}$$
$$effort = 50 \text{ lb}$$

Class 2 Lever In a class 2 lever, the load is between the effort and the fulcrum. The effort arm is as long as the whole lever, but the load arm is shorter. So a class 2 lever always has a mechanical advantage.

Example

The wheelbarrow shown is a class 2 lever. What is its mechanical advantage?

$$MA = effort \ distance/load \ distance = 3/1.5 = 2$$

Class 2 lever

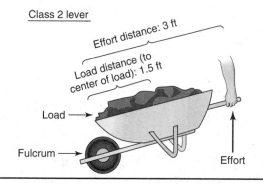

Class 3 Lever In a class 3 lever, the effort is between the load and the fulcrum. Tweezers and tongs are good examples of class 3 levers. The length of the effort arm and the load arm are calculated from the fulcrum, as with the class 2 lever.

Example

The figure shows a class 3 lever. What is the mechanical advantage?

$$MA = load/effort = 1/2 = 0.5$$

In other words, 2 pounds of effort would produce 1 pound of "squeeze" on the orange. We could call this a fractional mechanical advantage, or a mechanical disadvantage. But in return for reducing the squeezing force, each inch of effort movement produces 2 inches of load movement.

Class 3 lever

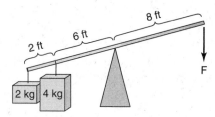

Balancing a Lever Some ASVAB problems may show you a diagram of a lever with various parts marked and ask you what force or weight is needed to balance the lever. To answer this kind of question, keep in mind that the moments of force (effort or load × distance) on either side of the fulcrum must be equal. Here is an example.

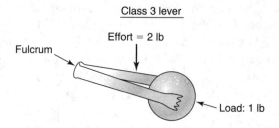

Example

What is the force F, in kilograms, needed to balance the lever? Add up the moments of force on either side of the fulcrum:

$$(2 \text{ kg} \times 8 \text{ ft}) + (4 \text{ kg} \times 6 \text{ ft}) = (8 \text{ ft} \times F)$$
$$16 + 24 = 8F$$
$$40 = 8F$$
$$F = 5 \text{ kg}$$

Pulleys Another kind of simple machine that helps you lift a heavy object is the pulley, also called a block and tackle. In pulleys, the mechanical advantage is found in either of the following two ways:

- MA = effort distance/load distance.
- MA = number of supporting strands. Supporting strands of rope or cable get shorter when you hoist the load. We'll return to this, but don't just count strands—some do not shorten as you hoist.

Example

The figure shows a pulley attached to a beam that is used to hoist a heavy crate. Each foot of pull on the rope lifts the crate 1 foot. Effort distance = load distance, so MA = 1. Although this pulley allows you to pull down instead of up, it gives no mechanical advantage.

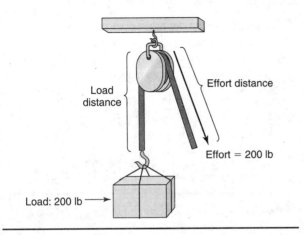

Load distance

Effort distance

Effort = 200 lb

Load: 200 lb →

Example

The figure shows two pulleys. When you hoist, two strands of the rope must be shortened. So for every 2 feet of pull (effort distance), you get 1 foot of lift (load distance).

MA = effort distance/load distance = 2/1
MA = 2

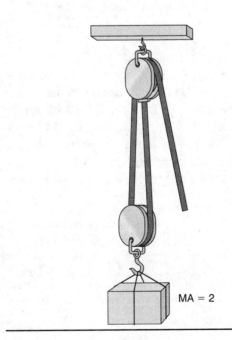

MA = 2

The simplest way to find the pulley MA is to count the strands of rope on the movable pulley (in

this case, the one attached to the load). MA = number of supporting strands.

Until now, we have ignored friction and the weight of the movable pulley and extra rope. As MA increases, these factors also increase, so there is a practical limit to the mechanical advantage of pulleys.

Gears Gears are a simple machine used to multiply rotating forces. Finding the MA of a gear is simplicity itself. Identify the driving gear (the one that supplies the force) and count the teeth. Count the teeth on the driven gear. Then use this formula:

Number of teeth on driven gear/number of teeth on driving gear = MA

Example

The figure shows a driving gear with 9 teeth and a driven gear with 36 teeth.

36/9 = 4
MA = 4

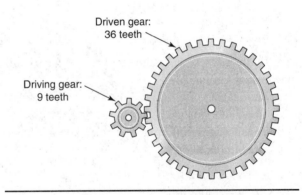

Driven gear:
36 teeth

Driving gear:
9 teeth

You will learn more about systems of driving and driven gears later in this chapter.

Sheaves Sheaves (often also called pulleys) and belts are a simple machine closely related to gears. To calculate the MA of a sheave system, divide the diameter of the driven sheave by the diameter of the drive sheave:

MA = driven diameter/drive diameter

Whenever the driven sheave is larger than the drive sheave, you get a mechanical advantage.

Example

The figure shows a sheave system. What is the MA?

9/3 = 3

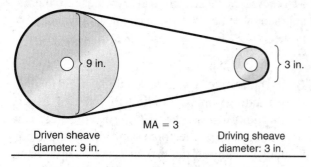

MA = 3

Driven sheave
diameter: 9 in.

Driving sheave
diameter: 3 in.

You will learn more about systems of sheaves (pulleys) later in this chapter.

Inclined Plane *Inclined plane* is a fancy term for "ramp." An inclined plane is another simple machine that is used to lift heavy objects. The formula for finding the mechanical advantage of an inclined plane is as follows:

MA = length of the slope/vertical rise

To find the mechanical advantage, measure vertically and diagonally along the ramp.

Example

The figure shows an inclined plane. What is the mechanical advantage?

MA = 12/3 = 4

If the load weighs 400 lb, how much force is needed to push it up the ramp?

MA = load/effort
4 = 400/effort
4 × effort = 400
Effort = 100 lb

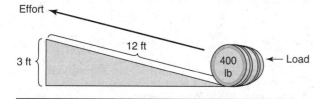

In real life, friction can play a huge role in ramps if the load is not on wheels. Most ASVAB problems will allow you to ignore friction in dealing with all simple machines.

Wedge The wedge is a type of inclined plane. It is one of the rarer simple machines. As always, MA = effort distance/load distance. The wedge is essentially two inclined planes, and the MA calculation also requires you to measure perpendicular to the long axis of the wedge.

Example

The figure shows a wedge. What is the MA?

Every time the wedge moves 5 inches, the load will move 2 inches. MA = 5/2 = 2.5. In reality, friction plays a major role in wedges.

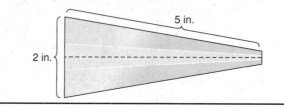

Screw Screws are some of the handiest simple machines, although we usually think of a screw as a fastener rather than as a way to multiply force. Finding mechanical advantage can be complicated because it comes from two sources: the threads and the wrench you use to tighten the screw. But if you consider effort distance and load distance, the calculation is simple.

MA = effort distance/load distance

Example

The figure shows an 8-inch wrench turning a screw with 8 threads per inch. This screw has a *pitch* (movement per turn of the screw) of 1/8 inch. The effort distance is π × diameter = 3.14 × 16 inches = about 50 inches. The load distance per turn of the wrench is 1/8 inch, so MA = 50/1/8 = 400. In reality, the MA is much less, because of friction and because you don't push on the

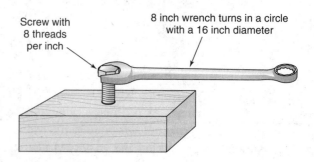

Screw with
8 threads
per inch

8 inch wrench turns in a circle
with a 16 inch diameter

absolute end of the wrench. But this still demonstrates the power of screws as simple machines!

Most ASVAB questions will not require this much calculation, but it never hurts to be prepared!

Wheel and Axle Wheels are a common and essential part of daily life, but most of these wheels are not simple machines. Instead, they are a way to reduce friction by the use of bearings. A wheel and axle is a simple machine only when the wheel and axle are fixed and rotate together.

A typical wheel-and-axle simple machine is the screwdriver. The screwdriver's handle is the wheel, and the screwdriver's blade is the axle. For wheel-and-axle machines, mechanical advantage is calculated as follows:

MA = effort distance (radius of the wheel)/load
distance (radius of the axle)

So for a screwdriver, MA = radius of the handle/radius of the blade.

Example

The figure shows a brace and bit, a kind of heavy-duty screwdriver that is an example of a wheel and axle as a simple machine. What is the MA?

Effort distance/load distance = MA
6 in./0.25 in. = 24
MA = 24

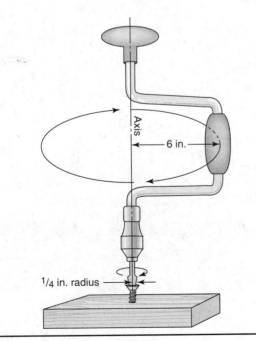

A wheel and axle can also give a mechanical disadvantage. In a car or a bicycle, where the axle drives the wheel instead of the wheel driving the axle, a small motion at the axle creates a large motion at the circumference of the rim. In these cases, you need a larger force, but you get more motion in return.

Compound Machines

A *compound machine* is one in which two or more simple machines work together. For example, a screwdriver (wheel and axle) driving a screw is a compound machine. To find the mechanical advantage of a compound machine, multiply the MA of the simple machines together.

Example

The axe shown is a compound machine. The handle is a lever, and the head is two inclined planes. Find the combined MA by multiplying the individual MA of each simple machine.

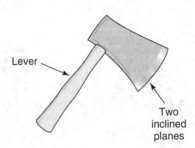

Structural Support

Some Mechanical Comprehension questions ask about the load carried by support structures such as beams or bridges—or sometimes by people. You will be given a diagram showing support structures and asked which one is strongest or weakest, or which support in the diagram is bearing the lesser or greater part of the load. To answer this kind of question, keep this in mind: When a load of any kind is supported by two support beams, posts, or people, the load is perfectly balanced if it is exactly centered. In that case, each beam, post, or person is bearing exactly half the load. However, if the load is not centered, then the beam, post, or person nearer to the load is bearing the greater part of the weight.

Example

The figure shows two people carrying a load. The load is centered. Each person is bearing half the load, or 75 lb.

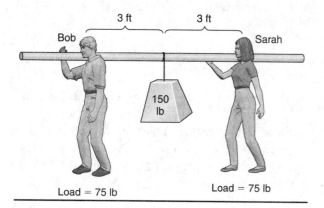

Example

The figure shows two people carrying a load. Which person is bearing the greater part of the load? Common sense tells you that it's the woman, because she is closer to the load.

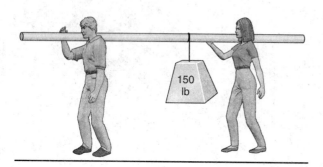

Questions about structural support may also ask you to decide which of several support structures is the strongest (or weakest). For this kind of question, you can often use your common sense. If a diagram shows several different structures, look for the one with the most brackets or other support elements. Also look for the one that leaves the least horizontal area (e.g., the front part of a flat bookshelf) unsupported. A bracket that runs the whole width of the shelf provides better support than one that does not.

Example

Which of these four shelves can bear the most weight?

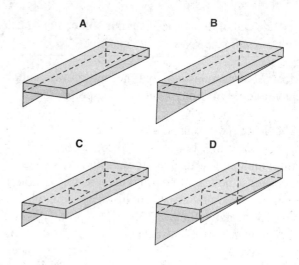

Choice D can bear the most weight. Of the four shelves, it is the strongest because it has the largest brackets, and because it has the most brackets.

Shape can also make a difference in the strength of a support structure. For example, a structure in which the support beams form rectangular shapes is not as strong as one in which the beams form triangular shapes. The reason is that while rectangular supports can easily bend out of shape, a triangle keeps its shape unless it falls apart entirely. That's why triangular shapes are used in support structures such as shelf brackets. That is also why you often see triangular shapes in the support structures of bridges, towers, and other buildings.

Example

Which bridge is the strongest?

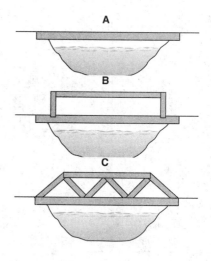

In the diagram, bridge C is the strongest because its framework is made of many triangles.

Properties of Materials

Tools, machines, and structures can be made of many different materials, such as wood, metal, plastic, and the like. Each material has its own *properties*, or characteristics. Some Mechanical Comprehension questions will ask you about those properties, or ask you to compare the properties of different materials. Most questions of this type can be answered with simple common sense, based on your everyday experience. For some, you may need to recall some of the basic ideas you learned in high school science.

One of the most important properties of materials is the ability to conduct heat. Here is an example of a question about this property.

Example

Three pots are cooking on a stove. Pot A has a wooden handle, pot B has a metal handle, and pot C has a plastic handle. Which handle will feel hottest to the touch?

A. A
B. B
C. C
D. All three will feel equally hot

To answer this question, think about your everyday experience. For which pot would you need a potholder? You might also remember from a high school science class that metal is a better conductor of heat than either wood or plastic. The correct answer is choice B.

Another important property of some materials is the ability to bend without breaking. This is called *flexibility*. The springs on a car can bend millions of times without breaking because they are made of spring steel. A ceramic cup, on other hand, cannot flex even once without breaking. Materials that are somewhat flexible include regular steel, copper, and wood. Some flexible materials will spring back to their original shape after bending. Others will stay in the new shape.

Example

Which object shown will resume its original shape after bending?

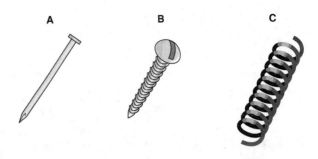

The spring (choice C) will resume its original shape. Nails, screws, and bolts are not made of spring steel, and while you can bend them, they will not spring back into their original shape.

Another important property of some materials is the ability to be formed into a new shape through repeated blows from a hammer. This property is called *malleability*. You will probably not need to be able to recall this term, but you should be able to identify substances that have this property. Metals like steel and copper are malleable and can thus be formed into the shapes of tools, containers, and other objects. Glass and ceramics, on the other hand, are not malleable.

Still another important property of some materials is the ability to stretch without breaking. Rubber can stretch in this way. Metal, glass, and plastic can stretch when heated. Wood and paper, on the other hand, will not stretch.

MECHANICAL MOTION

Many Mechanical Comprehension questions have to do with things that move or rotate: gears, pulleys, and other mechanisms.

Systems of Pulleys

You have already learned about pulleys as simple machines used to hoist heavy objects. Pulleys are also used as drive mechanisms. That is, systems of interconnected pulleys are used to transfer power, and often rotational speed, from one shaft to another. Two pulleys (sheaves) that are connected by a belt will run at the same speed if they are the same size. When the sheaves are of different sizes, the smaller one will run faster than the larger one because the smaller one must make more turns to move the belt the same distance.

In most cases on the ASVAB, you will simply be asked which of two or more interconnected pulleys runs fastest or slowest, and you can easily solve these problems by identifying the smallest or largest pulley. However, if you are asked to calculate the speed of a particular pulley in a system of pulleys, you can easily do so by using the following formula:

$$\text{Speed}_1 \times \text{diameter}_1 = \text{speed}_2 \times \text{diameter}_2$$

(Note that pulley speed is measured in revolutions per minute, or rpm.)

Example

Pulley 1 measures 9 inches in diameter. Pulley 2 measures 3 inches in diameter. If pulley 1 rotates at 1,200 rpm, how fast will pulley 2 rotate?

$$\text{Speed}_1 \times \text{diameter}_1 = \text{speed}_2 \times \text{diameter}_2$$
$$1,200 \times 9 = 10,800 = \text{speed}_2 \times 3$$
$$\text{Speed}_2 = 10,800/3 = 3,600$$

Another way to calculate the speed of pulley 2 is to look at the ratio between the two diameters. A ratio of 9:3, or 3:1, will multiply speed × 3. So 1,200 rpm × 3 = 3,600 rpm. Remember that the pulley with the smaller diameter is always the one that rotates faster!

Using these methods, you can calculate the speed of any pulley system as long as you know the diameters of both pulleys and the speed of either pulley.

Example

When pulley A runs at 400 rpm, what will be the speeds of pulleys B, C, and D?

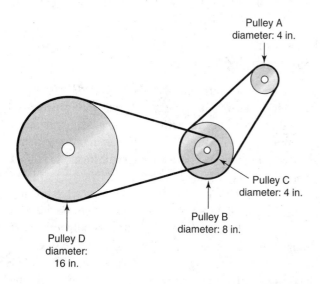

Pulley A
diameter: 4 in.

Pulley C
diameter: 4 in.

Pulley B
diameter: 8 in.

Pulley D
diameter: 16 in.

In this system, assume that the linked pulleys (B and C in the example) run at the same rpm, since they are

attached to the same shaft. Break the problem down into parts, and calculate them in order:

- Diameter of pulley A/diameter of pulley B = 4/8, so pulley B will run 1/2 as fast as pulley A.

  400/2 = 200 rpm

- You already know that pulley C runs at the same speed as pulley B.
- Diameter of pulley C/diameter of pulley D = 4/16 = 1/4, so pulley D will run 1/4 as fast as pulley C.

  200/4 = 50 rpm

Systems of Gears

Another way to transfer power between shafts is through systems of gears. The gears in a system typically have different diameters and different numbers of teeth per gear. The teeth of one gear mesh with the teeth of another, and as one gear (the *driving* gear) turns, it turns the other gear (the *driven* gear). When interlocking gears have different numbers of teeth, the gear with fewer teeth will rotate more times in a given period than the gear with more teeth. To see how this works, look at the following example.

Example

Gear A and gear B make up a system of gears. If gear A makes 6 revolutions, how many revolutions will gear B make?

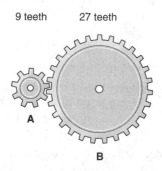

9 teeth   27 teeth

A

B

To solve this problem, use the picture and your common sense. Count the teeth on each gear. Gear A has 9 teeth. Gear B has 27 teeth. The ratio of the teeth on the two gears is 9:27 or 1:3. Common sense should tell you that gear A must rotate 3 times to make gear B rotate once. So if gear A rotates 6 times, gear B will rotate twice. Always keep in mind that in this kind of system, the gear with more teeth makes fewer rotations in the same period than the gear with fewer teeth.

Notice, too, that gears change the rotation direction, while pulleys usually do not. To rotate a gear in the same direction as the driving gear, you need a third gear, called an *idler* gear.

Example

In this system, gear A (the driving gear) is rotating clockwise. Gear B is the idler gear. Gear A makes gear B rotate counterclockwise. Gear B then makes gear C rotate clockwise, the same direction as gear A.

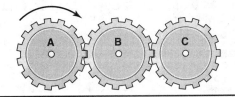

Rotating Wheels and Disks

Another way to drive shafts is to use what is called a *pin and slot arrangement*. In this arrangement, a pin is attached to a driving shaft, and a slotted disk is attached to a driven shaft. When the driving shaft rotates, the pin enters a slot on the disk and turns the driven shaft.

Example

In this pin and slot arrangement, each time the driving shaft turns one full revolution, the disk on the driven shaft will make 1/4 revolution. How far will the disk rotate when the pin turns three complete revolutions?

3 × 1/4 = 3/4 turn.

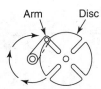

Arm   Disc

You may also be asked simple questions about what happens to points on a wheel when the wheel turns.

Example

How many rotations will point A make when point C makes 5 rotations? Point A will make 5 rotations

because point A and point C are both fixed on the same wheel.

Example

Which point will travel farthest as the wheel makes 10 rotations? Point B will travel farthest because it is farthest from the center. The distance it travels in each rotation is greater than the distance traveled by the other points.

Cams and Cam Followers

Cams are lobes attached to rotating shafts to push separate pieces, called *cam followers*. Cams are often found in engines, where they push intake and exhaust valves open when the engine turns. For every complete rotation of the camshaft, the cam follower will move away from and then back to its original position. A spring pushes the follower tight to the cam.

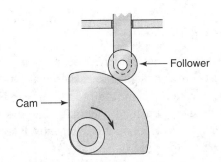

Cranks and Pistons

Cranks are used to change motion in a straight line to motion in a circle. You'll find cranks connected to pedals on a bike and to pistons in a car engine. When a crank makes one complete revolution, the piston must go up and down and return to its original position.

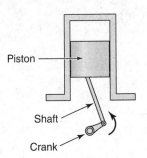

FLUID DYNAMICS

Fluids are substances that take the shape of their container. Gases and liquids are both fluids. The behavior of fluids is called *fluid dynamics*, and it can get rather complicated, but not on the ASVAB. Let's start with air, and move on to hydraulics—the engineering of liquids.

Air Pressure

Air pressure is measured in pounds per square inch. Atmospheric pressure at sea level is 14.7 lb/in.^{2}, which is actually quite a bit of pressure. Since it's present all around us, we don't notice it. However, if you create a vacuum inside a weak container, the container will be crushed by all that pressure.

Pneumatics and the Gas Laws Systems that use compressed air to do work are called *pneumatic systems*. Air is easily compressed, and the calculations are more complicated than they are with liquids, which usually can't be compressed. The larger the driven cylinder, the more air pressure it is exposed to, and the greater the force it can exert.

The "gas laws" apply to air as it is compressed and expanded.

- When a gas is compressed, it gains thermal energy—it warms up. The gas also gains potential energy, which is why compressed air can be used to drive nail guns and pneumatic hammers.
- When a given amount of gas expands, its pressure drops and the gas cools.
- When a gas cools without a change in outside pressure, it loses volume.

What happens when you increase the air pressure outside a balloon? The balloon shrinks until the pressure inside becomes great enough to balance the pressure outside.

Air pressure is also what keeps airplanes aloft. The bulge on the top of an airplane wing increases the speed of air passing over the wing, and that

causes a reduction in pressure. Because air pressure does not change below the wing, the result is an unbalanced upward force. This force lifts the airplane.

Refrigerators Refrigerators and air conditioners provide interesting applications of the gas laws. A compressor compresses a fluid, called a *refrigerant*. The refrigerant warms up, as predicted by the gas laws. Then the refrigerant loses heat (but not pressure) in the condenser. The refrigerant is piped into an evaporator, where it goes through a small hole and evaporates under reduced pressure. Expansion causes the temperature to drop, and the cold refrigerant can pick up heat from the surroundings. This is why the evaporator is placed in the area to be cooled. The condenser is placed where it's easy to get rid of excess heat—in the backyard for an air conditioner or in back for a refrigerator.

Water Pressure

On the ASVAB, water pressure questions often involve flow through pipes. Keep these principles in mind:

1. Total flow through a pipe system must be the same everywhere because water cannot be compressed.
2. When liquid speeds up, pressure falls.
3. When liquid slows down, pressure rises.

In the diagram, the same amount of water is flowing everywhere in the pipe system. For this to be true, water must be flowing faster at point B than at point A. That means that pressure is lower at point B.

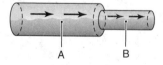

Water in a container also exerts pressure on the bottom of the container. The deeper the water, the greater the pressure. To find the amount of water pressure in a tank, calculate the total weight of the water and divide by the area of the base of the tank.

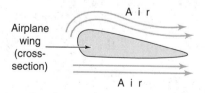

Example

A tank with a base that measures 2 feet × 4 feet holds 1,600 pounds of water. What is the water pressure at the base of the tank?

2 ft × 4 ft = 8 ft^2
1,600/8 = 200 lb/ft^2

Remember too that 1 ft^2 = 144 in^2. To convert pressure between pounds per square inch and pounds per square foot, divide or multiply by 144.

Filling and Emptying Tanks

The Mechanical Comprehension test often includes problems about filling and emptying tanks. Usually the filling and emptying take place at different rates, as in the following examples.

Example

Water is being piped into a tank at the rate of 2 gallons per second. At the same time, it is being piped out of the tank at the rate of 60 gallons per minute. How many gallons will be added in 5 minutes?

Convert the inflow rate so that you are working only with gallons per minute.

gal/sec × 60 = gal/min
2 gal/sec × 60 = 120 gal/min

Subtract:

120 gal/min inflow − 60 gal/min outflow
= 60 gal/min net gain

The net gain in 5 minutes is 5 × 60 = 300 gallons.

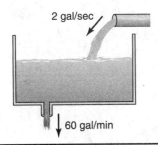

Example

A 100-gallon tank contains 10 gallons of water. Water is added through one pipe at the rate of 3 gallons per minute. It is drained away through another pipe at the rate of 2 gallons per minute. How long will it take to fill the tank?

Find the net gain of water per minute:

3 gal/min − 2 gal/min = 1 gal/min

It will take 100 − 10 = 90 gallons to fill the tank. At the rate of 1 gal/min, it will take 90 minutes to fill the tank.

CHAPTER GLOSSARY

chemical energy. Energy stored in chemicals or released in a chemical reaction

compound machine. A machine made up of two or more simple machines working together

compression. A force that pushes materials together

effort. In a lever, the point where you apply force

effort arm. In a lever, the distance from the force to the fulcrum

electrical energy. Energy in moving electrons

flexibility. The ability of a material to bend without breaking

friction. The force that resists the relative motion of two surfaces in contact

fulcrum. The stationary element that holds a lever but also allows it to rotate

gravity. An attractive force between objects

kinetic energy. Energy in a moving object

load. In a lever, the part where output force lifts or squeezes

load arm. In a lever, the distance from the load to the fulcrum

mechanical advantage. The amount by which a machine multiplies the force applied to it

potential energy. Energy that can be released under certain conditions

tension. A force that pulls materials apart

LAWS AND FORMULAS TO KNOW

How to calculate mechanical advantage (MA):

- *Lever*: MA = load/effort = effort distance/load distance
- *Pulley*: MA = load/effort = number of supporting strands
- *Gears*: MA = number of teeth on driven gear/ number of teeth on driving gear

- *Sheaves*: MA = driven diameter/drive diameter
- *Inclined plane*: MA = horizontal length/vertical rise
- *Wheel and axle*: MA = radius of wheel/radius of axle

Speed of pulleys in a system:

$$\text{Speed}_1 \times \text{diameter}_1 = \text{speed}_2 \times \text{diameter}_2$$

The gas laws:

- When a gas is compressed, it heats up.
- When a given amount of gas expands, its pressure drops and the gas cools.
- When a gas cools without a change in outside pressure, it loses volume.

Water pressure:

- Total flow through a pipe system is the same everywhere.
- When liquid speeds up, pressure falls.
- When liquid slows down, pressure rises.

MECHANICAL COMPREHENSION QUIZ

Take the following quiz to check your understanding of ASVAB Mechanical Comprehension topics. Circle each answer that you choose. There is no time limit. Answers and explanations are given immediately after the quiz.

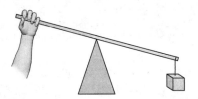

1. The diagram shows a class 1 lever. Which of the following is the same kind of lever?
   A. A pair of tweezers
   B. A pair of scissors
   C. A wheelbarrow
   D. A pair of tongs

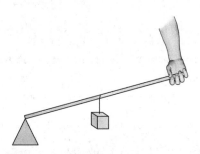

2. The diagram shows a class 2 lever. Which of the following is the same kind of lever?
   A. A seesaw
   B. A pair of scissors
   C. The human forearm
   D. A wheelbarrow

3. When a mass of air expands, which of the following is most likely to happen?
   A. The air warms up.
   B. The air cools down.
   C. The air stays at the same temperature.
   D. The air contracts.

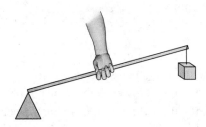

4. The diagram shows a class 3 lever. Which of the following is the same kind of lever?
   A. A pair of tweezers
   B. A wheelbarrow
   C. A seesaw
   D. A wedge

5. Which of the following would feel hottest to the touch if one end were placed in a pot of boiling water?
   A. A wooden spoon
   B. A metal fork
   C. A plastic knife
   D. A plastic cup

6. A lever lifts 300 lb with an effort of 60 lb. What is the mechanical advantage?
   A. 3
   B. 4
   C. 5
   D. 8

7. In a class 1 lever with an MA of 4, if the effort distance measures 8 ft, how long is the load distance?
   A. 2 ft
   B. 4 ft
   C. 8 ft
   D. 16 ft

8. Water is flowing into a tank at the rate of 180 gallons per minute. However, the tank has a leak, and water is leaking out at the rate of 2 gallons per second. How long will it take to add 120 gallons?

A. 1 min
B. 2 min
C. 3 min
D. 4 min

9. Water is flowing from a 3-inch pipe into a 1-inch pipe. Which statement is true?

A. Pressure is lower in the 3-inch pipe.
B. Pressure is higher in the 1-inch pipe.
C. More water is passing through the 3-inch pipe than through the 1-inch pipe.
D. The same amount of water is passing through every part of the piping system.

10. Gear A, with 18 teeth, meshes with gear B, which has 12 teeth. When gear A rotates 3 times, gear B will rotate

A. 3 times.
B. 4.5 times.
C. 6 times.
D. 9 times.

11. What is the mechanical advantage in this pulley system?

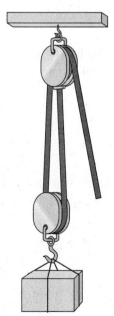

A. 2
B. 3
C. 4
D. 6

12. In the diagram, what can you tell about the load on posts A and B?

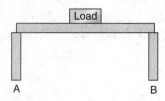

A. Post B carries more weight.
B. Post A carries more weight.
C. Post A carries no weight.
D. The load is equal on posts A and B.

13. Water is flowing through this pipe. Which statement is true?

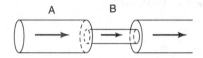

A. Water is moving faster at point A than at point B.
B. Water pressure is equal at points A and B.
C. Water pressure is greater at point A than at point B.
D. Water pressure is greater at point B than at point A.

14. What is the advantage of using triangle shapes in constructing a bridge?

A. Triangles are sturdier than other shapes.
B. Triangles are very flexible.
C. Triangles are inexpensive to manufacture.
D. Triangles are attractive to look at.

15. Which formula will calculate the mechanical advantage of a lever?

A. Load/effort
B. Effort distance/load distance
C. Both A and B
D. Load × effort

ANSWERS AND EXPLANATIONS

1. **B.** In a class 1 lever, the fulcrum is between the effort and the load. Of the choices, only the pair of scissors has the same arrangement of elements. The tweezers (choice A) and tongs (choice B) are examples of class 3 levers. The wheelbarrow (choice C) is a class 2 lever.

2. **D.** In a class 2 lever, the fulcrum is at one end, the effort is at the other end, and the load is between the fulcrum and the effort. Of the choices, only the wheelbarrow has this arrangement of elements. The seesaw (choice A) and the scissors (choice B) are class 1 levers. The human forearm (choice C) is a class 3 lever.

3. **B.** Air is a gas, so according to the gas laws, when a mass of warm air expands, it will most likely cool down.

4. **A.** In a class 3 lever, the fulcrum is at one end, the load is at the other, and the effort is between the fulcrum and the load. Of the choices, only the tweezers have this arrangement of elements. The wheelbarrow (choice B) is a class 2 lever. The seesaw (choice C) is a class 1 lever. The wedge (choice D) is not a lever at all but an inclined plane.

5. **B.** Metal is a better conductor of heat than wood or plastic, so the metal fork would feel hottest to the touch.

6. **C.** MA = load/effort. 300/60 = 5

7. **A.** MA = effort distance/load distance. 4 = 8/load distance. 4 = 8/2. The load distance is 2 ft.

8. **B.** The leak of 2 gal/sec = 120 gal/min. Net gain is 60 gal/min, so it will take 2 minutes to add 120 gallons to the tank.

9. **D.** Because water cannot be compressed, the same amount of water must be flowing through both parts of the pipe system. This makes choice C incorrect. If the same amount of water is flowing through all parts of the system, then the water must be flowing faster in the 1-inch section, reducing pressure in that section. That makes choices A and B both incorrect.

10. **B.** The ratio of teeth is 18 to 12. Each turn of gear A will turn gear B 1.5 times. 1.5 × 3 = 4.5.

11. **A.** Count the supporting strands (the strands of rope that will be shortened as the load is lifted). There are two supporting strands, so MA = 2.

12. **D.** Since the load is centered between the posts, each post carries equal weight.

13. **C.** When fluid speed increases, pressure decreases. Because the same amount of water must be in these pipes, water must be moving faster at point B. Therefore pressure is lower at point B.

14. **A.** The only way to distort the form of a triangular-shaped structure is to break the legs or angles. That makes triangles sturdier than other structural shapes, which is a good reason to use triangle shapes when constructing a bridge.

15. **C.** Either load/effort or effort distance/load distance will calculate the mechanical advantage.

Chapter 18

Assembling Objects

YOUR GOALS FOR THIS CHAPTER

- **Learn how to solve Assembling Objects questions.**
- **View sample Assembling Objects questions.**

INTRODUCTION: ASVAB ASSEMBLING OBJECTS QUESTIONS

The Assembling Objects (AO) subtest of the ASVAB was added to the battery of tests in 2002. This subtest is designed to test your skills in spatial orientation. It is not part of the Armed Forces Qualification Test (AFQT) and does not determine your eligibility for enlistment.

The types of questions you will see on the Assembling Objects (AO) subtest will likely not be familiar to you from tests you took in high school. They require the ability to comprehend the sizes and orientations of various shapes and to picture how those objects can be maneuvered to be joined together. In order to do well on this subtest, you will need a method to follow, common sense, and practice, in addition to your spatial ability.

There are two types of AO questions:

- **Puzzle:** In this question type, you are given several individual objects in a box. Then you will have four choices of what those objects could look like when they are assembled together into one figure. This is a bit like fitting together pieces in a jigsaw puzzle.

- **Connection:** In this question type, you are given objects that each have a point that is marked. You will again have four choices and you must choose which one shows the objects joined by a

line between the correct points. This is a bit like connect-the-dots.

Let's look at strategies for each type of question.

PUZZLE QUESTIONS

In this question type, you are given several individual objects and you choose the answer that shows what they could look like when assembled together into one figure.

For example, if you are given these objects:

The correct choice might look like this:

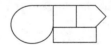

Rules

1. **The correct answer must have the exact same objects as the original (and no additional ones).** For example, this would be incorrect because it has an additional triangle that the original did not.

2. **The objects must remain the same size.**
For example, this would be incorrect because the rectangle at the bottom is smaller than the original one.

3. **The objects can be rotated.**
For example, this shape in the original:

might look like this in the answer:

4. **The objects must remain in their original orientation: they can rotate but not be flipped over.**
In other words, do not choose an answer that shows a mirror image of the original object.
For example, this shape in a question:

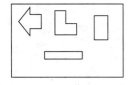

CANNOT look like this in the answer:

Strategies

1. First, count the number of objects and eliminate any answer choice that does not match. If the original figure has six objects, then the correct answer must also have six objects.
2. Look for one object at a time and eliminate any answer choice that does not include it.
3. Choose the most distinctive (unique) object to look for. For example, it is easier to look for a weird-looking object such as this

than a more common one like this

4. Keep choosing new objects until you have eliminated all but one answer choice. Once you are down to only one choice, you know you have found the correct answer.

Examples

1. *Which figure best shows how the objects in the box will appear if they are fit together?*

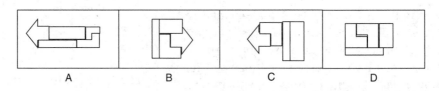

2. Which figure best shows how the objects in the box will appear if they are fit together?

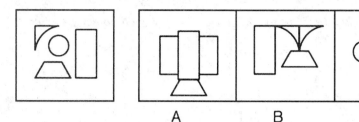

A   B   C   D

Explanations

1. **B.** First, count the number of objects. There are four in the original. Choice A has five, so eliminate A. The others each have four, so move on to the next strategy. Choose the most distinctive object to look for first—perhaps the arrow. Choice D does not have an arrow, so eliminate D. Now look for the L-shaped object. Choices B and C both have it, so choose another object. The long thin rectangle is in both choices as well. The rectangle at the top right of the original figure is the last object left. Choice C has another rectangle, but it is not the same size as the original, so eliminate C. That leaves only choice B. **The correct answer is B.**

2. **D.** First, count the number of objects. There are four in the original and in all the answer choices. Move on to the next strategy. Choose the most distinctive object to look for first: the object that looks like a triangle with one concave side. Choice A does not have it, so eliminate A. Choice B has two of them, so eliminate B as well. Choice C has a similar shape, but it is not the same as the original. Eliminate C. That leaves only choice D. **The correct answer is D.**

CONNECTION QUESTIONS

In this question type, you are given individual objects that each have a point labeled with a letter. You choose the answer that shows what the objects would look like when the labeled points are connected by a line.

For example, if you are given these objects:

   A

B

The correct choice might look like this:

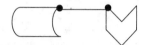

Rules

1. **Follow the same rules as Puzzle questions.**
   Plus
2. **Each object must connect at the exact same point that was labeled in the original.**
3. **Points may be located anywhere on the object— on the perimeter or within the area.**

Strategies

1. Look for one labeled point at a time.

2. Start with whichever shape has the easier point to find. For example, it is easier to look for a point in a distinctive location like a corner, such as this

   than one in a less distinct location such as this

3. Once you have eliminated choices that do not connect at the first point correctly, repeat the process for the other point. Both points must be connected at the place they were labeled in the original.

Examples

1. *Which figure best shows how the objects in the box will touch if the letters for each object are matched?*

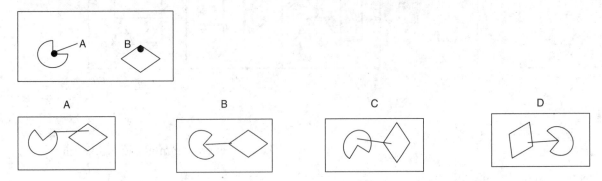

2. *Which figure best shows how the objects in the box will touch if the letters for each object are matched?*

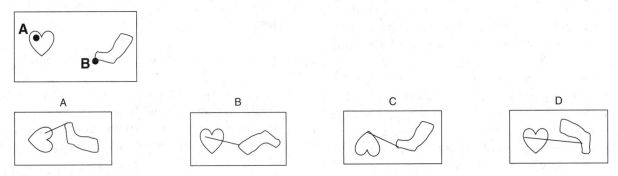

Explanations

1. **D.** The first object has point A located at the center of the ¾ circle. Choices A and C do not connect there, so eliminate them. The second object has point B located within the figure, in the wider angle of the rhombus. Choice B does not connect at that point. **The correct answer is D.**
2. **A.** The heart has point A located within the upper left rounded part. Choices C and D do not connect there, so eliminate them. Point B is located at the "toe" of the boot shape. Choice B does not connect at that point. That leaves only choice A. **The correct answer is A.**

ASVAB PAPER VERSION

Paper-and-pencil versions of the ASVAB are given at Military Entrance Test (MET) sites around the country or on high school and some college campuses.

On the paper-and-pencil version of the ASVAB, there are 25 questions and you will have 15 minutes to complete them.

Every candidate receives the same test questions, and questions range from very easy to very difficult.

Strategies for the Paper Version

- **Skip around:** You do not necessarily have to do the questions in order, as long as you fill in your answer sheet correctly. This will enable you to focus on your personal strengths. If, for example, you find that you perform better on the connection questions, you can seek those out and answer them first. That way you will not run out of time without doing all the questions you are best at. Save harder questions for last to maximize the number of correct answers you get.
- **Don't leave any blanks:** Since there is no penalty for guessing on the paper-and-pencil version of the test, be sure to keep an eye on the clock so that you can fill in an answer on any question you didn't get a chance to work. Do not leave anything blank. Wrong answers do not hurt you, but a blank question is a missed chance for points.
- **Use the process of elimination:** The easiest way to approach Assembling Objects questions is by

using the process of elimination. Cross out choices that you know cannot be correct and you will get to the correct answer more quickly. If you are taking a paper version of the test, this is easy to do.

- **Practice:** Working 25 questions in only 15 minutes is tough. The more you practice, the faster you will get. The Assembling Objects practice tests in this book are paper-and-pencil format, so you will have plenty of opportunity to work on your speed and accuracy. Answers and explanations are provided so that you can review your work and learn from your mistakes.

ASVAB COMPUTER VERSION (CAT)

Computer Adaptive Test (CAT) versions of the ASVAB are given at Military Entrance Processing Stations (MEPS) around the country.

On the CAT-ASVAB, there are 15 questions and you will have 17 minutes to complete them.

The CAT-ASVAB provides questions that adapt to each tester's ability. Questions are selected from a large pool of items that range in difficulty from very easy to very difficult. The first question presented will be of medium difficulty. If you answer correctly, you will get a slightly more difficult question next. If you answer incorrectly, you will be given a slightly easier question next. This process continues after each question, allowing the computer to quickly home in on your highest level of ability. For that reason, CATs generally have fewer questions than paper exams and provide more score-level precision.

Strategies for the CAT-ASVAB

- **Take your time:** You can't skip around on the CAT-ASVAB and earlier questions are very important since they determine the level of difficulty (and score level) of later questions. Do not rush. You have more time per question on the CAT than you would on the paper version, so use it wisely. Work carefully.

- **Use the process of elimination:** The easiest way to approach Assembling Objects questions is by using the process of elimination. Cross out choices that you know cannot be correct and you will get to the correct answer more quickly. This is easier to do on the paper version of the test, but it can be done on the CAT as well. You will need to use your scratch paper to write "ABCD," and then you can cross out choices as you eliminate them. Do this for each question.

- **Double-check:** Once you submit an answer, you cannot return to that question to change it. Be sure you get each question correct before you submit and confirm your answer.

- **Finish the test:** There are only 15 items and most testers are able to finish them in the time allowed. However, if you should happen to run out of time, there is a penalty. Any questions that were not completed are scored as though they were answered at random. This will have an effect on your score but not as much of an effect as would rushing through the test in order to finish and missing early questions. Work on your pacing while you practice so that this will not be an issue for you on test day.

- **Practice:** The more you practice, the faster you will get. The Assembling Objects practice tests in this book are paper-and-pencil format, but that will give you even more opportunity to work on your speed and accuracy. Answers and explanations are provided so that you can review your work and learn from your mistakes.

Part 4

THREE ASVAB PRACTICE TEST FORMS

ASVAB PRACTICE TEST FORM 1

Try to take this practice test under actual test conditions. Find a quiet place to work, and set aside a period of approximately 2½ hours when you will not be disturbed. Work on only one section at a time, and use your watch or a timer to keep track of the time limits for each section. Mark your answers on the answer sheet, just as you will when you take the real exam.

At the end of the test you'll find an answer key and explanations for every question. After you check your answers against the key, you can complete a chart that will show you how you did on the test and what test topics you might need to study more. Then review the explanations, paying particular attention to the ones for the questions that you answered incorrectly.

If you take the CAT-ASVAB, your test will be given by computer and the number of your question, the mode of administration, and the timing will be different. This process has been explained in previous chapters.

GENERAL DIRECTIONS

IF YOU ARE TAKING THE PAPER-AND-PENCIL VERSION OF THE ASVAB, THE TEST ADMINISTRATOR WILL READ THE FOLLOWING ALOUD TO YOU:

DO NOT WRITE YOUR NAME OR MAKE ANY MARKS in this booklet. Mark your answers on the separate answer sheet. Use the scratch paper which was given to you for any figuring you need to do. Return this scratch paper with your other papers when you finish the test.

If you need another pencil while taking this test, hold your pencil above your head. A proctor will bring you another one.

This booklet contains 9 tests. Each test has its own instructions and time limit. When you finish a test, you may check your work in that test ONLY. Do not go on to the next test until the examiner tells you to do so. Do not turn back to a previous test at any time.

For each question, be sure to pick the BEST ONE of the possible answers listed. Each test has its own instructions and time limit. When you have decided which one of the choices given is the best answer to the question, blacken the space on your answer sheet that has the same number and letter as your choice. Mark only in the answer space. BE CAREFUL NOT TO MAKE ANY STRAY MARKS ON YOUR ANSWER SHEET. Each test has a separate section on the answer sheet. Be sure you mark your answers for each test in the section that belongs to that test.

Here is an example of correct marking on an answer sheet.

S1. A triangle has

    A. 2 sides
    B. 3 sides
    C. 4 sides
    D. 5 sides

```
+-----------------------------+
|         PRACTICE            |
|  S1  (A) ● (C) (D)          |
|  S2  (A) (B) (C) (D)        |
|  S3  (A) (B) (C) (D)        |
+-----------------------------+
```

The correct answer to sample question S1 is B.

Next to the item, note how space B opposite number S1 has been blackened. Your marks should look just like this and be placed in the space with the same number and letter as the correct answer to the question. Remember, there is only ONE BEST ANSWER for each question. If you are not sure of the answer, make the BEST GUESS you can. If you want to change your answer, COMPLETELY ERASE your first answer mark.

Answer as many questions as possible. Do not spend too much time on any one question. Work QUICKLY, but work ACCURATELY. DO NOT TURN THE PAGE UNTIL TOLD TO DO SO. Are there any questions?

ASVAB PRACTICE TEST FORM 1

ANSWER SHEET

ANSWER SHEET   PAGE 3
ARMED SERVICES VOCATIONAL
APTITUDE BATTERY

PART 1. GENERAL SCIENCE

THE TEST ADMINISTRATOR WILL READ THE FOLLOWING ALOUD TO YOU:

Turn to Part 1 and read the directions for General Science silently while I read them aloud. This test has questions about science. Pick the BEST answer for each question and then blacken the space on your separate answer sheet that has the same number and letter as your choice. Here are three practice questions.

S1. A rose is a kind of
  A. animal.
  B. bird.
  C. flower.
  D. fish.

S2. A cat is a
  A. plant.
  B. mammal.
  C. reptile.
  D. mineral.

S3. The earth revolves around the
  A. sun.
  B. moon.
  C. meteorite.
  D. Mars.

Now look at the section of your answer sheet labeled "PRACTICE." Notice that answer space C has been marked for question S1. Now do practice questions S2 and S3 by yourself. Find the correct answer to the question, then mark the space on your answer sheet that has the same letter as the answer you picked. Do this now.

   You should have marked B for question S2 and A for question S3. If you make any mistakes, erase your mark carefully and blacken the correct answer space. Do this now.

   Your score on this test will be based on the number of questions you answer correctly. You should try to answer every question. If you finish before time is called, go back and check your work in this part <u>ONLY</u>. Now find the section of your answer sheet that is marked PART 1. When you are told to begin, start with question number 1 in Part 1 of your test booklet and answer space number 1 in Part 1 on your separate answer sheet. DO NOT TURN THE PAGE UNTIL TOLD TO DO SO. This part has 25 questions in it. You will have 11 minutes to complete it. Most of you should finish in this amount of time. Are there any questions?

   Turn the page and begin.

1. A major difference between a plant cell and an animal cell is
   A. plant cells do not reproduce.
   B. plant cells do not have cell walls.
   C. plant cells have chloroplasts.
   D. plant cells die after a year.

2. Sponges, worms, insects, and amphibians belong to which of the following kingdoms?
   A. Monera
   B. Protista
   C. Plantae
   D. Animalia

3. What name is given to the global ecosystem?
   A. Atmosphere
   B. Hydrosphere
   C. Lithosphere
   D. Biosphere

4. Which of the following parts of a plant is (are) not digestible by humans?
   A. Cell wall
   B. Stomata
   C. Root hairs
   D. Tap root

5. In fish, respiration takes place in which of the following body parts?
   A. Gills
   B. Spinal cord
   C. Tail
   D. Circulatory system

6. If a seed is planted upside down, the roots will still grow downward because of which of the following plant responses?
   A. Respiration
   B. Gravitropism
   C. Osmosis
   D. Turgidity

7. When a plant like mistletoe lives by stealing the nutrients from the tissue of living trees, the condition is called which of the following?
   A. Mutualism
   B. Commensalism
   C. Predation
   D. Parasitism

8. Which of the following organs contains smooth muscle cells?
   A. Biceps
   B. Heart
   C. Stomach
   D. Thyroid

9. Bacteria and viruses can invade your body from all except which of the following?
   A. Mouth
   B. Nose
   C. Eyes
   D. Skin

10. Most of the absorption of nutrients takes place in which part of the digestive system?
    A. Small intestine
    B. Mouth
    C. Large intestine
    D. Stomach

11. The autonomic system is not involved in controlling which of the following?
    A. Walking
    B. Digestion
    C. Heart rate
    D. Breathing

12. If one parent has two dominant genes for tallness (TT) and the other parent has two recessive genes for shortness (tt), what is the probability that the offspring will be tall?
    A. 25%
    B. 50%
    C. 75%
    D. 100%

13. If an object is more than two focal lengths away from a convex lens, which of the following best represents what the image would look like?
    A. The image is larger and inverted.
    B. The image is smaller and inverted.
    C. The image is larger and upright.
    D. The image is smaller and upright.

14. An element's atomic mass is determined by which of the following?
    A. The number of protons and the number of electrons
    B. The number of protons, electrons, and neutrons
    C. The number of protons only
    D. The number of protons and the number of neutrons

15. A rainbow is the outcome of which of the following behaviors of light?
    A. Diffusion
    B. Refraction
    C. Reflection
    D. Interference

16. Which of the following features indicate that continents such as Africa were once situated near the South Pole?
    A. Fault lines
    B. Glacial deposits
    C. Snow-capped mountains
    D. River deposits

17. Pumice formed from a volcano is considered to be which of the following types of rock?
    A. Intrusive igneous rock
    B. Extrusive igneous rock
    C. Intrusive metamorphic rock
    D. Extrusive metamorphic rock

18. Which of the following causes ocean waves?
    A. Persistent surface winds
    B. Excessive rainfall in one part of the ocean
    C. Steep slopes near the shoreline
    D. The relative alignment of the moon and the sun with Earth

19. Earth is closest to the sun during which of the following months?
    A. July
    B. June
    C. January
    D. September

20. Which planet is eighth in order from the sun?
    A. Venus
    B. Saturn
    C. Jupiter
    D. Neptune

21. Which of the following is NOT a possible effect of the global warming of Earth's atmosphere?
    A. Warmer weather might support an increase in tropical diseases such as malaria.
    B. Sea level might rise due to the melting of the polar ice caps.
    C. Earth's mesosphere might become more ionized, hindering space travel.
    D. The number of storms and hurricanes might increase.

22. Severe storms are most likely to be created under which of the following conditions?
    A. A cold front when the warm air it replaces rises quickly
    B. A warm front that is replacing cold air
    C. An occluded front where a fast-moving cold air mass replaces a slower warm air mass
    D. A stationary front that remains in place for several days

23. Which of the following are features associated with glaciers?
    A. Cirques, aràtes, moraines
    B. Sinkholes, groundwater flow, stalactites
    C. Sedimentary rocks, conglomerates, shale
    D. Vents, calderas, hot spots

24. A sea breeze is caused by which of the following conditions?
    A. The sea is warmer than the air.
    B. The sea is warmer than the land.
    C. The land is warmer than the sea.
    D. The sea is colder than the air.

25. An atom of the same element that has different numbers of neutrons is called a(n)
    A. ion.
    B. isotope.
    C. compound.
    D. mixture.

STOP! DO NOT TURN THIS PAGE UNTIL TIME IS UP FOR THIS TEST. IF YOU FINISH BEFORE TIME IS UP, CHECK OVER YOUR WORK ON THIS TEST ONLY.

PART 2. ARITHMETIC REASONING

THE TEST ADMINISTRATOR WILL READ THE FOLLOWING ALOUD TO YOU:

Turn to Part 2 and read the directions for Arithmetic Reasoning silently while I read them aloud. This is a test of arithmetic word problems. Each question is followed by four possible answers. Decide which answer is CORRECT, and then blacken the space on your answer sheet that has the same number and letter as your choice. Use your scratch paper for any figuring you wish to do.

Here is a sample question. DO NOT MARK your answer sheet for this or any further sample questions.

S1. A student buys a sandwich for 80 cents, milk for 20 cents, and pie for 30 cents. How much did the meal cost?
   A. $1.00
   B. $1.20
   C. $1.30
   D. $1.40

The total cost is $1.30; therefore, C is the right answer. Your score on this test will be based on the number of question you answer correctly. You should try to answer every question. DO NOT SPEND TOO MUCH TIME on any one question. If you finish before time is called, go back and check your work in this part ONLY.

Now find the section of your answer sheet that is marked PART 2. When you are told to begin, start with question number 1 in Part 2 of your test booklet and answer space number 1 in Part 2 on your separate answer sheet.

DO NOT TURN THIS PAGE UNTIL TOLD TO DO SO. You will have 36 minutes for the 30 questions. Are there any questions?

Begin.

1. At an electronics store, Sal bought a CD for $14.99, another CD for $19.99, a CD player for $49.75, and a software program for his computer for $56.59. What was his total bill?
   A. $121.45
   B. $125.42
   C. $139.72
   D. $141.32

2. Maurice placed several calls on his cell phone during the first half of the month. The calls were 15 minutes, 46 minutes, 59 minutes, 83 minutes, and 114 minutes. If his plan authorizes 400 minutes, how many minutes does he have left?
   A. 47
   B. 83
   C. 113
   D. 147

3. A math class has 32 students. Fourteen are females. What percent of the class is female?
   A. 24.22%
   B. 37.25%
   C. 43.75%
   D. 52.75%

4. Tim has run a 500-meter race four times. His times are 169 seconds, 212 seconds, 198 seconds, and 201 seconds. What is his average speed?
   A. 182 seconds
   B. 195 seconds
   C. 198 seconds
   D. 201 seconds

5. Trinidad earns $2,700 per month. If she spends $1,100 on the rent for her apartment, to the nearest percent what percent is spent on rent?
   A. 20%
   B. 41%
   C. 51%
   D. 65%

6. A restaurant had a delivery of 12 cartons of vegetables and 36 cartons of fruit. What is the ratio of vegetable cartons to fruit cartons?
   A. $\frac{1}{4}$
   B. $\frac{1}{8}$
   C. $\frac{1}{3}$
   D. $\frac{6}{12}$

7. A manufacturer can produce 112 shirts per hour. If a work day is 8 hours, how many shirts can be produced in 3.5 days?
   A. 392
   B. 1,125
   C. 1,236
   D. 3,136

8. Stella has a drawer full of headbands. Twenty are pink, 32 are blue, and 16 are black. If she reaches into the drawer and picks out a headband at random, what is the probability that the headband will not be black?
   A. $\frac{17}{32}$
   B. $\frac{13}{17}$
   C. $\frac{16}{32}$
   D. $\frac{1}{2}$

9. Five years ago, Ruth made a deposit of $500 into a savings account that pays simple interest. She made no further deposits, and now her account is worth $750. What was the rate of interest per year?
   A. 2%
   B. 4%
   C. 9%
   D. 10%

10. If Grayson earns $53/hour during a regular work-day of 8 hours and $57/hour for any work he does after 5 P.M., how much does he earn if he worked 5 regular work days and 6.5 overtime hours?
    A. $1,870.50
    B. $2,292.25
    C. $2,490.50
    D. $2,597.50

11. Ellen baked 112 cookies and shared them equally with her 23 classmates. How many whole cookies each can Ellen and her classmates have?
    A. 4
    B. 5
    C. 6
    D. 7

12. Leon wants to paint the four walls in his dining room. Each wall is 8 feet high by 30 feet wide. A quart of paint covers 40 ft². How many gallons of paint must he purchase?
    A. 6
    B. 8
    C. 12
    D. 15

13. A family goes to a restaurant, and the bill is $112.00. If the parents want to leave the wait-person a 15 percent tip, how much should they leave?
    A. $15.40
    B. $16.80
    C. $17.25
    D. $17.85

14. At the local department store, a flat screen television retails for $4,250. Manuel finds the same television for sale on the Internet for $450 less. What percent will Manuel save if he purchases the television via the Internet site?
    A. 5.2%
    B. 10.6%
    C. 12.8%
    D. 15%

15. Jenny is 34 years old. Two years ago she was twice as old as her cousin. How old is her cousin now?
    A. 15
    B. 17
    C. 18
    D. 21

16. Simon's garden is 12 by 24 feet. He will plant bushes that require 8 sq ft of space. How many bushes can he plant?
    A. 16
    B. 24
    C. 27
    D. 36

17. Jake is riding a bicycle on a circular track. The distance from the center of the track to the outside of the circle is 10 meters. How far will Jake ride if he goes around the track one time?
    A. 62.8 m
    B. 145.5 m
    C. 157.8 m
    D. 162.8 m

18. Robby is twice as old as Greg. If Robby is 46 years old, how old is Greg?
    A. 15
    B. 23
    C. 25
    D. 27

19. Karin is as old as her two cousins combined. If Karin is 54 and one cousin is 21, how old is the second cousin?
    A. 13
    B. 22
    C. 33
    D. 37

20. Ned works after school. When he cuts grass, he earns $6.25 per hour. When he delivers papers, he earns $5.95 per hour and when he runs errands for people, he earns $5.50 per errand. If in one week he cut grass for 6 hours, delivered papers for 11 hours, and ran 6 errands, how much money did he earn?
    A. $114.95
    B. $135.95
    C. $144.15
    D. $156.05

21. Gerard throws a discus distances of 10 meters, 14.5 meters, 14.8 meters, 16.7 meters, and 15.4 meters. On his last attempt he faults, which counts as 0 meters. What is his average throwing distance?
    A. 11.9 meters
    B. 12.8 meters
    C. 13.78 meters
    D. 14.28 meters

22. An aquarium is 12 inches high, 24 inches long, and 10 inches wide. A goldfish needs 144 cubic inches of space to live. How many goldfish can this aquarium support?
    A. 15
    B. 20
    C. 22
    D. 24

23. The movie *Gladiator* earned $6 million in the first week of its showing. All movies shown that week earned $114 million. Which of the following shows the ratio of the earnings of *Gladiator* to the earnings of all movies that week?
    A. 1:4
    B. 2:13
    C. 1:19
    D. 3:8

24. A soup can is 6 inches high with a diameter of 4 inches. How much soup can it hold?
    A. 44.6 in.^{3}
    B. 57.2 in.^{3}
    C. 75.4 in.^{3}
    D. 81.3 in.^{3}

25. Sylvia wants to plant flowers in a window box that measures 48 inches long, 10 inches high, and 18 inches wide. If a bag of potting soil holds 1,440 cubic inches of soil, how many bags should she purchase?
    A. 2
    B. 4
    C. 6
    D. 8

26. During the first year of life, a whale gains about 30 tons of weight. What is the average weight gain per month?
    A. 1.2 tons
    B. 2.5 tons
    C. 2.9 tons
    D. 3.1 tons

27. Sally's rent is $750.00 each month. After two years of paying rent, how much has she spent?
    A. $12,000
    B. $18,000
    C. $21,000
    D. $24,000

28. Jonathan's monthly take-home pay is $2,556.36. If 28 percent is routinely taken out of his pay for taxes and insurance, what is his original pay before the deductions?
    A. $3,550.50
    B. $3,665.72
    C. $4,175.50
    D. $4,278.50

29. Rita earns an annual salary of $30,000. Her boss promised her that next year she will earn $32,500. What percent increase is the new salary?
    A. 5.2%
    B. 6.3%
    C. 8.3%
    D. 9.1%

30. Elaine was born on December 15 and has lived in Minneapolis, Minnesota, all her life. Weather records show that over the last 100 years, it has snowed on December 15 60 percent of the time. If this trend continues and Elaine lives to be 85, on how many birthdays will she probably have snow?
    A. 47
    B. 51
    C. 54
    D. 62

STOP! DO NOT TURN THIS PAGE UNTIL TIME IS UP FOR THIS TEST. IF YOU FINISH BEFORE TIME IS UP, CHECK OVER YOUR WORK ON THIS TEST ONLY.

PART 3. WORD KNOWLEDGE

THE TEST ADMINISTRATOR WILL READ THE FOLLOWING ALOUD TO YOU:

Now turn to Part 3 and read the directions for Word Knowledge silently while I read them aloud.

This is a test of your knowledge of word meanings. These questions consist of a sentence or phrase with a word or phrase underlined. From the four choices given, you are to decide which one MEANS THE SAME OR MOST NEARLY THE SAME as the underlined word or phrase. Once you have made your choice, mark the space on your answer sheet that has the same number and letter as your choice.

Look at the sample question.

S1. The weather in this geographic area tends to be moderate.
  A. Severe
  B. Warm
  C. Mild
  D. Windy

The correct answer is "mild," which is choice C. Therefore, you would have blackened in space C on your answer sheet.

Your score on this test will be based on the number of questions you answer correctly. You should try to answer every question. DO NOT SPEND TOO MUCH TIME on any one question. If you finish before time is called, go back and check your work in this part ONLY.

Now find the section of your answer sheet that is marked PART 3. When you are told to begin, start with question number 1 in Part 3 of your test booklet and answer space number 1 in Part 3 on your separate answer sheet.

DO NOT TURN THE PAGE UNTIL TOLD TO DO SO. You will have 11 minutes to complete the 35 questions in this part. Are there any questions?

Begin.

1. At the time, the situation seemed <u>calamitous</u>.
   A. Ruthless
   B. Joyous
   C. Controlled
   D. Disastrous

2. <u>Lure</u> most nearly means
   A. prepare.
   B. drift.
   C. attract.
   D. repel.

3. Based on the facts, his solution was <u>judicious</u>.
   A. Mistaken
   B. Fantastic
   C. Misguided
   D. Sensible

4. He did not understand the <u>enormity</u> of his decision.
   A. Perfection
   B. Size
   C. Fluidity
   D. Logic

5. Before the storm, there was a definite feeling of <u>foreboding.</u>
   A. Serenity
   B. Calm
   C. Fear
   D. Distance

6. After his dog was lost for a week, Kevin was <u>forlorn</u>.
   A. Nervous
   B. Hopeless
   C. Happy
   D. Anxious

7. <u>Haughty</u> most nearly means
   A. proud.
   B. angry.
   C. naive.
   D. important.

8. Filling out those forms seemed to be an unnecessary bureaucratic <u>impediment</u>.
   A. Requirement
   B. Step
   C. Obstacle
   D. Challenge

9. She was <u>steadfast</u> in her belief about the causes of crime.
   A. Wrong
   B. Correct
   C. Wavering
   D. Firm

10. <u>Vilify</u> most nearly means
    A. respect.
    B. abuse.
    C. enhance.
    D. believe.

11. He was appreciated for his <u>wry</u> sense of humor.
    A. Clever
    B. Nasty
    C. Funny
    D. Tedious

12. Even though they had never won any championships, the team members were still <u>arrogant</u> about their skills.
    A. Forlorn
    B. Disappointed
    C. Conceited
    D. Ignorant

13. <u>Estranged</u> most nearly means
    A. together.
    B. strange.
    C. punished.
    D. separated.

14. Krista was <u>gratified</u> that the problem had been worked out.
    A. Hopeful
    B. Satisfied
    C. Thankful
    D. Disappointed

15. <u>Irascible</u> most nearly means
   A. smooth.
   B. heavy.
   C. testy.
   D. agreeable.

16. After the tornado, the town was <u>obliterated</u>.
   A. Destroyed
   B. Rebuilt
   C. Saved
   D. Warned

17. The apartment was <u>ransacked</u>.
   A. Decorated
   B. Furnished
   C. Plundered
   D. Rented

18. <u>Vendor</u> most nearly means
   A. outlaw.
   B. buyer.
   C. machine.
   D. salesperson.

19. <u>Scavenger</u> most nearly means
   A. employer.
   B. authority.
   C. gatherer.
   D. relative.

20. The speech given by the politician created <u>repercussions</u> around the world.
   A. Votes
   B. Consequences
   C. Applause
   D. Discontent

21. It took a while to <u>extricate</u> the victim from the burning house.
   A. Hear
   B. Find
   C. Elevate
   D. Remove

22. She <u>floundered</u> through the first reading of the play.
   A. Stumbled
   B. Rushed
   C. Mumbled
   D. Slept

23. <u>Mortify</u> most nearly means
   A. support.
   B. question.
   C. kill.
   D. humiliate.

24. Bob was simply <u>inept</u> with mechanical things.
   A. Superb
   B. Sloppy
   C. Incompetent
   D. Funny

25. Due to a busy schedule, Brad had to <u>forgo</u> his exercise.
   A. Increase
   B. Skip
   C. Reduce
   D. Change

26. <u>Foster</u> most nearly means
   A. encourage.
   B. abandon.
   C. scold.
   D. waver.

27. The day was <u>fraught</u> with excitement.
   A. Electrified
   B. Started
   C. Targeted
   D. Filled

28. The costume was <u>frivolous</u>.
   A. Skimpy
   B. Colorful
   C. Serious
   D. Playful

29. The boat tipped over in a gale.
    A. Lake
    B. Ocean
    C. Windstorm
    D. Panic

30. The new designer fabrics seemed gaudy.
    A. Skimpy
    B. Expensive
    C. Flashy
    D. Basic

31. The e-mail message contained gibberish.
    A. Photos
    B. Nonsense
    C. Warnings
    D. Errors

32. Glower most nearly means
    A. gleam.
    B. shirk.
    C. enrage.
    D. scowl.

33. Grapple most nearly means
    A. struggle.
    B. explain.
    C. interpret.
    D. define.

34. She filed a grievance over the way the issue was handled.
    A. Understanding
    B. Annoyance
    C. Complaint
    D. Question

35. To prevent them from communicating with each other, the brothers were sequestered.
    A. Praised
    B. Separated
    C. Questioned
    D. Encouraged

STOP! DO NOT TURN THIS PAGE UNTIL TIME IS UP FOR THIS TEST. IF YOU FINISH BEFORE TIME IS UP, CHECK OVER YOUR WORK ON THIS TEST ONLY.

PART 4. PARAGRAPH COMPREHENSION

THE TEST ADMINISTRATOR WILL READ THE FOLLOWING ALOUD TO YOU:

Turn to Part 4 and read the directions for Paragraph Comprehension silently while I read them aloud.

This is a test of your ability to understand what you read. In this section you will find one or more paragraphs of reading material followed by incomplete statements or questions. You are to read the paragraph and select one of four lettered choices which BEST completes the statement or answers the question. When you have selected your answer, blacken the space on your answer sheet that has the same number and letter as your answer.

Your score on this test will be based on the number of questions you answer correctly. You should try to answer every question. DO NOT SPEND TOO MUCH TIME on any one question. If you finish before time is called, go back and check your work in the part ONLY.

Now find the section of your answer sheet that is marked PART 4. When you are told to begin, start with question number 1 in Part 4 of your test booklet and answer space number 1 in Part 4 on your separate answer sheet.

DO NOT TURN THE PAGE UNTIL TOLD TO DO SO. You will have 13 minutes to complete the 15 questions in this part. Are there any questions?

Begin.

Anyone who pans for gold hopes to be rewarded by the glitter of colors in the fine material collected in the bottom of the pan. Although the exercise and outdoor activity experienced in prospecting are rewarding, there are few thrills comparable to finding gold. Even an assay report showing an appreciable content of gold in a sample obtained from a lode deposit is exciting. The would-be prospector hoping for financial gain, however, should carefully consider all the pertinent facts before deciding on a prospecting venture.

Only a few prospectors among the many thousands who searched the western part of the United States ever found a valuable deposit. Most of the gold mining districts in the West were located by pioneers, many of whom were experienced gold miners from the southern Appalachian region, but even in colonial times only a small proportion of gold seekers were successful. Over the past several centuries the country has been thoroughly searched by prospectors. During the depression of the 1930s, prospectors searched the better-known gold-producing areas throughout the nation, especially in the West, and the little-known areas as well. The results of their activities have never been fully documented, but incomplete records indicate that an extremely small percentage of the total number of active prospectors supported themselves by gold mining. Of the few significant discoveries reported, nearly all were made by prospectors of long experience who were familiar with the regions in which they were working.

The lack of outstanding success in spite of the great increase in prospecting during the depression of the 1930s confirms the opinion of those most familiar with the occurrence of gold and the development of gold mining districts that the best chances of success lie in systematic studies of known productive areas rather than in efforts to discover gold in hitherto unproductive areas. The development of new, highly sensitive, and relatively inexpensive methods of detecting gold, however, has greatly increased the possibility of discovering gold deposits which are too low grade to have been recognized earlier by the prospector using only a gold pan.

1. According to the passage, which of the following is true?
   A. Panning for gold in the western United States is a useful and profitable activity.
   B. Many prospectors got rich by panning for gold.
   C. There are many places in the United States where gold can be found.
   D. Prospectors who were successful knew the areas where they worked.

2. In this passage the word pertinent means
   A. confusing.
   B. contradictory.
   C. relevant.
   D. identifiable.

3. According to the passage, which of the following is true?
   A. Prospectors were hopeless people looking to strike it rich.
   B. Prospectors could not earn enough to support a family.
   C. Only a few prospectors found gold.
   D. The only successful prospectors were from the southern Appalachian region.

4. Which of the following would be the best title for the passage?
   A. Prospectors Strike It Rich
   B. Technology Finds Gold in the United States
   C. The Gold Rush—A Big Hoax
   D. Few Find It—Many Have Tried

5. With regard to the future of gold mining, the passage suggests which of the following?
   A. Gold mining will never be profitable.
   B. Only the prospectors from the southern Appalachian region seem to be able to find gold.
   C. There is no more gold to be found in the United States.
   D. New technologies will help to find more gold.

Organic food is produced by farmers who emphasize the use of renewable resources and the conservation of soil and water to enhance environmental quality for future generations. Organic meat, poultry, eggs, and dairy products come from animals that are given no antibiotics or growth hormones. Organic food is produced without using most conventional pesticides, petroleum-based fertilizers or sewage sludge-based fertilizers, bioengineering, or ionizing radiation. Before a product can be labeled "organic," a government-approved certifier inspects the farm where the food is grown to make sure the farmer is following all the rules necessary to meet federal organic standards. Companies that handle or process organic food before it gets to your local supermarket or restaurant must be certified, too.

No organization claims that organically produced food is safer or more nutritious than conventionally produced food. Organic food differs from conventionally produced food in the way it is grown, handled, and processed. At the supermarket, in order to distinguish organically produced food from conventionally produced food, consumers must look at package labels and watch for display signs. Along with the national organic standards, there are strict labeling rules to help consumers know the exact organic content of the food they buy. An official seal also tells you that a product is at least 95 percent organic.

The word "organic" and a small sticker version of the Official Organic seal will be on organic vegetables or pieces of fruit, or they may appear on the sign above the organic produce display. The word "organic" and the seal may also appear on packages of meat, cartons of milk or eggs, cheese, and other single-ingredient foods.

"Natural" foods are not necessarily organic foods. Truthful claims, such as "free-range," "hormone-free," and "natural," can still appear on food labels. However, this does not mean that they are "organic." Only food labeled "organic" has been certified as meeting government organic standards.

6. According to this passage, organic farming is helpful because
   A. food grown organically is generally healthier than conventional food.
   B. organic food uses no fertilizers or harmful rays.
   C. growing organic food is cheaper and easier for the farmer.
   D. organic farming tends to improve environmental quality for the future.

7. According to the passage, to be considered organic, a farmer must do which of the following?
   A. Use special fertilizers.
   B. Use special bioengineering techniques.
   C. Show that the foods produced on the farm are healthier than those produced using conventional methods.
   D. Follow certain rules and standards set by the federal government.

8. According to the passage, how can a person identify an organically produced food?
   A. There will be a label or a sign with the word "organic."
   B. There is no way to tell.
   C. Organic labels are prohibited.
   D. A label will indicate "Safe to Eat."

9. In this passage the word <u>renewable</u> most nearly means which of the following?
   A. Fresh
   B. New
   C. Replaceable
   D. Ignored

10. According to the passage, which of the following ensures that foods are organic?
    A. The foods are naturally grown and produced.
    B. The foods are hormone-free.
    C. The foods are grown without chemicals.
    D. The foods are grown according to government standards.

To understand what the Everglades is today, you need to know what it once was. The pristine Everglades was a wetland that spanned the state of Florida south of Lake Okeechobee, about 2.9 million acres of mostly peatland covered by tall saw grass growing in shallow water. When the lake was full, water overflowed into the northern Everglades and moved slowly to the south in a 50-mile-wide sheet, a foot deep. In the 1880s people began to drain the Kissimmee River–Lake Okeechobee–Everglades watershed. Drainage exposed the organic muck soil, which produced extraordinary crop yields.

Today more than 50 percent of the historic Everglades has been eliminated. More than 1,400 miles of drainage canals and levees have been constructed in and around the Everglades for flood control. Widespread population growth and land-use modification for agriculture and industry have altered the natural wetlands, affecting the quantity and quality of drinking water and increasing human exposure to hydrologic hazards such as floods. Chemicals used in farming, including fertilizers, insecticides, herbicides, and fungicides, now often leak into the groundwater or nearby surface waters. Storm-water runoff from urban areas commonly transports heavy metals and nutrients into canals and the Biscayne aquifer.

The last 100 years have seen tremendous change in the Everglades. Today the flow of water is controlled by a complex management system that includes canals, levees, and pumps. The Everglades has been called "the biggest artificial plumbing system in the world."

11. From this passage it can be inferred that
    A. people have harmed their own environment by changing it.
    B. the peatlands have been eliminated.
    C. people living near the Everglades need to use bottled water.
    D. the Everglades is overpopulated.

12. According to the passage, which of the following has been the cause of the shrinking of the Everglades?
    A. Loss of peatlands
    B. Draining of the watershed
    C. Home building in Florida
    D. Overgrowth of saw grass

13. Why is the Everglades called "the biggest artificial plumbing system in the world"?
   A. It is managed by a system of canals, levees, and pumps.
   B. Industrial fluids and chemicals flow into the aquifer.
   C. Insecticides and pesticides leak into the water.
   D. The extensive saw grass tends to block drainage.

14. According to the passage, what was the first event resulting in the elimination of much of the Everglades?
   A. Farming of the peatlands
   B. Flooding of Lake Okeechobee
   C. Draining of the Everglades
   D. Installation of pumps and levees

15. In this passage, what is the meaning of <u>hydrologic</u>?
   A. Catastrophic
   B. Chemical
   C. Water-based
   D. Electronic

STOP! DO NOT TURN THIS PAGE UNTIL TIME IS UP FOR THIS TEST. IF YOU FINISH BEFORE TIME IS UP, CHECK OVER YOUR WORK ON THIS TEST ONLY.

PART 5. MATHEMATICS KNOWLEDGE

THE TEST ADMINISTRATOR WILL READ THE FOLLOWING ALOUD TO YOU:

Now turn to Part 5 and read the directions for Mathematics Knowledge silently while I read them aloud.

This is a test of your ability to solve general mathematics problems. You are to select the correct response from the choices given. Then mark the space on your answer sheet that has the same number and letter as your choice. Use your scratch paper to do any figuring you wish to do.

Your score on this test will be based on the number of questions you answer correctly. You should try to answer every question. DO NOT SPEND TOO MUCH TIME on any one question. If you finish before time is called, go back and check your work in this part ONLY.

Now find the section of your answer sheet that is marked PART 5. When you are told to begin, start with question number 1 in Part 5 of your test booklet and answer space number 1 in Part 5 on your separate answer sheet.

DO NOT TURN THE PAGE UNTIL TOLD TO DO SO. You will have 24 minutes to complete the 25 questions in this part. Are there any questions?

Begin.

1. If $y - 12 = 24$, $y = ?$
   A. 12
   B. 15
   C. 17
   D. 36

2. $2^3 + 2(2 + 5) =$
   A. 12
   B. 17
   C. 19
   D. 22

3. $\dfrac{9}{12} \times 1\dfrac{1}{4} =$

   A. $\dfrac{12}{48}$

   B. $\dfrac{1}{8}$

   C. $\dfrac{15}{16}$

   D. $\dfrac{24}{5}$

4. $(2 \times 10^3)(4.3 \times 10^5) =$
   A. 6.3×10^8
   B. 6.3×10^{-2}
   C. 8.6×10^{-2}
   D. 8.6×10^8

5. Change the fraction $\dfrac{5}{2}$ to a percent.
   A. 2.5%
   B. 52%
   C. 250%
   D. 252%

6. Subtract:

   $24\dfrac{1}{8}$

   $-15\dfrac{3}{4}$

   A. $7\dfrac{1}{4}$

   B. $8\dfrac{3}{8}$

   C. $9\dfrac{2}{8}$

   D. $10\dfrac{7}{8}$

7. Subtract:

   $(4x^2 + 5y + 6)$
   $-(6x^2 + 3y + 8)$

   A. $10x^2 + 8y + 14$
   B. $2x^2 + 2y + 2$
   C. $10x^2 - 2y - 2$
   D. $-2x^2 + 2y + 2$

8. Solve for x.

   $7x = 4x - 12$

   A. 2
   B. −2
   C. −4
   D. 4

9. Solve for the two unknowns.

   $6c - 2d = 32$
   $3c + 2d = 22$

   A. $c = 2$; $d = 4$
   B. $c = 6$; $d = 3$
   C. $c = 6$; $d = 2$
   D. $c = 2$; $d = 6$

10. Divide h^6 by h^8.
    A. h^{-2}
    B. h^{14}
    C. h^2
    D. h^{48}

11. Solve for y.

    $y^2 - 25 = 0$

    A. $y = -2$
    B. $y = 5$
    C. $y = 6$
    D. $y = \dfrac{6}{5}$

12. Multiply:

    $\dfrac{8a}{5} \times \dfrac{3}{4y} =$

    A. $24a + 12$
    B. $\dfrac{6a}{5y}$
    C. $\dfrac{32ay}{15}$
    D. $\dfrac{27a}{20y}$

13. Divide:

    $\dfrac{6a^2}{5b} \div \dfrac{2b}{9a}$

    A. $\dfrac{12a^2b}{45ab}$
    B. $\dfrac{27a^3}{5b^2}$
    C. $\dfrac{27a^3}{10b^2}$
    D. $\dfrac{12a}{45b}$

14. $\sqrt[3]{512} =$
    A. 2
    B. 4
    C. 6
    D. 8

15. Solve for y.

    $gy - n = x$

    A. $y = g(nx)$
    B. $y = \dfrac{x - n}{g}$
    C. $y = \dfrac{g}{xn}$
    D. $y = \dfrac{x + n}{g}$

16. Which of the following is an isosceles triangle?
    A.
    B.
    C.
    D.

17. What is the perimeter of the following square?

    4 in.

    A. 8 in.
    B. 10 in.
    C. 14 in.
    D. 16 in.

18. What is the circumference of a circle with a radius of 8 in.?
    A. 12.25 in.
    B. 25.12 in.
    C. 30.24 in.
    D. 50.24 in.

19. In the following isosceles triangle, what is the measure of $\angle 1$?

    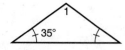

    A. 30°
    B. 85°
    C. 90°
    D. 110°

20. If ∠1 is 142°, what is the measure of ∠2?

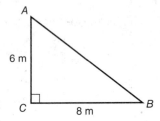

   A. 24°
   B. 38°
   C. 142°
   D. 218°

21. In the following right triangle, what is the length of side *AB*?

   A. 5 m
   B. 9 m
   C. 10 m
   D. 14 m

22. What is the area of the following triangle?

   A. 80 in.²
   B. 300 in.²
   C. 420 in.²
   D. 530 in.²

23. What is the area of the following circle?

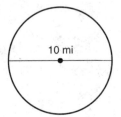

   A. 78.5 mi²
   B. 145.5 mi²
   C. 240.5 mi²
   D. 314 mi²

24. If lines *A* and *B* are parallel and are intersected by line *C* and ∠1 is 145°, what is the sum of the measures of angles 1, 2, 5, and 6?

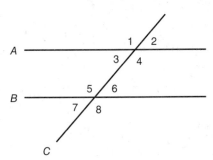

   A. 160°
   B. 180°
   C. 240°
   D. 360°

25. What is the volume of a cylinder with a radius of 4 cm and a height of 6 cm?
   A. 75.36 cm³
   B. 150.72 cm³
   C. 301.44 cm³
   D. 602.88 cm³

STOP! DO NOT TURN THIS PAGE UNTIL TIME IS UP FOR THIS TEST. IF YOU FINISH BEFORE TIME IS UP, CHECK OVER YOUR WORK ON THIS TEST ONLY.

PART 6. ELECTRONICS INFORMATION

THE TEST ADMINISTRATOR WILL READ THE FOLLOWING ALOUD TO YOU:

Now turn to Part 6 of your test booklet and read the directions for Electronics Information silently while I read them aloud.

This is a test of your knowledge of electrical, radio, and electronics information. You are to select the correct response from the choices given and then mark the space on your answer sheet that has the same number and letter as your choice.

Your score on this test will be based on the number of questions you answer correctly. You should try to answer every question. DO NOT SPEND TOO MUCH TIME on any one question. If you finish before time is called, go back and check your work in this part ONLY.

Now find the section of your answer sheet that is marked PART 6. When you are told to begin, start with question number 1 in Part 6 of your test booklet and answer space number 1 in Part 6 on your separate answer sheet.

DO NOT TURN THE PAGE UNTIL TOLD TO DO SO. You will have 9 minutes to complete the 20 questions in this part. Are there any questions?

Begin.

1. If you were installing lighting in a backyard, you might choose 12-volt equipment because it
   A. carries more energy than 120-volt equipment.
   B. does not need to be supplied through a transformer.
   C. would be less likely to "leak" through a wet insulator.
   D. would be more likely to "leak" through a wet insulator.

2. Of the following, the best conductor is
   A. copper.
   B. rubber.
   C. ceramic.
   D. glass.

3. On an electrical circuit diagram, the symbol shown below usually represents a

   A. resistor.
   B. capacitor.
   C. transformer.
   D. battery.

4. In residential wiring, you would use steel conduit to
   A. protect the wires.
   B. ground the system.
   C. protect the wires and ground the system.
   D. serve as a backup hot wire.

5. When you are hooking up an electric light, the switch should
   A. control the hot side of the circuit.
   B. control the ground side of the circuit.
   C. control the equipment grounding side of the circuit.
   D. be wired in parallel with the light.

6. In which kind of circuit is current always the same at all points?
   A. Series
   B. Parallel
   C. Series-parallel
   D. Series-series circuit

7. A battery makes which kind of current?
   A. Direct or alternating
   B. Alternating
   C. Direct
   D. No current

8. An electric current is made of moving
   A. ohms.
   B. neutrons.
   C. protons.
   D. electrons.

9. Which of the following materials would make the best insulator?
   A. Silver
   B. Gold
   C. Rubber
   D. Salt water

10. If you wanted to convert AC to DC, you might use
    A. an electromagnet.
    B. a transformer.
    C. a rectifier.
    D. a resistor.

11. On an electrical circuit diagram, the symbol shown below usually represents a

    ———⟋⟍⟋⟍⟋———

    A. resistor.
    B. capacitor.
    C. transformer.
    D. battery.

12. Which of these is a correct statement of Ohm's law?
    A. ohms = volts × amps
    B. volts = amps × ohms
    C. amps = volts × ohms
    D. volts × amps = ohms

13. If resistors R_1 and R_2 are both of equal size, what is the voltage at the arrow in the diagram below?

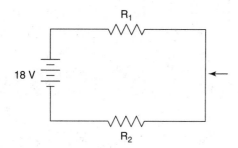

A. 6V
B. 9V
C. 18V
D. 24V

14. To find out if a circuit is live or shut off, you would connect
A. an ohmmeter between hot and ground.
B. a test light to the hot side.
C. a test light to the ground side.
D. a voltmeter between hot and ground.

15. How many amperes would flow in a 120-volt circuit with 12 ohms of resistance?
A. 0.5
B. 1.0
C. 5
D. 10

16. The purpose of a switch in a series circuit is to
A. shut off all current.
B. reduce the voltage.
C. increase the amperage.
D. reduce the amperage.

17. If you were choosing a fuse for a circuit supplying a large air compressor, you would most likely look for a fuse that would allow
A. high resistance.
B. low resistance.
C. a heavy surge of voltage.
D. heavy surge of current.

18. What is the reading on the voltmeter in this series circuit?

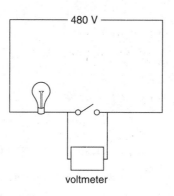

voltmeter

A. 0
B. 120
C. 240
D. 480

19. Which tells how much power an electric device requires?
A. Amperes/ohms
B. Volts/amperes
C. Volts × amperes
D. Amperes × ohms

20. In an electrical circuit diagram, the symbol shown below represents a

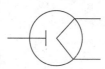

A. transistor.
B. diode.
C. rectifier.
D. transformer.

STOP! DO NOT TURN THIS PAGE UNTIL TIME IS UP FOR THIS TEST. IF YOU FINISH BEFORE TIME IS UP, CHECK OVER YOUR WORK ON THIS TEST ONLY.

PART 7. AUTO AND SHOP INFORMATION

THE TEST ADMINISTRATOR WILL READ THE FOLLOWING ALOUD TO YOU:

Now turn to Part 7 and read the directions for Auto and Shop Information silently while I read them aloud.

This test has questions about automobiles, shop practices, and the use of tools. Pick the BEST answer for each question and then mark the space on your answer sheet that has the same number and letter as your choice.

Your score on this test will be based on the number of questions you answer correctly. You should try to answer every question. DO NOT SPEND TOO MUCH TIME on any one question. If you finish before time is called, go back and check your work in this part ONLY.

Now find the section of your answer sheet that is marked PART 7. When you are told to begin, start with question number 1 in Part 7 of your test booklet and answer space number 1 in Part 7 on your separate answer sheet.

DO NOT TURN THE PAGE UNTIL TOLD TO DO SO. You will have 11 minutes to complete the 25 questions in this part. Are there any questions?

Begin.

1. If a diesel engine runs roughly, which device might be causing the problem?
   A. Carburetor
   B. Fuel injector
   C. Cooling system
   D. Spark plugs

2. If a tire label says it should be inflated to 32 lb/square inch,
   A. the pressure should be 32 lb/square inch when the tire is hot.
   B. the pressure should be 32 lb/square inch when the tire is cold.
   C. tire temperature does not matter; you can measure the pressure at any time.
   D. pressure should be within 10 lb/square inch.

3. When you step on the clutch pedal,
   A. the clutch engages.
   B. the clutch disengages.
   C. the throwout bearing engages.
   D. the throwout bearing disengages.

4. If an engine is overheating, the engine thermostat
   A. is allowing too much water to flow through the cooling system.
   B. is helping the heater work.
   C. may be preventing hot coolant from flowing through the cooling system.
   D. should not be inspected.

5. An engine control unit is basically a(n)
   A. linkage between the throttle and the camshaft.
   B. older method of controlling power output.
   C. substitute for the ignition switch.
   D. computer.

6. If black soot is gathering in the tailpipe,
   A. the fuel–air mix is not burning completely.
   B. the positive crankcase ventilator should be replaced.
   C. the fuel injectors must be replaced.
   D. the catalytic converter is working correctly.

7. If oil is puddling on the ground underneath a vehicle,
   A. check the piston rings immediately.
   B. it is time to change the oil.
   C. a gasket may be loose or defective.
   D. the oil filter is plugged.

8. If antilock brakes fail, the wheel sensor may not be
   A. sensing brake-pad position.
   B. detecting wheel rotation.
   C. opening the brake line.
   D. closing the brake line.

9. The torque converter shown below contains two parts that act as

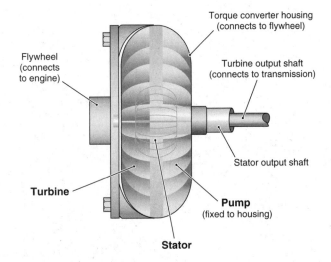

   A. fans.
   B. propellers.
   C. sieves.
   D. flywheels.

10. If a fan belt breaks, which of the following may stop functioning?
    A. Air-conditioning compressor, alternator, or power-steering pump
    B. Camshaft, generator, and distributor
    C. Generator and alternator
    D. Brakes

11. During periodic maintenance, which system(s) may need replacement oil?
    A. Air conditioner and carburetor
    B. Differential
    C. Fuel injection and brakes
    D. Brakes and power steering

12. If a valve stops opening, which of these parts should be inspected?
    A. Drive shaft
    B. Main bearing
    C. Cylinder head
    D. Camshaft, timing belt, and rocker arm

13. Which tool would you be most likely to use after sawing wood?
    A. Wood file
    B. Plane
    C. Belt sander
    D. Wood chisel

14. A countersink is
    A. a gauge used to check drill sharpness.
    B. a conical hole that allows a screw head to sit flat with a surface.
    C. a plumbing tool used to check flow through a sink drain.
    D. a specialized masonry drill bit.

15. The tool shown below is used for

    A. opening paint cans.
    B. chiseling metal.
    C. drilling metal.
    D. locating a drill hole in metal.

16. What is one advantage of a Phillips screwdriver?
    A. It is less likely to slip off the screw.
    B. It gets a stronger grip.
    C. It has a longer shaft.
    D. It is more symmetrical.

17. Which measuring device is generally not used for carpentry?
    A. Dividers
    B. Marking gauge
    C. Tape measure
    D. Dial gauge

18. Why is the end of a box wrench angled slightly to the body?
    A. To allow better access in tight quarters
    B. To grab square and hexagonal bolt heads
    C. To get a better grip on the bolt
    D. To allow the wrench to clear nearby obstructions

19. The tool shown below is a(n)

    A. arc-joint pliers.
    B. can-do clamp.
    C. Crescent® (adjustable) wrench.
    D. come-along.

20. Which of these tools are likely to be used together?
    A. Center punch and drill bit
    B. Phillips screwdriver and slot screwdriver
    C. Wood plane and coping saw
    D. Nail set and center punch

21. Which two tools are best for cutting plywood?
    A. Hand saw and coping saw
    B. Jigsaw and circular saw
    C. Hand saw and hacksaw
    D. Jigsaw and hacksaw

22. Rivets are used to
    A. make permanent connections.
    B. join wood to metal.
    C. replace glue.
    D. make joints that can be taken apart.

23. The tool shown below is used for

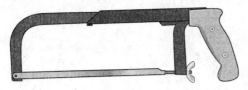

    A. cutting curves.
    B. aligning hinges.
    C. sawing molding.
    D. sawing metal.

24. Tightening the screw on the handle of a locking pliers will
    A. make the pliers stronger.
    B. open the jaws.
    C. adjust the pliers for smaller objects.
    D. allow the jaws to open wider.

25. The teeth on a hacksaw
    A. point away from the handle.
    B. must be sharpened frequently.
    C. are usually made of tungsten carbide.
    D. cut when you pull the saw toward you.

STOP! DO NOT TURN THIS PAGE UNTIL TIME IS UP FOR THIS TEST. IF YOU FINISH BEFORE TIME IS UP, CHECK OVER YOUR WORK ON THIS TEST ONLY.

PART 8. MECHANICAL COMPREHENSION

THE TEST ADMINISTRATOR WILL READ THE FOLLOWING ALOUD TO YOU:

Now turn to Part 8 and read the directions for Mechanical Comprehension silently while I read them aloud.

This test has questions about mechanical and physical principles. Study the picture and decide which answer is CORRECT and then mark the space on your separate answer sheet that has the same number and letter as your choice.

Your score on this test will be based on the number of questions you answer correctly. You should try to answer every question. DO NOT SPEND TOO MUCH TIME on any one question. If you finish before time is called, go back and check your work in this part ONLY.

Now find the section of your answer sheet that is marked PART 8. When you are told to begin, start with question number 1 in Part 8 of your test booklet and answer space number 1 in Part 8 on your separate answer sheet.

DO NOT TURN THE PAGE UNTIL TOLD TO DO SO. You will have 19 minutes to complete the 25 questions in this part. Are there any questions?

Begin.

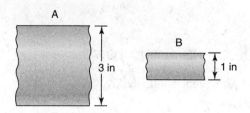

1. Pipes A and B are carrying water with the same amount of water pressure. Pipe A is carrying how much more water than pipe B?
   A. Ten times as much
   B. Nine times as much
   C. Three times as much
   D. Two times as much

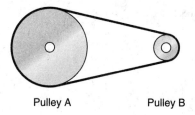

Pulley A   Pulley B

2. Pulleys A and B run on the same belt. Which pulley is rotating faster?
   A. Pulley A and pulley B rotate at the same rate.
   B. Pulley A rotates faster.
   C. Pulley B rotates faster.
   D. First pulley A rotates faster; then pulley B rotates faster.

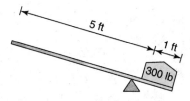

3. The lever shown is used to lift a 300-lb weight. What force is needed to lift the weight?
   A. 60 lb
   B. 50 lb
   C. 45 lb
   D. 30 lb

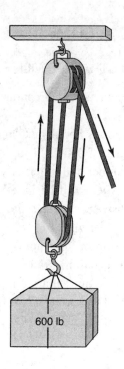

4. The rope shown must be able to lift at least
   A. 150 lb
   B. 200 lb
   C. 300 lb
   D. 1,200 lb

5. Which of the following is a type of lever?

A.

B.

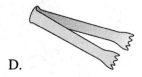

C.

D.

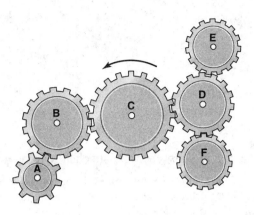

6. In the gear system shown, gear C is rotating in the direction shown. Which other gears are rotating in the same direction?
   A. A and D
   B. B, D, and F
   C. B and E
   D. A, E, and F

7. In a pulley system, the easiest way to find the theoretical mechanical advantage is to
   A. count the supporting strands.
   B. count the pulleys.
   C. count the number of rope strands you see.
   D. measure effort and load.

8. Gear A, with 24 teeth, is meshed with gear B, with 12 teeth. For every complete rotation of gear B, how many complete rotations will gear A make?
   A. ½
   B. 1
   C. 2
   D. 4

9. If all these items are the same temperature, which one will feel coldest to the touch?

   A.

   B.

   C.

   D.

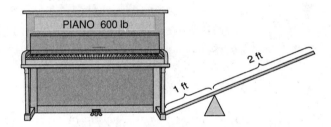

10. To lift one side of the piano, how much force must be applied to the lever?
    A. 150 lb
    B. 200 lb
    C. 250 lb
    D. 300 lb

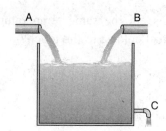

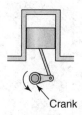

Crank

11. Water is flowing into the tank through pipe A at 6 gallons per minute and through pipe B at 7 gallons per minute. It is flowing out of the tank through pipe C at 8 gallons per minute. How much more water will be in the tank at the end of 5 minutes?
    A. 13 gallons
    B. 15 gallons
    C. 21 gallons
    D. 25 gallons

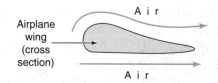

12. The diagram shows a cross section of an airplane wing in flight, with air passing above and beneath it. Lift is created because
    A. air pressure above the wing is less than air pressure below the wing.
    B. air pressure above the wing is equal to air pressure below the wing.
    C. the wing pushes air down, and the opposing force lifts the plane.
    D. air pressure above the wing is greater than air pressure below the wing.

13. The figure shows a crank attached to a rod and piston. When the crank turns, the piston moves in the cylinder. The piston will be at the base of the cylinder if the crank makes a
    A. ¼ turn
    B. ½ turn
    C. ¾ turn
    D. complete turn

14. A tank 4 feet wide by 4 feet long holds 800 lb of water. What is the water pressure at the bottom in lb/ft^2?
    A. 10
    B. 12.5
    C. 25
    D. 50

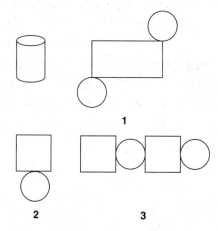

15. The flat sheet metal pattern that can be bent to form the completely closed cylinder shown is
    A. 1
    B. 2
    C. 3
    D. None of the Above

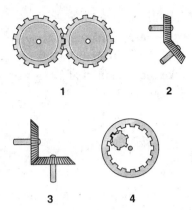

1   **2**

3   **4**

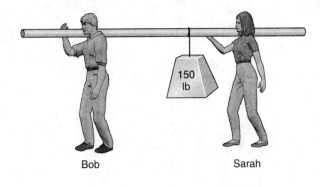

Bob   Sarah

16. Which of the gear types shown is used to transmit motion between shafts that form a 90° angle?
    A. 1
    B. 2
    C. 3
    D. 4

17. If gear A and gear B are meshed, and gear B rotates twice as fast as gear A, the gear ratio of A:B is
    A. 2:1
    B. 1:1
    C. 0.5:1
    D. 1:2

18. Oxygen gas stored in a metal cylinder exerts a pressure of 6 pounds per square inch on the inside of the cylinder. If you pump additional oxygen into the cylinder, which of the following could be the new pressure inside the cylinder?
    A. 4 pounds per square inch
    B. 5 pounds per square inch
    C. 6 pounds per square inch
    D. 8 pounds per square inch

19. In the diagram, who is carrying more of the load?
    A. Sarah
    B. Bob
    C. They are carrying equal loads.
    D. Sarah, then later Bob

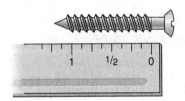

20. What happens when this screw is turned 8 complete turns tight?
    A. The screw moves in by 2/3 in.
    B. The screw moves out by 3/4 in.
    C. The screw moves in by 1/2 in.
    D. The screw moves in by 3/4 in.

23. A nut and a bolt exert compression on this board. You run the parts through a degreaser and tighten the nut to 100 ft/lbs. What happens if you take the assembly apart, grease the threads, and retighten to 100 ft/lbs?

A. Compression increases.
B. Compression decreases.
C. Compression does not change.
D. Compression decreases while tightening, then increases.

21. The mechanical advantage in the pulley system shown is
A. 2
B. 3
C. 4
D. 6

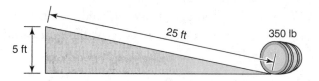

22. In the diagram, pressure in the main tank is 50 lb/in². What is the pressure in tank B in lb/in²?

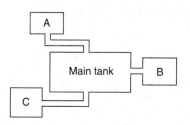

24. How much effort will be needed to roll a 350-lb drum up a ramp 25 feet long and 5 feet high? (Ignore the effect of friction.)
A. 35 lb
B. 60 lb
C. 70 lb
D. 100 lb

A. 25
B. 50
C. 100
D. 200

25. In the hydraulic system shown, how much will
the piston in cylinder B move when you move
the piston in cylinder A by 6 inches?

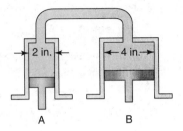

A. 1.5 inches
B. 2 inches
C. 3 inches
D. 6 inches

**STOP! DO NOT TURN THIS PAGE UNTIL TIME IS UP FOR THIS TEST. IF YOU FINISH
BEFORE TIME IS UP, CHECK OVER YOUR WORK ON THIS TEST ONLY.**

PART 9. ASSEMBLING OBJECTS

THE TEST ADMINISTRATOR WILL READ THE FOLLOWING ALOUD TO YOU:

Now turn to Part 9 and read the directions for Assembling Objects silently while I read them aloud.

This test has questions that will measure your spatial ability. Study the diagram, decide which answer is CORRECT, and then mark the space on your separate answer sheet that has the same number and letter as your choice.

Your score on this test will be based on the number of questions you answer correctly. You should try to answer every question. DO NOT SPEND TOO MUCH TIME on any one question. If you finish before time is called, go back and check your work in this part ONLY.

Now find the section of your answer sheet that is marked PART 9. When you are told to begin, start with question 1 in Part 9 of your test booklet and answer space number 1 in Part 9 on your separate answer sheet.

DO NOT TURN THE PAGE UNTIL TOLD TO DO SO. You will have 15 minutes to complete the 25 questions in this part. Are there any questions?

Begin.

For Questions 1–18, which figure best shows how the objects in the left box will appear if they are fit together?

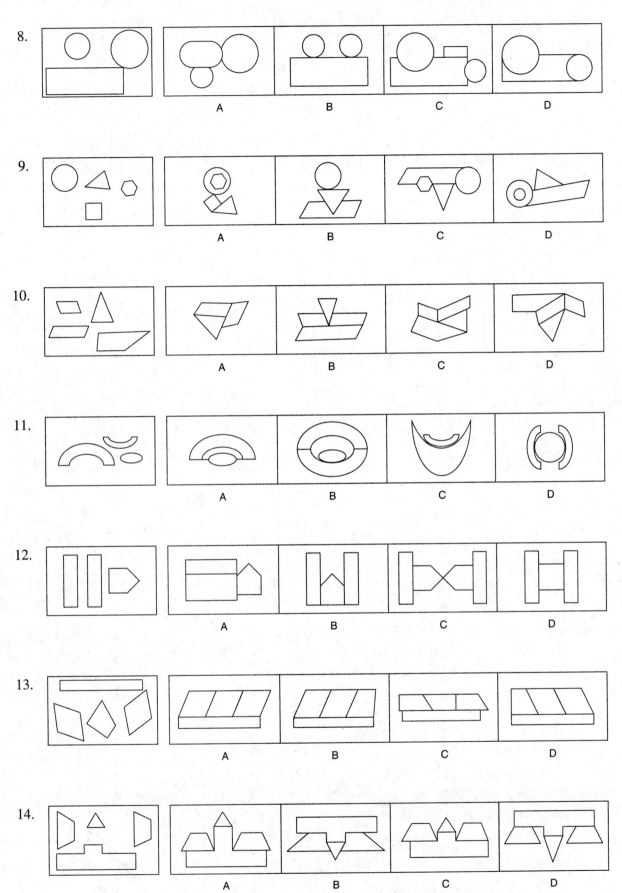

8.　A　B　C　D

9.　A　B　C　D

10.　A　B　C　D

11.　A　B　C　D

12.　A　B　C　D

13.　A　B　C　D

14.　A　B　C　D

15.

16.

17.

18.

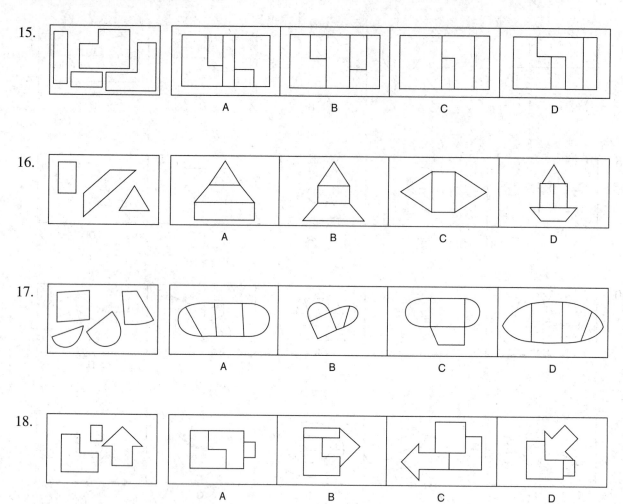

For Questions 19-25, which figure best shows how the objects in the left box will touch if the letters for each object are matched?

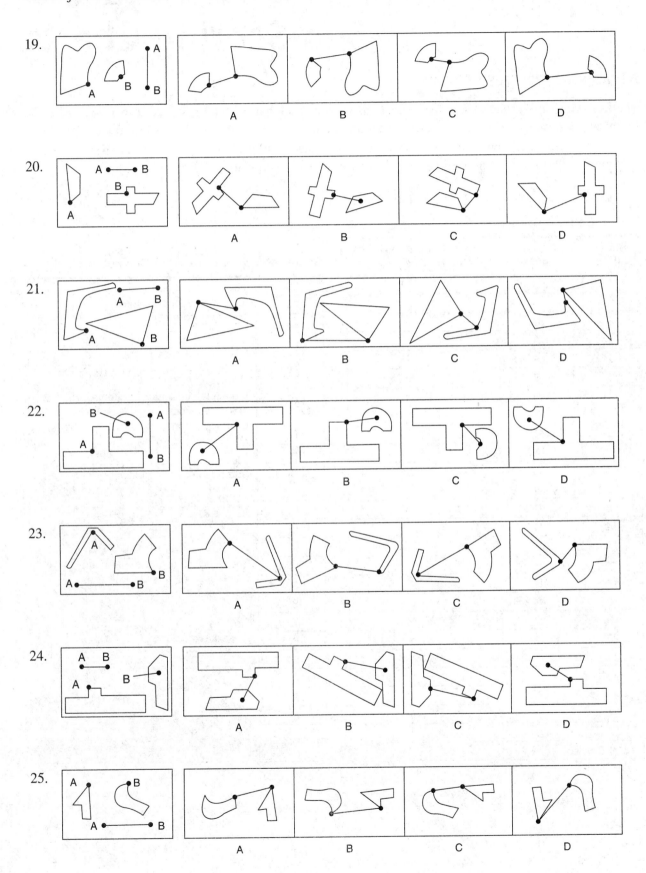

Answer Keys and Self-Scoring Charts

ASVAB PRACTICE TEST FORM 1

The following answer keys show the correct answers for each part of the practice test that you just took. For each question, compare your answer to the correct answer. Mark an × in the column to the right if you got the item correct. Then total the number correct for each part of the test. Find that number in the corresponding chart to the right of the answer key. See the suggestions listed for your performance.

Part 1. General Science

Item Number	Correct Answer	Mark X If You Picked the Correct Answer
1	C	
2	D	
3	D	
4	A	
5	A	
6	B	
7	D	
8	C	
9	D	
10	A	
11	A	
12	D	
13	B	
14	D	
15	B	
16	B	
17	B	
18	A	
19	C	
20	D	
21	C	
22	A	
23	A	
24	C	
25	B	
Total Correct		

Score Interpretation—Total Correct

25	This is pretty good work. Review the explanations for the answers you got incorrect.
24	
23	
22	
21	
20	You are doing pretty well. Review the explanations for the items you answered incorrectly. If you have time, review the other explanations and you will learn even more.
19	
18	
17	
16	
15	You need to keep studying. Pay close attention to the explanations for each item, even for the ones you got correct.
14	
13	
12	
11	
10	Spend time working on the General Science review in Part 3 of this book.
9	
8	
7	
6	If you have copies of your school science textbooks, it would be a good idea to study those.
5	
4	
3	
2	
1	
0	

Part 2. Arithmetic Reasoning

Item Number	Correct Answer	Mark X If You Picked the Correct Answer
1	D	
2	B	
3	C	
4	B	
5	B	
6	C	
7	D	
8	B	
9	D	
10	C	
11	A	
12	A	
13	B	
14	B	
15	C	
16	D	
17	A	
18	B	
19	C	
20	B	
21	A	
22	B	
23	C	
24	C	
25	C	
26	B	
27	B	
28	A	
29	C	
30	B	
Total Correct		

Score Interpretation—Total Correct

30	This is pretty good work. Review the explanations for the answers you got incorrect. This test is a part of the AFQT, and you must do well.
29	
28	
27	
26	
25	You are doing pretty well. Review the explanations for the items you answered incorrectly. If you have time, review the other explanations and you will learn even more.
24	
23	
22	
21	
20	You need to keep studying. Pay close attention to the explanations for each item, even for the ones you got correct.
19	
18	
17	
16	
15	Spend time working on the Arithmetic Reasoning reviews in Part 3 of this book.
14	
13	
12	
11	Keep working and reworking problems until you are comfortable with the processes.
10	
9	
8	
7	
6	
5	
4	
3	
2	
1	
0	

Part 3. Word Knowledge

Item Number	Correct Answer	Mark X If You Picked the Correct Answer
1	D	
2	C	
3	D	
4	B	
5	C	
6	B	
7	A	
8	C	
9	D	
10	B	
11	A	
12	C	
13	D	
14	C	
15	C	
16	A	
17	C	
18	D	
19	C	
20	B	
21	D	
22	A	
23	D	
24	C	
25	B	
26	A	
27	D	
28	D	
29	C	
30	C	
31	B	
32	D	
33	A	
34	C	
35	B	
Total Correct		

Score Interpretation—Total Correct

Score	Interpretation
35	This is pretty good work. Review the explanations for the answers you got incorrect. This is an important test that contributes to your AFQT.
34	
33	
32	
31	
30	You are doing pretty well. Review the explanations for the items you answered incorrectly. If you have time, review the other explanations and you will learn even more.
29	
28	
27	
26	
25	You need to keep studying. Pay close attention to the explanations for each item, even for the ones you got correct.
24	
23	
22	
21	
20	Spend time working on the Word Knowledge review in Part 3 of this book.
19	
18	
17	
16	Keep reading and identifying words you don't know.
15	
14	
13	
12	
11	
10	
9	
8	
7	
6	
5	
4	
3	
2	
1	
0	

Part 4. Paragraph Comprehension

Item Number	Correct Answer	Mark X If You Picked the Correct Answer
1	D	
2	C	
3	C	
4	D	
5	D	
6	D	
7	D	
8	A	
9	C	
10	D	
11	A	
12	B	
13	A	
14	C	
15	C	
Total Correct		

Score Interpretation—Total Correct

15	This is pretty good work. Review the explanations for the answers you got incorrect. This test is a part of the AFQT, and you must do well.
14	You are doing pretty well. Review the explanations for the items you got incorrect. This test is a part of the AFQT, and you must do well.
13	You need to keep studying. Pay close attention to the explanations for each item, even for the ones you got correct. This will help you understand why the answer is correct. Spend time reviewing the Paragraph Comprehension and Word Knowledge reviews in Part 3 of this book. Keep reading books and newspapers.
12	
11	
10	
9	
8	
7	
6	
5	
4	
3	
2	
1	
0	

Part 5. Mathematics Knowledge

Item Number	Correct Answer	Mark X If You Picked the Correct Answer
1	D	
2	D	
3	C	
4	D	
5	C	
6	B	
7	D	
8	C	
9	C	
10	A	
11	B	
12	B	
13	B	
14	D	
15	D	
16	C	
17	D	
18	D	
19	D	
20	B	
21	C	
22	B	
23	A	
24	D	
25	C	
Total Correct		

Score Interpretation—Total Correct

Score	Interpretation
25	This is pretty good work. Review the explanations for the answers you got incorrect.
24	
23	
22	You are doing pretty well. Review the explanations for the items you answered incorrectly. If you have time, review the other explanations and you will learn even more.
21	
20	
19	This test is part of the AFQT. You must perform well on this test.
18	
17	
16	You need to keep studying. Pay close attention to the explanations for each item, even for the ones you got correct.
15	
14	
13	
12	Spend time working on the Mathematics Knowledge reviews in Part 3 of this book, and work and rework the problems until you fully understand the processes.
11	
10	
9	
8	
7	
6	
5	
4	
3	
2	
1	
0	

Part 6. Electronics Information

Item Number	Correct Answer	Mark X If You Picked the Correct Answer
1	C	
2	A	
3	D	
4	C	
5	A	
6	A	
7	C	
8	D	
9	C	
10	C	
11	A	
12	B	
13	B	
14	D	
15	D	
16	A	
17	D	
18	A	
19	C	
20	A	
Total Correct		

Score Interpretation—Total Correct

Score	Interpretation
20	This is pretty good work. Review the explanations for the answers you got incorrect.
19	
18	
17	You are doing pretty well. Review the explanations for the items you answered incorrectly. If you have time, review the other explanations and you will learn even more.
16	
15	
14	
13	
12	You need to keep studying. Pay close attention to the explanations for each item, even for the ones you got correct.

Spend time working on the Electronics Information review in Part 3 of this book.

Your school electronics or physics books may have some helpful information as well. |
11	
10	
9	
8	
7	
6	
5	
4	
3	
2	
1	
0	

Part 7. Auto and Shop Information

Item Number	Correct Answer	Mark X If You Picked the Correct Answer
1	B	
2	B	
3	B	
4	C	
5	D	
6	A	
7	C	
8	B	
9	B	
10	A	
11	B	
12	D	
13	B	
14	B	
15	D	
16	A	
17	D	
18	D	
19	C	
20	A	
21	B	
22	A	
23	D	
24	C	
25	A	
Total Correct		

Score Interpretation—Total Correct

25	This is pretty good work. Review the explanations for the answers you got incorrect.
24	
23	
22	
21	
20	You are doing pretty well. Review the explanations for the items you answered incorrectly. If you have time, review the other explanations and you will learn even more.
19	
18	
17	
16	
15	
14	You need to keep studying. Pay close attention to the explanations for each item, even for the ones you got correct.
13	
12	
11	
10	Spend time working on the Auto and Shop Information reviews in Part 3 of this book.
9	
8	
7	
6	
5	
4	
3	
2	
1	
0	

Part 8. Mechanical Comprehension

Item Number	Correct Answer	Mark X If You Picked the Correct Answer
1	B	
2	C	
3	A	
4	B	
5	D	
6	D	
7	A	
8	A	
9	B	
10	A	
11	D	
12	A	
13	C	
14	D	
15	A	
16	C	
17	A	
18	D	
19	A	
20	A	
21	A	
22	B	
23	A	
24	C	
25	A	
Total Correct		

Score Interpretation—Total Correct

25	This is pretty good work. Review the explanations for the answers you got incorrect.
24	
23	
22	
21	
20	You are doing pretty well. Review the explanations for the items you answered incorrectly. If you have time, review the other explanations and you will learn even more.
19	
18	
17	
16	
15	You need to keep studying. Pay close attention to the explanations for each item, even for the ones you got correct.
14	
13	
12	
11	Spend time working on the Mechanical Comprehension review in Part 3 of this book
10	
9	
8	
7	
6	
5	
4	
3	
2	
1	
0	

Part 9. Assembling Objects

Item Number	Correct Answer	Mark X if You Picked the Correct Answer
1	C	
2	B	
3	A	
4	B	
5	C	
6	B	
7	A	
8	D	
9	C	
10	C	
11	A	
12	B	
13	B	
14	C	
15	D	
16	B	
17	A	
18	D	
19	D	
20	D	
21	A	
22	D	
23	C	
24	A	
25	C	
Total Correct		

Score Interpretation–Total Correct

Score	Interpretation
25	This is pretty good work. Review the explanations of the answers you got incorrect.
24	
23	
22	
21	
20	You are doing pretty well. Review the explanations for the items you answered incorrectly. If you have time, review the other explanations and you will learn even more.
19	
18	
17	
16	
15	Keep studying. Pay close attention to the explanations for each item, even for the ones you got correct. Spend time working on the Assembling Objects review in Part 3 of this book.
14	
13	
12	
11	
10	
9	
8	
7	
6	
5	
4	
3	
2	
1	
0	

Answers and Explanations

PART 1. GENERAL SCIENCE

1. **C.** Plant cells do reproduce and they do have cell walls. Plant cells do not necessarily die after one year. Plant cells do contain chloroplasts, and animal cells do not. The correct answer is therefore choice C.

2. **D.** All these creatures are animals and are part of the kingdom Animalia.

3. **D.** An ecosystem is the collection of organisms, living and nonliving, along with the factors that affect them such as heat, light, landforms, gases, and the like. The atmosphere (choice A) is the air that surrounds Earth, the hydrosphere (choice B) is the water found in and on Earth, and the lithosphere (choice C) is the land on Earth. Biosphere is the name given to the global ecosystem. Choice D is the correct answer.

4. **A.** Stomata (choice B) are parts of a leaf and should be digestible in most plants. Root hairs (choice C) and roots (choice D) should be digestible as well. The cell wall (choice A) is made of cellulose, which is not digestible by humans. Choice A is the correct answer.

5. **A.** The process of respiration takes place in the gills of a fish. Choice A is the correct answer.

6. **B.** Respiration (choice A) is the series of chemical reactions that release energy that is stored in food molecules. Osmosis (choice C) is the process by which water diffuses through a cell membrane. Turgidity (choice D) is the swelling of a plant stem as it fills with water and stands up. A plant that is not turgid is limp. Choice B, gravitropism, is a plant's response to gravity. Roots have a positive reaction to gravity and grow downward regardless of how the plant is situated.

7. **D.** A predator is an animal that seeks out another animal for food. Choice C is not correct. Mutualism is a condition in which both organisms benefit from their cooperation. Choice A is not the correct answer. Commensalism is a situation in which one organism benefits by living off another, but without harming the other organism. Choice B is not the correct answer. Mistletoe is a parasite, an organism that steals nutrients from another organism. Choice D is the correct answer.

8. **C.** The biceps and the heart contain certain muscle cells, but the stomach has specialized, smooth muscle cells. Choice C is the correct answer.

9. **D.** Pathogens can enter your body from areas that are open and related to mucus like the mouth (choice A), nose (choice B), and eyes (choice C). The skin (choice D) is the first line of defense against viruses and bacteria, unless it is cut in some way. Choice D is the correct answer.

10. **A.** The mouth (choice B) is the starting place of the digestive system where food begins to break down, but it is not the place where nutrients are absorbed. The large intestine (choice C) absorbs a lot of the liquid, but not much in the way of nutrients. The stomach (choice D) is important for the chemical breakdown of food into absorbable particles that can be absorbed by the small intestine (choice A). Choice A is the correct answer.

11. **A.** The autonomic system controls functions that are involuntary, such as breathing (choice D), heart rate (choice C), and digestion (choice B). Voluntary movements, such as walking (choice A) and other purposeful actions, are not controlled by the autonomic system. Choice A is the correct answer.

12. **D.** You can use Punnett Squares to determine the probability of offspring having certain characteristics. The dominant trait or allele is shown in capital letters, and the recessive allele is in lower case letters.

		Father	
		T	T
Mother	t	Tt	Tt
	t	Tt	Tt

From this it can be seen that every offspring will be tall. Choice D, 100%, is the correct answer.

13. **B.** If an object is more than two focal lengths away from a convex lens, the image would be smaller and inverted. The correct answer is choice B.

14. **D.** An element's atomic mass is determined by the number of protons and neutrons it contains. Choice D is the correct answer.

15. **B.** Light has many characteristics; one is that it bends when it changes speed as it passes through various materials. The degree of bending depends on the original frequency and the medium through which the light is passing. When a light wave moves through water (for example, through a raindrop), it forms a spectrum of colors we know as a rainbow. Although light diffuses (choice A) or spreads and can be reflected (choice C), these do not cause a rainbow. Bending of light is refraction, so choice B is the correct answer.

16. **B.** In the past, when the continents were connected in a megacontinent called Pangaea, areas that are currently tropical or temperate, like Africa, South America, and India were closer to the South Pole. Parts of these continents were covered by glaciers where there are no glaciers now. Evidence of glacial deposits, along with the scouring of rocks typical of glaciers, is present. Choice B, glacial deposits, is the correct answer.

17. **B.** Pumice is formed when molten rock is shot out of a volcano by high concentrations of gases. It is extrusive because it is above the surface of the earth and igneous because it is formed from volcanic activity. The correct answer is choice B, extrusive igneous rock.

18. **A.** The alignment of Earth, the moon, and the sun (choice D) causes oceanic tides, not waves. Excessive rainfall (choice B) does not cause waves. Steep slopes near the shoreline (choice C) can cause waves to crash forming breakers, but the slopes are not the cause of wave formation. Persistent surface winds over the ocean are the chief cause of wave formation. Choice A is the correct answer.

19. **C.** Earth's orbit around the sun is elliptical, not circular. As a result there are times of the year when Earth is closer to the sun than at other times. It is January when Earth is closest to the sun. For people in the northern hemisphere, this may seem odd because it is winter in the northern hemisphere in January. In the southern hemisphere, however, January is in the middle of the summer. Choice C is the correct answer.

20. **D.** Know your planetary order: Mercury, Venus, Earth, Mars, Jupiter, Saturn, Uranus, and Neptune. The eighth from the sun is Neptune (choice D).

21. **C.** If the atmosphere heats up over time to the point where the climate is changed (called global warming), that could have severe environmental effects. The warmer weather could indeed help increase the amount of area considered tropical, thus spreading tropical diseases such as malaria (choice A). If the temperature rises high enough, polar ice caps and glaciers would melt, thus raising the level of the oceans and seas (choice B). Since hurricanes are initiated by low pressure systems over warm waters, the number and severity of hurricanes could increase (choice D). All these are possibilities. We are seeing some evidence of global warming already. Scientists are debating whether the increase in global temperature is

part of a natural cycle or whether some human actions are causing this to happen. There is no known relationship between Earth's mesosphere being ionized and global warming. Choice C is the correct answer.

22. **A.** A stationary front (choice D) means a front is not moving or not moving very fast. When a warm air mass is replacing cold air (choice B) or when an occluded front overtakes a warm air mass (choice C), the weather tends to be rainy across large areas around the front. The weather is not necessarily severe. A cold front that replaces warm air causing the warm air to uplift rapidly generally causes severe weather. Choice A is the correct answer.

23. **A.** Sinkholes, groundwater flow, and stalactites (choice B) are features related to groundwater. Sedimentary rocks, conglomerates, and shale (choice C) are related to a category of rocks called *sedimentary rocks*. Vents, calderas, and hot spots (choice D) are all associated with volcanism. Choice A, cirques, aràtes, and moraines, are features of glaciers. Choice A is the correct answer.

24. **C.** Land heats faster than water. As the land heats up, it warms the air above it. The air expands and rises, creating an upward current. The rising warm air is replaced by cooler air flowing in from over the water. This process creates a wind current that goes from the water to the land, called a *sea breeze*. In the evening the reverse takes place because the land cools faster than the water. The air current moves in the opposite direction, from the land to the sea, and is called a *land breeze*. Choice C is the correct answer.

25. **B.** An atom of the same element that has a different number of neutrons is called an *isotope*. Choice B is the correct answer.

PART 2. ARITHMETIC REASONING

1. **D.** This is a simple addition problem. Add the amounts to get the total of $141.32.

2. **B.** This problem has two steps. First you need to add the number of minutes Maurice used. Then you need to subtract that amount from his allocation of 400 minutes. The total number of minutes used was 317, so subtracting that from 400 leaves 83 minutes.

3. **C.** In order to calculate the percent, you need to divide the number of female students by the total number of students. In this case you would divide 14 by 32 to get 0.4375. To change that to a percent, move the decimal point two places to the right and add the percent sign to get the correct answer of 43.75%.

4. **B.** To find the average, add the numbers and divide by the number of numbers. In this problem Tim's times add up to 780 seconds. Dividing that number by 4 gives an average of 195 seconds.

5. **B.** To determine the percent, you need to divide the amount Trinidad spends on rent by the amount that she earns. In this problem, $1,100 is divided by $2,700 with the result coming to 40.7%. Rounding to the nearest percent gives the correct answer of 41%.

6. **C.** To find the ratio of one number to another, create a fraction to show the relationship. In this problem 12 cartons of vegetables are compared to 36 cartons of fruit. The resulting fraction is $\frac{12}{36}$. Simplifying this fraction results in a ratio of $\frac{1}{3}$. You might see this as 1:3 on the ASVAB test. $\frac{1}{3}$ and 1:3 mean the same thing.

7. **D.** Set this up as a proportion. If 112 shirts can be produced in 1 hour, then [x] shirts can be produced in 8 hours. The proportion would look like this: $\frac{112}{1} = \frac{x}{8}$. Cross-multiply and solve for x. 1x = 896. So 896 shirts can be produced in one 8-hour day. The question asks how many can be produced in 3.5 days, so multiply 896 by 3.5 to get the final answer of 3,136.

8. **B.** Probability = number of favorable outcomes/ number of possible outcomes. In this problem the number of possible outcomes equals $20 + 32 + 16 = 68$. The number of favorable outcomes is the sum of pink and blue or $20 + 32 = 52$. $\frac{52}{68} = \frac{13}{17}$, the correct answer when simplified.

9. **D.** Simple interest problems use the formula $I = prt$, where I is the amount of interest, p is the principal or the amount saved or invested, r is the rate of interest, and t is the amount of time in which the interest is accruing. In this problem you are asked to find the rate of interest given that Ruth has earned $250 in interest over 5 years. Substitute the information into the formula to get $250 = $500(r)5$ and solve for r. $250 = $2,500r$; $r = 0.10$ or 10%.

10. **C.** Multiply the number of hours worked by the regular rate of $53 per hour and add that to the overtime hours multiplied by the overtime rate. The number of regular hours worked is 5×8 or 40.

$$40($53) + 6.5($57) = $2,490.50$$

11. **A.** If Ellen baked 112 cookies for herself and her 23 classmates, divide 24 into 112 to get 4.67. The question asks for the number of whole cookies, so there are 4 per person plus some fraction of a cookie left over. The correct answer is 4.

12. **A.** Leon has four walls that are 8 ft by 30 ft. That's an area of 240 ft^2 for each of four walls. For all four walls, multiply 240 by 4, giving a total square footage of 960 ft^2. If a quart of paint covers 40 ft^2, then you need to divide 960 by 40, which results in 24 quarts of paint. The question asks for the number of gallons. Since there are 4 quarts in a gallon, divide 24 by 4 to get 6 gallons.

13. **B.** If the bill is $112.00, a tip of 15% is calculated by multiplying $112.00 by 0.15.

$$\$112.00 \times 0.15 = \$16.80$$

14. **B.** Use the following formula:

$$\text{percent of change} = \frac{\text{amount of change}}{\text{original amount}}.$$

In this problem we know that the amount of change is $450 if Manuel purchases the TV online. The starting point is the cost of the TV at the local department store, $4,250. Substitute that information into the formula:

$$\text{percent of change} = \frac{\$450.00}{\$4,250.00} = 0.106.$$

Change the result to a percent by moving the decimal place two places to the right and adding the percent sign.

$$0.106 = 10.6\%$$

15. **C.** Two years ago Jenny was 32 years old. She was twice the age of her cousin. Set this up as an equation.

$$2c = 32.$$

Solve for c.

$$c = 16$$

Since we want the cousin's age now, which is two years later, it would be 18 $(16 + 2)$.

16. **D.** To calculate the area of Simon's garden, use the formula $A = lw$.

$$A = 12 \text{ ft} \times 24 \text{ ft}$$
$$A = 288 \text{ ft}^2$$

If each bush needs 8 ft^2 of space, divide 288 by 8.

$$\frac{288 \text{ ft}^2}{8} = 36 \text{ bushes}$$

17. **A.** To find the circumference of a circle, use the formula $C = 2\pi r$. The radius is given as 10 meters. Substitute that information into the formula $C = 2(3.14)10$.
Solve for C. $C = 62.8$ meters.

18. **B.** Create an equation to solve this problem: Robby = 2(Greg).

You are told that Robby is 46. Substitute that information into the formula: 46 = 2(Greg).

Solve for Greg by dividing both sides by 2.

23 = Greg, so Greg is 23 years old.

19. **C.** Create an equation to solve this problem.

$$Karin = cousin\ 1 + cousin\ 2.$$

You are told that Karin is 54 and one cousin is 21. Substitute that information into the formula.

$$54 = 21 + cousin\ 2$$

Solve for cousin 2 by subtracting 21 from both sides.

$$33 = cousin\ 2$$

20. **B.** Create an equation to solve this problem. This problem requires both multiplication and addition.

6($6.25) + 11($5.95) + 6($5.50)
= $37.50 + $65.45 + $33.00 = $135.95

21. **A.** To find the average, add the numbers and divide by the number of numbers. In this problem add 10, 14.5, 14.8, 16.7, 15.4, and 0 to get 71.4. Divide this by 6.

$$\frac{71.4}{6} = 11.9$$

Caution: Don't forget that the zero counts as a number, so be sure to divide by 6, not 5.

22. **B.** To solve this problem, use the formula $V = lwh$. You have been given the dimensions of $l = 24$ inches, $w = 10$ inches, and $h = 12$ inches. Substitute that information into the formula.

$$V = (24)(10)(12) = 2,880\ in^3$$

If a goldfish needs 144 in^3 to exist in the aquarium, divide that number into the total amount of volume.

2880 in.^{3} ÷ 144 in.^{3} = 20 goldfish.

23. **C.** The ratio would be stated as 6:114 or $\frac{6}{114}$. Simplify this fraction to get $\frac{1}{19}$.

24. **C.** To calculate the volume of a cylinder, use the formula $V = \pi r^2 h$. In this problem the diameter is 4 in., so the radius is 2 in. The height is given as 6 in. Substitute that information into the formula.

$$V = (3.14)(2^2)(6)$$
$$V = (3.14)(4)(6)$$
$$V = 75.36\ or\ 75.4\ in.^3$$

Be sure to pay attention to the information you are given. You could have made an error by using the diameter rather than the radius because the diameter was given in the problem.

25. **C.** To calculate the volume of a rectangular solid, use the formula $V = lwh$. In this problem you are given a length of 48 inches, a width of 18 inches, and a height of 10 in.

$$V = (48)(18)(10)$$
$$V = 8,640\ in^3$$

Next you need to determine how many bags of soil measuring 1,440 in.^{3} are needed to fill the container. Calculate the answer by dividing 8,640 in.^{3} by 1,440 in.^{3}.

$$\frac{8,640}{1,440} = 6\ Bags\ of\ Soil$$

26. **B.** To calculate the average, take the number of tons and divide by the number of months.

$$\frac{30}{12} = 2.5\ tons\ per\ month$$

27. **B.** Multiply the monthly rent by the number of months. There are 24 months in two years.

$$24 \times \$750 = \$18,000$$

28. **A.** Create an equation.

$$\$2,556.36 = x - 0.28x$$

Solve for x.

$$\$2{,}556.36 = 0.72x$$

$$\frac{\$2{,}556.36}{0.72} = \$3{,}550.50$$

29. **C.** Use the formula percent of change = $\frac{\text{amount of change}}{\text{starting point}}$. The amount. of change is the difference between $30,000 and $32,500 or $2,500. The starting point is the original salary of $30,000. Substitute the information into the formula.

$$\text{percent of change} = \frac{\$2{,}500}{\$30{,}000} = 0.083$$

or an 8.3% raise

30. **B.** Multiply 85 years by 0.60 to get 51 years with snow.

PART 3. WORD KNOWLEDGE

1. **D.** *Calamitous* means "relating to misfortune or disaster." The words *ruthless*, *joyous*, and *controlled* do not match the definition. *Disastrous*, choice D, is the correct answer.

2. **C.** To *lure* means "to tempt or entice." The word *repel* is the opposite of this concept. *Drift* also means moving away, not attracting or enticing. *Prepare* is not related to the definition. *Attract*, choice C, is the correct answer.

3. **D.** *Judicious* means "having sound judgment." The words *mistaken* and *misguided* are opposite to that meaning. *Fantastic* is unrelated. *Sensible*, choice D, is the correct answer.

4. **B.** *Enormity* means "great size, scale, or impact." *Perfection*, *fluidity*, and *logic* are unrelated. *Size*, choice B, is the correct answer.

5. **C.** A *foreboding* is a feeling that something bad will happen. The words *serenity* and *calm* are somewhat opposite to this definition. *Distance* is unrelated. *Fear* is closest in meaning to *foreboding*. Choice C is the correct answer.

6. **B.** Kevin probably would not be *happy* if his dog was lost for a week, so clearly that is not correct. He might be *anxious* or *nervous* about the situation, but since *forlorn* means "miserable and wretched and without hope," choice B, *hopeless*, is the best answer.

7. **A.** *Haughty* means "being full of oneself, feeling self-important, or being arrogant." The words *correct*, *wavering*, and *fixed* do not seem related to that idea. *Proud*, choice A, is the closest word to the correct meaning.

8. **C.** To *impede* means "to block or obstruct." An *impediment* is something that blocks or discourages movement or progress. The words *requirement* and *step* do not match the definition. An impediment could be a *challenge*, but the word that most closely reflects the idea of blocking or obstructing is *obstacle*. The correct answer is choice C.

9. **D.** To be *steadfast* means "to be unwavering and sure of one's beliefs or actions." *Wavering* is the opposite of this idea, so it is clearly the wrong answer. The words *correct* and *wrong* seem unrelated, so *firm*, choice D, is the correct answer.

10. **B.** *Vilify* relates to the word *vile*, which means "morally evil, wicked, and degrading." To *vilify* is to say wicked or degrading things about a person or situation. *Respect* and *enhance* are somewhat the opposite in meaning. *Believe* does not relate to the definition, but *abuse* does. Choice B is the correct answer.

11. **A.** *Wry* means "ironic, dry, or sarcastic." *Tedious*, *funny*, and *nasty* do not relate to the definition, so *clever*, choice A, is the best answer.

12. **C.** *Arrogant* means "proud and full of oneself." *Forlorn*, *disappointed*, and *ignorant* have no relation to the definition. *Conceited*, choice C, is the correct answer.

13. **D.** *Estranged* means "changed in one's view from affectionate to indifferent" or "at odds

with another person." The word *together* conveys the opposite meaning. The words *strange* and *punished* do not match the definition. *Separated* is the word that is closest to the meaning. Choice D is the correct answer.

14. **C.** *Gratified* means "pleased or be grateful." The words *hopeful*, *satisfied*, and *disappointed* are not descriptions of being pleased or grateful. *Thankful* is the most appropriate answer. Choice C is correct.

15. **C.** *Irascible* means "easily angered or hot-tempered." *Smooth*, *heavy*, and *agreeable* are clearly not correct. Choice C, *testy*, is correct.

16. **A.** *Obliterated* means "blotted out or destroyed." The words *rebuilt* and *saved* are somewhat opposite to this definition. *Warned* has no relationship to the correct definition. Choice A, *destroyed*, is the correct answer.

17. **C.** *Ransacked* means "searched thoroughly" or "plundered." The words *decorated*, *furnished*, and *rented* do not fit that meaning. *Plundered*, choice C, is the correct answer.

18. **D.** *Vend* means "to sell." A *vendor* is a person or organization that sells. The word *buyer* is opposite to the word *vendor*. *Outlaw* and *machine* are not related. *Salesperson* is the most correct answer from among the options presented. Choice D is the best answer.

19. **C.** A *scavenger* collects or gathers things that are discarded by others. The words *employer*, *authority*, and *relative* are not related to this definition. The word most closely related to this concept is *gatherer*. Choice C is the correct answer.

20. **B.** *Repercussions* are far-reaching effects or reactions to some event. *Applause* and *discontent* are certainly reactions, but they are not necessarily far-reaching. The word *votes* doesn't seem to fit the definition of *repercussion*s. The correct

answer, *consequences*, matches the meaning of a far-reaching effect. Choice B is the correct answer.

21. **D.** *Ex-* is a prefix that means "out of." *Extricate* means "to remove or set free." The words *hear*, *find*, and *elevate* do not relate to the stated definition. *Remove*, choice D, is clearly the correct answer.

22. **A.** *Floundered* means "struggled" or "acted in an awkward way." The words *rushed* and *slept* do not relate to the definition. The word *mumbled* could be a result of floundering, but choice A, *stumbled*, is the correct answer.

23. **D.** To *mortify* means "to humiliate." The words *support*, *question*, and *kill* do not relate to the definition, so choice D, *humiliate*, is the correct answer.

24. **C.** *Inept* means "clumsy or foolish." *Superb* seems to be the opposite of that meaning, and *sloppy* and *funny* do not relate to the definition. *Incompetent*, choice C, is the correct answer.

25. **B.** To *forgo* means "to relinquish, miss, or give up on something." The words *increase*, *reduce*, and *change* do not carry that meaning. Choice B, *skip*, is the best answer.

26. **A.** To *foster* means "to nourish, help develop, or promote." The word *abandon* means somewhat the opposite of *foster*. The words *scold* and *waver* do not match the definition. Choice A, *encourage*, has a similar meaning and is the correct answer.

27. **D.** *Fraught* means "full of or loaded with something." The words *electrified*, *started*, and *targeted* do not relate to the definition. *Filled*, choice D, is the correct answer.

28. **D.** *Frivolous* means "silly, giddy, or of little value." The words *skimpy* and *colorful* do not match the definition. The word *serious* conveys an idea opposite the meaning of *frivolous*. *Playful*, choice D, is the correct answer.

29. **C.** A *gale* is a strong wind. The words *lake*, *ocean*, and *panic* do not match the definition, so *windstorm*, choice C, is the correct answer

30. **C.** *Gaudy* means "bright and showy but lacking in good taste." The words *skimpy*, *expensive*, and *basic* do not share that meaning, so choice C, *flashy*, is the correct answer.

31. **B.** *Gibberish* is spoken or written communication that is unintelligible. The words *photos*, *warnings*, and *errors* do not share that idea, so *nonsense*, choice B, is the correct answer.

32. **D.** To *glower* means "to look angrily or glare at someone." Don't be fooled by the word *enrage*. *Glower* does not mean "to make someone angry," but to look angrily at someone. *Gleam* and *shirk* are clearly incorrect, so *scowl*, choice D, is the correct answer.

33. **A.** To *grapple* means "to wrestle with or to try to cope with a situation or condition." *Struggle*, choice A, has the same meaning and is the correct answer.

34. **C.** A *grievance* is an accusation that something is thought to be unjust or unfair. *Understanding*, *annoyance*, and *question* do not relate to this idea, so choice C, *complaint*, is the correct answer.

35. **B.** *Sequestered* means "set apart or isolated." The only word that shares that meaning is choice B, *separated*.

PART 4. PARAGRAPH COMPREHENSION

1. **D.** The passage indicates that panning for gold was profitable for only a few people, so choices A and B cannot be correct. Only the western United States is mentioned, and the passage does not indicate that gold could be found in many places, so choice C is not correct. The last sentence of the passage states that discoveries of gold were made by people who were familiar with the regions in which they were working. That makes choice D the correct answer.

2. **C.** The word *pertinent* means "having some connection with the subject mentioned." The correct answer is choice C, *relevant*.

3. **C.** There is no indication that prospectors were hopeless people (choice A) or that they could not afford to support a family (choice B). Though some successful prospectors were from the southern Appalachian region (choice D), there is nothing in the passage to suggest that they were the only successful prospectors. Choice C, that only a few prospectors were successful, is the correct answer.

4. **D.** The major idea of the passage is that a lot of people have tried to find gold in the United States, but few people have been successful. The only title that reflects that idea is choice D, "Few Find It—Many Have Tried."

5. **D.** The passage suggests that new, highly sensitive, and relatively inexpensive methods of detecting gold have greatly increased the possibility of discovering gold deposits. Choice D is the correct answer.

6. **D.** The first sentence indicates that farmers who emphasize the use of renewable resources and the conservation of soil and water do so to enhance environmental quality for future generations. So choice D is the correct answer. There is no mention that the food is healthier (choice A) or cheaper (choice C), so those cannot be correct. Choice B is true, but it does not refer to an aspect of organic farming that is helpful. So choice B cannot be correct.

7. **D.** The passage clearly indicates that to be considered organic, food must be grown according to all the rules that meet government standards. Choice D is the correct answer. Choice A is partly true, but it is only one aspect of organic farming. Choice B, using bioengineering, is clearly forbidden by the standards, and there is no claim that the food is healthier (choice C).

8. **A.** The passage indicates that any organic product is required to have a label stating that it is an organic product. The label appears on the product or on a sign above the product display. Choice A is the correct answer.

9. **C.** *Renewable* means "replaceable." Choice C is the correct answer.

10. **D.** Foods that are naturally grown and produced (choice A), that are hormone-free (choice B), and that are grown without chemicals (choice C) might be organic, but the passage indicates that much more is involved. Only foods that are grown according to specific government standards can be labeled organic. Choice D is the correct answer.

11. **A.** The passage does not suggest that the peatlands have been eliminated (choice B) or that people near the Everglades need to use bottled water (choice C). Overpopulation in Florida may have contributed to the problem (choice D), but the passage mostly suggests that activities such as draining and land-use changes have caused the problem. These are human activities, so choice A is the best and correct answer.

12. **B.** The passage indicates that the Everglades is shrinking because of the draining of the water. Choice B is the correct answer.

13. **A.** The passage clearly indicates that the Everglades is controlled by a complex management system that includes canals, levees, and pumps, making it the world's largest artificial plumbing system. Choice A is the correct answer.

14. **C.** Although all the answer options have taken place, the first one mentioned in the passage is the draining of the Everglades. Choice C is the correct answer.

15. **C.** The passage gives you a direct hint at the meaning of *hydrologic*. The passage mentions "hydrologic hazards such as floods." This suggests that *hydrologic* has to do with water. Choice C is the correct answer.

PART 5. MATHEMATICS KNOWLEDGE

1. **D.** Isolate the unknown on one side. Add 12 to each side.

$$y = 24 + 12$$
$$y = 36$$

2. **D.** Perform operations in the parentheses first, then the exponents, and then the remaining operations.

$$8 + 2(7)$$
$$8 + 14 = 22$$

3. **C.** Change the mixed numbers to improper fractions. Carry out the operation by multiplying the two numerators and the two denominators. In this case you can reduce the $\frac{9}{12}$ to $\frac{3}{4}$.

$$\frac{9}{12} \times \frac{5}{4}$$
$$= \frac{3}{4} \times \frac{5}{4}$$
$$= \frac{15}{16}$$

4. **D.** When multiplying numbers in scientific notation, multiply the numbers and add the exponents.

$$2 \times 4.3 = 8.6$$

Add the two exponents.

$$3 + 5 = 8$$

The answer is 8.6×10^8

5. **C.** Change the fraction to a decimal, which is 2.50. Move the decimal point two places to the right and add the percent sign. The result is 250%.

6. **B.** In order to subtract the fractions, they need to be changed to the least common denominator. In this instance the least common denominator is 8, so the fraction $\frac{3}{4}$ becomes $\frac{6}{8}$. But $\frac{6}{8}$ is larger than $\frac{1}{8}$, so, using the notion of regrouping, we can change the 24 to a 23 and add the $\frac{8}{8}$ to the $\frac{1}{8}$, making the fraction $\frac{9}{8}$. Now the $\frac{6}{8}$ can be subtracted from $\frac{9}{8}$ leaving $\frac{3}{8}$. Now subtract the whole numbers, 15 from 23, leaving $8\frac{3}{8}$ as the correct answer.

7. **D.** When subtracting, change the minus sign to a plus sign and change the sign of ALL of the terms/numbers that follow. In this problem the signs for $-(6x^2 + 3y + 8)$ becomes: $+ (-6x^2 - 3y - 8)$.

$$4x^2 + 5y + 6$$
$$+ (-6x^2 - 3y - 8)$$

Now add like terms and get: $-2x^2 + 2y - 2$.

8. **C.** To solve such problems, move all the terms with an unknown to one side of the equals sign and the numbers to the other. In this problem, subtract $4x$ from both sides. This results in $7x - 4x = -12$. Combine the terms to give $3x = -12$, so $x = -4$.

9. **C.** Set each equation to equal zero. Next, arrange the equations so that when one is subtracted from or added to the other, one of the terms results in a zero and drops out of the result. Next, solve for the remaining unknown.

In this problem,

$$6c - 2d - 32$$
$$3c + 2d - 22$$

Set each equation to zero.

$$6c - 2d - 32 = 0$$
$$3c + 2d - 22 = 0$$

Note that if you add the two equations, the d term drops out, leaving $9c - 54 = 0$. Solve for c: $9c = 54$; $c = 6$

Substitute for c in one of the equations to solve for d.

$$6(6) - 2d - 32 = 0$$
$$36 - 2d - 32 = 0$$
$$-2d = 32 - 36$$
$$-2d = -4$$
$$d = 2$$

The correct answer is $c = 6$; $d = 2$.

10. **A.** To divide numbers with exponents, subtract the exponents. In this instance, subtracting 8 from 6 leaves -2. So the correct answer is h^{-2}. This also can be written as $\dfrac{1}{h^2}$.

11. **B.** In this problem you should see two squares. Solve for y:

$$y^2 = 25$$
$$y = 5.$$

12. **B.** To multiply fractions, multiply the numerators and denominators. Simplify where possible.

$$\frac{8a}{5} \times \frac{3}{4y} = \frac{24a}{20y}$$

Simplify to $\dfrac{6a}{5y}$.

13. **B.** To divide fractions, invert the second term and multiply. Simplify where possible.

$$\frac{6a^2}{5b} \div \frac{2b}{9a} \text{ becomes } \frac{6a^2}{5b} \times \frac{9a}{2b}.$$

Multiply the numerators and denominators.

$$\frac{54a^3}{10b^2}$$

Simplify.

$$\frac{27a^3}{5b^2}$$

14. **D.** The cube root is the number that, when multiplied by itself three times, results in the answer. In this instance, $8 \times 8 \times 8 = 512$, so the cube root of $512 = 8$.

15. **D.** Isolate the variable of interest on one side of the equation. In this problem the variable of interest is y. First add n to both sides, resulting in $gy = x + n$. Next, divide each side by g to isolate y. The correct answer is $y = \dfrac{x + n}{g}$.

16. **C.** An isosceles triangle has at least two equal sides and two equal angles, so C is the correct answer.

17. **D.** To find the perimeter of a quadrilateral, add the lengths of the four sides. In this problem, each side measures 4 inches, so the correct answer is 16 in. For a square, just multiply the length of one side by 4 to obtain the correct answer.

18. **D.** The formula for the circumference of a circle is πd. The problem gives the radius, which is $\frac{1}{2}$ the length of the diameter. The diameter is 16, so the correct answer is 16π, or approximately 50.24 inches.

19. **D.** The two known angles total 70°. Subtracting that from 180° gives the correct answer of 110°.

20. **B.** A straight line or straight angle is 180°. If $\angle 1$ is 142°, then $\angle 2$ must be the difference between 180° and 142° or 38°.

21. **C.** $\angle C$ is a right angle, so you can employ the Pythagorean Theorem $a^2 + b^2 = c^2$ to get your answer. $6^2 + 8^2 = c^2$. So $36 + 64 = 100$. $c = 10$, the square root of 100.

22. **B.** In an isosceles triangle, the base is the unequal side and the height is the length of a line from the base to the opposite angle. In this problem the base is 30 inches and the height is 20 inches. The indication that the other two sides are 25 inches each, is irrelevant to this problem.

    The formula for the area of a triangle is $\frac{1}{2}bh$. Substituting the information into the formula gives $\frac{1}{2}(30)(20)$, making the correct answer 300 in.^{2}.

23. **A.** The formula for the area of a circle is πr^2. In this problem you are given the diameter, 10 miles. The radius is half the diameter, 5 miles. Substituting the information into the formula gives $3.14\,(5)^2$, or 78.5 mi^2.

24. **D.** $\angle 1$ and $\angle 2$ make up a straight line of 180°. $\angle 5$ and $\angle 6$ also make up a straight line of 180°. That adds up to 360°.

25. **C.** The formula for finding the volume of a cylinder is $V = \pi r^2 h$. In this problem you are given a radius of 4 cm and a height of 6 cm. Substituting that information into the formula gives $\pi(4)^2\,(6)$, or 301.44 cm^3.

PART 6. ELECTRONICS INFORMATION

1. **C.** Higher voltage is more likely to "push" electricity across air gaps and faulty insulation.

2. **A.** Copper is an excellent conductor; the other materials are all insulators.

3. **D.** This is the symbol for a battery.

4. **C.** Steel conduit both protects wires and supplies the backup ground.

5. **A.** Switches should turn the hot side on and off. They should never control the ground or equipment grounding side.

6. **A.** Current must flow through the entire series circuit, so it must be the same at all points.

7. **C.** Batteries make direct current.

8. **D.** When electrons move in a circuit, an electric current exists.

9. **C.** Rubber is a good insulator; the other materials listed are conductors.

10. **C.** A rectifier converts alternating current to direct current.

11. **A.** This is the symbol for a resistor.

12. **B.** Ohm's law is important, and you should memorize it before taking the ASVAB.

13. **B.** The total voltage drop in the circuit must equal the voltage across the battery, 18V. Because the resistors are the same size, each creates half that voltage drop, or 9V apiece. Thus the voltage at the arrow is 9V.

14. **D.** If the meter detects voltage when connected between hot and ground, the circuit is live.

15. **D.** According to Ohm's law, amperes = volts/ohms = 120/12 = 10.

16. **A.** A switch shuts off all current in a series circuit.

17. **D.** Motors often require a brief surge of current when they start; special motor fuses can handle this surge without blowing.

18. **A.** The switch is open, so the potential across the voltmeter is 0V.

19. **C.** Volts × amps = watts. The watt is the unit that measures electrical power.

20. **A.** This is the symbol for a transistor.

PART 7. AUTO AND SHOP INFORMATION

1. **B.** Diesel engines contain fuel injectors and neither carburetor nor spark plugs. In a diesel engine, the piston compresses the fuel–air mixture hot enough to cause ignition. A cooling system problem is unlikely to cause roughness.

2. **B.** Tire pressure should always be measured when the tire is cold.

3. **B.** Stepping on the clutch disengages the clutch, so you can shift gears.

4. **C.** The thermostat must open after the coolant warms up. If the thermostat is stuck closed, coolant cannot circulate through the engine and radiator to cool the engine, and overheating will result.

5. **D.** The control unit gets input from sensors about gas pedal position and engine conditions. Then it controls the ignition and fuel systems for optimum performance and minimum pollution.

6. **A.** Soot is a sign of incomplete combustion. Why are the other answers incorrect? Choice C: the problem may be with the fuel injectors or ECM, but incomplete combustion is the immediate cause of soot. Choice D: The converter may not be working correctly, but it would not be the first thing to check.

7. **C.** Leaking gaskets are the primary cause of leaks in a stationary vehicle.

8. **B.** The wheel sensor detects wheel rotation and "tells" the antilock brake control that one wheel has stopped turning. The control immediately eases off on that brake, to prevent skidding.

9. **B.** The wheels in a torque converter act like propellers. One wheel moves the transmission fluid, and the other catches the moving fluid and is forced to spin. Choice A is the second-best answer; choice B is better because fans move air, while propellers move water.

10. **A.** Why are the other answers incorrect? Choice B: the camshaft and distributor are not driven by belts. Choice C: a car would not have both a generator and an alternator. Choice D: a belt does not drive the brakes.

11. **B.** The differential is lubricated with a high-viscosity oil.

12. **D.** The camshaft, timing belt, and rocker arm all work together to lift the valve off its seat. Why are the other answers incorrect? Choices A and B: The drive shaft and main bearing have nothing to do with the valves. Choice D: The cylinder head is not a moving part.

13. **B.** Planes remove minor roughness from board edges. Why are the other answers incorrect? Choice A: A wood file leaves a rougher surface. Choice C: A belt sander is hard to use on an edge. Choice D: A wood chisel is not used to smooth wood.

14. **B.** A countersink bit makes a conical hole (a countersink) so a screw head can sit flush.

15. **D.** A center punch prevents the drill bit from wandering before you drill.

16. **A.** Phillips screwdrivers are no stronger than slotted screwdrivers, but they do stay in place on the screw head.

17. **D.** Machinists use a dial gauge to measure surface irregularities; the gauge is too accurate for carpentry.

18. **D.** The wrench is angled so that it does not lie flat against a surface. This improves access in tight quarters and protects your fingers.

19. **C.** This wrench can be adjusted to fit a bolt or nut.

20. **A.** A center punch is used to mark metal before drilling.

21. **B.** A hand saw is awkward for cutting plywood. A coping saw will cut only near the edge. A hacksaw is for cutting metal, not wood.

22. **A.** Rivets can be taken apart only by destroying them.

23. **D.** Hacksaws have fine teeth, and they are only good for cutting metal.

24. **C.** Tightening the screw makes the jaws close tighter.

25. **A.** Hacksaws cut on the down (away) stroke. Why are the other answers incorrect? Choice B: Hacksaw blades are discarded, not sharpened. Choice C: They are made of steel, not tungsten carbide. Choice D: They cut on the away stroke.

PART 8. MECHANICAL COMPREHENSION

1. **B.** The volume of water depends on the cross-sectional area of the pipe. For the 1-inch pipe, the cross-sectional area is $\pi r^2 = \pi(1^2) = \pi$. For the 3-inch pipe, the cross sectional area is $\pi r^2 = \pi(3^2) = 9\pi$. So the 3-inch pipe can carry nine times as much water as the 1-inch pipe.

2. **C.** Pulley B is smaller, so it must rotate faster than Pulley A.

3. **A.** MA = effort arm/load arm = 5/1 = 5. So the force needed to move the 300-lb weight is 300/5 = 60 lb.

4. **B.** The pulley has three working strands and a mechanical advantage of 3. Each strand must support 200 lb.

5. **D.** The tongs shown in choice D are a type of class 3 lever. The screwdriver (choice A) is a type of wheel and axle. The bolt (choice B) is a type of inclined plane. The eggbeater (choice C) is a system of meshed gears.

6. **D.** In this gear system, gear C is rotating counterclockwise. That means that gears B and D are rotating clockwise, and gears A, E, and F are rotating counterclockwise.

7. **A.** Count the supporting strands—the ones that are shortened as you hoist.

8. **A.** If gear A has 24 teeth and gear B has 12 teeth, every time gear B makes one complete revolution, gear A will make 12/24 = ½ of a complete revolution.

9. **B.** Of the items shown, only the pliers are made of metal. Since metal conducts heat better than other substances, when you touch the pliers, heat will be conducted away from your fingers. As a result, the metal pliers will feel colder to the touch than the other items pictured, even though all are actually at the same temperature.

10. **A.** The piano weighs 600 lb so one side carries 300 lb The mechanical advantage is 2, so a 150-lb effort will lift 300 lb.

11. **D.** The amount of water flowing into the tank each minute is 6 + 7 = 13 gallons. At the same time, 8 gallons are flowing out each minute, so the net gain per minute is 13 − 8 = 5 gallons. So in 5 minutes the tank will gain 5 × 5 = 25 gallons of water.

12. **A.** The curved upper surface of the wing forces the air going over it to travel a longer distance.

The air speeds up, moving faster than the air that is going under the wing. This makes the air pressure pressing down on the wing from above less than the air pressure pressing up on the wing from below. The result is the lift force that keeps the airplane airborne.

13. **C.** If the crank makes a ¼ turn, the piston will be at the top of the cylinder. If the crank makes a ½ turn, it will return to the position shown. If the crank makes a ¾ turn, the piston will be at the base of the cylinder. If the crank makes one complete turn, the cylinder will return to the position shown.

14. **D.** Pressure = force/area. Area = $4 \times 4 = 16 \text{ ft}^2$. Pressure = $800/16 = 50 \text{ lb/ft}^2$.

15. **A.** Of the patterns shown, the only one that could form a completely closed cylinder is pattern number 1. The rectangle can be rolled to form the one continuous side of the cylinder, and the two circles form its top and bottom.

16. **C.** The gear labeled 3 is called a *miter gear*. This kind of gear is used to transmit motion between shafts that form a 90° angle.

17. **A.** If gear B rotates twice as fast as gear A, then gear B must have fewer teeth than gear A. The gear ratio of A:B is 2:1.

18. **D.** Pressure = force/area. When more gas is added, it creates more force. Therefore, choice D is the only possible answer.

19. **A.** The load is closer to Sarah, so she is carrying more of it.

20. **A.** This screw has 12 threads per inch, so each turn tight moves it 1/12 in. in the "in" direction.

$$1/12 \times 8 = 8/12 = 2/3 \text{ in.}$$

21. **A.** There are two supporting strands, so the MA is 2. To find the number of supporting strands,

look at how many parts of the rope will get shorter as you tighten the rope.

22. **B.** Pressure is the same at all points in a closed system, so the pressure is 50 lb/in.² in tank B.

23. **A.** Friction causes the simple machine to be less efficient, so less of the rotating force is transferred into compressive force on the board.

24. **C.** The ramp is an inclined plane. The mechanical advantage is 25/5 = 5. So the effort required is 350/5 = 70 lb.

25. **A.** Volume of a cylinder equals height $\times \pi \times$ radius². The same volume of fluid that leaves cylinder A must enter cylinder B. So $h \times \pi \times r^2$ for A = $h \times \pi \times r^2$ for B. Dividing both sides by π gives $h \times r^2 = h \times r^2$ (radius = diameter/2).

    Now substitute:

$$6 \times 1 \times 1 = h \times 2 \times 2$$
$$6 = 4h$$
$$1.5 = h$$

Notice that you can just look at the ratio of the diameters to quickly get this answer: $2^2/4^2 = 1/4$. The large piston moves 1/4 as far as the small piston.

PART 9. ASSEMBLING OBJECTS

1. **C.** The original figure has four objects and choice B has five, so eliminate B. Focus on the most distinctive object, perhaps the long triangle with one jagged side at the bottom right of the original. Choice A does not have the squared-off jagged object. Look for the right triangle. Choice D does not have it. Choice C is the correct answer.

2. **B.** The original figure has three objects and choice C has only 2, so eliminate C. Look for the two identical teardrop shapes. Choices A and D have circles instead. Choice B is the correct answer.

3. **A.** The original has four objects and choice B has five, so eliminate it. Look for the small triangle. Choice D does not have a triangle. Look for the two curved objects. Choice C only has one and it is bigger than either of the original objects. Choice A is the correct answer.

4. **B.** The original has five objects and choice A has four, so eliminate it. Look for the large object with a curved side and a right angle. Choice C does not have that object. Choices B and D have that object, but in choice D the curved object is a mirror image of the original. Choice B is the correct answer.

5. **C.** The original has four objects and choice D has five, so eliminate D. Now find a distinct shape. The large curved object should stand out. In choice A, this object does not have two straight-edged ends. Look for the two equal circles. Choice B does not have circles of equal size. Choice C is the correct answer.

6. **B.** There are five objects in the original and seven in choice D, so eliminate it. Look at the large dome-shaped object on the right. It is not present in choice A. In choice C, the dome is too thin. Choice B is the correct answer.

7. **A.** There are five objects in the original and six in choice B, so eliminate it. Look for the small right triangle. Choice C has an equilateral triangle and choice D does not have a triangle at all. Choice A is the correct answer.

8. **D.** There are three objects in the original and four in choice C, so eliminate C. Look for the long rectangle. It is not present in choice A. Look for the large circle. Choice B has two circles of equal size. Choice D is the correct answer.

9. **C.** The original has four objects, while choice B has only three. Look for the small hexagon. It is not present in choice D. Look for the parallelogram. Choice A has a rectangle. Choice C is the correct answer.

10. **C.** The original figure contains four objects, so eliminate choice A, which has only three. Now focus on one shape at a time. Look for the largest object: the quadrilateral at the lower right. It is not present in choice B. Look for the isosceles triangle. The one in choice D is a right triangle. Choice C is the correct answer.

11. **A.** There are three objects in the original and two in choice C, so eliminate C. Look at the large arch. There are two of them in choice B and it is not present in choice D. Choice A is the correct answer.

12. **B.** The original figure has three objects and choice C has four, so eliminate C. Focus on the most distinctive object: the five-sided object on the right. In choice D there is a square instead. Look for the two identical thin rectangles. Choice A has one that is too wide. Choice B is the correct answer.

13. **B.** Since each figure has four objects, look for each shape. There is one long thin rectangle, but each figure has it. There are two identical rhombuses. Choice C does not have them, so eliminate it. Choice A has three identical rhombuses, so eliminate it as well. Choices B and D have the correct objects, but in choice D the irregular quadrilateral is a mirror image of the original. Choice B is the correct answer.

14. **C.** Since each figure has four objects, look for the most distinctive shape. The object at the bottom of the figure looks like a rectangle with a bump in the middle. Choice A has a similar shape, but the bump is much taller. Look for the small equilateral triangle. Choice D has a tall isosceles triangle instead. Finally, look for the two trapezoids. Choice B has quadrilaterals that are shaped differently. Choice C is the correct answer.

15. **D.** The original has four objects, but choices A and B have five. Look for the large L-shaped object at the lower right. Choice C does not have the same shape. Choice D is the correct answer.

16. **B.** There are three objects in the original and four in choice D, so eliminate D. Look for the most distinctive shape: the isosceles trapezoid. It is not present in choice C. Look for the rectangle. Choice A has a rectangle, but it is much too long. Choice B is the correct answer.

17. **A.** Since each choice has the same number of objects as the original, focus on one object at a time. There is a rounded dome shape in the original that is not the same in choice D. Choice B has a dome, but it is much too small. Look for the quadrilateral with the slanted side at the top right. Choice C has a quadrilateral, but it is not the same shape. Choice A is the correct answer.

18. **D.** Since all the choices have three objects, look for the most distinctive object: the arrow. Choice A does not have it. Look for the small rectangle. Choices B and C do not have one of the proper size. Choice D is the correct answer.

19. **D.** Start with the smaller second object because it is easier to orient. The small object has point B on the center of the smaller curved line. Choices A and B do not join at that point. Choice C is joined at the correct spots, but the larger object is a mirror image of the original. Choice D is the correct answer.

20. **D.** If you picture the first object lying on its larger base, then point A is at the lower right corner. Choice A connects on the lower left corner of that object, and Choice B connects in the middle of that object. On the second object point B is at an internal angle. Choice C has point B at the end of the t-shaped object. Choice D is the correct answer.

21. **A.** With the first object in its original orientation, point A is at the bottom right. This is not true in choices B and C. Point B is at the right angle of the triangle. Choice D does not connect the triangle at that point. Choice A is the correct answer.

22. **D.** Point A is on the inside left corner of the T-shaped object. This point is in the wrong location in choice A (here it is in the inside right corner) and choice B. Point B is in the center of the arch object. Choice C places point B on the perimeter of that object. Choice D is the correct answer.

23. **C.** Point A is at the center of the inside edge on the first object. Point A is not at the correct spot in choices B or D, so eliminate them. Point B is at the lower right point of the second object (in its current orientation). Point B does not connect at this spot in choice A. Choice C is the correct answer.

24. **A.** Since point B is quite distinctive, look for it first. It should be in the center of the large end of the second object. Choice C does not connect at this spot. Now look for point A. In choice B, it is on the wrong corner of the bump on the rectangle. Choice D may appear to connect the objects in the correct spots, but the object with point A is a mirror image of the original. Choice A is the correct answer.

25. **C.** Each object has its designated point at the top tip. Choice A does not connect the second object at that point. Choice B does not connect the first object at the correct point. Choice D may appear to connect the objects in the correct spots, but the second object is a mirror image of the original. Choice C is the correct answer.

ASVAB PRACTICE TEST FORM 2

> ### YOUR GOALS FOR THIS TEST:
>
> * **Take a complete practice ASVAB under actual test conditions.**
> * **Mark your answers on a sample ASVAB answer sheet.**
> * **Check your test answers and read explanations for every question.**
> * **Use your results to check your progress.**

Try to take this practice test under actual test conditions. Find a quiet place to work, and set aside a period of approximately 2½ hours when you will not be disturbed. Work on only one section at a time, and use your watch or a timer to keep track of the time limits for each section. Mark your answers on the answer sheet, just as you will when you take the real exam.

At the end of the test you'll find an answer key and explanations for every question. After you check your answers against the key, you can complete a chart that will show you how you did on the test and what test topics you might need to study more. Then review the explanations, paying particular attention to the ones for the questions that you answered incorrectly.

If you take the CAT-ASVAB, your test will be given by computer and the number of your question, the mode of administration, and the timing will be different. This has been explained in previous chapters.

GENERAL DIRECTIONS

IF YOU ARE TAKING THE PAPER-AND-PENCIL VERSION OF THE ASVAB, THE TEST ADMINISTRATOR WILL READ THE FOLLOWING ALOUD TO YOU:

DO NOT WRITE YOUR NAME OR MAKE ANY MARKS in this booklet. Mark your answers on the separate answer sheet. Use the scratch paper which was given to you for any figuring you need to do. Return this scratch paper with your other papers when you finish the test.

If you need another pencil while taking this test, hold your pencil above your head. A proctor will bring you another one.

This booklet contains 9 tests. Each test has its own instructions and time limit. When you finish a test, you may check your work in that test ONLY. Do not go on to the next test until the examiner tells you to do so. Do not turn back to a previous test at any time.

For each question, be sure to pick the BEST ONE of the possible answers listed. Each test has its own instructions and time limit. When you have decided which one of the choices given is the best answer to the question, blacken the space on your answer sheet that has the same number and letter as your choice. Mark only in the answer space. BE CAREFUL NOT TO MAKE ANY STRAY MARKS ON YOUR ANSWER SHEET. Each test has a separate section on the answer sheet. Be sure you mark your answers for each test in the section that belongs to that test.

Here is an example of correct marking on an answer sheet.

S1. A triangle has

    A. 2 sides

    B. 3 sides

    C. 4 sides

    D. 5 sides

PRACTICE

S1 Ⓐ ● Ⓒ Ⓓ

S2 Ⓐ Ⓑ Ⓒ Ⓓ

S3 Ⓐ Ⓑ Ⓒ Ⓓ

The correct answer to sample question S1 is B.

Next to the item, note how space B opposite number S1 has been blackened. Your marks should look just like this and be placed in the space with the same number and letter as the correct answer to the question. Remember, there is only ONE BEST ANSWER for each question. If you are not sure of the answer, make the BEST GUESS you can. If you want to change your answer, COMPLETELY ERASE your first answer mark.

Answer as many questions as possible. Do not spend too much time on any one question. Work QUICKLY, but work ACCURATELY. DO NOT TURN THE PAGE UNTIL TOLD TO DO SO. Are there any questions?

ASVAB PRACTICE TEST FORM 2

ANSWER SHEET

LAST   FIRST   MI   SSN

PRACTICE

S1 Ⓐ Ⓑ ● Ⓓ   Ⓐ Ⓑ Ⓒ Ⓓ
S2 Ⓐ Ⓑ Ⓒ Ⓓ   Ⓐ Ⓑ Ⓒ Ⓓ
S3 Ⓐ Ⓑ Ⓒ Ⓓ

PART 1-GS
PART 2-AR
PART 3-WK
PART 4-PC
PART 5-MK
PART 6-EI
PART 7-AS
PART 8-MC
PART 9-AO

PART 1. GENERAL SCIENCE

THE TEST ADMINISTRATOR WILL READ THE FOLLOWING ALOUD TO YOU:

Turn to Part 1 and read the directions for General Science silently while I read them aloud. This test has questions about science. Pick the BEST answer for each question and then blacken the space on your separate answer sheet that has the same number and letter as your choice. Here are three practice questions.

S1. A rose is a kind of
   A. animal.
   B. bird.
   C. flower.
   D. fish.

S2. A cat is a
   A. plant.
   B. mammal.
   C. reptile.
   D. mineral.

S3. The earth revolves around the
   A. sun.
   B. moon.
   C. meteorite.
   D. Mars.

Now look at the section of your answer sheet labeled "PRACTICE." Notice that answer space C has been marked for question S1. Now do practice questions S2 and S3 by yourself. Find the correct answer to the question, then mark the space on your answer sheet that has the same letter as the answer you picked. Do this now.

You should have marked B for question S2 and A for question S3. If you make any mistakes, erase your mark carefully and blacken the correct answer space. Do this now.

Your score on this test will be based on the number of questions you answer correctly. You should try to answer every question. If you finish before time is called, go back and check your work in this part <u>ONLY</u>. Now find the section of your answer sheet that is marked PART 1. When you are told to begin, start with question number 1 in Part 1 of your test booklet and answer space number 1 in Part 1 on your separate answer sheet. DO NOT TURN THE PAGE UNTIL TOLD TO DO SO. This part has 25 questions in it. You will have 11 minutes to complete it. Most of you should finish in this amount of time. Are there any questions?

Turn the page and begin.

1. Parts of a cell that store food, waste, and/or water are called
   A. vacuoles.
   B. alleles.
   C. DNA.
   D. bacteria.

2. The organelle responsible for holding hereditary information is the
   A. cell wall.
   B. chloroplasts.
   C. nucleus.
   D. vacuole.

3. Which of the following is not a function of the plant root?
   A. Conduct photosynthesis
   B. Absorb nutrients
   C. Anchor the plant to the soil
   D. Store food

4. Which of the following parts of a plant controls the amount of water within the plant?
   A. Stem
   B. Xylem
   C. Stomata
   D. Flower

5. The eating habits of humans would classify them as
   A. herbivores.
   B. carnivores.
   C. omnivores.
   D. arachnids.

6. Kangaroos and other animals with pouches are considered to be
   A. placentals.
   B. amphibians.
   C. marsupials.
   D. endotherms.

7. Which of the following is not a function of the human skeletal system?
   A. Produce red blood cells
   B. Protect body organs
   C. Create nutrients
   D. Store minerals for later use

8. When you lift a weight with your biceps, which of the following occurs?
   A. Your bicep sends electrical charges to your bone marrow.
   B. Your biceps contract, and your triceps relax.
   C. Your muscle increases in weight.
   D. Your abdominals tighten and relax alternatively.

9. Sunlight promotes the development of which of the following vitamins important for bone development?
   A. Vitamin A
   B. Vitamin C
   C. Vitamin D
   D. Vitamin E

10. Which of the following are the vessels through which blood moves from your heart to other parts of your body?
    A. Veins
    B. Capillaries
    C. Atriums
    D. Arteries

11. The part of the eye that responds to light intensity and color is called the
    A. retina.
    B. iris.
    C. vitreous humor.
    D. lens.

12. What a person actually looks like as a result of his or her genetic makeup is called which of the following?
    A. Genotype
    B. Phenotype
    C. Mutation
    D. Variation

13. Which of the following reasons makes the laser such a good surgical cutting tool?
    A. It makes incisions less prone to bacterial infection.
    B. It makes an incision quickly.
    C. A lot of energy can be directed to a small area.
    D. The light is made of different wavelengths.

14. Which of the following is required in order to change a solid to a liquid?
    A. Chemical compounds
    B. Neutrons
    C. Thermal energy
    D. Extra molecules

15. The tendency of Earth's winds and surface water currents to flow in a clockwise direction in the northern hemisphere is a result of which of the following?
    A. Doppler effect
    B. Convection effect
    C. Coriolis effect
    D. Compression effect

16. Which type of volcano is the steepest, forming from alternating layers of lava and volcanic ash, cinders, and other materials from volcanic explosions?
    A. Composite volcano
    B. Fissure eruptions
    C. Cinder cone
    D. Shield volcano

17. Which of the following represents the way the oceans were formed?
    A. Water flowed from the land to the lower areas of Earth's surface.
    B. Volcanoes released water vapor that eventually condensed to form rain that filled the ocean basins.
    C. Organisms during photosynthesis released oxygen, which then combined with hydrogen creating water that filled the ocean basins.
    D. Water trapped underground was released to the ocean basins after Earth's crust cooled.

18. The strongest and highest tides occur under which of the following conditions?
    A. Hurricane force winds and heavy rains
    B. The perpendicular alignment of the sun, Earth, and moon
    C. The vertical alignment of the sun, Earth, and moon
    D. An earthquake caused by faulting in the ocean floor

19. The term that describes the time when the sun's rays are directly over the equator is
    A. season.
    B. solstice.
    C. equatorial tilt.
    D. equinox.

20. The space between Mars and Jupiter is occupied by which of the following?
    A. Uranus
    B. Asteroids
    C. Comets
    D. Gas nebulas

21. If there were no greenhouse effect, which would likely happen to Earth?
    A. Pollution would decrease.
    B. Earth would get colder.
    C. The ozone layer would increase.
    D. Plant and animal diversity would increase.

22. Which of the following conditions is necessary to create a hurricane?
    A. A cold front in the tropics
    B. A low-pressure area over the tropical ocean
    C. A downdraft caused by air circulation
    D. Surface winds that form over the ocean and move toward land

23. The process by which moisture from a lake enters the atmosphere is called
    A. transpiration.
    B. condensation.
    C. evaporation.
    D. precipitation.

24. The south pole of a magnet is attracted to which of the following?
   A. The geographic south pole
   B. The magnetic north pole
   C. The magnetic south pole
   D. The geographic north pole

25. Which of the following is a substance that speeds up chemical reactions?
   A. Catalyst
   B. Ionic converter
   C. Inhibitor
   D. Rising temperature

STOP! DO NOT TURN THIS PAGE UNTIL TIME IS UP FOR THIS TEST. IF YOU FINISH BEFORE TIME IS UP, CHECK OVER YOUR WORK ON THIS TEST ONLY.

PART 2. ARITHMETIC REASONING

THE TEST ADMINISTRATOR WILL READ THE FOLLOWING ALOUD TO YOU:

Turn to Part 2 and read the directions for Arithmetic Reasoning silently while I read them aloud. This is a test of arithmetic word problems. Each question is followed by four possible answers. Decide which answer is CORRECT, and then blacken the space on your answer sheet that has the same number and letter as your choice. Use your scratch paper for any figuring you wish to do.

Here is a sample question. DO NOT MARK your answer sheet for this or any further sample questions.

S1. A student buys a sandwich for 80 cents, milk for 20 cents, and pie for 30 cents. How much did the meal cost?

    A. $1.00
    B. $1.20
    C. $1.30
    D. $1.40

The total cost is $1.30; therefore, C is the right answer. Your score on this test will be based on the number of questions you answer correctly. You should try to answer every question. DO NOT SPEND TOO MUCH TIME on any one question. If you finish before time is called, go back and check your work in this part ONLY.

Now find the section of your answer sheet that is marked PART 2. When you are told to begin, start with question number 1 in Part 2 of your test booklet and answer space number 1 in Part 2 on your separate answer sheet.

DO NOT TURN THIS PAGE UNTIL TOLD TO DO SO. You will have 36 minutes for the 30 questions. Are there any questions?

Begin.

1. Lyn had a birthday party and spent $9.98 on balloons, $47.23 on party favors, $16.97 on a cake, $21.77 on ice cream, and $15.15 on invitations. How much did Lyn spend on the party?
   A. $97.20
   B. $111.10
   C. $125.20
   D. $127.30

2. Ned's physical fitness program includes a goal of 175 minutes of jogging per week. He ran 35 minutes on Monday, 44 minutes on Wednesday, 17 minutes on Thursday, and 62 minutes on Friday, how many more minutes does he need to run to reach his goal?
   A. 17
   B. 29
   C. 37
   D. 42

3. Bob's CD collection totals 200. If 30 CDs contain jazz music, what percent of his collection is jazz?
   A. 2%
   B. 15%
   C. 27%
   D. 35%

4. Dave is on a kayaking trip. On the first day he travels 15 miles. On the second day he travels 35 miles, and the third day he rests. On the fourth day he travels 57 miles, and on the fifth day he travels 43 miles. What is the average number of miles that he travels per day?
   A. 25
   B. 27
   C. 30
   D. 32

5. Blair plays football. In the last game he threw completed passes of 3 yards, 22 yards, 45 yards, 19 yards, and 16 yards. What is his average?
   A. 17 yards
   B. 19 yards
   C. 21 yards
   D. 22 yards

6. A volleyball team won 25 games during a season of 55 games. What is the ratio of wins to losses?
   A. $\dfrac{5}{11}$
   B. $\dfrac{5}{6}$
   C. $\dfrac{1}{3}$
   D. $\dfrac{1}{2}$

7. A certain relay team is made up of eight members. If each team member runs 1.2 kilometers, how many meters is the race?
   A. 9.6
   B. 116.6
   C. 5,280
   D. 9,600

8. The Ice Cream Manufacturers Association is conducting a survey of favorite ice cream flavors. The results so far are as follow:

Flavor	Number of Persons
Chocolate	602
Rocky Road	411
Vanilla	589
Strawberry	214
Peach	78
Pistachio	514

If these numbers are representative, what is the probability that the next randomly selected person will say that their favorite flavor is chocolate?
   A. $\dfrac{1}{4}$
   B. $\dfrac{1}{3}$
   C. $\dfrac{4}{5}$
   D. $\dfrac{6}{7}$

9. Shanice and Trevon are saving to buy a car. They deposit $10,000 into a savings account that pays simple interest at a rate of 15% per year. If a car costs $17,500, in how many years will they have enough money to buy the car?
   A. 3
   B. 5
   C. 6
   D. 7

10. A manufacturing plant in Detroit makes 80 cars on Monday, 78 on Tuesday, 84 on Wednesday, 77 on Thursday, and 86 on Friday. What is the average number of cars made each day?
    A. 81
    B. 82
    C. 83
    D. 84

11. Wayne has 256 pieces of candy that he needs to pack into boxes that hold 30 pieces each. How many pieces of candy are left over after he fills as many boxes as he can?
    A. 4
    B. 16
    C. 18
    D. 22

12. Leroy wants to tile the floor in his kitchen. The floor is 10 feet by 12 feet. Each tile is 6 by 8 inches. How many tiles will he need to finish the job?
    A. 124 tiles
    B. 360 tiles
    C. 412 tiles
    D. 514 tiles

13. The Metro Fitness Center has a joining fee of $125.00 per person. The Center runs a special promotion every December with a discount of 25% off the usual fee. What is the special joining fee in December?
    A. $80.00
    B. $93.75
    C. $100.00
    D. $112.25

14. Blythe sees a computer that she wants to buy. It costs $1,600. The store manager says that next week the computer will be on sale for $350 less. What percent will Blythe save if she waits a week before purchasing the computer?
    A. 12.8%
    B. 15.3%
    C. 21.9%
    D. 31.3%

15. Two numbers add to 1,500. One number is 4 times the size of the other. What are the two numbers?
    A. 300; 1,200
    B. 200; 1,300
    C. 500; 1,000
    D. 750; 750

16. Miles is starting a tree farm. His plot of land is triangular with one side 36 feet and the other two sides 30 feet each. The height of this triangle-shaped plot is 24 feet. If each tree needs 8 square feet of space to grow, how many trees can Miles plant?
    A. 35
    B. 42
    C. 54
    D. 76

17. A meteor made a circular crater with a circumference of 56.52 meters. What is the diameter of the crater?
    A. 9 m
    B. 12 m
    C. 15 m
    D. 18 m

18. Jamie is three times as old as Francis. If Francis is 14, how old is Jamie?
    A. 39
    B. 42
    C. 44
    D. 47

19. In 1968 the population of the Chicago metropolitan area was 8,435,978. In 1998, the population was 10,544,972. By about what percent did the population increase in 30 years?
    A. 13%
    B. 18%
    C. 21%
    D. 25%

20. Sandra has a piece of ribbon that is 5 yards long. She needs to cut the ribbon into pieces of 18 inches each in order to wrap presents. How many presents can she wrap with the ribbon that she has?
    A. 8
    B. 10
    C. 12
    D. 18

21. Randy is training to run a 12-kilometer race. His training program starts at him running 3 kilometers per day for the first week and increasing that amount by 15% each week. How many kilometers will he be running per day in the fourth week?
    A. 3.45 km
    B. 3.97 km
    C. 4.12 km
    D. 4.57 km

22. A swimming pool is 3 meters deep, 50 meters long, and 25 meters wide. If it takes 0.25 pounds of chlorine for every 15 cubic meters of water to keep the pool clean and healthy, how many pounds of chlorine are needed?
    A. 27.2 pounds
    B. 35.7 pounds
    C. 45.4 pounds
    D. 62.5 pounds

23. A person needs 1,500 milligrams of calcium each day. For breakfast, Joel eats a bowl of cereal with milk that has 250 milligrams of calcium, a bagel with cream cheese that has 100 milligrams of calcium, and a small container of yogurt that has 225 milligrams of calcium. What percent of the daily calcium requirement does he get at breakfast?
    A. 24%
    B. 38%
    C. 44%
    D. 49%

24. A circular baking pan is 4 inches high with a diameter of 16 inches. How much cake batter can it hold?
    A. 404.2 in.^{3}
    B. 512.7 in.^{3}
    C. 788.8 in.^{3}
    D. 803.8 in.^{3}

25. A basketball has a diameter of 12 inches. What is the volume?
    A. 76.2 in.^{3}
    B. 89.52 in.^{3}
    C. 904.32 in.^{3}
    D. 1068.12 in.^{3}

26. A box of cereal contains 45.6 oz of cereal. If the cost for one ounce is 12 cents, how much does the box of cereal cost?
    A. $4.12
    B. $4.56
    C. $5.47
    D. $5.97

27. Melanie's credit card bill was $4,000 in February. If her monthly interest is 18%, what will her bill be in March if she makes no payments and no new purchases?
    A. $4,440
    B. $4,570
    C. $4,720
    D. $4,890

28. In a jar, there are 45 green marbles, 32 red marbles, 13 purple marbles, 16 orange marbles, and 29 yellow marbles. If Duke selects one marble at random, what is the probability that the marble will be red or purple?
    A. 1:3
    B. 2:5
    C. 3:5
    D. 3:7

29. A corner table is the shape of a right triangle. One side is 3 feet, another side is 4 feet, and the third side is 5 feet. What is the area of the table top?
    A. 6 ft^2
    B. 7.2 ft^2
    C. 8 ft^2
    D. 8.5 ft^2

30. Arnold signs an agreement with the A-1 car rental company for a car at $25 per day plus 15 cents per mile driven. If Arnold rents the car for 2 days and drives 200 miles, how much does he owe the company?
    A. $55
    B. $75
    C. $80
    D. $95

STOP! DO NOT TURN THIS PAGE UNTIL TIME IS UP FOR THIS TEST. IF YOU FINISH BEFORE TIME IS UP, CHECK OVER YOUR WORK ON THIS TEST ONLY.

PART 3. WORD KNOWLEDGE

THE TEST ADMINISTRATOR WILL READ THE FOLLOWING ALOUD TO YOU:

Now turn to Part 3 and read the directions for Word Knowledge silently while I read them aloud.

This is a test of your knowledge of word meanings. These questions consist of a sentence or phrase with a word or phrase underlined. From the four choices given, you are to decide which one MEANS THE SAME OR MOST NEARLY THE SAME as the underlined word or phrase. Once you have made your choice, mark the space on your answer sheet that has the same number and letter as your choice.

Look at the sample question.

S1. The weather in this geographic area tends to be <u>moderate</u>.

    A. Severe
    B. Warm
    C. Mild
    D. Windy

The correct answer is "mild," which is choice C. Therefore, you would have blackened in space C on your answer sheet.

Your score on this test will be based on the number of questions you answer correctly. You should try to answer every question. DO NOT SPEND TOO MUCH TIME on any one question. If you finish before time is called, go back and check your work in this part <u>ONLY</u>.

Now find the section of your answer sheet that is marked PART 3. When you are told to begin, start with question number 1 in Part 3 of your test booklet and answer space number 1 in Part 3 on your separate answer sheet.

DO NOT TURN THE PAGE UNTIL TOLD TO DO SO. You will have 11 minutes to complete the 35 questions in this part. Are there any questions?

Begin.

1. The comedy was <u>hilarious</u>.
   A. Confusing
   B. Repellent
   C. Serious
   D. Funny

2. Richard was <u>wary</u> about what he was being told.
   A. Enthusiastic
   B. Cautious
   C. Nervous
   D. Happy

3. He felt a <u>tremor</u> in his hand.
   A. Trembling
   B. Pain
   C. Warmth
   D. Object

4. <u>Ingenious</u> most nearly means
   A. destructive.
   B. threatening.
   C. clever.
   D. sorrowful.

5. <u>Illusory</u> most nearly means
   A. real.
   B. first.
   C. deceptive.
   D. bright.

6. <u>Thrive</u> most nearly means
   A. move.
   B. separate.
   C. think.
   D. flourish.

7. He was able to <u>articulate</u> the meaning of the paragraph.
   A. Explain
   B. Change
   C. Contradict
   D. Support

8. Tom felt <u>reticent</u> about expressing his opinion.
   A. Hopeful
   B. Encouraged
   C. Restrained
   D. Hopeless

9. Carrying those packages was <u>cumbersome</u> for Julie.
   A. Clumsy
   B. Easy
   C. Difficult
   D. Annoying

10. The argument was <u>coherent</u>.
    A. Misguided
    B. Logical
    C. Mixed
    D. Unfavorable

11. The discussion between the two was <u>contentious</u>.
    A. Quarrelsome
    B. Agreeable
    C. Laughable
    D. Interesting

12. The data started to <u>converge</u> into a possible theory.
    A. Dissolve
    B. Join
    C. Form
    D. Slide

13. <u>Invert</u> most nearly means
    A. rebuke.
    B. lecture.
    C. overturn.
    D. break.

14. Judy <u>feigned</u> interest in the subject.
    A. Lost
    B. Showed
    C. Pretended
    D. Disregarded

15. The new measurements <u>refute</u> the original hypothesis.
    A. Disprove
    B. Support
    C. Question
    D. Direct

16. <u>Impetuous</u> most nearly means
    A. natural.
    B. impulsive.
    C. slow.
    D. severe.

17. He had a <u>propensity</u> for taking criticism to heart.
    A. Fault
    B. Awareness
    C. Aversion
    D. Tendency

18. He <u>spurned</u> the accusation.
    A. Moved
    B. Rejected
    C. Scattered
    D. Accepted

19. During the fall, squirrels tend to <u>hoard</u> acorns.
    A. Eat
    B. Stockpile
    C. Share
    D. Hide

20. Their relationship was <u>tempestuous</u>.
    A. Friendly
    B. Adversarial
    C. Stormy
    D. Calm

21. <u>Charade</u> most nearly means
    A. pretense.
    B. nervousness.
    C. subtlety.
    D. contamination.

22. The speech he made was <u>incoherent</u>.
    A. Jumbled
    B. Nasty
    C. Simple
    D. Incorrect

23. His attitude <u>inhibited</u> him from moving forward.
    A. Pushed
    B. Prevented
    C. Scared
    D. Questioned

24. The atmosphere in the boardroom was one of <u>levity</u>.
    A. Seriousness
    B. Lightheartedness
    C. Anger
    D. Distrust

25. <u>Malady</u> most nearly means
    A. illness.
    B. heart.
    C. comedy.
    D. strength.

26. In the rainstorm, the umbrella was <u>unwieldy</u>.
    A. Heavy
    B. Useless
    C. Cumbersome
    D. Colorful

27. His action was a serious <u>transgression</u> of their agreement.
    A. Mistake
    B. Misspelling
    C. Diversion
    D. Violation

28. The paragraph was a <u>preamble</u> to the story.
    A. Summary
    B. Preface
    C. Ending
    D. Interruption

29. <u>Artificial</u> most nearly means
    A. false.
    B. creative.
    C. faithful.
    D. late.

30. The new rules caused a <u>rebellion</u> in the office.
    A. Meeting
    B. Parade
    C. Revolt
    D. Celebration

31. <u>Benefactor</u> most nearly means
    A. supporter.
    B. partner.
    C. employer.
    D. counselor.

32. The beach was <u>secluded</u>.
    A. Windy
    B. Rocky
    C. Narrow
    D. Hidden

33. Good health is a <u>prerequisite</u> for a physically active life.
    A. Requirement
    B. Outcome
    C. Cause
    D. Desire

34. The grade was a <u>blemish</u> on her perfect record.
    A. Notation
    B. Inspiration
    C. Defect
    D. Entry

35. His argument was based on <u>hearsay</u>.
    A. Facts
    B. Statements
    C. Rumor
    D. Measurements

STOP! DO NOT TURN THIS PAGE UNTIL TIME IS UP FOR THIS TEST. IF YOU FINISH BEFORE TIME IS UP, CHECK OVER YOUR WORK ON THIS TEST ONLY.

PART 4. PARAGRAPH COMPREHENSION

THE TEST ADMINISTRATOR WILL READ THE FOLLOWING ALOUD TO YOU:

Turn to Part 4 and read the directions for Paragraph Comprehension silently while I read them aloud.

This is a test of your ability to understand what you read. In this section you will find one or more paragraphs of reading material followed by incomplete statements or questions. You are to read the paragraph and select one of four lettered choices which BEST completes the statement or answers the question. When you have selected your answer, blacken the space on your answer sheet that has the same number and letter as your answer.

Your score on this test will be based on the number of questions you answer correctly. You should try to answer every question. DO NOT SPEND TOO MUCH TIME on any one question. If you finish before time is called, go back and check your work in this part ONLY.

Now find the section of your answer sheet that is marked PART 4. When you are told to begin, start with question number 1 in Part 4 of your test booklet and answer space number 1 in Part 4 on your separate answer sheet.

DO NOT TURN THE PAGE UNTIL TOLD TO DO SO. You will have 13 minutes to complete the 15 questions in this part. Are there any questions?

Begin.

Everyone wants to protect children. Yet when it comes to the sensitive issue of childhood obesity, too often we fall silent. We need to break the silence and lay open for the country the hard facts and necessary choices we need to make to deal with what has become a quiet epidemic in America. The simple fact is that more people die in the United States of too much food than of too little, and the habits that lead to this epidemic become ingrained at an early age.

Everyone knows the statistics: obesity and excessive weight affect 10 million U.S. children. That's a record, and there's no real sign that it won't be broken again soon. In the past 20 years, the number of obese children has doubled, placing more Americans at risk of high cholesterol, high blood pressure, heart disease, diabetes, arthritis, and cancer—all at an earlier age. Obesity contributes to 300,000 deaths each year. That's close to 1,000 lives lost each day at a cost to our healthcare system of $70 billion a year, or 8 percent of all medical bills.

We need to take this issue seriously. For at least one in five kids, excessive weight is not a cute phase that will be outgrown. It's the start of a lifetime of serious health problems. It is time we elevate this issue to its rightful place near the top of the public health agenda alongside cancer, heart disease, and other leading killers of Americans today.

Solving the problem is a complex issue because it involves some very sensitive areas: personal choice, culture, and economic status. Children soak up the wrong lessons not just from TV, but also from the one in three adults who is overweight. The apple isn't falling far from the tree, here. So we can't blame only fast-food restaurants or the media for the problem. We need to have a strong educational program so that children and their parents know the facts.

The only way kids will succeed is if they have easier access to healthy foods, fewer temptations, and role models that set the right example both with eating habits and exercise. Success almost always hinges on changing the whole family's eating habits, and that is good for everyone.

1. In this passage, the author is attempting to point out which of the following?
   A. Reasons to combat childhood obesity
   B. Statistics about obesity
   C. The causes of obesity
   D. The definition of the word *obesity*

2. Which of the following does the passage suggest as a solution to the obesity problem?
   A. Education of parents and children
   B. Mass media campaigns
   C. Elimination of school vending machines
   D. Personal choice

3. According to the passage, childhood obesity
   A. is influenced mostly by parents and schools.
   B. will probably continue to set records.
   C. will not be a problem in the future.
   D. can be eliminated with exercise.

4. The passage indicates that obesity results in
   A. 300,000 deaths per day.
   B. a society that does not exercise.
   C. public admiration for cute children.
   D. $70 billion in healthcare costs.

5. The passage indicates that the number of adults who are overweight is
   A. one in three.
   B. one in four.
   C. one in five.
   D. one in ten.

Pocahontas was the daughter of Powhatan, an important chief of the Algonquian Native Americans (the Powhatans) who lived in the Virginia region. Her real name was Matoaka. Pocahontas was a nickname meaning "playful" or "mischievous one."

Pocahontas was only about 10 years old when her world changed forever. English settlers arrived from far across the ocean and created a settlement at Jamestown, Virginia. These new English settlers looked and acted very differently from Powhatan's tribe. Some of Pocahontas's people were afraid or even hateful of the newcomers. But the chief's daughter had a curious mind and a friendly manner. She wanted to know more about these newcomers.

Pocahontas got to know and make friends with the new colonists. Her warm nature and natural curiosity led the English to like and trust her as well. One of the colonists, Captain John Smith, said that her appearance, intelligence, and friendly personality "much exceedeth any of the rest of Powhatan's people." But not all of Powhatan's people were so curious and friendly. In December 1607, Captain Smith was captured and held at Chief Powhatan's capital, Werowocomoco.

Pocahontas is most famous for reportedly saving the life of English Captain John Smith. Throughout her short life (she died at the age of 22), however, she was important in other ways as well. Pocahontas tried to promote peace between the Powhatans and the English colonists. She even converted to Christianity and married John Rolfe, a Jamestown colonist, a union which helped bring the two groups together. Her untimely death in England hurt the chance for continued peace in Virginia between the Algonquians and the colonists.

6. According to the passage, Pocahontas might best be remembered as a
   A. warrior.
   B. mediator.
   C. daughter.
   D. child.

7. According to the passage, the nickname "Pocahontas" means
   A. beautiful child.
   B. white rose.
   C. playful cloud.
   D. mischievous one.

8. Pocahontas can be described by which of the following words?
   A. Baby
   B. Friendly
   C. Superior
   D. Hateful

9. According to the passage, what was one of the major reasons Pocahontas married John Rolfe?
   A. To promote peace
   B. To bear children
   C. To increase her influence
   D. To ensure her status in her tribe

10. According to the passage, John Rolfe was which of the following?
    A. A messenger
    B. A warrior
    C. A trader
    D. A colonist

Conflict over issues of how much control the federal government should have over the states, industrialization, trade, and especially slavery had increased tension between Northern and Southern states. After Abraham Lincoln was elected president in 1860, turmoil arose. One after another, Southern states were seceding from the United States of America.

On January 10, 1861, Florida delegates who were meeting in the state capital, Tallahassee, voted to secede from the United States. Florida then became one of six Southern states that formed a new independent government, the Confederate States of America. Five more Southern states would soon join the original six, making a total of 11 Confederate states. These events led to the outbreak of the Civil War—a brutal, bloody, four-year conflict that left the South defeated and ended slavery at the cost of more than half a million lives.

Though Florida had the smallest population of the Southern states, some 16,000 Floridians fought in the Civil War. While this was a small number when compared to the forces provided by other Southern states, it was the highest percentage of available men of military age from any Confederate state. The state also provided resources valuable to the Confederate cause. Florida's coastline had important harbors, and its products, such as sugar, pork, and salt, helped to feed the Southern soldiers.

Very little military action took place in Florida, except in a couple of coastal cities. When Confederate General Robert E. Lee surrendered in 1865, Tallahassee was the only Southern capital held by rebel forces. What do you and your family think of when you think of Florida? The Sunshine State? Oranges? Beaches? Walt Disney World? You might not think of Florida's Civil War history.

11. Which of the following would be the best title for this passage?
    A. Sunny State—Confederate History
    B. Florida Secedes
    C. Florida Leaves the Union
    D. Florida Feeds the Soldiers

12. According to the passage, what was the major reason Florida seceded from the Union?
    A. Territorial issues related to its borders
    B. Issues related to feeding Confederate soldiers
    C. Florida's weather
    D. Disagreements over slavery

13. According to the passage, how many Floridians fought in the Civil War?
    A. 5,000
    B. 7,500
    C. 11,000
    D. 16,000

14. How many states eventually left the Union?
    A. 6
    B. 9
    C. 10
    D. 11

15. According to the passage, which of the following is true?
    A. Florida was the first state to leave the Union.
    B. Florida provided the highest percentage of soldiers of any Confederate state.
    C. Most of the battles in the South took place in Florida's coastal cities.
    D. Florida's position on slavery and government control contributed to the end of the Civil War.

STOP! DO NOT TURN THIS PAGE UNTIL TIME IS UP FOR THIS TEST. IF YOU FINISH BEFORE TIME IS UP, CHECK OVER YOUR WORK ON THIS TEST ONLY.

PART 5. MATHEMATICS KNOWLEDGE

THE TEST ADMINISTRATOR WILL READ THE FOLLOWING ALOUD TO YOU:

Now turn to Part 5 and read the directions for Mathematics Knowledge silently while I read them aloud.

This is a test of your ability to solve general mathematics problems. You are to select the correct response from the choices given. Then mark the space on your answer sheet that has the same number and letter as your choice. Use your scratch paper to do any figuring you wish to do.

Your score on this test will be based on the number of questions you answer correctly. You should try to answer every question. DO NOT SPEND TOO MUCH TIME on any one question. If you finish before time is called, go back and check your work in this part ONLY.

Now find the section of your answer sheet that is marked PART 5. When you are told to begin, start with question number 1 in Part 5 of your test booklet and answer space number 1 in Part 5 on your separate answer sheet.

DO NOT TURN THE PAGE UNTIL TOLD TO DO SO. You will have 24 minutes to complete the 25 questions in this part. Are there any questions?

Begin.

1. If $z + 12 = 24$, $z = ?$
   A. 6
   B. 12
   C. 24
   D. 36

2. $4^2 + 3(2 - 5) =$
   A. 7
   B. 12
   C. -4
   D. -12

3. 30 is what percent of 90?
   A. 30
   B. 33
   C. 45
   D. 50

4. What is the mean of the following numbers?

   14, 24, 12, 16, 34, 104

   A. 14
   B. 24
   C. 34
   D. 44

5. Divide $\dfrac{1}{3}$ by $\dfrac{5}{6}$.

   A. $\dfrac{2}{5}$

   B. $\dfrac{5}{18}$

   C. $\dfrac{15}{6}$

   D. $2\dfrac{1}{2}$

6. Factor the following expression:

   $x^2 - 100$

   A. $(x + 10)(x + 10)$
   B. $2x(100)$
   C. $(x - 10)(x - 10)$
   D. $(x - 10)(x + 10)$

7. Subtract:

   $$12x^2 + 6y - 3$$
   $$-(3x^2 - 2y + 12)$$

   A. $9x^2 + 8y - 15$
   B. $15x^4 + 4y + 9$
   C. $15x^2 + 4y + 9$
   D. $9x^2 + 4y + 9$

8. Solve for x.

   $$\frac{r}{j} = \frac{x}{y}$$

   A. $\dfrac{yr}{j} = x$

   B. $x = \dfrac{rx}{jy}$

   C. $x = \dfrac{rx}{jy}$

   D. $jy = x$

9. Solve for the two unknowns.

   $3a + 4b = 12$
   $2a + 2b = 15$

   A. $a = 12$; $b = 10$
   B. $a = -10$; $b = 12$
   C. $a = -4$; $b = 9.5$
   D. $a = 18$; $b = -10.5$

10. $\dfrac{x^4 y^6}{x^7 y^2} =$

    A. $x^{-3}y^8$

    B. $x^3 y^8$

    C. $\dfrac{y^4}{x^3}$

    D. $\dfrac{y^4}{x^4}$

11. Solve for g.

    $3g^2 + 4g + 1 = 0$

    A. $g = 4$; $g = -1$
    B. $g = 4$; $g = 1$
    C. $g = 1/3$; $g = -1$
    D. $g = -1/3$; $g = -1$

12. Multiply:

$$\frac{6}{x+1} \times \frac{3x+3}{5}$$

A. $\dfrac{18x+18}{5x+1}$

B. $3\dfrac{3}{5}$

C. $30(3x^2+1)$

D. $18x^2+30$

13. Divide:

$$\frac{5}{x+1} \div \frac{6}{3x+3}$$

A. $\dfrac{30}{3x^2+6x+3}$

B. $\dfrac{30}{(x+1)(3x+3)}$

C. $\dfrac{15x}{6x^2}$

D. $2\dfrac{1}{2}$

14. $\sqrt[2]{121} =$

A. 3

B. 5

C. 6

D. 11

15. Solve for y.

$$\frac{v}{y} = \frac{s}{t}$$

A. $y = \dfrac{vs}{yt}$

B. $y = \dfrac{s}{vt}$

C. $y = \dfrac{vt}{s}$

D. $y = vst$

16. Which of the following is a hexagon?

A.

B.

C.

D.

17. What is the perimeter of the following parallelogram?

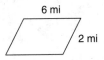

A. 9 mi

B. 12 mi

C. 16 mi

D. 20 mi

18. Round the following number to the nearest hundredth.

1,478.964

A. 1,478.99

B. 1,500.00

C. 1,478.96

D. 1,478.970

19. If the following is a right triangle and $\angle 2$ is 35°, what is the measure of $\angle 1$?

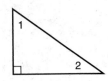

A. 25°

B. 55°

C. 65°

D. 75°

20. In the following figure, if ∠2 is 29°, what is the measure of ∠4?

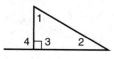

 A. 61°
 B. 85°
 C. 90°
 D. 119°

21. In the following right triangle, what is the length of side *AC*?

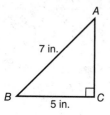

 A. 6 in.
 B. 12 in.
 C. √24 in.
 D. √35 in.

22. What is the area of the following rectangle?

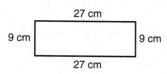

 A. 36 cm²
 B. 99 cm²
 C. 121.5 cm²
 D. 243 cm²

23. What is the area of the following circle?

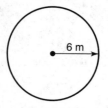

 A. 18.84 m²
 B. 97.02 m²
 C. 113.04 m²
 D. 452.16 m²

24. What is 1% of 5,000?
 A. 5
 B. 10
 C. 50
 D. 105

25. What is the volume of a sphere that has a radius of 3 meters?
 A. 12.56 m³
 B. 37.68 m³
 C. 75.36 m³
 D. 113.04 m³

STOP! DO NOT TURN THIS PAGE UNTIL TIME IS UP FOR THIS TEST. IF YOU FINISH BEFORE TIME IS UP, CHECK OVER YOUR WORK ON THIS TEST ONLY.

PART 6. ELECTRONICS INFORMATION

THE TEST ADMINISTRATOR WILL READ THE FOLLOWING ALOUD TO YOU:

Now turn to Part 6 of your test booklet and read the directions for Electronics Information silently while I read them aloud.

This is a test of your knowledge of electrical, radio, and electronics information. You are to select the correct response from the choices given and then mark the space on your answer sheet that has the same number and letter as your choice.

Your score on this test will be based on the number of questions you answer correctly. You should try to answer every question. DO NOT SPEND TOO MUCH TIME on any one question. If you finish before time is called, go back and check your work in this part ONLY.

Now find the section of your answer sheet that is marked PART 6. When you are told to begin, start with question number 1 in Part 6 of your test booklet and answer space number 1 in Part 6 on your separate answer sheet.

DO NOT TURN THE PAGE UNTIL TOLD TO DO SO. You will have 9 minutes to complete the 20 questions in this part. Are there any questions?

Begin.

1. When an air compressor starts up, other lights on the circuit often dim because
   A. motors need large amperage to start.
   B. motors need large voltage when starting.
   C. motors need large resistance when starting.
   D. most circuits are not designed for motors.

2. A capacitor is used to store
   A. voltage.
   B. current.
   C. capacitance.
   D. resistance.

3. A 480-volt circuit has a resistance of 24 ohms. How much current is flowing?
   A. 0.05 amperes
   B. 20 amperes
   C. 48 amperes
   D. Can't tell from this information

4. The electrical property that is usually referred to as pressure is
   A. current.
   B. resistance.
   C. amperage.
   D. voltage.

5. A meter at point A will read

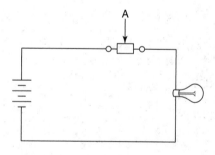

   A. voltage.
   B. resistance.
   C. current.
   D. capacitance.

6. In an electrical diagram, the following symbol represents a(n)

   A. transformer.
   B. electromagnet.
   C. battery.
   D. capacitor.

7. How much power is being carried in a 120-volt circuit when the current is 12 amps?
   A. 10 watts
   B. 240 watts
   C. 1,200 watts
   D. 1,440 watts

8. A dishwasher on a 120-volt circuit is consuming 1,800 watts of power. What is the amperage?
   A. 1.5
   B. 15
   C. 120
   D. 1,680

9. An electrical wire is usually made by
   A. stretching a conductor until it's thin enough to make wire.
   B. wrapping a conductor around an insulator.
   C. wrapping an insulator around a conductor.
   D. transforming a conductor into an insulator.

10. Many computer parts use gold-plated contacts because gold
    A. resists corrosion, ensuring a good contact.
    B. has a high impedance.
    C. adds prestige to the computer.
    D. is heat-resistant.

11. Which of these devices would you use to control the flow of current?
    A. Generator
    B. Transistor
    C. Alternator
    D. Capacitor

12. A material that allows free flow of electrons is called a(n)
    A. insulator.
    B. resistor.
    C. semiconductor.
    D. conductor.

13. A lead-acid storage battery presents an explosion hazard if you
    A. undercharge it.
    B. add too much water.
    C. add too little water.
    D. get a spark near it.

14. One kilowatt hour is the same as
    A. 1,000 watts for 1 hour
    B. 10 watts for 10 hours
    C. 1,000 watts for 1,000 hours
    D. 1 watt for 1 hour

15. 12-gauge wire is _____ than 16-gauge wire.
    A. more flexible
    B. longer
    C. thicker
    D. thinner

16. What is the total resistance in this circuit?

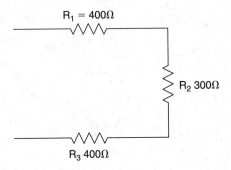

    A. 0.11 ohm
    B. 11 ohms
    C. 425 ohms
    D. 1,100 ohms

17. What is the approximate total resistance in this circuit?

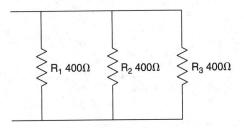

    A. 0.133 ohm
    B. 33 ohms
    C. 133.3 ohms
    D. 3,300 ohms

18. When the primary side of a transformer has more turns than the secondary side, the output voltage will be _____ the input voltage.
    A. lower than
    B. higher than
    C. the same as
    D. a multiple of

19. Which of these devices contains an electromagnet?
    A. Incandescent light bulb
    B. Fluorescent light bulb
    C. Light switch
    D. Doorbell

20. Current that changes polarity many times per second is called
    A. direct current.
    B. alternating current.
    C. AC/DC current.
    D. reverse-polarity current.

STOP! DO NOT TURN THIS PAGE UNTIL TIME IS UP FOR THIS TEST. IF YOU FINISH BEFORE TIME IS UP, CHECK OVER YOUR WORK ON THIS TEST ONLY.

PART 7. AUTO AND SHOP INFORMATION

THE TEST ADMINISTRATOR WILL READ THE FOLLOWING ALOUD TO YOU:

Now turn to Part 7 and read the directions for Auto and Shop Information silently while I read them aloud.

This test has questions about automobiles, shop practices, and the use of tools. Pick the BEST answer for each question and then mark the space on your answer sheet that has the same number and letter as your choice.

Your score on this test will be based on the number of questions you answer correctly. You should try to answer every question. DO NOT SPEND TOO MUCH TIME on any one question. If you finish before time is called, go back and check your work in this part ONLY.

Now find the section of your answer sheet that is marked PART 7. When you are told to begin, start with question number 1 in Part 7 of your test booklet and answer space number 1 in Part 7 on your separate answer sheet.

DO NOT TURN THE PAGE UNTIL TOLD TO DO SO. You will have 11 minutes to complete the 25 questions in this part. Are there any questions?

Begin.

1. If you hear a "ping" or knocking sound on acceleration, what is the most likely cause?
   A. The fuel–air mixture is too rich.
   B. The fuel–air mixture is too lean.
   C. The gasoline octane is too low.
   D. The gasoline octane is too high.

2. If the antifreeze is defective,
   A. the engine may overheat.
   B. the engine block or cylinder head may crack.
   C. the engine may not warm up enough.
   D. the defroster will stop working.

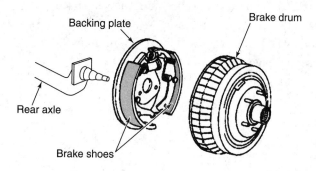

Backing plate
Brake drum
Rear axle
Brake shoes

3. Problems with this system
   A. can prevent a vehicle from stopping fast enough.
   B. should be checked out when convenient.
   C. may cause engine overheating.
   D. will always cause steering problems.

4. What is not an advantage of front-wheel drive?
   A. Improved traction in mud, rain, and snow
   B. Greater traction on the driving wheels
   C. More interior room
   D. Better tire life on the front wheels

5. If engine oil gets too low, you can expect more
   A. corrosion.
   B. friction.
   C. noise.
   D. exhaust.

6. If there are problems with the valve train, you can expect
   A. overheating.
   B. loss of power.
   C. engine roughness.
   D. B and C.

7. If a vehicle loses acceleration,
   A. check the fuel.
   B. horsepower is falling off for some reason.
   C. torque is falling off for some reason.
   D. press harder on the gas pedal.

8. All other things being equal, a heavier car will
   A. have more headroom.
   B. be harder to start in winter.
   C. use more gasoline per mile.
   D. have more gears.

9. If an engine turns over too slowly in the winter,
   A. the starter motor may need replacement.
   B. the battery may be losing its charge.
   C. the oil may be too heavy.
   D. B and C.

10. When jump starting a car, be sure to
    A. connect the ground cable to the front axle.
    B. remove all the battery caps.
    C. connect the terminals positive to positive, and negative to negative.
    D. keep the transmission in drive.

11. If a manual transmission "grinds" during a normal gearshift, check the
    A. synchronizers.
    B. clutch adjustment.
    C. fluid in the torque converter.
    D. A and B.

12. If an engine seizes up (cannot turn), you might suspect problems with the
    A. lubrication system.
    B. cooling system.
    C. main bearings.
    D. all the above.

13. What type of screw head would best hold a flat plate to a joist?
    A. Round head or pan head
    B. Phillips head
    C. Flat head
    D. Any head

14. The tool shown below is a

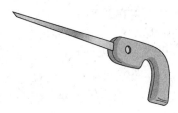

    A. hacksaw.
    B. prybar saw.
    C. dowel saw.
    D. keyhole saw.

15. When applying a glossy paint to wood, what is the correct order of operations?
    A. Prime, paint, repaint
    B. Prime, sand, paint
    C. Sand, prime, sand, paint
    D. Sand, prime, paint

16. When fastening drywall,
    A. all joints should occur on top of a stud.
    B. vertical joints should occur on top of a stud.
    C. joints should overlap slightly.
    D. joints should be avoided.

17. When sanding drywall, always
    A. use a belt sander.
    B. start with medium sandpaper.
    C. sand with the grain.
    D. wear a dust mask.

18. The tool shown below is a(n)

    A. box wrench.
    B. open-end wrench.
    C. arc-joint wrench.
    D. combination wrench.

19. When an electric grinder spins at 4,500 rpm,
    A. it should not use fine-grained wheels.
    B. it would be smart to use a 3,000 rpm wheel, which would cut even faster at 4,500 rpm.
    C. you must use water to cool the metal you are grinding.
    D. the metal you are grinding will heat up quickly.

20. For which group of operations would you use a wire brush?
    A. Removing paint from metal, removing rust, cleaning bolt threads
    B. Stripping paint, dulling paint before repainting, cleaning a wood deck
    C. Preparing drywall for painting, removing rust
    D. Cleaning threads inside a nut, removing paint from metal, dulling paint before repainting

21. The tool shown below is best used for

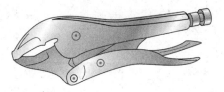

    A. turning bolts.
    B. clamping.
    C. cutting sheet metal.
    D. pounding a chisel.

22. The standard size of a sheet of plywood, drywall, or oriented-strand board
    A. varies
    B. is 4×8 ft
    C. is 1/2 in. thick
    D. is 5/8 in. $\times 4$ ft $\times 8$ ft

23. What is the purpose of galvanizing?
    A. To prevent oxidation on steel
    B. To smooth a steel surface to reduce friction
    C. To harden steel
    D. To sharpen steel

24. Which items have a similar mechanical role?
    A. Sheet-metal screw, screen door operator
    B. Deadbolt in lock, toggle in wall switch
    C. Door hinge pin, auto axle
    D. Door weather-strip, door lock

25. Which group of tools has the same function?
    A. Rasp, plane, sandpaper
    B. Screwdriver, awl, prybar
    C. Drill, auger, jigsaw
    D. Band saw, hole saw, jigsaw

STOP! DO NOT TURN THIS PAGE UNTIL TIME IS UP FOR THIS TEST. IF YOU FINISH BEFORE TIME IS UP, CHECK OVER YOUR WORK ON THIS TEST ONLY.

PART 8. MECHANICAL COMPREHENSION

THE TEST ADMINISTRATOR WILL READ THE FOLLOWING ALOUD TO YOU:

Now turn to Part 8 and read the directions for Mechanical Comprehension silently while I read them aloud.

This test has questions about mechanical and physical principles. Study the picture and decide which answer is CORRECT and then mark the space on your separate answer sheet that has the same number and letter as your choice.

Your score on this test will be based on the number of questions you answer correctly. You should try to answer every question. DO NOT SPEND TOO MUCH TIME on any one question. If you finish before time is called, go back and check your work in this part ONLY.

Now find the section of your answer sheet that is marked PART 8. When you are told to begin, start with question number 1 in Part 8 of your test booklet and answer space number 1 in Part 8 on your separate answer sheet.

DO NOT TURN THE PAGE UNTIL TOLD TO DO SO. You will have 19 minutes to complete the 25 questions in this part. Are there any questions?

Begin.

1. A tank holds 480 gallons of fuel. If it is drained at a rate of 120 gallons per minute, it will be empty in _____ seconds.
   A. 60
   B. 120
   C. 240
   D. 480

2. If the lever balances as shown, what is the mechanical advantage?

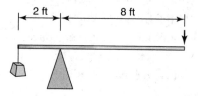

   A. 2
   B. 3
   C. 4
   D. 8

3. When a pump compresses air, the air
   A. gets hotter.
   B. gets colder.
   C. becomes a liquid.
   D. becomes a solid.

4. In a hydraulic system, the driving cylinder has a radius of 4 inches and the driven cylinder has a radius of 8 inches. The mechanical advantage is
   A. 2
   B. 4
   C. 6
   D. 8

5. For two gears to mesh properly, they must
   A. have the same number of teeth.
   B. rotate at the same speed.
   C. have teeth that are the same size.
   D. be made of the same material.

6. A screwdriver is used to turn a screw with a pitch of ¼ inch. After 10 complete turns of the screwdriver, the screw will have moved
   A. 1 inch
   B. 1 1/2 inches
   C. 2 inches
   D. 2 1/2 inches

7. Five gears are shown below. If gear A turns as shown, the other gears that turn in the same direction are

   A. Gears A and B
   B. Gears B and C
   C. Gears C and D
   D. Gears A and D

8. Which bookshelf is supported most securely?

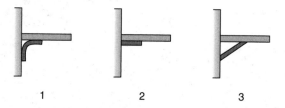

   A. 1
   B. 2
   C. 3
   D. 2 and 3 are equally secure.

9. Which flat cardboard pattern can be folded along the dotted lines to form the complete, totally enclosed, three-sided pyramid shown?

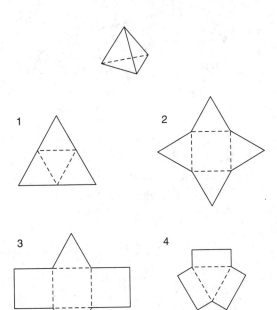

A. 1
B. 2
C. 3
D. 4

10. Which of these is not a simple machine?
    A. Lever
    B. Inclined plane
    C. Pulley
    D. Electric motor

11. A tank 2 feet wide by 6 feet long holds 600 pounds of water. What is the water pressure at the bottom in lb/ft²?
    A. 25
    B. 35
    C. 40
    D. 50

12. The figure shows a crank attached to a rod and piston. When the crank turns, the piston moves in the cylinder. The piston will be at the top of the cylinder if the crank makes

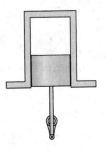

A. a ¼ turn
B. a ½ turn
C. a ¾ turn
D. one complete turn

13. How much effort is needed to roll a 120-lb drum up a ramp 30 ft long and 3 ft high at its high end? (Ignore the effect of friction.)
    A. 6 lb
    B. 9 lb
    C. 10 lb
    D. 12 lb

14. If you pulled the following objects from a pot of hot water, which would seem hottest to your hand?
    A. wooden spoon
    B. metal spoon
    C. plastic spoon
    D. paper straw

15. A gear and pinion have a ratio of 6 to 1. If the gear is rotating at a speed of 150 revolutions per minute (rpm), the speed of the pinion is most nearly
    A. 900 rpm
    B. 750 rpm
    C. 150 rpm
    D. 25 rpm

16. A spring, a device that can resume its original shape after bending, is best made from which of the following materials?
    A. ceramic
    B. paper
    C. steel
    D. wood

17. What is the mechanical advantage in this pulley system?

    A. 2
    B. 3
    C. 4
    D. 6

18. If pulley A rotates as shown, pulley C will rotate

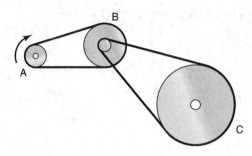

    A. faster, in the same direction.
    B. slower, in the same direction.
    C. faster, in the opposite direction.
    D. slower, in the opposite direction.

19. If you wanted to hit a heavy nail, you might choose a hammer with a
    A. lighter head.
    B. longer handle.
    C. lead head.
    D. shiny striking surface.

20. A sheet metal screw is likely to be _____ than a wood screw.
    A. longer
    B. shorter
    C. harder
    D. softer

21. The tank shown below holds 100 gallons of water. The pipe is delivering 5 gallons/second at the top, and the pipe at the bottom is removing 100 gallons/minute. How many gallons will remain in the tank after 2 minutes?

    A. 200
    B. 300
    C. 500
    D. 700

22. On the bicycle wheel shown, A and B are fixed points. When the wheel turns, which of the following is true?

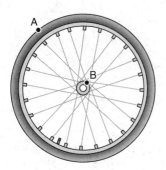

A. Point A travels farther than point B in each turn of the wheel.
B. Point A makes fewer complete turns per minute than does point B.
C. Point B travels farther than point A in each turn of the wheel.
D. Point A turns in complete circles, but point B does not turn.

23. If gear A is rotating at 200 rpm, how fast will gear C be rotating?

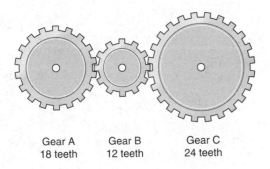

Gear A 18 teeth   Gear B 12 teeth   Gear C 24 teeth

A. 100 rpm
B. 150 rpm
C. 300 rpm
D. 400 rpm

24. When water flows in the direction shown, pressure at point B will be

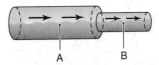

A B

A. higher, and water velocity will be higher.
B. higher, and water velocity will be lower.
C. lower, and the water velocity will be higher.
D. lower, but velocity will not change.

25. Gear X, which has 60 teeth, meshes with gear Y, which has 12 teeth. Each time gear X makes one complete rotation, how many rotations does gear Y make?
A. 6
B. 5
C. 4
D. 3

STOP! DO NOT TURN THIS PAGE UNTIL TIME IS UP FOR THIS TEST. IF YOU FINISH BEFORE TIME IS UP, CHECK OVER YOUR WORK ON THIS TEST ONLY.

PART 9. ASSEMBLING OBJECTS

THE TEST ADMINISTRATOR WILL READ THE FOLLOWING ALOUD TO YOU:

Now turn to Part 9 and read the directions for Assembling Objects silently while I read them aloud.

This test has questions that will measure your spatial ability. Study the diagram, decide which answer is CORRECT, and then mark the space on your separate answer sheet that has the same number and letter as your choice.

Your score on this test will be based on the number of questions you answer correctly. You should try to answer every question. DO NOT SPEND TOO MUCH TIME on any one question. If you finish before time is called, go back and check your work in this part ONLY.

Now find the section of your answer sheet that is marked PART 9. When you are told to begin, start with question 1 in Part 9 of your test booklet and answer space number 1 in Part 9 on your separate answer sheet.

DO NOT TURN THE PAGE UNTIL TOLD TO DO SO. You will have 15 minutes to complete the 25 questions in this part. Are there any questions?

Begin.

For Questions 1–17, which figure best shows how the objects in the left box will appear if they are fit together?

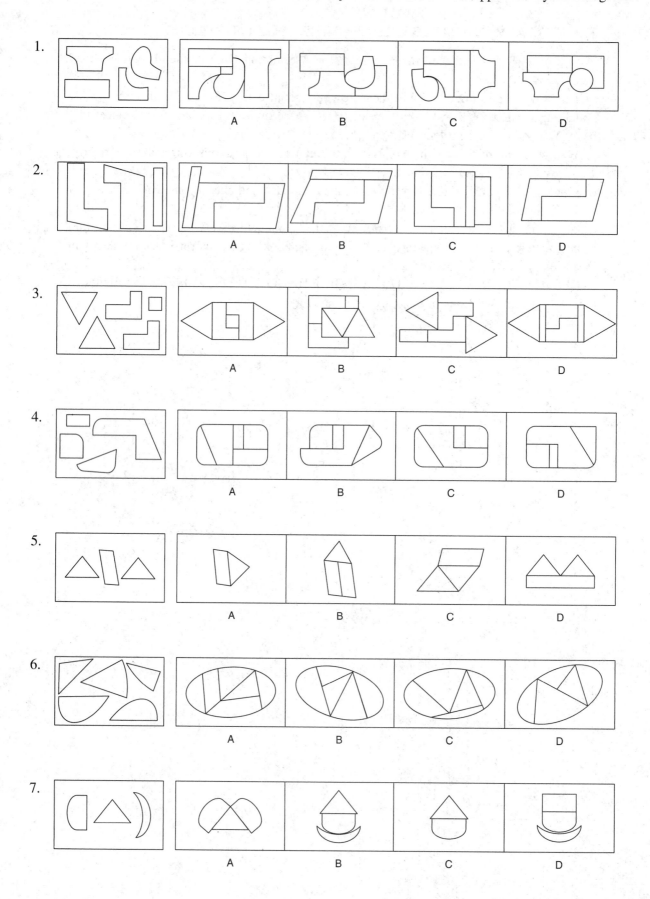

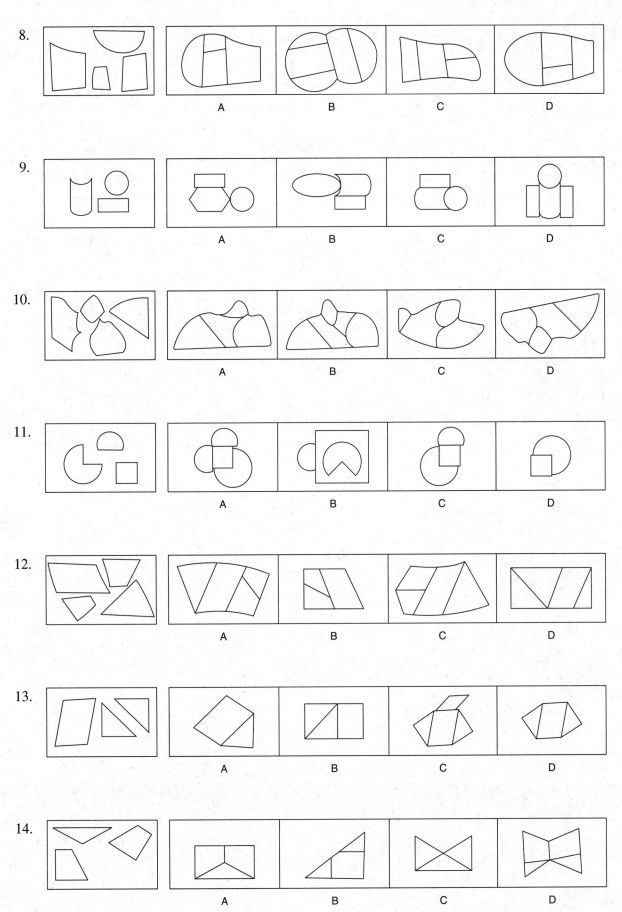

15.

16.

17.

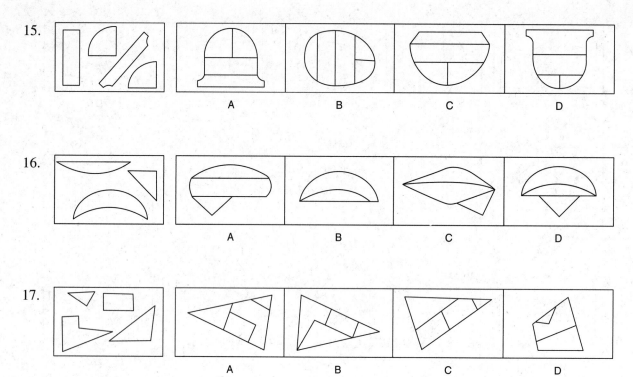

For Questions 18-25, which figure best shows how the objects in the left box will touch if the letters for each object are matched?

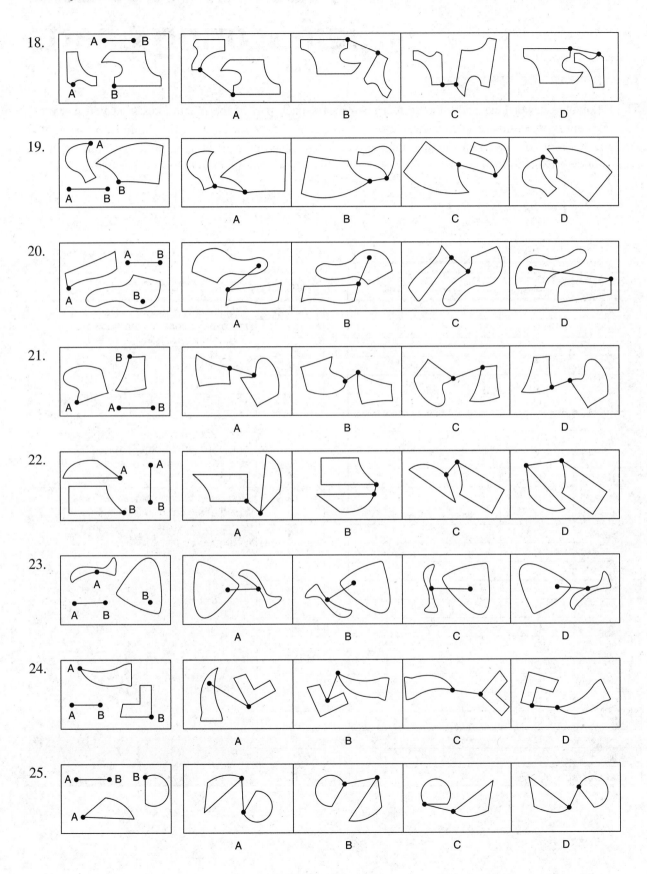

Answer Keys and Self-Scoring Charts

ASVAB PRACTICE TEST FORM 2

The following answer keys show the correct answers for each part of the practice test that you just took. For each question, compare your answer to the correct answer. Mark an X in the column to the right if you got the item correct. Then total the number correct for each part of the test. Find that number in the corresponding chart to the right of the answer key. See the suggestions listed for your performance.

Part 1. General Science

Item Number	Correct Answer	Mark X If You Picked the Correct Answer
1	A	
2	C	
3	A	
4	C	
5	C	
6	C	
7	C	
8	B	
9	C	
10	D	
11	A	
12	B	
13	C	
14	C	
15	C	
16	A	
17	B	
18	C	
19	D	
20	B	
21	B	
22	B	
23	C	
24	B	
25	A	
Total Correct		

Score Interpretation—Total Correct

25	This is pretty good work. Review the explanations for the answers you got incorrect.
24	
23	
22	
21	
20	You are doing pretty well. Review the explanations for the items you answered incorrectly. If you have time, review the other explanations and you will learn even more
19	
18	
17	
16	
15	You need to keep studying. Pay close attention to the explanations for each item, even for the ones you got correct.
14	
13	
12	
11	
10	Spend time working on the General Science review in Part 3 of this book.
9	
8	
7	
6	If you have copies of your school science textbooks, it would be a good idea to study those.
5	
4	
3	
2	
1	
0	

Part 2. Arithmetic Reasoning

Item Number	Correct Answer	Mark X If You Picked the Correct Answer
1	B	
2	A	
3	B	
4	C	
5	C	
6	B	
7	D	
8	A	
9	B	
10	A	
11	B	
12	B	
13	B	
14	C	
15	A	
16	C	
17	D	
18	B	
19	D	
20	B	
21	D	
22	D	
23	B	
24	D	
25	C	
26	C	
27	C	
28	A	
29	A	
30	C	
Total Correct		

Score Interpretation—Total Correct

30	This is pretty good work. Review the explanations for the answers you got incorrect. This test is a part of the AFQT, and you must do well
29	
28	
27	
26	
25	You are doing pretty well. Review the explanations for the items you answered incorrectly. If you have time, review the other explanations and you will learn even more.
24	
23	
22	
21	
20	You need to keep studying. Pay close attention to the explanations for each item, even for the ones you got correct.
19	
18	
17	
16	
15	Spend time working on the Arithmetic Reasoning reviews in Part 3 of this book.
14	
13	
12	
11	Keep working and reworking problems until you are comfortable with the processes.
10	
9	
8	
7	
6	
5	
4	
3	
2	
1	
0	

Part 3. Word Knowledge

Item Number	Correct Answer	Mark X If You Picked the Correct Answer
1	D	
2	B	
3	A	
4	C	
5	C	
6	D	
7	A	
8	C	
9	A	
10	B	
11	A	
12	B	
13	C	
14	C	
15	A	
16	B	
17	D	
18	B	
19	B	
20	C	
21	A	
22	A	
23	B	
24	B	
25	A	
26	C	
27	D	
28	B	
29	A	
30	C	
31	A	
32	D	
33	A	
34	C	
35	C	
Total Correct		

Score Interpretation—Total Correct

35	This is pretty good work. Review the explanations for the answers you got incorrect. This is an important test that contributes to your AFQT.
34	
33	
32	
31	
30	You are doing pretty well. Review the explanations for the items you answered incorrectly. If you have time, review the other explanations and you will learn even more.
29	
28	
27	
26	
25	You need to keep studying. Pay close attention to the explanations for each item, even for the ones you got correct.
24	
23	
22	
21	
20	
19	Spend time working on the Word Knowledge review in Part 3 of this book.
18	
17	
16	
15	Keep reading and identifying words you don't know.
14	
13	
12	
11	
10	
9	
8	
7	
6	
5	
4	
3	
2	
1	
0	

Part 4. Paragraph Comprehension

Item Number	Correct Answer	Mark X If You Picked the Correct Answer
1	A	
2	A	
3	B	
4	D	
5	A	
6	B	
7	D	
8	B	
9	A	
10	D	
11	A	
12	D	
13	D	
14	D	
15	B	
Total Correct		

Score Interpretation—Total Correct

15	This is pretty good work. Review the explanations for the answers you got incorrect. This test is a part of the AFQT, and you must do well.
14	You are doing pretty well. Review the explanations for the items you got incorrect. This test is a part of the AFQT, and you must do well.
13	You need to keep studying. Pay close attention to the explanations for each item, even for the ones you got correct. This will help you understand why the answer is correct.

Spend time reviewing the Paragraph Comprehension and Word Knowledge reviews in Part 3 of this book. Keep reading books and newspapers. |
12	
11	
10	
9	
8	
7	
6	
5	
4	
3	
2	
1	
0	

Part 5. Mathematics Knowledge

Item Number	Correct Answer	Mark X If You Picked the Correct Answer
1	B	
2	A	
3	B	
4	C	
5	A	
6	D	
7	A	
8	A	
9	D	
10	C	
11	D	
12	B	
13	D	
14	D	
15	C	
16	A	
17	C	
18	C	
19	B	
20	C	
21	C	
22	D	
23	C	
24	C	
25	D	
Total Correct		

Score Interpretation—Total Correct

25	This is pretty good work. Review the explanations for the answers you got incorrect.
24	
23	
22	You are doing pretty well. Review the explanations for the items you answered incorrectly. If you have time, review the other explanations and you will learn even more.
21	
20	
19	This test is part of the AFQT. You must perform well on this test.
18	
17	
16	You need to keep studying. Pay close attention to the explanations for each item, even for the ones you got correct.
15	
14	
13	
12	Spend time working on the Mathematics Knowledge reviews in Part 3 of this book, and work and rework the problems until you fully understand the processes.
11	
10	
9	
8	
7	
6	
5	
4	
3	
2	
1	
0	

Part 6. Electronics Information

Item Number	Correct Answer	Mark X If You Picked the Correct Answer
1	A	
2	B	
3	B	
4	D	
5	C	
6	A	
7	D	
8	B	
9	C	
10	A	
11	B	
12	D	
13	D	
14	A	
15	C	
16	D	
17	C	
18	A	
19	D	
20	B	
Total Correct		

Score Interpretation—Total Correct

20	This is pretty good work. Review the explanations for the answers you got incorrect.
19	
18	
17	You are doing pretty well. Review the explanations for the items you answered incorrectly. If you have time, review the other explanations and you will learn even more.
16	
15	
14	
13	
12	You need to keep studying. Pay close attention to the explanations for each item, even for the ones you got correct.
11	
10	
9	
8	Spend time working on the Electronics Information review in Part 3 of this book.
7	
6	
5	Your school electronics or physics books may have some helpful information as well.
4	
3	
2	
1	
0	

Part 7. Auto and Shop Information

Item Number	Correct Answer	Mark X If You Picked the Correct Answer
1	C	
2	B	
3	A	
4	D	
5	B	
6	D	
7	C	
8	C	
9	D	
10	C	
11	D	
12	D	
13	A	
14	D	
15	C	
16	B	
17	D	
18	D	
19	D	
20	A	
21	B	
22	B	
23	A	
24	C	
25	A	
Total Correct		

Score Interpretation—Total Correct

Score	Interpretation
25	This is pretty good work. Review the explanations for the answers you got incorrect.
24	
23	
22	
21	
20	You are doing pretty well. Review the explanations for the items you answered incorrectly. If you have time, review the other explanations and you will learn even more.
19	
18	
17	
16	
15	
14	You need to keep studying. Pay close attention to the explanations for each item, even for the ones you got correct.
13	
12	
11	
10	Spend time working on the Auto and Shop Information reviews in Part 3 of this book.
9	
8	
7	
6	
5	
4	
3	
2	
1	
0	

Part 8. Mechanical Comprehension

Item Number	Correct Answer	Mark X If You Picked the Correct Answer
1	C	
2	C	
3	A	
4	B	
5	C	
6	D	
7	C	
8	C	
9	A	
10	D	
11	D	
12	B	
13	D	
14	B	
15	A	
16	C	
17	A	
18	B	
19	B	
20	C	
21	C	
22	A	
23	B	
24	C	
25	B	
Total Correct		

Score Interpretation—Total Correct

25	This is pretty good work. Review the explanations for the answers you got incorrect.
24	
23	
22	
21	
20	You are doing pretty well. Review the explanations for the items you answered incorrectly. If you have time, review the other explanations and you will learn even more.
19	
18	
17	
16	
15	You need to keep studying. Pay close attention to the explanations for each item, even for the ones you got correct.
14	
13	
12	
11	Spend time working on the Mechanical Comprehension review in Part 3 of this book
10	
9	
8	
7	
6	
5	
4	
3	
2	
1	
0	

Part 9. Assembling Objects

Item Number	Correct Answer	Mark X if You Picked the Correct Answer
1	B	
2	A	
3	A	
4	C	
5	C	
6	B	
7	B	
8	A	
9	C	
10	D	
11	C	
12	A	
13	D	
14	A	
15	A	
16	D	
17	B	
18	D	
19	C	
20	D	
21	C	
22	D	
23	B	
24	C	
25	D	
Total Correct		

Score Interpretation–Total Correct

Score	Interpretation
25	This is pretty good work. Review the explanations of the answers you got incorrect.
24	
23	
22	
21	
20	You are doing pretty well. Review the explanations for the items you answered incorrectly. If you have time, review the other explanations and you will learn even more.
19	
18	
17	
16	
15	Keep studying. Pay close attention to the explanations for each item, even for the ones you got correct. Spend time working on the Assembling Objects review in Part 3 of this book.
14	
13	
12	
11	
10	
9	
8	
7	
6	
5	
4	
3	
2	
1	
0	

Answers and Explanations

PART 1. GENERAL SCIENCE

1. **A.** Vacuoles are the parts of a cell that are the storage containers, holding water, waste, and food. The correct answer is choice A.

2. **C.** The nucleus contains the hereditary information for a cell. Choice C is the correct answer.

3. **A.** Root hairs have several functions including absorbing nutrients from the soil (choice B), helping the root system to anchor the plant (choice C), and storing food (choice D). The root hairs do not conduct photosynthesis because they do not contain chloroplasts. Choice A is the correct answer.

4. **C.** The flower (choice D) is not associated with controlling water. The stem (choice A) helps to transport water, but does not really control it. Xylem (choice B) is a tissue that makes up vessels in plants that help to support the plant and transport water; however, those vessels do not control the amount of water. The stomata are holes in the leaf that are surrounded by guard cells. The guard cells open and close the stomata, depending on how much water is contained in the plant and how much humidity there is in the atmosphere, thus controlling the amount of water that is released to the atmosphere. Choice C is the correct answer.

5. **C.** Arachnids, choice D, are spiders and this does not relate to the question. Herbivores (choice A) are plant eaters, carnivores (choice B) are meat eaters, and omnivores (choice C) eat both plants and animals as do humans. Choice C is the correct answer.

6. **C.** Kangaroos are marsupials. Choice C is the correct answer.

7. **C.** Your bones do indeed produce red blood cells (choice A), protect your body organs (choice B), and store minerals like calcium (choice D). Bones do not create nutrients. Choice C is the correct answer.

8. **B.** Any time you lift a weight or move a muscle, one muscle contracts, and the opposing muscle relaxes. So when you contract your biceps, your triceps relax. Choice B is the correct answer.

9. **C.** Sunlight promotes the development of vitamin D, which is important for the development of bones. Choice C is the correct answer.

10. **D.** Arteries (choice D) are structures that move the blood from the heart to other parts of the body. Veins (choice A) bring blood into the heart. Atriums (choice C) are not part of the circulatory system. Capillaries (choice B) are the tiny blood vessels found throughout the body. Choice D is the correct answer.

11. **A.** The retina (choice A) contains rods and cones that respond to light intensity and color. Choice A is the correct answer.

12. **B.** What a person actually ends up looking like based on his or her genetic makeup is called the *phenotype*. Choice B is the correct answer.

13. **C.** A laser emits light rays that are all the same wavelength; the waves are in phase. As a result, the light can travel great distances with very high energy without spreading. A laser can direct a lot of energy to a very small area and do it accurately, making it a great cutting tool. Choice C is the correct answer.

14. **C.** In order to change a solid to a liquid, thermal energy is necessary. Choice C is the correct answer.

15. **C.** The Doppler effect (choice A) relates to the change in the frequency of wavelengths. It is not the correct answer. Winds and surface water currents are affected by the Earth's rotation, which bends them in the clockwise direction in the northern hemisphere and counterclockwise in the southern hemisphere. This condition is called the Coriolis effect. Choice C is the correct answer.

16. **A.** A fissure eruption (choice B) is merely magma that flows out from a crack in the Earth's surface and can flood a large part of the land with basalt. Cinder cones (choice C) are volcanoes formed by explosive events, throwing cinders, lava, and ash into the air. When these materials drop back to the ground, they form a small cone. A shield volcano (choice D) is formed by lava flows of layers of basalt. A shield volcano is not very steep. The volcano type that forms from alternating layers of explosive materials and lava flows and is the steepest of all the volcanoes is a composite volcano. Choice A is the correct answer.

17. **B.** Earlier in the history of the Earth there was much more volcanic activity than there is now. The volcanoes spewed molten material and also gases. The gases included water vapor. When the atmosphere cooled to a certain point, it began to rain. It rained heavily and for a very long time. The water collected in the low areas which are now the oceans. The best answer is choice B.

18. **C.** Tides are caused by the relative positions of the Earth, the moon, and the sun. When the moon and the sun are in a straight line with Earth, the gravitational pull is the strongest. Choice C is the correct answer.

19. **D.** When the sun's rays are directly over latitude $23\frac{1}{2}°$ N (the Tropic of Cancer), it is the summer solstice for the northern hemisphere and the winter solstice for the southern hemisphere. When the sun's rays are directly over latitude $23\frac{1}{2}°$ S (the Tropic of Capricorn), it is the winter solstice for the northern hemisphere and the summer solstice for the southern hemisphere. When the sun's rays are directly over the equator, it is

the equinox, when days and nights are nearly equal all over the Earth. Choice D, equinox, is the correct answer.

20. **B.** Asteroids are small rocky bodies that revolve around the sun between the orbits of Mars and Jupiter. Scientists speculate that Jupiter's strong gravitational pull prevented a larger planet from forming in that plane. Choice B is the correct answer.

21. **B.** Without the greenhouse effect, Earth would get colder. Choice B is the correct answer.

22. **B.** A low-pressure area in the tropics is the cause of hurricanes. Choice B is the correct answer.

23. **C.** Water enters the atmosphere through transpiration from plants and evaporation from bodies of water like rivers, lakes, and oceans. Choice C is the correct answer.

24. **B.** The geographic pole is the axis point around which Earth rotates. The magnetic and geographic poles differ by about $11\frac{1}{2}$ degrees at the current time. The magnetic poles have reversed many times during Earth's long history. Opposite magnetic poles attract. The magnetic south pole of a magnet is attracted to the magnetic north pole. Choice B is the correct answer.

25. **A.** A substance that speeds up chemical reactions is called a catalyst. Choice A is the correct answer.

PART 2. ARITHMETIC REASONING

1. **B.** This is a simple addition problem. Add the amounts to get the total of $111.10.

2. **A.** This problem has two steps. First you need to add the number of minutes that Ned ran and then subtract that amount from the goal of 175 minutes. He ran 158 minutes, so he needs to run 17 minutes more to reach 175.

3. **B.** In order to calculate the percent, you need to divide the total number of CDs into the number of jazz CDs. In this case you would divide 200

into 30 to get 0.15. To change that to a percent, move the decimal point two places to the right and add the percent sign to get the correct answer of 15%.

4. **C.** To find the average, add the numbers and divide by the number of numbers. In this instance the number of miles Dave has traveled add up to 150. The number of numbers is 5, so the average number of miles per day he has traveled is 30. Don't make the mistake of dividing by 4 because on one day his number of miles was 0. That number needs to be used in calculating the average.

5. **C.** To find the average, add up the numbers and divide by the number of numbers. In this problem the number of yards in Blair's completed passes adds to 105 yards. Since he has thrown 5 times, his average pass is 21 yards.

6. **B.** To find the ratio of one number to another, create a fraction to show the relationship. In this problem, there are 25 wins in a total of 55 games. The question asks for the ratio of wins to losses, so you need to calculate the number of losses. Subtracting 25 from 55 gives the result of 30 losses. Representing this as a fraction would be $\frac{25}{30}$. Simplifying this results in the fraction $\frac{5}{6}$.

7. **D.** Set this up as a proportion. If one member runs 1.2 km, how many km would eight members run? The proportion would look like this: $\frac{1}{1.2} = \frac{8}{x}$. Cross-multiply and solve.

$$x = 8(1.2) \text{ or } x = 9.6 \text{ km}$$

However, the problem asks for the number of meters. Realizing that 1 km = 1,000 meters, change 9.6 km to 9,600 meters.

8. **A.**

$$\text{Probability} = \frac{\text{number of favorable outcomes}}{\text{number of possible outcomes}}$$

In this problem the number of favorable outcomes is the number of people selecting chocolate as a favorite flavor. The number of possible outcomes is the number of all those surveyed or $602 + 411 + 589 + 214 + 78 + 514 = 2,408$.

So substitute into the formula: $\frac{602}{2,408}$. This simplifies to $\frac{1}{4}$, the correct ratio.

9. **B.** Simple interest problems use the formula $I = prt$, where I is the amount of interest, p is the principal or the amount saved or invested, r is the rate of interest, and t is the amount of time that the interest is accruing. In this problem, the interest needed is the difference between what was invested and the amount needed to buy the car: $\$17,500 - \$10,000 = \$7,500$. The principal is $\$10,000$ and the rate is 15%. Substitute that information into the formula to get $\$7,500 = \$10,000(0.15)t$ and solve for t. $\$7,500 = \$1500t$; $t = 5$ years.

10. **A.** To find the average, add the numbers and divide by the number of numbers. In this case you would add $80 + 78 + 84 + 77 + 86$, and then divide by 5 to get an average of 81 cars per day.

11. **B.** If Wayne has 256 pieces of candy that he packs into boxes of 30 each, he can fill $256 \div 30 = 8$ boxes with 16 pieces left over— not enough to fill another box.

12. **B.** The formula for calculating area is $A = lw$. Since we are working with the result in tiles that are measured in inches, calculate the area of the floor in square inches. 10 feet $\times$ 12 inches = 120 inches. 12 feet $\times$ 12 inches = 144 inches. So the area of the floor is 17,280 in.^{2}. Each tile has an area of 6×8 inches or 48 in.^{2}. Dividing 17,280 in.^{2} by 48 in.^{2} = 360 tiles.

13. **B.** The cost of the joining fee is $\$125.00$. Reduce that amount by 25% by multiplying the total cost by 0.25.

$$\$125.00 \times 0.25 = \$31.25$$

Subtract $\$31.25$ from the original cost.

$$\$125.00 - \$31.25 = \$93.75.$$

A faster way to do this (and speed is important on the ASVAB) is to multiply the original cost by 75% or 0.75 because that is the cost of the items after the discount.

$$\$125.00 \times 0.75 = \$93.75$$

14. **C.** Use the following formula:

$$\text{Percent of Change} = \frac{\text{Amount of Change}}{\text{Starting Point}}$$

In this problem we know that the amount of change is $350 and that the starting price of the computer is $1,600. Substitute that information into the formula:

$$\text{Percent of Change} = \frac{\$350.00}{\$1,600.00} = 0.219.$$

Change that number to a percent by moving the decimal place two places to the right and add the percent sign: $0.219 = 21.9\%$

15. **A.** You are told that two numbers add up to 1,500, with one number being 4 times the size of other. Create an equation.

$$x + 4x = 1,500$$

Solve for x.

$$5x = 1,500$$

$$x = 300$$

$$4x = 1,200$$

So the two numbers are 300 and 1,200.

16. **C.** To calculate the area of a triangle, use the formula $A = \frac{1}{2}bh$. In this instance the base is 36 feet and the height is 24 feet.

$$A = \frac{1}{2}(36)(24)$$

$$A = \frac{1}{2}864$$

$$A = 432 \text{ ft}^2$$

If each tree needs 8 ft², divide that into the area.

$$\frac{432 \text{ ft}^2}{8 \text{ ft}^2} = 54$$

17. **D.** Use the formula $C = 2\pi r$ or $C = \pi d$.

If the circumference is 56.52 meters, substitute that into the formula: $56.52 = \pi d$

Solve for d by dividing both sides by π.

$$\frac{56.52}{3.14} = d$$

$$d = 18 \text{ m}$$

18. **B.** Create an equation to solve this problem: Jamie = 3(Francis). You are told that Francis is 14. Substitute that information into the formula.

$$\text{Jamie} = 3(14)$$

Solve for Jamie.

$$\text{Jamie} = 42$$

19. **D.** Use the following formula:

$$\text{Percent of Change} = \frac{\text{Amount of Change}}{\text{Starting Point}}.$$

The starting point is 8,435,978. The amount of change is the difference between 10,544,972 and 8,435,978. You find the difference by subtracting:

$$10,544,972 - 8,435,978 = 2,108,994$$

Substitute that information into the equation.

$$\text{Percent of Change} = \frac{2,108,994}{8,435,978} = 0.2499 \text{ or } 25\%$$

20. **B.** Watch the units in this problem. Change the yards to inches.

$$5 \text{ yards} = 15 \text{ feet} = 180 \text{ inches.}$$

The problem asks for the number of 18-in.-long pieces. To find the answer, you need to divide 180 by 18.

$$\frac{180}{18} = 10 \text{ pieces}$$

21. **D.** Treat this like a compound interest problem where you calculate the increase in kilometers

each week, using the increased amount as the amount to use in the calculations the next week.

Week One

3 km

Week Two

3 km + 0.15(3) = 3.45 km

Week Three

3.45 km + 0.15(3.45) = 3.97 km

Week Four

3.97 km + 0.15(3.97) = 4.57 km

22. **D.** To solve this problem, use the formula $V = lwh$. You have been given the dimensions of $l = 50$, $w = 25$ m, and $h = 3$ m. Substitute that information into the formula.

$$V = (50)(25)(3)$$
$$V = 3,750 \text{ m}^3$$

If 0.25 lb of chlorine are required for every 15 m^3, calculate how many units of 15 m^3 there are in the total volume of 3,750 m^3.

$$3,750 \text{ m}^3 \div 15 \text{ m}^3 = 250 \text{ quarter-pound units}$$

The question asks for pounds, so divide that number by 4.

$$250 \div 4 = 62.5 \text{ lb}$$

23. **B.** In this problem you need to add the amounts of calcium that Joel eats at breakfast. Add 250, 100, and 225 to get his calcium intake of 575 mg. To calculate the percentage, divide the required amount into 575.

$$\frac{575}{1,500} = 38\%$$

24. **D.** To calculate the volume of a cylinder, use the formula $V = \pi r^2 h$. In this problem the diameter is 16 in. so the radius is 8 in. The height is given as 4 in. Substitute that information into the formula.

$$V = (3.14)(8^2)(4)$$
$$V = (3.14)(64)(4)$$
$$V = 803.84 \text{ or } 803.8 \text{ in.}^3$$

Be sure to pay attention to the information you are given. You could have made an error by using the diameter rather than the radius, because the diameter was given in the problem.

25. **C.** To calculate the volume of a sphere, use the formula $V = \frac{4}{3}\pi r^3$. You are given a diameter of 12 inches, so the radius is 6 in. Substitute that information into the formula.

$$V = \frac{4}{3}\pi(6)^3$$
$$V = \frac{4(3.14)(216)}{3}$$
$$V = \frac{2,712.96}{3} = 904.32 \text{ in}^3$$

26. **C.** To calculate the answer, multiply 45.6 by 0.12.

$$45.5 \times 0.12 = \$5.47$$

27. **C.** Melanie's charge will be \$4,000 plus interest of 18%. Multiply \$4,000 by 1.18 to get \$4,720. Her bill will be \$4,720 if she makes no payments and no new purchases.

28. **A.**

$$\text{Probability} = \frac{\text{number of favorable outcomes}}{\text{number of possible outcomes}}$$

Favorable outcomes would be red or purple.

$$\text{Probability} = \frac{32+13}{45+32+13+16+29}$$
$$= \frac{45}{135} = \frac{1}{3} = 1:3$$

29. **A.** Use the following formula for finding the area of a right triangle: $A = \frac{1}{2}bh$. The height of the triangle is 3 feet, and the base is 4 feet. Substitute that into the formula.

$$A = \frac{1}{2}(4)(3)$$
$$A = 6 \text{ ft}^2$$

30. **C.** The basic rental is $25 \times 2 = $50. For the mileage, multiply 200 miles times 15 cents. $200 \times 0.15 = $30. Combined the costs are $50 + $30 = $80.

PART 3. WORD KNOWLEDGE

1. **D.** *Hilarious* means "highly amusing." *Serious* has the opposite meaning and is clearly incorrect. The words *confusing* and *repellent* don't share the meaning, so choice D, *funny*, is the best answer.

2. **B.** To be *wary* means "to be guarded or mistrustful of something or someone." To mistrust something might make someone *nervous*, but that is not a synonym for *wary*. *Enthusiastic* and *happy* are not related to the meaning of *wary*, so choice B, *cautious*, is the correct answer.

3. **A.** A *tremor* is a shaking, quivering, or vibration. The words *pain*, *warmth*, and *object* are not related to the definition. The word that most closely relates to these is *trembling*. Choice A is the correct answer.

4. **C.** To be *ingenious* is to be smart and resourceful. The words *destructive*, *threatening*, and *sorrowful* do not seem like good matches. *Clever* is the word closest to the meaning of *ingenious*. Choice C is the correct answer.

5. **C.** *Illusory* means "misleading or false." *Real* is somewhat the opposite of this meaning, while the words *first* and *bright* are unrelated to the definition. *Deceptive*, choice C, is the correct answer.

6. **D.** To *thrive* means "to prosper or be successful." The words *move*, *separate*, and *think* do not seem to fit this meaning. *Flourish*, choice D, is the correct answer.

7. **A.** To *articulate* means "to speak clearly or to clarify." The words *change*, *contradict*, and *support* do not fit this meaning. *Explain*, choice A, is the word that most closely fits the meaning of *articulate*.

8. **C.** *Reticent* means "reserved, silent, or hesitant." The words *hopeful* and *hopeless* do not relate to the definition of *reticent*. *Encouraged* is somewhat the opposite in meaning. Choice C, *restrained*, is the word that most closely resembles the meaning of *reticent*.

9. **A.** *Cumbersome* means "unwieldy or awkward." That might make carrying packages *difficult*, but the best word to replace *cumbersome* is *clumsy*, choice A.

10. **B.** The prefix *co-* means "together." To be *coherent* means "to bring together ideas in a clear and logical way." The ideas "stick together" well. The words *misguided*, *mixed*, and *unfavorable* are not related to the definition of *coherent*. *Logical*, choice B, is the only answer that conveys this idea.

11. **A.** *Contentious* relates to strife or arguments—people in disagreement. The word *agreeable* suggests the opposite of that meaning and cannot be correct. Although contentious discussions might be *laughable* and even *interesting*, the best synonym for *contentious* is *quarrelsome*, choice A.

12. **B.** The prefix *con-* means "together or together with." To *converge* means "to come together at a point or place." The words *dissolve* and *slide* seem somewhat opposite of the concept of bringing together. The word *form* relates somewhat to the meaning of *converge*, but *join*, choice B, is the best answer.

13. **C.** To *invert* something means "to turn it over, turn it upside down, or reverse it." *Rebuke*, *lecture*, and *break* do not share the same meaning. *Overturn*, choice C, is the correct answer.

14. **C.** *Feigned* means "made up, faked, or invented." *Showed*, *disregarded*, and *lost* are not related to this idea. *Pretended*, choice C, is the best synonym for *feigned*.

15. **A.** To *refute* means "to prove something false or wrong." The word *support* conveys the opposite

meaning. *Direct* is also incorrect. To *question* the hypothesis seems to come close, but the best answer is choice A, *disprove*.

16. **B.** *Impetuous* means "hasty, spontaneous, or reckless." The words *natural*, *slow*, and *severe* are not related to the definition of *impetuous*. Only choice B, *impulsive*, conveys the correct meaning and is the best answer.

17. **D.** *Propensity* means "a leaning toward or a tendency to do something." *Aversion* conveys the opposite message as that word means "to move away from something." *Fault* and *awareness* are unrelated to the meaning of *propensity*. *Tendency*, choice D, is the best answer.

18. **B.** *Spurned* means "scorned, snubbed, or rebuffed." *Accepted* conveys the opposite meaning and cannot be correct. *Scattered* and *moved* do not relate to the definition of *spurned*. *Rejected*, choice B, is the correct answer.

19. **B.** To *hoard* means "to store or accumulate." The words *eat* and *share* do not relate to the meaning of *hoard*. In order to store or accumulate acorns, a squirrel probably does have to *hide* them, but the best answer is choice B, *stockpile*.

20. **C.** A *tempest* is a storm, so to be *tempestuous* certainly cannot mean *calm* or *friendly*. An *adversarial* relationship certainly can be tempestuous, but *stormy*, choice C, is the best synonym for *tempestuous*.

21. **A.** *Charade* means "a farce or fake." The words *nervousness*, *subtlety*, and *contamination* are not related to that idea. *Pretense*, choice A, is the best answer.

22. **A.** The prefix *in-* means "not." To be "not coherent" means to be illogical, disjointed, or not logically connected. The words *nasty*, *simple*, and *incorrect* are not related to the idea of being illogical. Choice A, *jumbled*, is the best answer.

23. **B.** To be *inhibited* means "to be repressed, held back, hindered, or restrained." The words *pushed*, *scared*, and *questioned* do not relate to this definition. *Prevented*, choice B, is the best answer.

24. **B.** *Levity* means "gaiety or frivolity." An atmosphere of levity would have laughter, jokes, and humorous comments. It would not be characterized by *seriousness*, *anger*, and *distrust*. *Lightheartedness*, choice B, is the best answer.

25. **A.** The prefix *mal-* means "bad, ill, or wrong." The words *comedy*, *heart*, and *strength* do not relate to this idea. A *malady* is an *illness*. Choice A is the correct answer.

26. **C.** *Unwieldy* means "hard to manage, awkward, or clumsy." *Colorful* does not relate to this idea. Though a *heavy* and *useless* umbrella may be unwieldy, the best answer is choice C, *cumbersome*.

27. **D.** *Transgression* means "a crime or offense." The words *misspelling* and *diversion* do not seem to relate to this definition. Although a transgression may be a *mistake*, choice D, *violation*, is the best synonym.

28. **B.** The prefix *pre-* means "before." *Preamble* means "a statement before or introduction to a longer narrative." It is certainly not an *ending* or an *interruption*. A preamble may well be a *summary*, but the word closest in meaning is choice B, *preface*.

29. **A.** *Artificial* means "not made by natural means; an imitation or a fake." *Creative*, *faithful*, and *late* do not relate to this idea. Choice A, *false*, is the best answer.

30. **C.** A *rebellion* is an uprising or upheaval. The words *celebration*, *parade*, and *meeting* do not share the same meaning of *uprising* or *upheaval*. Choice C, *revolt*, is the best answer.

31. **A.** The prefix *bene-* means "good or well." A *benefactor* is one who does something good or helpful for another. A *partner*, an *employer*, or a *counselor* might do good deeds for another person, but choice A, *supporter*, is the best answer.

32. **D.** *Secluded* means "shut off from others or isolated." The words *windy*, *rocky*, and *narrow* do not relate to this definition. *Hidden*, choice D, is the best answer.

33. **A.** The prefix *pre-* means "before." *Requisite* means "necessary or indispensable." Thus a *prerequisite* is a condition that needs to be fulfilled before another step or action can be taken. Choice A, *requirement*, is the best answer

34. **C.** A *blemish* is a spot, mark, or imperfection. The words *entry* and *inspiration* are clearly not correct answers. A blemish on the record could be in the form of a *notation*, but that word does not convey the meaning of being an imperfection. *Defect*, choice C, is the best answer.

35. **C.** *Hearsay* means "based on unsubstantiated statements or information, or on gossip." As a result, *facts*, *statements*, and *measurements* cannot be correct. In this case, the best answer is choice C, *rumor*.

PART 4. PARAGRAPH COMPREHENSION

1. **A.** In this passage the author's major intent is not to define obesity or point out obesity statistics or even the causes of obesity. The author spends most of the passage listing all the detrimental effects of obesity on health. Choice A is the correct answer.

2. **A.** The passage indicates that mass media might be a possible cause of obesity, but not part of the solution. Choice B is not the correct answer. The passage does not mention school vending machines, so choice C is not the correct answer. The passage suggests that personal choice is a factor in suggesting solutions, but it is not a solution by itself, so Choice D is not the correct answer. In several places, the passage indicates that knowledge of the facts by parents and children is the solution to the problem. Choice A is the correct answer.

3. **B.** The passage indicates that obesity is likely to be a problem in the future, so choice C cannot be correct. The passage does not suggest that exercise will eliminate obesity, so choice D cannot be correct. Although it is suggested that parents and schools have an influence on children's health (choice A), the best answer is stated directly in the passage—childhood obesity will likely continue to set records. Choice B is the correct answer.

4. **D.** The passage indicates that there are 300,000 deaths per year, not per day, from obesity-related causes, so choice A is not correct. Nothing in the passage suggests that obesity results in a society that does not exercise, although the passage does suggest that exercise is part of the solution. Choice B is not correct. Nothing indicates that obesity results in public admiration of cute children, so choice C is not correct. The passage specifically says that obesity is associated with $70 billion in healthcare costs, so choice D is the correct answer.

5. **A.** The passage says that "children soak up the wrong lessons not just from TV, but also from the one in three adults who is overweight." So choice A is the correct answer.

6. **B.** There is no indication that Pocahontas was a warrior (choice A). She certainly was a daughter (choice C), and she was only a child (choice D) when the English colonists first arrived. The passage, however, indicates that Pocahontas took various actions, such as promoting peace between the Powhatans and the English colonists and actually marrying one of the colonists, in hopes of bringing the colonists and Algonquians together. These actions made her a mediator, a person who tries to bring together people who disagree on some issue. Choice B is the correct answer.

7. **D.** The passage specifies that "Pocahontas" means "mischievous one." Choice D is the correct answer.

8. **B.** Pocahontas was young, but she is not described as a baby (choice A), as "superior" (choice C) or as "hateful" (choice D). The correct answer is "friendly," choice B.

9. **A.** There is no mention that Pocahontas married to have children (choice B), to increase her influence (choice C), or to ensure her status in her tribe (choice D). The correct answer is choice A, to promote peace.

10. **D.** Not much is said in the passage about John Rolfe. He is mentioned only as a colonist. Choice D is the correct answer.

11. **A.** Although Florida seceded from the United States (choice B), left the Union (choice C), and helped to feed the soldiers (choice D), these are not the major ideas in the passage. The passage indicates that we think of Florida as a sunny vacation state, but it has a history regarding the issues between the northern and southern states. This would lead to the best title: Sunny State—Confederate History. Choice A is the best answer.

12. **D.** There is no mention that issues about Florida's borders (choice A), the weather (choice C), or feeding soldiers (choice B) led to Florida's secession from the Union. These are all incorrect. The major sticking point between Florida and the Union was a disagreement about slavery. Choice D is the correct answer.

13. **D.** The passage indicates that 16,000 Floridians fought in the Civil War. The correct answer is choice D.

14. **D.** The passage states that 11 states left the Union. Choice D is the correct answer.

15. **B.** Although Florida did leave the Union, the passage does not state that it was the first to do so. Choice A is not correct. Although the passage states that there were some battles on the Florida coast, it does not state that most of the battles in the South took place there. Choice C is not correct. Clearly, Florida's position on government control and slavery were the cause of the beginning, not the end, of the Civil War. Choice D is not correct. The correct answer is B, that although Florida's contribution in total numbers was small, its percentage of soldiers among available men of military age was the highest of any Confederate state.

PART 5. MATHEMATICS KNOWLEDGE

1. **B.** Isolate the unknown on one side. Subtract 12 from both sides.

$$z = 24 - 12$$
$$z = 12$$

2. **A.** Perform operations in the parentheses first, then the exponents, and then the remaining operations.

$$16 + 3(-3)$$
$$16 + (-9) = 7$$

3. **B.** $30 = x\%$ of 90.

$$30 = x90$$

Solve for x.

$$x = \frac{30}{90}$$
$$x = \frac{1}{3} \text{ or } 33\%$$

4. **C.** When calculating the average or mean, add the numbers and then divide by the number of numbers. The sum of 14, 24, 12, 16, 34 and 104 is 204 divided by 6 which comes to 34.

5. **A.** When dividing fractions, invert and multiply. Then see if you can simplify the fractions.

$$\frac{1}{3} \div \frac{5}{6} = \frac{1}{3} \times \frac{6}{5} = \frac{6}{15} = \frac{2}{5}$$

6. **D.** In this problem you should note three things. First, x times x gives x^2. Second, 10×10 gives 100. Last, note that there are only two terms in the expression and that there is a minus sign. The minus sign tells you that the term in one factor needs to be a minus sign. This results in two factors: $(x - 10)(x + 10)$.

7. **A.** When subtracting, change the sign and add. In this problem, $3x^2 - 2y + 12$ is changed to $-3x^2 + 2y - 12$ and then added to the first term.

$$12x^2 + 6y - 3$$
$$-3x^2 + 2y - 12$$

Add the terms. This results in the correct answer of $9x^2 + 8y - 15$.

8. **A.** For this problem you want to isolate the x on one side of the equals sign. To do this, multiply each side by y, giving the correct answer: $\frac{yr}{j} = x$.

9. **D.** Set each equation to equal zero. Next, arrange the equations so that when one is subtracted from or added to the other, one of the terms results in a zero and drops out of the result. Next, solve for the remaining unknown. In this problem,

$$3a + 4b = 12$$
$$2a + 2b = 15$$

Set the equations to zero:

$$3a + 4b - 12 = 0$$
$$2a + 2b - 15 = 0$$

Note that neither adding nor subtracting will eliminate a term, but if you multiply one equation by an appropriate term to make two terms equal, you can add or subtract as necessary to eliminate a term. For this problem, if you multiply the second equation by -2 and then add the equations, the b term can be eliminated.

$$3a + 4b - 12 = 0$$
$$-2(2a + 2b - 15) = 0$$

$$3a + 4b - 12 = 0$$
$$-4a - 4b + 30 = 0$$

Add the two equations:

$$-a + 18 = 0$$
$$-a = -18$$
$$a = 18$$

Substitute the value of a into one of the original equations.

$$2(18) + 2b - 15 = 0$$
$$36 + 2b - 15 = 0$$
$$21 + 2b = 0$$
$$2b = -21$$
$$b = -10.5$$

The correct answer is choice D, $a = 18$, $b = -10.5$.

10. **C.** To divide numbers with exponents, subtract the exponents. In this problem, subtract 7 from 4 giving -3 as the exponent for x, and 2 from 6, giving 4 as the exponent for y. The term thus becomes $x^{-3} y^4$. This also can be written as $\frac{y^4}{x^3}$, which is choice C.

11. **D.** Factor this equation, resulting in $(3g + 1)(g + 1)$. Set each factor to equal zero and solve for g.

$$3g = -1; \ g = -\frac{1}{3}$$
$$g = -1$$

The correct answer is $g = -\frac{1}{3}; \ g = -1$.

12. **B.** To multiply fractions, multiply the numerators and denominators. Simplify where possible.

$$\frac{6}{x+1} \times \frac{3x+3}{5}$$
$$= \frac{18x+18}{5x+5}$$

Factor this fraction to get:

$$\frac{18\,(\cancel{x+1})}{5\,(\cancel{x+1})}$$

To simplify, cancel the $x + 1$ terms, leaving $\frac{18}{5}$ or $3\frac{3}{5}$.

13. **D.** To divide fractions, invert the second term and multiply. Simplify where possible.

$$\frac{5}{x+1} \div \frac{6}{3x+3} \text{ becomes } \frac{5}{x+1} \times \frac{3x+3}{6}.$$

Multiply the numerators and denominators.

$$\frac{15x+15}{6x+6}$$

Simplify by factoring.

$$\frac{15\,(\cancel{x+1})}{6\,(\cancel{x+1})}$$

Simplify by canceling.

$$\frac{15}{6} = \frac{5}{2} = 2\frac{1}{2}$$

14. **D.** The square root is the number that, when multiplied by itself, results in the original number. In this case, $11 \times 11 = 121$, so 11 is the correct answer.

15. **C.** In this type of problem, where two fractions are separated by an equals sign, you can cross multiply and then solve for the variable of interest. Cross multiply to get $sy = vt$. Solving for y gives $\frac{vt}{s}$, the correct answer.

16. **A.** A hexagon has six sides, so A is the correct answer.

17. **C.** A parallelogram has two equal lengths and two equal widths. The perimeter is the sum of the widths and lengths, or $2(l) + 2(w)$. In this problem, the perimeter is 16 miles.

18. **C.** The hundredths place is two places to the right of the decimal point. Underline the hundredths place. If the number to the right of it is less than 5, keep the number in that place the same. If it is 5 or greater, increase the number in the hundredths place by one. In this problem, the number to the right of the hundredths place is 4 so the number is rounded to 1,478.96, which is the correct answer.

19. **B.** A right triangle has one angle that measures 90°. If a second angle measures 35°, that accounts for 125°. Subtracting that total from 180° (the total number of degrees in a triangle) gives the result of 55°.

20. **C.** In this problem, information on angles 1 and 2 is irrelevant. Since you know that $\angle 3$ is a right angle, $\angle 4$ must be $180° - 90°$ or 90°, another right angle.

21. **C.** $\angle C$ is a right angle, so you can employ the Pythagorean Theorem $a^2 + b^2 = c^2$ to get your answer. In this problem, you know that side c, the side opposite the right angle, has a length of 7. $c = 7$ and $c^2 = 49$. You also know that one other side has a length of 5.

$$a^2 + 5^2 = 7^2$$
$$a^2 + 25 = 49.$$

Solve for a:

$$a^2 = 49 - 25$$
$$a^2 = 24$$
$$a = \sqrt{24} \text{ in.}$$

22. **D.** The area of a rectangle is length × width. In the figure, the length is 27 cm, and the width is 9 cm. Multiplying 27×9 gives the correct answer of 243 cm².

23. **C.** The formula for the area of a circle is πr^2. You are given the radius of 6 meters. Substituting the information into the formula gives $3.14\,(6)^2$, or 113.04 m².

24. **C.** 1% is the equivalent of 0.01.

$$0.01 \times 5,000 = 50$$

25. **D.** The volume of a sphere is found by using the formula $\frac{4}{3}\pi r^3$. Substituting the information into the formula gives $\frac{4}{3}(3.14)(3)^3$ or $4 \times 3.14 \times 27 \div 3$, making the correct answer 113.04 m^3.

PART 6. ELECTRONICS INFORMATION

1. **A.** An electric motor makes a short but heavy current demand while starting. Current is measured in amperes.

2. **B.** Capacitors store current.

3. **B.** Current = volts/resistance = 480/24 = 20 amperes.

4. **D.** Voltage measures electrical "pressure."

5. **C.** To measure current, the meter must become part of the circuit, as it is at point A.

6. **A.** This is the standard symbol for a transformer.

7. **D.** Watts = amperes × volts = 120 × 12 = 1,440 watts.

8. **B.** Watts = volts × amperes. 1,800 = 120 × amperes. Amperes = 1,800/120 = 15 amperes.

9. **C.** An electrical wire is usually a copper conductor wrapped with a plastic insulator.

10. **A.** Corrosion (oxidation) is usually a good insulator. By resisting corrosion, gold ensures a low-resistance contact.

11. **B.** A transistor can slow or stop current.

12. **D.** Current is the flow of electrons, and current flows through a conductor.

13. **D.** Under certain conditions, lead-acid batteries generate hydrogen gas, which is explosive. Keep cigarettes and sparks away from these batteries!

14. **A.** In the metric system, "kilo" means 1,000, so 1 kilowatt hour translates into 1,000 watt hours, or 1,000 watts for 1 hour.

15. **C.** Wire gauge is sized backwards; the higher the gauge, the smaller the diameter.

16. **D.** In a series circuit like this, total resistance is the sum of the individual resistances. 400 + 400 + 300 = 1,100.

17. **C.** In a parallel circuit like this, $1/R_{total} = 1/R_1 + 1/R_2 + 1/R_3 = 1/400 + 1/400 + 1/400 = 3/400$. That equals $1/R_{total}$, so invert : 400/3 = 133.3 ohms.

18. **A.** The change in voltage is determined by the ratio of windings. If the output side has fewer windings, the transformer reduces the voltage.

19. **D.** An electromagnet moves the clapper that bangs against the bell in a doorbell (and in older telephones).

20. **B.** The polarity of alternating current goes through 60 complete cycles per second in 60-hertz current.

PART 7. AUTO AND SHOP INFORMATION

1. **C.** Knocking or "pinging" usually means that the octane of your gasoline is too low. The noise indicates that the fuel–air mixture is igniting too soon inside the cylinders.

2. **B.** Water expands as it freezes, and that could destroy the engine block and cylinder head.

3. **A.** Brake problems reduce stopping ability. Why are the other answers incorrect? Choice B: it's vital to check brake problems right away, not when it's convenient. Choices C and D: these can result from brake problems, but are less complete answers than choice A.

4. **D.** Front tire life is lower with front-wheel drive because the wheels are driving, steering, and braking. However, front-wheel drive pulls the car in the direction you steer, and it is especially helpful in slippery conditions. Because the engine is above the driving wheels, the wheels get better traction. Because there is no driveshaft, there is more room in the interior.

5. **B.** The primary job of oil is to reduce friction. More friction also increases noise and temperature, but without oil, an engine would barely rotate at all.

6. **D.** Correct valve operation helps an engine run smoothly, with maximum power.

7. **C.** Torque, not horsepower, is what accelerates a vehicle, and so it's the most likely problem. Torque is defined as a twisting force, and it is measured in foot-pounds; one pound of force acting through one foot of rotary motion.

8. **C.** Because a heavier car needs more energy, it also needs more gasoline, which is its source of energy.

9. **D.** Batteries lose some of their power when they get cold, and high-viscosity oil gets thick in winter, requiring more force to turn the engine over. Choice A: starter motor, may be a problem, but either choice B or choice C is much more likely to be the cause.

10. **C.** When jump-starting a car, the battery cables must be connected positive to positive, and negative to negative. Why are the other answers incorrect? Choice A: the ground cable should be connected to the engine block or the ground terminal on the battery. Choice B: battery caps should remain in place. Choice D: the transmission should be in neutral or park.

11. **D.** Problems with the synchronizers or clutch adjustment can cause poor shifting. Choice C is incorrect: a torque converter is found in automatic, not manual, transmissions.

12. **D.** Problems with lubrication, cooling, or main bearings can all cause an engine to seize.

13. **A.** Both round and pan heads are flat on the bottom, so they will securely fasten a flat plate to a joist or other flat piece of wood. Why are the other answers incorrect? Choice B: a Phillips head is not necessarily flat on the bottom. Choice C: a flat head screw is cone-shaped on the bottom and would not hold the plate securely. Choice D: as explained earlier, the best screw head will be flat on the bottom.

14. **D.** The tool shown is a keyhole saw.

15. **C.** Sanding smoothes the wood, priming prevents it from absorbing too much paint, sanding again smoothes the primer, making it ready for painting.

16. **B.** Why are the other answers incorrect? Choice A: it's impossible to make vertical and horizontal joints occur on top of studs. Choice C: joints cannot overlap. D: joints are needed in any wall larger than one sheet of drywall.

17. **D.** Sanding drywall creates lots of dust. Why are the other answers incorrect? Choice A: a belt sander is much too aggressive for drywall. Choice B: medium sandpaper would gouge drywall; use fine paper. Choice C: drywall has no grain.

18. **D.** The tool shown is a combination wrench.

19. **D.** Metal always heats quickly on an electric grinder. Why are the other answers incorrect? Choice A: a fine-grained wheel would be acceptable if it's rated at 4,500 rpm. Choice B: it is *unsafe* to use a 3,000 rpm wheel on a 4,500 rpm grinder. Choice C: use water only on a grinder designed for water cooling.

20. **A.** A wire brush will remove paint from metal, remove rust, and clean bolt threads. It will not clean a wood deck, prepare drywall for painting, clean threads inside a nut, or dull paint before repainting.

21. **B.** The tool shown is best used for clamping.

22. **B.** Almost all sheets of plywood, drywall, and oriented-strand board are 4 feet by 8 feet. The thickness can vary.

23. **A.** Galvanizing is a zinc coating designed to prevent rusting (oxidation) on steel. Why are the other answers incorrect? Galvanizing does not smooth, harden, or sharpen steel.

24. **C.** A door hinge pin and auto axle both hold rotating objects in place.

25. **A.** A rasp, a plane, and sandpaper are all used to smooth wood surfaces.

PART 8. MECHANICAL COMPREHENSION

1. **C.** 480/120 = 4 minutes until drained. $4 \times 60 =$ 240 seconds.

2. **C.** MA = load/effort = 8/2 = 4.

3. **A.** When a gas is compressed, it warms up because the energy that runs the pump is converted into heat.

4. **B.** Compare the areas of the bases of the cylinders to find mechanical advantage of a hydraulic system. Cylinder base area = $\pi \times$ radius^2. So the area of two cylinders is proportional to the square of their radii (plural of radius). $8^2 = 64$, $4^2 = 16$. 64/16 = 4.

5. **C.** For two gears to mesh properly, they must have teeth that are the same size. The gears can be of different sizes, they can rotate at different speeds, and they can be made of different materials.

6. **D.** If the screw has a pitch of ¼ inch, then with 4 complete turns of the screwdriver it will move $4 \times ¼ = 1$ inch. After 10 complete turns it will have moved $10 \times ¼ = 10/4 = 2¼ =$ 2½ inches.

7. **C.** Gear A is turning clockwise. That means that gear B is turning counterclockwise. Gears C and D turn clockwise, the same direction as gear A. Gear E turns counterclockwise, the same as gear B.

8. **C.** Of the support brackets shown, choice C, which creates a triangle with the vertical wall, provides the most secure support.

9. **A.** Of the patterns shown, choice A is the only one that, when folded on the dotted lines, will make a three-sided pyramid.

10. **D.** Levers, inclined planes, and pulleys are all simple machines.

11. **D.** To find the water pressure in lb/ft^2, first find the area of the bottom of the tank: $2 \times 6 =$ 12 ft^2. Then divide the total weight of the water by the area of the bottom: $600 \div 12 = 50$ lb/ft.^{2}

12. **B.** If the crank makes 1/2 turn, the piston will move up to the top of the cylinder.

13. **D.** The ramp (inclined plane) is a simple machine. To calculate its mechanical advantage, divide the length by the height: $30 \div 3 = 10$. To find how much effort is needed, divide the weight of the drum by the mechanical advantage: $120 \div 10 = 12$ lb of effort.

14. **B.** Metal conducts heat better than does wood or plastic, so it would feel hottest.

15. **A.** The gear is rotating at a speed of 150 rpm. At a ratio of 6:1, the pinion makes 6 revolutions for each revolution of the gear. So the speed of the pinion is $150 \times 6 = 900$ rpm.

16. **C.** A spring is a device that can be twisted or stretched by some force applied to it, and it will resume its original shape when the force is released. Of the choices, the only one that can be made to act in this manner is choice C, steel.

17. **A.** The system has 2 supporting strands, so MA = 2.

18. **B.** The larger pulley in a system will rotate slower than the smaller one, but both pulleys will rotate in the same direction.

19. **B.** A longer handle allows you to move the hammer head faster, producing more striking force.

20. **C.** Sheet metal screws must cut threads in metal, so they must be made of hard steel. They may be longer or shorter than wood screws.

21. **C.** If water is flowing into the tank at a rate of 5 gal/sec, then the amount flowing in per minute is $5 \times 60 = 300$ gallons. The amount flowing out per minute is 100 gallons, so the net gain per minute is $300 - 100 = 200$ gallons. In 2 minutes, the tank will gain 400 gallons more than the 100 gallons already present, so the total amount in the tank after 2 minutes will be $400 + 100 = 500$ gallons.

22. **A.** Point A, located on the outside of the wheel, travels farther than point B each time the wheel makes a complete turn.

23. **B.** In gear systems with an idler gear, you can discount the idler gear and just look at the number of teeth on the drive gear (A) and the driven gear (C). Here, the ratio is 18 to 24, or 3 to 4. So the driven gear will be rotating slower than the drive gear. $3/4 \times 200 = 150$.

24. **C.** The same quantity of water must be flowing past every point in the system, but the water must speed up through the smaller pipe. Fluid pressure falls when velocity increases.

25. **B.** Gear X has 60 teeth and gear Y has 12 teeth. $60 \div 12 = 5$, so for each complete rotation by gear X, gear Y makes 5 rotations.

PART 9. ASSEMBLING OBJECTS

1. **B.** The original figure has four objects, while choices A and C have five. Look for the object shaped like a large comma. Choice D does not have this object. Choice B is the correct answer.

2. **A.** The original figure has three objects, while choice C has four and choice D has only two. Look for the long thin rectangle. It is not present in choice B (there is a longer parallelogram there). Choice A is the correct answer.

3. **A.** The original figure has five objects, while choices B and D have six. Look for the small square. There is no square in choice C. Choice A is the correct answer.

4. **C.** The original figure has four objects, while choice A has five. Look for the largest piece. Choice B has a mirror image of this piece, so eliminate B. Look for the piece that looks like a right triangle with the right angle rounded off. Choice D has an actual right triangle instead. Choice C is the correct answer.

5. **C.** The original figure has three objects, while choice A has only two. Look for the pair of identical triangles. Choice B has only one triangle. Choice D has a rectangle rather than a parallelogram. Choice C is the correct answer.

6. **B.** The original figure has five objects, while choice A has six. Look for the wedge shape at the top middle of the original box. Choices C and D have wedges that are much too large. Choice B is the correct answer.

7 **B.** The original figure has three objects, while choice C has only two. Look for the moon shape. Choice A does not have it. Look for the triangle. Choice D does not have it. Choice B is the correct answer.

8. **A.** The original figure has four objects, while choice B has five. Look for the object shaped like a segment of a circle. Choice C does not have it. Look at the large shape on the left in the original. It is too small in choice D. Choice A is the correct answer.

9. **C.** The original figure has three objects and choice D has four, so eliminate D. Look for the distinctive object at the left. Choice A does not have it. Look for the circle. Choice B has an oval instead. Choice C is the correct answer.

10. **D.** The original figure contains four objects. Choice B has five, so eliminate it. Look for the distinctive shape on the left of the original figure. Choices A and C do not have that shape. Choice D is the correct answer.

11. **C.** The original figure has three objects, while choice A has four and choice D has two. Choice B has a square that is much larger than the original. Choice C is the correct answer.

12. **A.** The original figure has four objects, while choice B has only three. Look for the most distinctive object: the top left object looks like a parallelogram but with a curved right side. Choice D has a similar object, but it does not have this curved side. That leaves choices A and C. Look for the triangle. The one in choice C may be too large, but if you can't tell, look at the two quadrilaterals. In choice C, they are not the same shapes. Choice A is the correct answer.

13. **D.** The original figure contains three objects. Choice A has two and choice C has four, so eliminate them. Look for the parallelogram. Choice B does not have a parallelogram. Choice D is the correct answer.

14. **A.** The original figure has three objects, while choice D has four. None of the objects are really distinctive, but choose one to look for first. The isosceles triangle at the top left might be a good choice. Choice B does not have it. Look for the pair of quadrilaterals. Choice C has only triangles. Choice A is the correct answer.

15. **A.** The original figure has four objects, while choice B has five. Look for the most distinctive object: the one that looks like the base of a column. Choice C does not have it and in choice D it is much too tall. Choice A is the correct answer.

16. **D.** The original figure has three objects and choice B only has two, so eliminate B. Look for the most distinctive object—perhaps the one that looks like a banana. Choices A and C do not have that object. Choice D is the correct answer.

17. **B.** The original figure has four objects and choice D has three, so eliminate D. Look for the most distinctive object—perhaps the one at the lower left that looks like a pointed boot. Choice C does not have an object like that. There is a vaguely similar object in choice A, but it is not the same pointed-boot shape. Choice B is the correct answer.

18. **D.** P oint A is in the center of the bottom left base of the first object in its original orientation. Choices A, B, and C do not connect the first object at that point. Choice D is the correct answer.

19. **C.** Point A is at the top right corner in the original orientation. In choice A this point is at the bottom right corner. Point B is at the bottom left corner in the original orientation. In choice B it is at the pointed corner. In choice D it is not at a corner at all. Choice C is the correct answer.

20. **D.** In this set of objects, the larger object with point B is more distinctive, so start with that one. The object looks like the head of a duck (upside down) with point B as the eye. Choices A and C do not have point B at this spot. Choice B has point B in the correct spot, but the object is a mirror image of the original, so eliminate it. Choice D is the correct answer.

21. **C.** The first object looks like a mitten with point A at the thumb. Choices A and B do not have point A at the correct spot. Point B is at the top left of the second object in its original orientation. In choice D, point B is at the bottom right of the object. Choice C is the correct answer.

22. **D.** In this set of objects, the object on bottom with point B is more distinctive, so start with that one. Point B is at the sharp tip of this object. Choice A does not connect at that point. The object on top has point A at the sharpest point as well. Choices B and C do not connect at that point. Choice D is the correct answer.

23. **B.** Point A is at the middle of the base of the first object (in its original orientation). Choice D places point A on the top side of the object. Picture the second object sitting on the flat base. In this orientation, point B is located in the center of the bottom right corner, not on the edge, but within the object. Choice A has point B located in the center of the bottom left corner. Choice C has point B located in the center of the object (plus the object with point A is shown in mirror image). Choice B is the correct answer.

24. **C.** Point A is at the pointed tip of the first object. Choice A does not connect at that point. Point B is at the pointed outside corner of the second object. Choice B has point B at the inside corner. Choice D has point B on one leg of the object. Choice C is the correct answer.

25. **D.** Point A is at the sharpest of the three corners. Choices A and C do not connect at that point. If you orient the second object with the flat base on the bottom, then point B is at the left bottom corner. Choice B has point B at the bottom right corner. Choice D is the correct answer.

ASVAB PRACTICE TEST FORM 3

Try to take this practice test under actual test conditions. Find a quiet place to work and set aside a period of approximately $2\frac{1}{2}$ hours when you will not be disturbed. Work on only one section at a time, and use your watch or a timer to keep track of the time limits for each section. Mark your answers on the answer sheet, just as you will when you take the real exam.

At the end of the test, you'll find an answer key and explanations for every question. After you check your answers against the key, you can complete a chart that will show you how you did on the test and what test topics you might need to study more. Then review the explanations, paying particular attention to the ones for the questions that you answered incorrectly.

If you take the CAT-ASVAB, your test will be given by computer and the number of your question, the mode of administration, and the timing will be different. This process has been explained in previous chapters.

GENERAL DIRECTIONS

IF YOU ARE TAKING THE PAPER AND PENCIL VERSION OF THE ASVAB, THE TEST ADMINISTRATOR WILL READ THE FOLLOWING ALOUD TO YOU:

DO NOT WRITE YOUR NAME OR MAKE ANY MARKS in this booklet. Mark your answers on the separate answer sheet. Use the scratch paper which was given to you for any figuring you need to do. Return this scratch paper with your other papers when you finish the test.

If you need another pencil while taking this test, hold your pencil above your head. A proctor will bring you another one.

This booklet contains 9 tests. Each test has its own instructions and time limit. When you finish a test, you may check your work in that test ONLY. Do not go on to the next test until the examiner tells you to do so. Do not turn back to a previous test at any time.

For each question, be sure to pick the BEST ONE of the possible answers listed. Each test has its own instructions and time limit. When you have decided which one of the choices given is the best answer to the question, blacken the space on your answer sheet that has the same number and letter as your choice. Mark only in the answer space. BE CAREFUL NOT TO MAKE ANY STRAY MARKS ON YOUR ANSWER SHEET. Each test has a separate section on the answer sheet. Be sure you mark your answers for each test in the section that belongs to that test.

Here is an example of correct marking on an answer sheet.

S1. A triangle has

    A. 2 sides
    B. 3 sides
    C. 4 sides
    D. 5 sides

PRACTICE

S1 Ⓐ ● Ⓒ Ⓓ
S2 Ⓐ Ⓑ Ⓒ Ⓓ
S3 Ⓐ Ⓑ Ⓒ Ⓓ

The correct answer to Sample Question S1 is B.

Next to the item, note how space B opposite number S1 has been blackened. Your marks should look just like this and be placed in the space with the same number and letter as the correct answer to the question. Remember, there is only ONE BEST ANSWER for each question. If you are not sure of the answer, make the BEST GUESS you can. If you want to change your answer, COMPLETELY ERASE your first answer mark.

Answer as many questions as possible. Do not spend too much time on any one question. Work QUICKLY, but work ACCURATELY. DO NOT TURN THE PAGE UNTIL TOLD TO DO SO. Are there any questions?

ASVAB PRACTICE TEST FORM 3

ANSWER SHEET

ANSWER SHEET PAGE 3
ARMED SERVICES VOCATIONAL
APTITUDE BATTERY

DD FORM 1304-5AS   FOR OFFICIAL USE ONLY (WHEN COMPLETED)

PART 1. GENERAL SCIENCE

THE TEST ADMINISTRATOR WILL READ THE FOLLOWING ALOUD TO YOU:

Turn to Part 1 and read the directions for General Science silently while I read them aloud.

This test has questions about science. Pick the BEST answer for each question and then blacken the space on your separate answer sheet that has the same number and letter as your choice. Here are three practice questions.

S1. A rose is a kind of
   A. animal.
   B. bird.
   C. flower.
   D. fish.

S2. A cat is a
   A. plant.
   B. mammal.
   C. reptile.
   D. mineral.

S3. The earth revolves around the
   A. sun.
   B. moon.
   C. meteorite.
   D. Mars.

Now look at the section of your answer sheet labeled "PRACTICE." Notice that answer space C has been marked for question S1. Now do practice questions S2 and S3 by yourself. Find the correct answer to the question, then mark the space on your answer sheet that has the same letter as the answer you picked. Do this now.

You should have marked B for question S2 and A for question S3. If you make any mistakes, erase your mark carefully and blacken the correct answer space. Do this now.

Your score on this test will be based on the number of questions you answer correctly. You should try to answer every question. If you finish before time is called, go back and check your work in this part ONLY. Now find the section of your answer sheet that is marked PART 1. When you are told to begin, start with question number 1 in Part 1 of your test booklet and answer space number 1 in Part 1 on you separate answer sheet. DO NOT TURN THE PAGE UNTIL TOLD TO DO SO. This part has 25 questions in it. You will have 11 minutes to complete it. Most of you should finish in this amount of time. Are there any questions?

Turn the page and begin.

1. The part of a cell that determines what enters and exits the cell is called the
   A. cell membrane.
   B. vacuoles.
   C. mitochondria.
   D. chlorophyll.

2. Roses, apple trees, and other plants that produce flowers are called
   A. angiosperms.
   B. annuals.
   C. deciduous.
   D. vascular.

3. A stalk of green celery has been placed in a glass with 2 inches of water with purple food coloring. Two hours later parts of the stalk and leaves of the celery are purple. Which of the following plant systems does this situation demonstrate?
   A. Vascular system
   B. Reproductive system
   C. Hereditary system
   D. Lymphatic system

4. Which of the following terms best represents a whale?
   A. Exoskeleton, cold-blooded
   B. Vertebrate, mammal
   C. Warm-blooded, invertebrate
   D. Carnivore, invertebrate

5. Which of the following is classified as a mammal?
   A. Fish
   B. Snake
   C. Crocodile
   D. Ape

6. Which of the following is considered to be a food producer?
   A. Frog
   B. Tree
   C. Parrot
   D. Manatee

7. The joints in bones are held together by which of the following?
   A. Cartilage
   B. Ligaments
   C. Hinges
   D. Marrow

8. Which of the following is the largest organ in your body?
   A. Skin
   B. Liver
   C. Kidney
   D. Lymph

9. Which of the following nutrients contain the most energy?
   A. Carbohydrates
   B. Proteins
   C. Minerals
   D. Fats

10. Most of your blood is made up of which of the following?
    A. Plasma
    B. White blood cells
    C. Red blood cells
    D. Platelets

11. Which part of your ear is responsible for your balance?
    A. Outer ear
    B. Middle ear
    C. Inner ear
    D. Olfactory membrane

12. Which of the following represents the basic difference between mitosis and meiosis?
    A. Cells divide twice as fast in mitosis as in meiosis.
    B. In mitosis each cell produced has 46 chromosomes, but in meiosis each cell produced has only 23 chromosomes.
    C. More nutrients are produced in mitosis than in meiosis.
    D. Mitosis occurs only with sexual reproduction, while meiosis can occur with sexual or asexual reproduction.

13. Which of the following characteristics describe ultraviolet radiation as compared to visible light?
    A. Less energy; lower frequency
    B. More energy; lower frequency
    C. Less energy; higher frequency
    D. More energy; higher frequency

14. A candle burns for 2 hours. What happens to the total mass as a result of the burning?
    A. It decreases.
    B. It increases.
    C. It stays the same.
    D. It first decreases and then increases.

15. Which of the following layers of Earth takes up the least amount of Earth's volume?
    A. Crust
    B. Mantle
    C. Lithosphere
    D. Core

16. Tectonic plates that are moving away from one another create which of the following features?
    A. Rifts
    B. Ridges
    C. Currents
    D. Land bridges

17. Sinking water that forms density currents is likely to have which of the following characteristics?
    A. Warm temperature and many animal inhabitants
    B. Swift-moving surface currents
    C. Cold temperatures and high salt content
    D. A high concentration of dissolved gases

18. The tilt of Earth's axis as the planet orbits around the sun causes which of the following?
    A. Tides
    B. Day and night
    C. Bulging of Earth at the equator
    D. Seasons

19. What causes the same side of the moon to always face Earth?
    A. Earth's gravitational pull on the moon
    B. The fast rotation of Earth in comparison to the moon
    C. The phases of the moon in comparison to the sun
    D. The friction of the moon in space

20. Which of the following elements is not naturally a gas at room temperature?
    A. Sulfur
    B. Oxygen
    C. Neon
    D. Fluorine

21. Most of Earth's weather takes place in which of the following atmospheric zones?
    A. Troposphere
    B. Stratosphere
    C. Thermosphere
    D. Exosphere

22. Which layer in the atmosphere reflects radio waves and allows long-distance communication on Earth?
    A. Troposphere
    B. Stratosphere
    C. Thermosphere
    D. Ionosphere

23. Which of the following is not an agent of erosion?
    A. Freezing and thawing
    B. Rivers
    C. Lava flows
    D. Lichen and mosses

24. The sun's energy comes from which of the following processes?
    A. Sunspot formation
    B. Magnetic fields
    C. Fusion of hydrogen to make helium
    D. Prominences that eject highly charged solar particles

25. When the nucleus of an atom gives up a particle and its energy in order to become more stable, the process is called which of the following?
    A. Ionization
    B. Refraction
    C. Radioactive decay
    D. Half-life

STOP! DO NOT TURN THIS PAGE UNTIL TIME IS UP FOR THIS TEST. IF YOU FINISH BEFORE TIME IS UP, CHECK OVER YOUR WORK ON THIS TEST ONLY.

PART 2. ARITHMETIC REASONING

THE TEST ADMINISTRATOR WILL READ THE FOLLOWING ALOUD TO YOU:

Turn to Part 2 and read the directions for Arithmetic Reasoning silently while I read them aloud.

This is a test of arithmetic word problems. Each question is followed by four possible answers. Decide which answer is CORRECT, and then blacken the space on your answer sheet that has the same number and letter as your choice. Use your scratch paper for any figuring you wish to do.

Here is a sample question. DO NOT MARK your answer sheet for this or any further sample questions.

S1. A student buys a sandwich for 80 cents, milk for 20 cents, and pie for 30 cents. How much did the meal cost?

    A. $1.00
    B. $1.20
    C. $1.30
    D. $1.40

The total cost is $1.30; therefore, C is the right answer. Your score on this test will be based on the number of question you answer correctly. You should try to answer every question. DO NOT SPEND TOO MUCH TIME on any one question. If you finish before time is called, go back and check your work in this part ONLY.

Now find the section of your answer sheet that is marked PART 2. When you are told to begin, start with question number 1 in Part 2 of your test booklet and answer space number 1 in Part 2 on your separate answer sheet.

DO NOT TURN THIS PAGE UNTIL TOLD TO DO SO. You will have 36 minutes for the 30 questions. Are there any questions?

Begin.

1. Len bought several books for school at $25.16, $28.99, $35.15, $42.27, and $62.12. How much did Len spend?
   A. $147.16
   B. $154.29
   C. $193.69
   D. $199.39

2. Marty is driving from Nashville, Tennessee to Mobile, Alabama. The distance is 527 miles. If Marty drives 111 miles on Monday, 67 miles on Tuesday, and 211 miles on Wednesday, how many miles must he drive if he wants to reach Nashville on Thursday?
   A. 28 miles
   B. 118 miles
   C. 138 miles
   D. 168 miles

3. Jamal was up to bat 72 times during the baseball season. If he got 54 hits, what percent of the time was he successful?
   A. 56%
   B. 75%
   C. 79%
   D. 81%

4. Phil is monitoring his calorie consumption. On Monday he eats meals totaling 1,700 calories. On Tuesday he eats meals totaling 3,000 calories. On Wednesday he eats meals totaling 2,500 calories, and on Thursday his meals total 3,200 calories. What is his average calorie consumption per day?
   A. 1,600 calories
   B. 1,885 calories
   C. 2,100 calories
   D. 2,600 calories

5. If there are 450 members of a sports club and 60 percent are males, how many members are males?
   A. 210
   B. 225
   C. 255
   D. 270

6. On a map of the United States, 1 inch represents a distance of 50 miles. If the distance between New Town and Old Town is 450 miles, how far apart are they on the map?
   A. 2.5 inches
   B. 8.3 inches
   C. 9 inches
   D. 12 inches

7. Phil bought two hamburgers and a milkshake for $10.30. Janice bought three hamburgers and a milkshake for $14.80. What is the cost of the two milkshake?
   A. $1.30
   B. $1.45
   C. $2.60
   D. $4.50

8. A box contains 35 green balls, 75 red balls, and 45 orange balls. If Nikki picks out a ball at random, what is the probability that the ball will be green or orange?
   A. $\dfrac{5}{12}$
   B. $\dfrac{6}{11}$
   C. $\dfrac{16}{31}$
   D. $\dfrac{18}{47}$

9. Jose bought a $3,000 savings bond that earns 5 percent compounded interest per year. How much will his bond be worth after 3 years?
   A. $3,150.00
   B. $3,307.50
   C. $3,348.88
   D. $3,472.88

10. After five science tests, Sonya's average score is 89. If she gets a 95 on the sixth test, what is her new average?
    A. 89.50
    B. 89.90
    C. 90.00
    D. 90.50

11. A pasta recipe calls for $\frac{2}{3}$ cup of tomato paste for each batch of lasagna. If Victor has 18 cups of tomato paste, how many batches of lasagna can he make?
    A. 12
    B. 21
    C. 26
    D. 27

12. Madeline wants to carpet her living room floor. The floor measures 12 feet by 24 feet. How many square yards of carpeting must she purchase?
    A. 12
    B. 22
    C. 24
    D. 32

13. Joanne earns $3,000 per month at her job at the department store. If her taxes are 28 percent of her pay, what is her monthly take-home pay?
    A. $2,160
    B. $2,280
    C. $2,370
    D. $2,780

14. On his 16th birthday Gene was 60 inches tall. On his 19th birthday he was 69 inches tall. What percent increase in height did Gene experience in the three years from his 16th to his 19th birthday?
    A. 10%
    B. 12%
    C. 13%
    D. 15%

15. Patrick took a trip on his motorcycle. He drove at a steady speed of 50 mph, and the trip took him 3 hours. How long would the trip have taken if his speed was 40 mph?
    A. 3.75 hr
    B. 4.00 hr
    C. 4.25 hr
    D. 4.45 hr

16. Troy is building a patio. He will make it 360 inches by 480 inches. If one bag of cement makes 16 sq feet of patio, how many bags will Troy need to complete the job?
    A. 32.5 bags
    B. 75 bags
    C. 95 bags
    D. 97.5 bags

17. A sprinkler sprays water in a circular region with a radius of 20 feet. What area can it cover?
    A. 1,256 ft^2
    B. 1,596 ft^2
    C. 1,996 ft^2
    D. 2,056 ft^2

18. Suzie is twice as old as Will, plus 4 years. If Will is 9, how old is Suzie?
    A. 19
    B. 21
    C. 22
    D. 29

19. Milo is saving quarters in a jar so that he can buy a computer. The computer costs $950. How many quarters must Barry save in order to purchase the computer?
    A. 1,200
    B. 3,800
    C. 4,200
    D. 4,276

20. Suzie has a piece of ribbon 6 yards long. She needs to cut the ribbon into lengths of 1½ feet. How many pieces will she be able to cut?
    A. 10
    B. 11
    C. 12
    D. 13

21. Randy is training to cycle in a 15-kilometer race. His training program starts at cycling 5 kilometers per day for the first week and increases that amount by 20 percent each week. How many kilometers will he be cycling per day in the fourth week?
    A. 7.20 km
    B. 8.64 km
    C. 9.33 km
    D. 9.98 km

22. A mother is sending a box of brownies to her daughter. The box is 6 inches high, 6 inches wide, and 15 inches long. If each brownie is 2 inches wide, ½ inch high, and 3 inches long, how many brownies will fit into the box?
    A. 25
    B. 55
    C. 90
    D. 180

23. If a pound of potatoes make 24 cups of hash browns, how many pounds of potatoes would it take to make 36 cups of hash browns?
    A. 1.5
    B. 2.0
    C. 3.4
    D. 4.4

24. A can of tomato paste is 10 cm high and has a diameter of 8 cm. How much tomato paste can the cans hold?
    A. 202.8 cm^3
    B. 347.0 cm^3
    C. 502.4 cm^3
    D. 538.4 cm^3

25. A circular balloon has a diameter of 18 inches. What is the volume?
    A. 114.24 in^3
    B. 3,052.08 in^3
    C. 5,624.10 in^3
    D. 8,416.34 in^3

26. Christy orders 1 pound of meat from the butcher. She wants it cut into slices each weighing 0.25 oz. About how many slices will Christy receive?
    A. 24
    B. 44
    C. 64
    D. 84

27. If Kent reads an average of 55 pages per day, how many days will it take him to read a book of 880 pages?
    A. 16
    B. 18
    C. 22
    D. 25

28. James took a trip of 448 miles in 8 hours. What was his average speed?
    A. 45 miles per hour
    B. 56 miles per hour
    C. 62 miles per hour
    D. 65 miles per hour

29. Kiki has been asked to decorate a square room with an area of 576 sq feet. What is the length of each side of the room?
    A. 17 feet
    B. 24 feet
    C. 26 feet
    D. 29 feet

30. A diver's score is determined by the degree of difficulty of each dive multiplied by the points awarded for that dive. The table below shows the points Wayne earned for five dives and the degree of difficulty for each dive. What was Wayne's total score?

Points	Difficulty
8.2	1.2
9.0	1.4
8.3	1.5
9.5	1.4
9.5	1.2

    A. 47.59
    B. 59.59
    C. 64.29
    D. 66.29

STOP! DO NOT TURN THIS PAGE UNTIL TIME IS UP FOR THIS TEST. IF YOU FINISH BEFORE TIME IS UP, CHECK OVER YOUR WORK ON THIS TEST ONLY.

PART 3. WORD KNOWLEDGE

THE TEST ADMINISTRATOR WILL READ THE FOLLOWING ALOUD TO YOU:

Now turn to Part 3 and read the directions for Word Knowledge silently while I read them aloud.

This is a test of your knowledge of word meanings. These questions consist of a sentence or phrase with a word or phrase underlined. From the four choices given, you are to decide which one MEANS THE SAME OR MOST NEARLY THE SAME as the underlined word or phrase. Once you have made your choice, mark the space on your answer sheet that has the same number and letter as your choice.

Look at the sample question.

S1. The weather in this geographic area tends to be <u>moderate</u>.

    A. severe
    B. warm
    C. mild
    D. windy

The correct answer is "mild" which is choice C. Therefore, you would have blackened in space C on your answer sheet.

Your score on this test will be based on the number of questions you answer correctly. You should try to answer every question. DO NOT SPEND TOO MUCH TIME on any one question. If you finish before time is called, go back and check your work in this part <u>ONLY</u>.

Now find the section of your answer sheet that is marked PART 3. When you are told to begin, start with question number 1 in Part 3 of your test booklet and answer space number 1 in Part 3 on your separate answer sheet.

DO NOT TURN THE PAGE UNTIL TOLD TO DO SO. You will have 11 minutes to complete the 35 questions in this part. Are there any questions?

Begin.

1. The hot afternoon sun made them <u>listless</u>.
   A. Active
   B. Energized
   C. Limp
   D. Thirsty

2. It was a <u>potent</u> combination of chemicals.
   A. Movable
   B. Powerful
   C. Colorful
   D. Explosive

3. They felt <u>compelled</u> to fulfill the request.
   A. Forced
   B. Happy
   C. Honored
   D. Free

4. He had a <u>vivid</u> imagination.
   A. Poor
   B. Stagnant
   C. Dull
   D. Colorful

5. His attempt to escape was <u>futile</u>.
   A. Successful
   B. Useless
   C. Easy
   D. Powerful

6. The patient <u>solicited</u> the advice of the doctor.
   A. Ignored
   B. Followed
   C. Requested
   D. Rejected

7. <u>Appease</u> most nearly means
   A. satisfy.
   B. hurt.
   C. scare.
   D. wonder.

8. <u>Abhor</u> most nearly means
   A. adore.
   B. worship.
   C. hate.
   D. reject.

9. Sally was <u>prudent</u> in the way she spent her allowance.
   A. Reckless
   B. Careful
   C. Stingy
   D. Wasteful

10. The family was a <u>staunch</u> supporter of the candidate.
    A. Steadfast
    B. Part-time
    C. Talkative
    D. Periodic

11. Her <u>zealous</u> beliefs contradicted the norm.
    A. Whimsical
    B. Incorrect
    C. Unusual
    D. Passionate

12. He had a <u>volatile</u> temper.
    A. Motivated
    B. Humorous
    C. Explosive
    D. Acceptable

13. The note she received was <u>terse</u> in its explanation.
    A. Expansive
    B. Concise
    C. Confusing
    D. Wordy

14. The living quarters were <u>austere</u>.
    A. Extravagant
    B. Fancy
    C. Simple
    D. Crowded

15. Darius was <u>meticulous</u> about his finances.
    A. Painstaking
    B. Careless
    C. Doubtful
    D. Persistent

16. Her contribution was <u>vital</u> to the success of the project.
    A. Unimportant
    B. Unrelated
    C. Harmful
    D. Essential

17. John was <u>timid</u> when asked to answer in class.
    A. Talkative
    B. Hesitant
    C. Upset
    D. Present

18. Her words were <u>inaudible.</u>
    A. Unheard
    B. Loud
    C. Funny
    D. Suspicious

19. The best time to plant bushes is when they are <u>dormant</u>.
    A. Growing
    B. Inactive
    C. Dry
    D. Seedless

20. I <u>relinquished</u> my place in line because I had to go back to work.
    A. Saved
    B. Transferred
    C. Surrendered
    D. Enjoyed

21. <u>Delegated</u> most nearly means
    A. absolved.
    B. assigned.
    C. withheld.
    D. promoted.

22. The new administrative assistant proved to be <u>indispensable</u> to the law firm.
    A. Hurtful
    B. Annoying
    C. Crucial
    D. Energizing

23. The employee gave a <u>plausible</u> explanation for the missing equipment.
    A. Ridiculous
    B. Lengthy
    C. Lofty
    D. Credible

24. The tax cut was <u>advantageous</u> to the middle class.
    A. Beneficial
    B. Harmful
    C. Useless
    D. Compromising

25. The court decision <u>aroused</u> the anger of many parents and teachers.
    A. Reduced
    B. Increased
    C. Appeased
    D. Awakened

26. The board gave <u>provisional</u> approval for the new airport.
    A. Initial
    B. Lasting
    C. Tentative
    D. Final

27. Marty's musical ability was <u>unparalleled</u>.
    A. Equaled
    B. Unmatched
    C. Appreciated
    D. Misunderstood

28. <u>Intermittent</u> most nearly means
    A. periodic.
    B. constant.
    C. disruptive.
    D. annoying.

29. Joan was <u>obstinate</u> about her decision to enroll in a creative writing class.
    A. Positive
    B. Unsure
    C. Hopeful
    D. Stubborn

30. Ramon was <u>tactful</u> when he told her the news.
    A. Unruly
    B. Thoughtless
    C. Diplomatic
    D. Forceful

31. Han was hired because she was <u>proficient</u> in computers.
    A. Superior
    B. Unschooled
    C. Trained
    D. Skilled

32. <u>Augment</u> most nearly means
    A. reduce.
    B. supplement.
    C. invent.
    D. forget.

33. The person's critique of the movie was <u>vindictive</u>.
    A. Gentle
    B. Accurate
    C. Mistaken
    D. Spiteful

34. The computer system was <u>obsolete</u>.
    A. Vibrant
    B. State of the art
    C. Outdated
    D. Broken

35. Because of his victory, he was <u>inundated</u> with phone calls.
    A. Awakened
    B. Flooded
    C. Interrupted
    D. Greeted

STOP! DO NOT TURN THIS PAGE UNTIL TIME IS UP FOR THIS TEST. IF YOU FINISH BEFORE TIME IS UP, CHECK OVER YOUR WORK ON THIS TEST ONLY.

PART 4. PARAGRAPH COMPREHENSION

THE TEST ADMINISTRATOR WILL READ THE FOLLOWING ALOUD TO YOU:

Turn to Part 4 and read the directions for Paragraph Comprehension silently while I read them aloud.

This is a test of your ability to understand what you read. In this section you will find one or more paragraphs of reading material followed by incomplete statements or questions. You are to read the paragraph and select one of four lettered choices which BEST completes the statement or answers the question. When you have selected your answer, blacken the space on your answer sheet that has the same number and letter as your answer.

Your score on this test will be based on the number of questions you answer correctly. You should try to answer every question. DO NOT SPEND TOO MUCH TIME on any one question. If you finish before time is called, go back and check your work in this part ONLY.

Now find the section of your answer sheet that is marked PART 4. When you are told to begin, start with question number 1 in Part 4 of your test booklet and answer space number 1 in Part 4 on your separate answer sheet.

DO NOT TURN THE PAGE UNTIL TOLD TO DO SO. You will have 13 minutes to complete the 15 questions in this part. Are there any questions?

Begin.

More than 50 volcanoes in the United States have erupted one or more times in the past 200 years. The most volcanically active regions of the nation are in Alaska, Hawaii, California, Oregon, and Washington. Volcanoes produce a wide variety of hazards that can kill people and destroy property. Large explosive eruptions can endanger people and property hundreds of miles away and even affect global climate. Some volcanic hazards can occur even when a volcano is not erupting.

An explosive eruption blasts solid and molten rock fragments and volcanic gases into the air with tremendous force. The largest rock fragments, called *bombs*, usually fall back to the ground within 2 miles of the vent. Small fragments (less than about 0.1 inch across) of volcanic glass, minerals, and ash rise high into the air, forming a huge, billowing eruption column.

Eruption columns can grow rapidly and reach more than 12 miles above a volcano in less than 30 minutes, forming an eruption cloud. The volcanic ash in the cloud can pose a serious hazard to aviation. During the past 15 years, about 80 commercial jets have been damaged by inadvertently flying into ash clouds, and several have nearly crashed because of engine failure. Large eruption clouds can extend hundreds of miles downwind, resulting in ash fall over enormous areas; the wind carries the smallest ash particles the farthest. Ash from the May 18, 1980, eruption of Mount St. Helens, Washington, fell over an area of 22,000 square miles in the western United States. Heavy ash fall can collapse buildings, and even minor ash fall can damage crops, electronics, and machinery.

Volcanoes emit gases during eruptions. Even when a volcano is not erupting, cracks in the ground allow gases to reach the surface through small openings. Common volcanic gases are carbon dioxide, sulfur dioxide, hydrogen sulfide, hydrogen, and fluorine. Sulfur dioxide gas can react with water droplets in the atmosphere to create acid rain, which causes corrosion and harms vegetation. Carbon dioxide is heavier than air and can be trapped in low areas in concentrations that are deadly to people and animals.

1. According to this passage, fragments that fall close to the volcano itself during an eruption are referred to as
   A. bombs.
   B. vents.
   C. volcanic ash.
   D. rocks.

2. From the passage you can infer that
   A. areas a long distance from the volcano can be damaged by an eruption.
   B. volcanoes hurt the farming industry in many ways.
   C. it's best for airplanes to fly around volcanoes even when they are not erupting.
   D. volcanic ash has caused a few airlines to go bankrupt.

3. According to the passage, most volcanoes in the United States are located in which part of the country?
   A. The central states
   B. The western states
   C. Along the East Coast
   D. The southern states

4. According to the passage, which of the following creates acid rain?
   A. Volcanic glass
   B. Ash combined with water
   C. Fluorine
   D. Sulfur dioxide

5. In this passage, the word *billowing* most nearly means which of the following?
   A. Tall
   B. Dense
   C. Imploding
   D. Swelling

Langston Hughes was born in Joplin, Missouri, to two bookkeepers. His parents separated when he was very young. His father moved to Mexico, and his mother left him for long periods while she searched for steady employment. Hughes's grandmother raised him in Lawrence, Kansas, until he was 12, when he moved to Illinois to live with his mother and stepfather. The family later moved to Ohio. From these humble origins, Langston developed a deep admiration for those he called "low-down folks," poor people who had a strong sense of emotion and pride.

To express his feeling and emotions, Hughes began writing poetry in high school. He gained some early recognition and support among important black intellectuals such as James Weldon Johnson and W. E. B. DuBois. While working as a busboy at the Wardman Park Hotel in Washington, D.C., Hughes gave three of his poems to Vachel Lindsay, a famous critic. Lindsay's enthusiastic praise won Hughes an even wider audience.

Hughes spent the summers of 1919 and 1920 with his father in Mexico. While on a train on his second trip, he wrote his first great poem, "The Negro Speaks of Rivers." The poem was published in *The Crisis*, a magazine of the National Association for the Advancement of Colored People.

Langston Hughes became one of the most important writers and thinkers of the Harlem Renaissance, which was the African American artistic movement in the 1920s that celebrated black life and culture. Hughes's creative genius was influenced by his life in New York City's Harlem, a primarily African American neighborhood. His literary works helped shape American literature and politics. Hughes, like others active in the Harlem Renaissance, had a strong sense of racial pride. Through his poetry, novels, plays, essays, and children's books, he promoted equality, condemned racism and injustice, and celebrated African American culture, humor, and spirituality.

6. Of the following, which is the best title for the passage?
   A. A Kid from Missouri
   B. An Extraordinary Poet
   C. From Busboy to Poet
   D. A Poet Is Praised

7. Who was Vachel Lindsay?
   A. A restaurant worker
   B. A family friend
   C. One of the "low-down folks"
   D. A literary critic

8. According to the passage, what was the Harlem Renaissance?
   A. The neighborhood were Langston Hughes was born
   B. His grandmother's home in Illinois
   C. An African American artistic movement
   D. The topic of Hughes' first great poem

9. Why was Langston Hughes considered a great poet?
   A. His literary works helped shape American politics and literature.
   B. He came from a humble background to reach fame and fortune.
   C. His poems were published in many countries around the world.
   D. His poems were praised by Vachel Lindsay.

10. According to the passage, why did Langston Hughes live with his grandmother?
    A. He didn't want to stay in Missouri.
    B. He didn't want to move to Ohio.
    C. He wanted to continue to write his poems.
    D. His parents were not able to care for him.

Laura Ingalls Wilder was born on February 7, 1867, in the Big Woods of Wisconsin. Reflecting the pioneer spirit of the era, Laura's family moved several times throughout her childhood. She lived for periods in Kansas, Minnesota, Iowa, and South Dakota. Life on the frontier was difficult, and members of Laura's family experienced many hardships that threatened their survival. For example, grasshoppers destroyed their crops two seasons in a row, and a winter of continuous blizzards threatened their supplies. Laura's brother died nine months after his birth, and Laura's sister Mary lost her eyesight at the young age of 15 because of a stroke. The lessons of life on the prairie in America's heartland made a lasting impression on Laura.

Laura and her sisters attended nearby schools during their childhood, and Laura's love for learning sparked her interest in teaching. At age 15 while living in De Smet, South Dakota, Laura earned her teaching certificate. In a rural area just 12 miles from De Smet, she began her first teaching job. During this time, she met and married Almanzo Wilder.

Following in the pioneer footsteps of their families, Laura and Almanzo struggled to establish homes first in South Dakota and later in Minnesota. They, along with their daughter, Rose, suffered through droughts, hail storms, fires, and diseases. Laura and Almanzo eventually settled on a farm in Missouri called "Rocky Ridge."

While living at Rocky Ridge, Laura developed her writing abilities, and for 12 years she edited the *Missouri Ruralist*. As she grew older, Laura became interested in sharing her pioneer experiences with the next generation. Unable to find a publisher for her autobiography called *Pioneer Girl*, Laura rewrote a section of the book and called it *Little House in the Big Woods*, which was published in 1932 when she was 65 years old. Soon after the book was released, *Little House* became a success. Children everywhere wanted to read more stories about the Ingalls family. Laura continued to write books about her family and finished her 18-volume series in 1943. Her desire to share her family's history painted a picture of life on the frontier for generations of children.

11. According to this passage what profession did Laura choose?
    A. Farmer
    B. Daughter
    C. Teacher
    D. Student

12. According to this passage, why did Laura choose to write about her childhood?
    A. She was otherwise unoccupied.
    B. She was too old to be considered a credible author.
    C. She wanted to share her experiences with others.
    D. She wanted to become a success.

13. According to the passage, where did Laura and Almanzo finally settle down?
    A. South Dakota
    B. Missouri
    C. Minnesota
    D. Iowa

14. Which of the following was *not* one of the hardships that Laura faced while growing up?
    A. Death of her brother
    B. Invasion of grasshoppers
    C. Blizzards
    D. Volcanic eruptions

15. Which of the following is the best title for this passage?
    A. Laura Ingalls Wilder—a Great Author
    B. Pioneer Woman and Writer
    C. Hardships of Farmers
    D. Little House in the Big Woods

STOP! DO NOT TURN THIS PAGE UNTIL TIME IS UP FOR THIS TEST. IF YOU FINISH BEFORE TIME IS UP, CHECK OVER YOUR WORK ON THIS TEST ONLY.

PART 5. MATHEMATICS KNOWLEDGE

THE TEST ADMINISTRATOR WILL READ THE FOLLOWING ALOUD TO YOU:

Now turn to Part 5 and read the directions for Mathematics Knowledge silently while I read them aloud.

This is a test of your ability to solve general mathematics problems. You are to select the correct response from the choices given. Then mark the space on your answer sheet that has the same number and letter as your choice. Use your scratch paper to do any figuring you wish to do.

Your score on this test will be based on the number of questions you answer correctly. You should try to answer every question. DO NOT SPEND TOO MUCH TIME on any one question. If you finish before time is called, go back and check your work in this part ONLY.

Now find the section of your answer sheet that is marked PART 5. When you are told to begin, start with question number 1 in Part 5 of your test booklet and answer space number 1 in Part 5 on your separate answer sheet.

DO NOT TURN THE PAGE UNTIL TOLD TO DO SO. You will have 24 minutes to complete the 25 questions in this part. Are there any questions?

Begin.

1. $k - 37 = -10$

   $k = ?$

   A. 27
   B. 29
   C. −29
   D. −12

2. $(4 - 7)(6 - 2)^2 =$
   A. −6
   B. −12
   C. −48
   D. −65

3. 45 is what percent of 180?
   A. 12
   B. 16
   C. 21
   D. 25

4. What percent is equivalent to $\frac{1}{8}$?
   A. 8%
   B. 12.5%
   C. 15%
   D. 17%

5. Divide $3\frac{1}{2}$ by $\frac{3}{4}$.

   A. $\frac{21}{8}$

   B. $2\frac{5}{8}$

   C. $4\frac{2}{3}$

   D. $5\frac{1}{8}$

6. Factor the following expression:

   $y^2 + 8y + 15$

   A. $(y + 1)(y + 15)$
   B. $2y^2 (y + 4)$
   C. $(y + 3)(y + 5)$
   D. $(y + 2)(y + 6)$

7. Subtract:

   $4x^3 - 2x^2 + 6$
   $-(4x^3 + 5x^2 - 12)$

   A. $8 + 3x^2 - 12$
   B. $3x^2 + 12$
   C. $8x^3 - 6$
   D. $-7x^2 + 18$

8. Solve for h.

   $\dfrac{3}{h} = \dfrac{1}{11}$

   A. $h = \dfrac{3}{11h}$

   B. $h = 33$
   C. $3h = 33$
   D. $h = 11$

9. Solve for the two unknowns.

   $3m + 3p = 24$
   $2m + p = 13$

   A. $m = 5; p = 3$
   B. $m = 2; p = 1\frac{1}{2}$
   C. $m = 5; p = 2$
   D. $m = 3; p = 5$

10. $\dfrac{3r^5 s^3}{12r^3 s^2} =$

    A. $4r^2 s$

    B. $\dfrac{1r^8 s^5}{4}$

    C. $\dfrac{r^2 s}{4}$

    D. $0.25r^2 s^5$

11. Solve for x.

    $x^2 - 18x = -45$

    A. $x = 9; x = 5$
    B. $x = 15; x = 3$
    C. $x = -3; x = -15$
    D. $x = 8; x = 2$

12. Multiply:

$$\frac{6g}{2} \times \frac{3}{4y}$$

A. $\dfrac{18g}{8g}$

B. $\dfrac{6g}{6y}$

C. $\dfrac{24gy}{6}$

D. $\dfrac{9g}{4y}$

13. Divide:

$$\frac{x^2}{3y} \div \frac{3x}{2y}$$

A. $\dfrac{2x}{9}$

B. $\dfrac{3x}{12y}$

C. $-\dfrac{3x^3}{12y^2}$

D. $\dfrac{x^3}{4}$

14. $\sqrt[3]{1,000} =$

A. 5
B. 10
C. 20
D. 25

15. Solve for g.

$$\frac{g}{8} = \frac{1}{4}$$

A. $g = 2$

B. $g = \dfrac{2}{4}$

C. $g = \dfrac{1}{2}$

D. $g = \dfrac{1}{32}$

16. Which of the following is a quadrilateral?

A. ▱ (triangle)

B. (cylinder)

C. (pentagon)

D. (parallelogram)

17. What is the perimeter of the following isosceles triangle?

24m

3m

A. 27 m
B. 30 m
C. 51 m
D. 72 m

18. Multiply −4.5 by −6.25.
A. −28.13
B. 24.23
C. 28.13
D. −24.23

19. In the following triangle, $\angle 1$ is 30°, and $\angle 2$ is 45°. What is the measure of $\angle 3$?

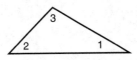

A. 5°
B. 25°
C. 95°
D. 105°

20. Which of the following is true about the sides of the following triangle?

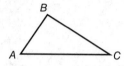

   A. $AB + BC > AC$
   B. $BC + AC < AB$
   C. $AB + AC = BC$
   D. $AC - BC = AB$

21. If the following figure is an isosceles right triangle, what is the measure of $\angle A$?

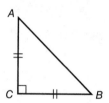

   A. 30°
   B. 45°
   C. 65°
   D. 90°

22. What is the area of the following parallelogram?

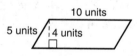

   A. 30 units²
   B. 40 units²
   C. 50 units²
   D. 200 units²

23. If lines A and B are parallel and are intersected by line C and $\angle 2$ is 25°, what is the measure of $\angle 4$?

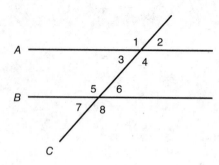

   A. 25°
   B. 45°
   C. 110°
   D. 155°

24. The edges of the following cube each measure 3 cm. What is its volume?

   A. 3 cm³
   B. 9 cm³
   C. 27 cm³
   D. 32 cm³

25. What is the volume of a sphere that has a diameter of 10 cm?
   A. 104.67 cm³
   B. 140 cm³
   C. 418.67 cm³
   D. 523.33 cm³

STOP! DO NOT TURN THIS PAGE UNTIL TIME IS UP FOR THIS TEST. IF YOU FINISH BEFORE TIME IS UP, CHECK OVER YOUR WORK ON THIS TEST ONLY.

PART 6. ELECTRONICS INFORMATION

THE TEST ADMINISTRATOR WILL READ THE FOLLOWING ALOUD TO YOU:

Now turn to Part 6 of your test booklet and read the directions for Electronics Information silently while I read them aloud.

This is a test of your knowledge of electrical, radio, and electronics information. You are to select the correct response from the choices given and then mark the space on your answer sheet that has the same number and letter as your choice.

Your score on this test will be based on the number of questions you answer correctly. You should try to answer every question. DO NOT SPEND TOO MUCH TIME on any one question. If you finish before time is called, go back and check your work in this part ONLY.

Now find the section of your answer sheet that is marked PART 6. When you are told to begin, start with question number 1 in Part 6 of your test booklet and answer space number 1 in Part 6 on your separate answer sheet.

DO NOT TURN THE PAGE UNTIL TOLD TO DO SO. You will have 9 minutes to complete the 20 questions in this part. Are there any questions?

Begin.

1. Which of these materials is a good insulator?
   A. Silver
   B. Copper
   C. Aluminum
   D. Neoprene

2. Voltage measures electrical
   A. current.
   B. pressure.
   C. resistance.
   D. frequency.

3. Which formula calculates the amount of power in an electrical circuit?
   A. Amps^2
   B. Watts/ohms
   C. Amps × volts
   D. Watts × amps

4. Four amps are flowing in a 12-volt circuit. How much power is being used?
   A. 1/3 watt
   B. 3 watts
   C. 8 watts
   D. 48 watts

5. A transformer changes the _____ in a circuit.
   A. voltage
   B. ohms
   C. capacitance
   D. polarity

6. A semiconductor gets its name because it can be used as a(n)
   A. generator or an alternator.
   B. switch or a transformer.
   C. insulator or a conductor.
   D. insulator or a capacitor.

7. Electric motors depend on
   A. capacitance.
   B. inductance.
   C. resistance.
   D. magnetic repulsion.

8. The purpose of a generator is to
   A. convert rotary motion into electric current.
   B. convert electric current into rotary motion.
   C. convert potential energy into kinetic energy.
   D. None of the above.

9. All circuits must have a
   A. load.
   B. current.
   C. voltage.
   D. switch.

10. If the electrochemical reaction in a battery is reversible, the battery is
    A. fully charged.
    B. dead.
    C. rechargeable.
    D. not rechargeable.

11. In the circuit shown, the total resistance is:

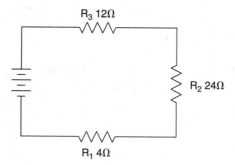

    A. 9/24 ohm
    B. 24/9 ohms
    C. 32 ohms
    D. 40 ohms

12. What is the total resistance in the circuit shown?

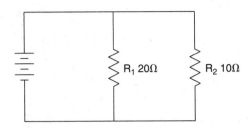

A. 0.066 ohm
B. 6 2/3 ohms
C. 30 ohms
D. 66 2/3 ohms

13. The type of circuit shown is

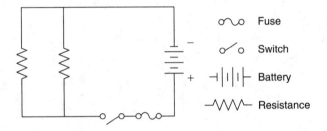

A. series.
B. parallel.
C. series-parallel.
D. parallel-parallel.

14. Silicon is commonly used in electronics because it is a
A. good conductor.
B. good insulator.
C. semiconductor.
D. resistor.

15. If you needed to ensure that current was flowing in one direction only, you would use a(n)
A. diode.
B. resistor.
C. capacitor.
D. alternator.

16. If the equipment grounding leg of a home circuit is not hooked up, what could happen?
A. Devices could get twice as much voltage as required because the ground is absent.
B. You could get a shock if something goes wrong with the wiring.
C. The circuit will be disconnected.
D. No significant change will be seen.

17. The major use of a simple circuit tester is to test
A. the voltage in a circuit.
B. the current in a circuit.
C. the resistance in a circuit.
D. whether a circuit is hot or cold.

18. How many three-way switches would you use to control a hallway light from both ends of the hallway?
A. 2
B. 3
C. 2 or 3
D. None of the above

19. In a circuit breaker, a(n) _____ shuts off the current when too much current flows.
A. rheostat
B. thermostat
C. electromagnet
D. resistor

20. In a circuit with 40 ohms of resistance, how much voltage would be present when the current is 10 amperes?
A. 0.25 V
B. 25 V
C. 200 V
D. 400 V

STOP! DO NOT TURN THIS PAGE UNTIL TIME IS UP FOR THIS TEST. IF YOU FINISH BEFORE TIME IS UP, CHECK OVER YOUR WORK ON THIS TEST ONLY.

PART 7. AUTO AND SHOP INFORMATION

THE TEST ADMINISTRATOR WILL READ THE FOLLOWING ALOUD TO YOU:

Now turn to Part 7 and read the directions for Auto and Shop Information silently while I read them aloud.

This test has questions about automobiles, shop practices, and the use of tools. Pick the BEST answer for each question and then mark the space on your answer sheet that has the same number and letter as your choice.

Your score on this test will be based on the number of questions you answer correctly. You should try to answer every question. DO NOT SPEND TOO MUCH TIME on any one question. If you finish before time is called, go back and check your work in this part ONLY.

Now find the section of your answer sheet that is marked PART 7. When you are told to begin, start with question number 1 in Part 7 of your test booklet and answer space number 1 in Part 7 on your separate answer sheet.

DO NOT TURN THE PAGE UNTIL TOLD TO DO SO. You will have 11 minutes to complete the 25 questions in this part. Are there any questions?

Begin.

1. If you hear a squealing noise that rises in pitch as the engine turns faster, what would you inspect?
   A. Distributor shaft
   B. Timing chain
   C. Fan belt
   D. Flywheel

2. The job of a spark plug is to
   A. make a spark every time the piston reaches top dead center.
   B. spark the catalytic converter.
   C. plug a hole in the cylinder head.
   D. ignite the fuel mixture.

3. If you smell a sickly sweet odor after the engine warms up, what would you suspect is the problem?
   A. Oil leak
   B. Transmission fluid leak
   C. Antifreeze leak
   D. Gasoline leak

4. The purpose of a flywheel is to
   A. smooth out crankshaft motion.
   B. help the brakes when the car is on a rough road.
   C. seal the wheel cylinder.
   D. increase friction between the tires and the road.

5. Gaskets are used to
   A. seal between two rigid surfaces.
   B. prevent overpressure in the crankcase.
   C. seal the two halves of the gas tank.
   D. seal between the piston and the cylinder wall.

6. When an ignition coil goes bad,
   A. the engine will overheat.
   B. the heater will not work.
   C. the timing chain is probably broken.
   D. cylinders may misfire because the spark plugs are not getting enough voltage.

7. The purpose of the connecting rod and crankshaft is to
   A. start the engine during cold weather.
   B. enable the power steering.
   C. change linear motion into rotary motion.
   D. transfer energy from the engine to the clutch.

8. When you step on the brake pedal,
   A. slave cylinders press on the master cylinder, causing the wheel cylinders to press against the brake pads.
   B. the brake lights come on, and an electrical signal is passed to the brakes, which stop the car.
   C. a servo-actuator in the hydraulic system creates pressure on the brake hoses, which press the brake pads against the brake drums.
   D. the master cylinder pushes brake fluid through the brake lines, causing the wheel cylinders to expand and push the brake pads against the disk or drum.

9. When replacing a cylinder head,
   A. tighten the bolts as much as possible.
   B. tighten the bolts in any order, as long as you tighten them with the correct torque.
   C. tighten the bolts to the correct torque, in the correct order.
   D. use the original bolts.

10. A fan belt should be tightened until
    A. the belt is slightly stretched.
    B. the bolts read 48 foot-pounds.
    C. it resists any deflection.
    D. it deflects slightly when you press midway between the pulleys.

11. If the tube feeding a fuel injector leaks,
    A. you may see a spray of gasoline.
    B. air will leak into the engine.
    C. air and gasoline will leak into the engine.
    D. methane will leak into the cylinder.

12. In a four-cylinder, four-stroke engine, how many explosions occur in one minute when the engine is running at 2,000 rpm?
    A. 1,000
    B. 2,000
    C. 4,000
    D. 8,000

13. Which list of fasteners is organized from weakest to strongest?
    A. Carriage bolt, lag bolt, screw
    B. Nail, screw, lag bolt, carriage bolt
    C. Screw, nail, lag bolt, carriage bolt
    D. Staple, carriage bolt, nail, lag bolt

14. The tool shown below is a(n)

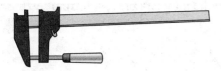

    A. eight-point wrench.
    B. box wrench.
    C. open-end wrench.
    D. bar clamp.

15. When using carpenter's glue, it is necessary to
    A. wait 30 minutes for the glue to dry.
    B. clamp the parts for 30 minutes.
    C. clamp the parts overnight.
    D. put primer on the wood before gluing.

16. Flux is needed for soldering because it
    A. transfers heat to the metal.
    B. melts the solder.
    C. prevents oxidation during heating.
    D. holds the joint tight.

17. What kind of force does this tool create?

    A. Shearing
    B. Grinding
    C. Slicing
    D. Twisting

18. Unreinforced concrete
    A. has great compressive strength but little tensile strength.
    B. has great compressive strength and great tensile strength.
    C. has little compressive strength and little tensile strength.
    D. has little compressive strength but great tensile strength.

19. The tool shown below is a(n)

    A. punch.
    B. awl.
    C. Phillips screwdriver.
    D. slot screwdriver.

20. A center punch
    A. finds the center of a hole.
    B. replaces a drill when drilling wood.
    C. prevents a drill from wandering as it starts in metal.
    D. lines up hinge holes.

21. In which building material does actual size equal nominal size?
    A. 2 × 4
    B. 1 × 4
    C. 8-inch concrete block
    D. plywood

22. The tool shown below is a

    A. putty knife.
    B. paint scraper.
    C. wood chisel.
    D. tang chisel.

23. When using a lag-shield anchor in concrete, it is necessary to
    A. use a power-actuated tool.
    B. drill a hole first.
    C. set the anchor in wet cement.
    D. use construction adhesive.

24. A pipe wrench
    A. has two fixed jaws.
    B. has one fixed jaw and one movable jaw.
    C. can be used on pipe, but not on pipe fittings.
    D. is always 16 inches long.

25. The purpose of pipe-joint compound is to
    A. seal the joint only.
    B. hold Teflon tape in place while the tape dries.
    C. prevent rust in steel threads.
    D. seal the joint and prevent it from seizing up.

STOP! DO NOT TURN THIS PAGE UNTIL TIME IS UP FOR THIS TEST. IF YOU FINISH BEFORE TIME IS UP, CHECK OVER YOUR WORK ON THIS TEST ONLY.

PART 8. MECHANICAL COMPREHENSION

THE TEST ADMINISTRATOR WILL READ THE FOLLOWING ALOUD TO YOU:

Now turn to Part 8 and read the directions for Mechanical Comprehension silently while I read them aloud.

This test has questions about mechanical and physical principles. Study the picture and decide which answer is CORRECT and then mark the space on your separate answer sheet that has the same number and letter as your choice.

Your score on this test will be based on the number of questions you answer correctly. You should try to answer every question. DO NOT SPEND TOO MUCH TIME on any one question. If you finish before time is called, go back and check your work in this part ONLY.

Now find the section of your answer sheet that is marked PART 8. When you are told to begin, start with question number 1 in Part 8 of your test booklet and answer space number 1 in Part 8 on your separate answer sheet.

DO NOT TURN THE PAGE UNTIL TOLD TO DO SO. You will have 19 minutes to complete the 25 questions in this part. Are there any questions?

Begin.

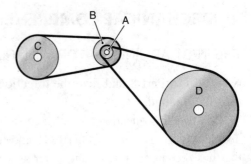

1. The mechanical advantage in the pulley system shown is
   A. 1
   B. 2
   C. 3
   D. 4

5. Which sheave would be rotating the slowest in this system?
   A. A
   B. B
   C. C
   D. D

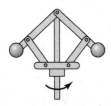

2. This object is a mechanical governor for an engine. What would cause the balls to move outward?
   A. Spring tension
   B. Slower rotation
   C. Faster rotation
   D. Linear momentum

6. A fuel tank is receiving fuel at the rate of 1 gallon per minute. It is supplying an engine that uses 3 gallons per hour. How many gallons of fuel will be in the tank after 1/2 hour, assuming the tank was empty at the start?
   A. 18.5
   B. 28.5
   C. 31.5
   D. 57

3. When you drag a concrete block across the pavement, one way to reduce the amount of pulling force needed would be to
   A. get a heavier block.
   B. use a better rope.
   C. pull harder.
   D. lubricate the pavement.

4. What type of simple machine is shown?
   A. Class 1 lever
   B. Class 2 lever
   C. Class 3 lever
   D. Two inclined planes

7. What amount of effort would be required to pull on this rope?
   A. 25 lb
   B. 50 lb
   C. 100 lb
   D. 200 lb

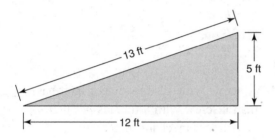

100 lb

400 lb

8. Using the weights given, what mechanical advantage would allow a 100 lb effort to raise the 400 lb weight? (Note: fhe figure is not drawn to scale.)
   A. 2
   B. 4
   C. 8
   D. 16

9. What is the mechanical advantage of this inclined plane?
   A. 13/12
   B. 12/13
   C. 13/5
   D. 12/5

10. Gear A, with 24 teeth, is meshed with gear B, with 36 teeth. When gear A rotates 3 times, how many times will gear B rotate?
    A. 2
    B. 3
    C. 4
    D. 6

11. This tool combines which simple machines?
    A. Lever and wheel and axle
    B. Inclined plane and pulley
    C. Lever and wedge
    D. Pulley and wheel and axle

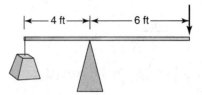

12. What is the mechanical advantage of this lever?
    A. 0.5
    B. 1.5
    C. 2
    D. 3

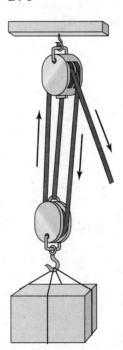

13. In this figure, pulling 12 feet of rope will raise the weight
    A. 2 feet.
    B. 3 feet.
    C. 4 feet.
    D. 8 feet.

14. When two gears are engaged,
    A. the driving gear must be smaller than the driven gear.
    B. one must rotate faster than the other.
    C. they always rotate at the same speed.
    D. they always rotate in opposite directions.

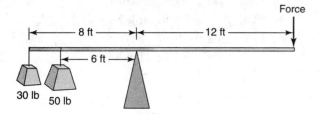

15. How much force would be needed to balance this lever?
    A. 12 lb
    B. 25 lb
    C. 35 lb
    D. 45 lb

16. Which of these items would return to its original shape after being deformed?

A.

B.

C.

D.

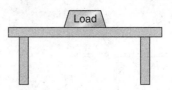

17. What would help the beam carry a heavier load?
    A. Split the beam in two.
    B. Add a post under the load.
    C. Coat the beam with oil.
    D. Remove one post and turn the other into a fulcrum.

18. A bolt has 8 threads per inch. How far does it move if you tighten it five turns, and then loosen it three turns?
    A. 1/8 in.
    B. 1/4 in.
    C. 3/8 in.
    D. 5/8 in.

19. A screw with 12 threads per inch has _____ than a screw with 8 threads per inch.
    A. a larger mechanical advantage
    B. a smaller mechanical advantage
    C. more strength
    D. less strength

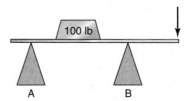

20. What would happen if you added a force at the arrow?
    A. Load would not change at A, and it would increase at B.
    B. Load would not change at A, but it would increase at B.
    C. Load would increase at A and decrease at B.
    D. Load would decrease at A and increase at B.

21. When you add water to a tank, water pressure on the bottom will
    A. stay the same.
    B. decrease.
    C. increase.
    D. depend on temperature.

22. If a bicycle rider changes to a smaller gear on the rear of the bike and wants to go the same speed,
    A. The rider must place more force on the pedals.
    B. The rider can put less force on the pedals.
    C. The rider need not change the force on the pedals.
    D. The rider should put more force on the left pedal and less force on the right pedal.

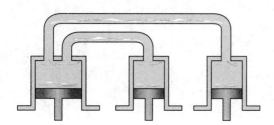

23. In the hydraulic system shown, where is the greatest pressure?
    A. A
    B. B
    C. C
    D. Pressure is the same at all points.

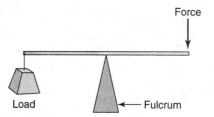

24. Moving the fulcrum closer to the load will
    A. increase the mechanical advantage.
    B. decrease the mechanical advantage.
    C. not change the mechanical advantage.
    D. allow you to lift farther with the lever.

25. What would you use to prevent a nut from moving on a bolt?
    A. Lock washer
    B. Lock nut
    C. Degreaser
    D. A and B

STOP! DO NOT TURN THIS PAGE UNTIL TIME IS UP FOR THIS TEST. IF YOU FINISH BEFORE TIME IS UP, CHECK OVER YOUR WORK ON THIS TEST ONLY.

PART 9. ASSEMBLING OBJECTS

THE TEST ADMINISTRATOR WILL READ THE FOLLOWING ALOUD TO YOU:

Now turn to Part 9 and read the directions for Assembling Objects silently while I read them aloud.

This test has questions that will measure your spatial ability. Study the diagram, decide which answer is CORRECT, and then mark the space on your separate answer sheet that has the same number and letter as your choice.

Your score on this test will be based on the number of questions you answer correctly. You should try to answer every question. DO NOT SPEND TOO MUCH TIME on any one question. If you finish before time is called, go back and check your work in this part ONLY.

Now find the section of your answer sheet that is marked PART 9. When you are told to begin, start with question 1 in Part 9 of your test booklet and answer space number 1 in Part 9 on your separate answer sheet.

DO NOT TURN THE PAGE UNTIL TOLD TO DO SO. You will have 15 minutes to complete the 25 questions in this part. Are there any questions?

Begin.

For Questions 1–17, which figure best shows how the objects in the left box will appear if they are fit together?

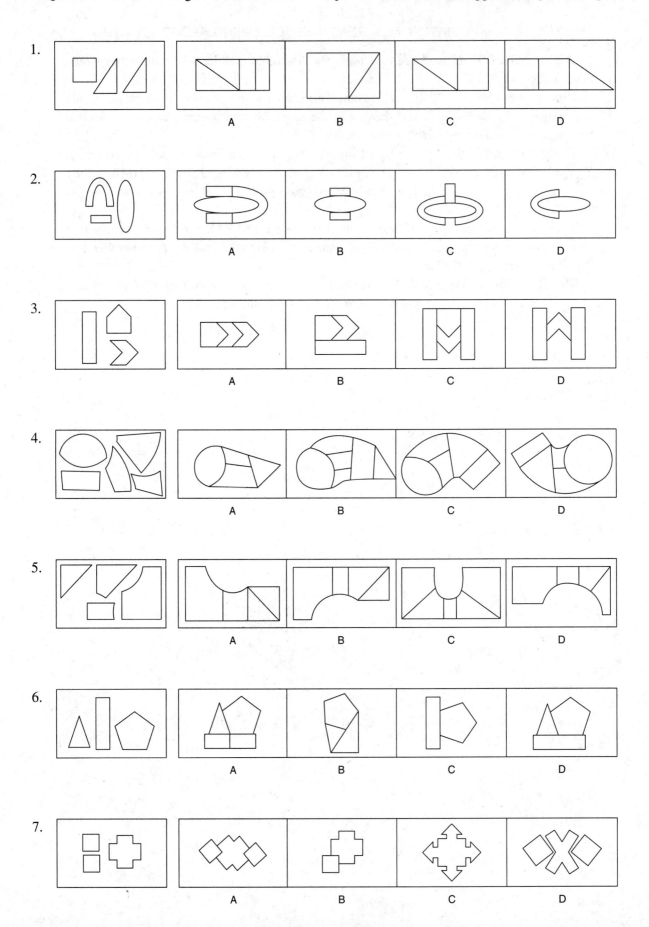

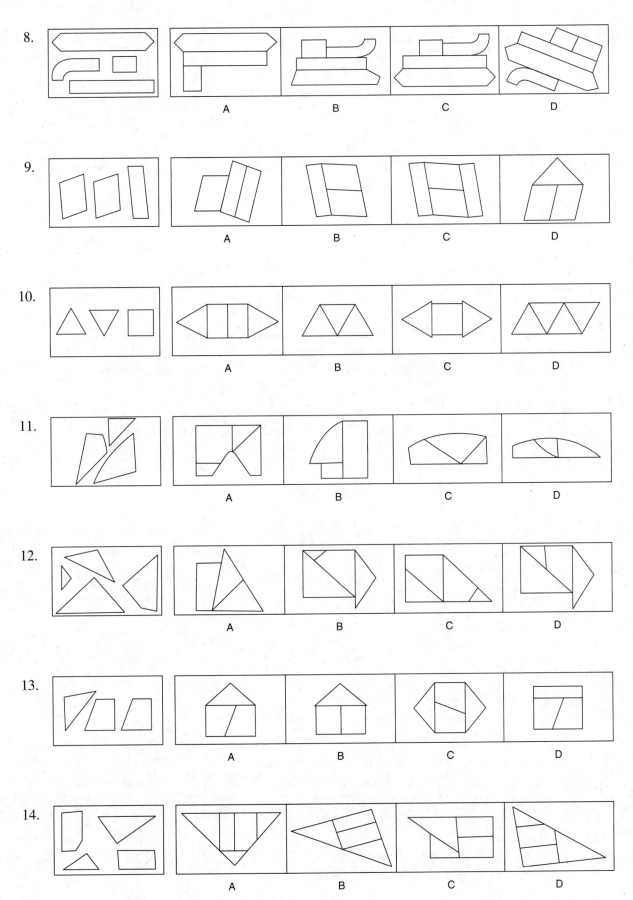

15.

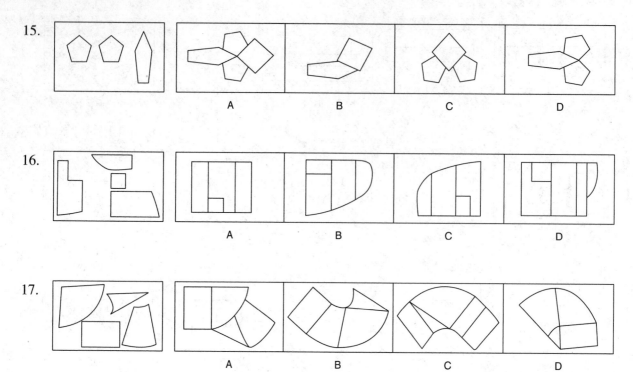

16.

17.

For Questions 18–25, which figure best shows how the objects in the left box will touch if the letters for each object are matched?

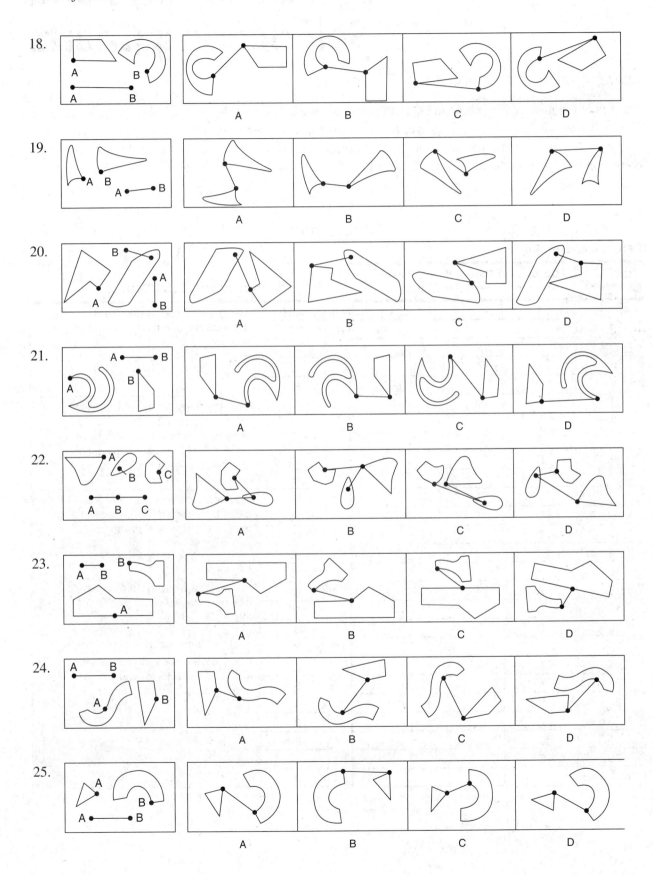

Answer Keys and Self-Scoring Charts

ASVAB PRACTICE TEST FORM 3

The following answer keys show the correct answers for each part of the practice test that you just took. For each question, compare your answer to the correct answer. Mark an X in the column to the right if you got the item correct. Then total the number correct for each part of the test. Find that number in the corresponding chart to the right of the answer key. See the suggestions listed for your performance.

Part 1. General Science

Item Number	Correct Answer	Mark X If You Picked the Correct Answer
1	A	
2	A	
3	A	
4	B	
5	D	
6	B	
7	B	
8	A	
9	D	
10	A	
11	C	
12	B	
13	D	
14	C	
15	A	
16	A	
17	C	
18	D	
19	A	
20	A	
21	A	
22	D	
23	C	
24	C	
25	C	
Total Correct		

Score Interpretation—Total Correct

25	This is pretty good work. Review the explanations for the answers you got incorrect.
24	
23	
22	
21	
20	You are doing pretty well. Review the explanations for the items you answered incorrectly. If you have time, review the other explanations and you will learn even more.
19	
18	
17	
16	
15	You need to keep studying. Pay close attention to the explanations for each item, even for the ones you got correct.
14	
13	
12	
11	
10	Spend time working on the General Science review in Part 3 of this book.
9	
8	
7	
6	If you have copies of your school science textbooks, it would be a good idea to study those.
5	
4	
3	
2	
1	
0	

Part 2. Arithmetic Reasoning

Item Number	Correct Answer	Mark X If You Picked the Correct Answer
1	C	
2	C	
3	B	
4	D	
5	D	
6	C	
7	C	
8	C	
9	D	
10	C	
11	D	
12	D	
13	A	
14	D	
15	A	
16	B	
17	A	
18	C	
19	B	
20	C	
21	B	
22	D	
23	A	
24	C	
25	B	
26	C	
27	A	
28	B	
29	B	
30	B	
Total Correct		

Score Interpretation—Total Correct

Score	Interpretation
30	This is pretty good work. Review the explanations for the answers you got incorrect. This test is a part of the AFQT, and you must do well
29	
28	
27	
26	
25	You are doing pretty well. Review the explanations for the items you answered incorrectly. If you have time, review the other explanations and you will learn even more.
24	
23	
22	
21	
20	You need to keep studying. Pay close attention to the explanations for each item, even for the ones you got correct.
19	
18	
17	
16	
15	Spend time working on the Arithmetic Reasoning reviews in Part 3 of this book.
14	
13	
12	
11	Keep working and reworking problems until you are comfortable with the processes.
10	
9	
8	
7	
6	
5	
4	
3	
2	
1	
0	

Part 3. Word Knowledge

Item Number	Correct Answer	Mark X If You Picked the Correct Answer
1	C	
2	B	
3	A	
4	D	
5	B	
6	C	
7	A	
8	C	
9	B	
10	A	
11	D	
12	C	
13	B	
14	C	
15	A	
16	D	
17	B	
18	A	
19	B	
20	C	
21	B	
22	C	
23	D	
24	A	
25	D	
26	C	
27	B	
28	A	
29	D	
30	C	
31	D	
32	B	
33	D	
34	C	
35	B	
Total Correct		

Score Interpretation—Total Correct

Score	Interpretation
35	
34	This is pretty good work. Review the explanations for the answers you got incorrect. This is an important test that contributes to your AFQT.
33	
32	
31	
30	You are doing pretty well. Review the explanations for the items you answered incorrectly. If you have time, review the other explanations and you will learn even more.
29	
28	
27	
26	
25	You need to keep studying. Pay close attention to the explanations for each item, even for the ones you got correct.
24	
23	
22	
21	Spend time working on the Word Knowledge review in Part 3 of this book.
20	
19	
18	
17	Keep reading and identifying words you don't know.
16	
15	
14	
13	
12	
11	
10	
9	
8	
7	
6	
5	
4	
3	
2	
1	
0	

Part 4. Paragraph Comprehension

Item Number	Correct Answer	Mark X If You Picked the Correct Answer
1	A	
2	A	
3	B	
4	D	
5	D	
6	C	
7	D	
8	C	
9	A	
10	D	
11	C	
12	C	
13	B	
14	D	
15	B	
Total Correct		

Score Interpretation—Total Correct

15	This is pretty good work. Review the explanations for the answers you got incorrect. This test is a part of the AFQT, and you must do well.
14	You are doing pretty well. Review the explanations for the items you got incorrect. This test is a part of the AFQT, and you must do well.
13	You need to keep studying. Pay close attention to the explanations for each item, even for the ones you got correct. This will help you understand why the answer is correct.
12	
11	
10	
9	
8	Spend time reviewing the Paragraph Comprehension and Word Knowledge reviews in Part 3 of this book. Keep reading books and newspapers.
7	
6	
5	
4	
3	
2	
1	
0	

Part 5. Mathematics Knowledge

Item Number	Correct Answer	Mark X If You Picked the Correct Answer
1	A	
2	C	
3	D	
4	B	
5	C	
6	C	
7	D	
8	B	
9	A	
10	C	
11	B	
12	D	
13	A	
14	B	
15	A	
16	D	
17	C	
18	C	
19	D	
20	A	
21	B	
22	B	
23	D	
24	C	
25	D	
Total Correct		

Score Interpretation—Total Correct

25	This is pretty good work. Review the explanations for the answers you got incorrect.
24	
23	
22	You are doing pretty well. Review the explanations for the items you answered incorrectly. If you have time, review the other explanations and you will learn even more.
21	
20	
19	This test is part of the AFQT. You must perform well on this test.
18	
17	
16	You need to keep studying. Pay close attention to the explanations for each item, even for the ones you got correct.
15	
14	
13	
12	Spend time working on the Mathematics Knowledge reviews in Part 3 of this book, and work and rework the problems until you fully understand the processes.
11	
10	
9	
8	
7	
6	
5	
4	
3	
2	
1	
0	

Part 6. Electronics Information

Item Number	Correct Answer	Mark X If You Picked the Correct Answer
1	D	
2	B	
3	C	
4	D	
5	A	
6	C	
7	D	
8	A	
9	A	
10	C	
11	D	
12	B	
13	C	
14	C	
15	A	
16	B	
17	D	
18	A	
19	C	
20	D	
Total Correct		

Score Interpretation—Total Correct

20	This is pretty good work. Review the explanations for the answers you got incorrect.
19	
18	
17	You are doing pretty well. Review the explanations for the items you answered incorrectly. If you have time, review the other explanations and you will learn even more.
16	
15	
14	
13	
12	You need to keep studying. Pay close attention to the explanations for each item, even for the ones you got correct.
11	
10	
9	
8	Spend time working on the Electronics Information review in Part 3 of this book.
7	
6	
5	Your school electronics or physics books may have some helpful information as well.
4	
3	
2	
1	
0	

Part 7. Auto and Shop Information

Item Number	Correct Answer	Mark X If You Picked the Correct Answer
1	C	
2	D	
3	C	
4	A	
5	A	
6	D	
7	C	
8	D	
9	C	
10	D	
11	A	
12	C	
13	B	
14	D	
15	B	
16	C	
17	A	
18	A	
19	C	
20	C	
21	D	
22	C	
23	B	
24	B	
25	D	
Total Correct		

Score Interpretation—Total Correct

Score	Interpretation
25	This is pretty good work. Review the explanations for the answers you got incorrect.
24	
23	
22	
21	
20	You are doing pretty well. Review the explanations for the items you answered incorrectly. If you have time, review the other explanations and you will learn even more.
19	
18	
17	
16	
15	
14	You need to keep studying. Pay close attention to the explanations for each item, even for the ones you got correct. Spend time working on the Auto and Shop Information reviews in Part 3 of this book.
13	
12	
11	
10	
9	
8	
7	
6	
5	
4	
3	
2	
1	
0	

Part 8. Mechanical Comprehension

Item Number	Correct Answer	Mark X If You Picked the Correct Answer
1	A	
2	C	
3	D	
4	C	
5	D	
6	B	
7	C	
8	B	
9	C	
10	A	
11	C	
12	B	
13	C	
14	D	
15	D	
16	C	
17	B	
18	B	
19	A	
20	D	
21	C	
22	A	
23	D	
24	A	
25	D	
Total Correct		

Score Interpretation—Total Correct

Score	Interpretation
25	This is pretty good work. Review the explanations for the answers you got incorrect.
24	
23	
22	
21	
20	You are doing pretty well. Review the explanations for the items you answered incorrectly. If you have time, review the other explanations and you will learn even more.
19	
18	
17	
16	
15	You need to keep studying. Pay close attention to the explanations for each item, even for the ones you got correct.
14	
13	
12	
11	Spend time working on the Mechanical Comprehension review in Part 3 of this book
10	
9	
8	
7	
6	
5	
4	
3	
2	
1	
0	

Part 9. Assembling Objects

Item Number	Correct Answer	Mark X if You Picked the Correct Answer
1	C	
2	A	
3	B	
4	C	
5	B	
6	D	
7	A	
8	C	
9	B	
10	C	
11	C	
12	B	
13	A	
14	D	
15	D	
16	C	
17	B	
18	D	
19	A	
20	D	
21	A	
22	A	
23	D	
24	B	
25	A	
Total Correct		

Score Interpretation–Total Correct

Score	Interpretation
25	This is pretty good work. Review the explanations of the answers you got incorrect.
24	
23	
22	
21	
20	You are doing pretty well. Review the explanations for the items you answered incorrectly. If you have time, review the other explanations and you will learn even more.
19	
18	
17	
16	
15	Keep studying. Pay close attention to the explanations for each item, even for the ones you got correct.
14	
13	
12	
11	Spend time working on the Assembling Objects review in Part 3 of this book.
10	
9	
8	
7	
6	
5	
4	
3	
2	
1	
0	

Answers and Explanations

PART 1. GENERAL SCIENCE

1. **A.** The cell membrane is the part of the cell that determines what enters and exits a cell. The correct answer is choice A.

2. **A.** *Vascular* is the name given to the systems that transport water and nutrients within certain plants, so choice D is not correct. *Annuals* are plants that undergo a full growing cycle in one year, so choice B is not correct. *Deciduous* is a term used for trees that shed their leaves in the winter, so choice C is not correct. Choice A, *angiosperms*, is the term given to describe plants that flower.

3. **A.** The hereditary system (choice C) and the reproductive system (choice B) are clearly incorrect answers. The lymphatic system (choice D) is one of the human organ systems and does not occur in plants. Choice A, the vascular system, is the correct answer. The vascular system takes fluid up from the water and transports it through the stem into the leaves.

4. **B.** The whale is a vertebrate mammal because it has an internal skeletal system and nurses its young with milk. Choice B is the correct answer.

5. **D.** A mammal is an animal that has live births and feeds its young with milk-secreting glands called mammary glands. Fish (choice A), snakes (choice B), and crocodiles (choice C) do not fit this group. Of this group, only apes (choice D) are mammals. Choice D is the correct answer.

6. **B.** Animals are food consumers, while plants are food producers. Frogs (choice A), parrots (choice C), and the manatee (choice D) are all animals, so they are consumers. The tree (choice B) is the only plant on the list and is the food producer.

7. **B.** Bones are held together by fibrous bands called ligaments. Choice B is the correct answer.

8. **A.** The skin is clearly the largest organ of your body, covering your entire skeletal and muscular system. Choice A is the correct answer.

9. **D.** Per gram, fat has much more energy than carbohydrates, proteins and minerals. Fat cells store the most energy. Choice D is the correct answer.

10. **A.** Platelets (choice D), red blood cells (choice C), and white blood cells (choice B) are all part of blood, but the substance that makes up most of the blood's volume is plasma (choice A).

11. **C.** The inner ear contains structures and fluids that move and swirl when you move. These structures send signals to the brain that help you understand your relative position. Choice C is the correct answer.

12. **B.** The basic difference between mitosis and meiosis is in the mechanism of cell division. The result of the division of mitosis is that each cell has the full complement of 46 chromosomes. In sexual reproduction, the process of meiosis produces cells with 23 chromosomes, so that when fertilization takes place between an egg and sperm, the full complement of 46 chromosomes will be present. Choice B is the correct answer.

13. **D.** Looking at the electromagnetic spectrum, ultraviolet radiation is to the right of visible light, which indicates more energy and a higher frequency. Choice D is the correct answer.

14. **C.** It is a fundamental principle that matter cannot be created or destroyed. Although a candle would look smaller, if we would trap

the gases, the total matter would end up the same. The total mass stays the same, so choice C is the correct answer.

15. **A.** The crust is a very thin layer (although very important to our daily lives) and takes up very little of Earth's volume. Choice A, crust, is the right answer.

16. **A.** Oceanic features formed by tectonic plates moving away from one another are called *rifts*. A rift is much like a huge valley across a continent or on the ocean floor.

17. **C.** Warm water is lighter or less dense than cold water. Choice A is not the correct answer. Currents on the surface are not density currents. Choice B is not the correct answer. Water with a high concentration of dissolved gases would be lighter and less dense, so choice D is not the correct answer. Water that is cold and saline or salty would be more dense and would be much more prone to sink, forming density currents. These types of currents form near the cold poles and generally flow slowly along the ocean floor. So choice C is correct.

18. **D.** Tides (choice A) are caused by the gravity of the moon. Day and night (choice B) are caused by the spinning or rotation of Earth as it revolves around the sun. Earth does bulge at the equator (choice C), but Earth's tilt does not cause this. The tilt of Earth's axis as it orbits around the sun causes the sun's rays to fall more directly on the northern hemisphere during the months of June, July, and August, and in the southern hemisphere during the months of December, January, and February. More direct sunlight makes for warmer temperatures and longer days, creating summer. Less direct sunlight and shorter days create winter weather. The tilt causes the seasons, so choice D is the correct answer.

19. **A.** Earth's gravitational pull is so strong that is locks the moon into having revolution and rotation periods that are nearly the same. The moon makes one rotation in about every 27.3 days.

It also revolves around Earth in about that same amount of time. As a result, the moon never shows most of its far side to Earth. We are able to see a small part of the far side of the moon because the rotation and revolution times of the moon are not exactly the same, but most of the far side is never visible by anyone on Earth. Choice A is the correct answer.

20. **A.** Substances can be in a solid, liquid, or gas form, the three states of matter. For matter to change from one form to another, there needs to be a change in thermal energy. At room temperature some substances are naturally gases, some are naturally liquids, and some are naturally solids. Of the options given, the only substance that is naturally a solid at room temperature is sulfur. The other three are gases, so choice A is the correct answer. However, if sulfur is heated to a high enough temperature, it too can be a gas. This temperature, however, is well beyond what we consider room temperature. You usually can smell sulfur gases near volcanic activity and thermal activity such as hot springs.

21. **A.** Earth's weather takes place in the atmospheric layer closest to the ground. That is choice A, the troposphere.

22. **D.** The sun's radiation causes a part of our atmosphere called the *ionosphere* to be electrically charged. AM radio waves from Earth's surface bounce off the ionosphere and are reflected back to ground level. Broadcasters use this property of radio waves to make the waves travel long distances. When solar radiation is overactive, it can cause the ionosphere to become highly charged. When that happens, radio waves can be disturbed, causing static. Choice D is the correct answer.

23. **C.** Erosion is the breaking up and transport of rock. Freezing and thawing (choice A) can cause rocks to break up into smaller pieces. The actions of water in rivers (choice B) break up pieces of rock and can also erode river banks

and create deep valleys and canyons. Lichen and mosses (choice D) use chemical action to break up rock. Lava flows (choice C) are considered to be builders of land rather than erosion agents. Choice C is the correct answer.

24. **C.** The core of the sun is so hot that it promotes the fusion of hydrogen atoms into helium, thereby releasing energy. Choice C is the correct answer.

25. **C.** A charged particle that has either more electrons than protons or fewer electrons than protons is called an *ion*. The process of creating charged particles is called *ionization*. Choice A is not the correct answer. *Refraction* is the bending of light rays. Choice B is not the correct answer. The rate of decay of an atomic nucleus is called its *half-life*. Choice D is not the correct answer. When the nucleus of an atom gives up a particle and its energy in order to become more stable, the process is called *radioactive decay*. Choice C is the correct answer.

PART 2. ARITHMETIC REASONING

1. **C.** This is a simple addition problem. Add the amounts to get the total of $193.69.

2. **C.** This problem has two steps. First you need to add the number of miles driven and then determine how many more miles are left in the trip. Marty has driven a total of 389 miles, so he has 138 to go.

3. **B.** In order to calculate the percent, you need to divide the total number of times at bat into the number of hits. In this case you would divide 72 into 54 to get 0.75. To change that to a percent, move the decimal point two places to the right and add the percent sign to get the correct answer of 75%.

4. **D.** To find the average, add the numbers and divide by the number of numbers. In this problem the total number of calories eaten by Phil is 10,400. Dividing this by the number of days (4) gives the average of 2,600 calories.

5. **D.** For this problem, you need to find out what number is 60% of 450. Change the 60% to 0.60 and multiply that by 450. Doing that calculation results in the number 270, the correct answer.

6. **C.** The basic question is that if 1 inch = 50 miles, how many inches = 450 miles? Set this up as a proportion. $\dfrac{1}{50} = \dfrac{x}{450}$. Cross-multiply and solve for x. $50x = 450$. $x = 9$. So the distance on the map between New Town and Old Town is 9 inches.

7. **C.** You need to find either the cost of one burger or the cost of one milkshake. It is possible to find the cost of one burger by subtracting Phil's cost from Janice's cost because the only difference is the cost of one burger. So $14.80 minus $10.30 = $4.50. Therefore, the cost of one burger is $4.50. If Phil's two burgers = $9.00, then one milkshake = $10.30 − $9.00 = $1.30. The problem asks for the cost of two milkshakes, which would be 2($1.30) or $2.60. Therefore, the correct answer is $2.60.

8. **C.** Probability = number of favorable outcomes/ number of possible outcomes. In this problem, the number of favorable outcomes is a combination of green and orange balls or 35 + 45 = 80. The number of possible outcomes is 35 + 75 + 45 = 155. So substituting into the formula, the probability is $\dfrac{80}{155}$ or $\dfrac{16}{31}$.

9. **D.** Simple interest problems use the formula $I = prt$, where I is the amount of interest, p is the principal or the amount saved or invested, r is the rate of interest, and t is the amount of time that the interest is accruing. For compounded interest problems, you need to calculate the simple interest for each year, add it to the principal and use that figure as the new principal for the next year. The process is repeated for each year. In this problem, the interest is compounded for 3 years, so you will calculate the interest three times, each time using the new principal.

Year One
Principal = $3,000
Rate = 0.05
Interest = $150
Add the interest to the principal to get $3,150.

Year Two
Principal = $3,150
Rate = 0.05
Interest = $157.50
Add the interest to the principal to get $3,307.50.

Year Three
Principal = $3,307.50
Rate = 0.05
Interest = $165.38
Add the interest to get the principal of $3,472.88, the amount Jose would have after three years.

10. **C.** If Sonya's average over five tests is 89, she has earned a total of 445 points for the five tests. If she received a 95 on the sixth test, that brings her total points to 540. Dividing that by the six tests gives a new average of 90.

11. **D.** In this problem you need to find out how many $\frac{2}{3}$ cups are in 18 cups. So $18 \div \frac{2}{3} = ?$ is the equation. When you divide by a fraction, you invert and multiply, so the equation becomes $18 \times \frac{3}{2} = \frac{18 \times 3}{2} = \frac{54}{2} = 27$. Victor can make 27 batches of lasagna with 18 cups of tomato paste.

12. **D.** Since the final result needs to be in square yards, take the original measurements and convert them to yards. 12 feet = 4 yards, and 24 feet = 8 yards. Since area = $l \times w$, $4 \times 8 = 32$ yd². That's how many square yards of carpeting are needed.

13. **A.** Joanne's salary is $3,000 per month. If she pays 28% of that amount on taxes, reduce that amount by 28% by multiplying the total salary by 0.28.

$$\$3,000 \times 0.28 = \$840.00$$

$840.00 represents the amount paid in taxes. To find out how much of Joanne's salary is left, subtract $840.00 from the salary:

$$\$3,000 - \$840.00 = \$2,160.00$$

A faster way to do this (and speed is important on the ASVAB) is to multiply the original cost by 72% or 0.72 because that is the amount of salary after taxes.

$$\$3,000 \times 0.72 = \$2,160.00$$

14. **D.** Use the following formula:

$$\text{Percent of Change} = \frac{\text{Amount of Change}}{\text{Starting Point}}.$$

In this problem we know that Gene grew from 60 to 69 inches. So the amount of change is the difference between 69 and 60 or 9 inches. The starting point is his original height of 60 inches. Substitute that information into the formula.

$$\text{Percent of Change} = \frac{9}{60} = 0.15.$$

Change that number to a percent by moving the decimal place two places to the right and add the percent sign. $0.15 = 15\%$

15. **A.** If Patrick traveled at 50 mph for 3 hours, the trip was 3(50) or 150 miles. Recall the formula $D = rt$, where D = distance, r = rate, and t = time. Substitute what you know. $D = 150$ miles; $r = 40$. $150 = 40t$. Solve for t. $t = 3.75$ hours.

16. **B.** To calculate the area of the patio, use the formula $A = lw$. In this problem, notice that the dimensions of the patio are given in inches, but the cement bag information is in square feet. Change the inches to feet.

$$360 \text{ inches} = 30 \text{ feet}$$
$$480 \text{ inches} = 40 \text{ feet}$$

Substitute that information into the formula.

$$A = (30)(40)$$
$$A = 1200 \text{ ft}^2$$

If one bag of cement is needed for every 16 square feet, you need to divide that into the total area.

$$\frac{1200 \text{ ft}^2}{16 \text{ ft}^2} = 75 \text{ bags}$$

17. **A.** To calculate the area of a circle, use the formula $A = \pi r^2$. In this problem the radius is 20 feet. Substitute the radius into the formula and solve for A.

$$A = 3.14(20^2)$$
$$A = 3.14(400)$$
$$A = 1,256 \text{ ft}^2$$

18. **C.** Create an equation to solve this problem.

$$\text{Suzie} = 2(\text{Will}) + 4$$

You are told that Will is 9. Substitute that information into the formula. Suzie = 2(9) + 4. Solve for Suzie. Suzie = 22

19. **B.** If there are four quarters in every dollar, Milo needs to save 4×950 quarters or 3,800 quarters.

20. **C.** Watch the units in this problem. Change the yards to feet.

$$6 \text{ yards} = 18 \text{ feet}$$

To find out how many $1\frac{1}{2}$-foot pieces there are in 18 feet, divide 18 by $1\frac{1}{2}$. Change the mixed number to a fraction. $18 \div \frac{3}{2}$. When dividing by a fraction, invert and multiply.

$$18 \times \frac{2}{3}$$
$$= \frac{18 \times 2}{3} = \frac{36}{3} = 12 \text{ pieces}$$

21. **B.** Treat this like a compound interest problem where you calculate the increase in kilometers each week, using the increased amount as the amount to use in the calculations the next week.

Week One
$$5 \text{ km}$$

Week Two
$$5 \text{ km} + 0.20(5) = 6 \text{ km}$$

Week Three
$$6 \text{ km} + 0.20(6) = 7.2 \text{ km}$$

Week Four
$$7.2 + 0.20(7.2) = 8.64 \text{ km}$$

22. **D.** To solve this problem, use the formula $V = lwh$. First you need to calculate the volume of the box. You have been given the dimensions of $l = 15$ in, $w = 6$ in, and $h = 6$ in. Substitute that information into the formula.

$$V = (15)(6)(6)$$
$$V = 540 \text{ in}^3$$

Now you need to calculate the volume of one brownie. You have been given the dimensions of $l = 3$ inches, $w = 2$ inches, and $h = \frac{1}{2}$ inches. Substitute that information into the formula.

$$V = (3)(2)\left(\frac{1}{2}\right)$$
$$V = 3 \text{ in.}^3$$

The last step is to find out how many brownies of 3 in.^{3} fit into 540 in.^{3}

Divide 540 in.^{3} by 3 in.^{3} to get 180 brownies.

23. **A.** Set this up as a proportion.

$$\frac{1 \text{ lb}}{24 \text{ cups}} = \frac{x}{36 \text{ cups}}$$

Cross-multiply and solve for x.

$$24x = 36$$
$$x = 1.5 \text{ lb of potatoes}$$

24. **C.** To calculate the volume of a cylinder, use the formula $V = \pi r^2 h$. In this problem the diameter is 8 cm, so the radius is 4 cm. The height is given as 10 cm. Substitute that information into the formula.

$$V = (3.14)(4^2)(10)$$
$$V = (3.14)(16)(10)$$
$$V = 502.4 \text{ cm}^3$$

Be sure to pay attention to the information you are given. You could have made an error by using the diameter rather than the radius because the diameter was given in the problem.

25. **B.** To calculate the volume of a sphere, use the formula $V = \frac{4}{3}\pi r^3$. You are given a diameter of 18 inches, so the radius is 9. Substitute that information into the formula.

$$V = \frac{4}{3}\pi(9)^3$$
$$V = \frac{4(3.14)(729)}{3}$$
$$V = \frac{9,156.24}{3} = 3,052.08 \text{ in.}^3$$

26. **C.** There are 16 ounces in a pound. Divide 16 by 0.25, the weight of one slice of meat, to get the number of slices.

$$\frac{16}{0.25} = 64 \text{ slices.}$$

27. **A.** Set this up as a proportion.

$$\frac{55 \text{ pages}}{1 \text{ day}} = \frac{880 \text{ pages}}{x \text{ days}}$$

Cross-multiply and solve for x.

$$55x = 880$$
$$x = 16 \text{ days}$$

28. **B.** Use the formula $D = rt$. You are given the distance of 448 miles and the time of 8 hours. Substitute those into the formula. $448 = r8$. Solve for r.

$$r = \frac{448}{8} = 56 \text{ miles per hour}$$

29. **B.** The area of a square is calculated by using the formula $A = s^2$ or $s \times s$. Given the area of 576 ft^2, you need to find a number that, when multiplied by itself, equals 576. That number is 24. For problems like this, the fastest way to find the correct answer is to multiply each answer choice by itself and see which produces 576 ft^2.

30. **B.** For this problem, multiply the points by the difficulty and add the scores for each dive.

$$(8.2 \times 1.2) + (9.0 \times 1.4) + (8.3 \times 1.5)$$
$$+ (9.5 \times 1.4) + (9.5 \times 1.2)$$
$$= 9.84 + 12.60 + 12.45 + 13.30 + 11.4$$
$$= 59.59 \text{ pts}$$

PART 3. WORD KNOWLEDGE

1. **C.** *Listless* means "not caring, indifferent, or without energy." *Thirsty* is clearly wrong. The words *active* and *energized* are opposite in meaning. *Limp*, choice C, is the best answer.

2. **B.** *Potent* means "having authority; strong, effective, and forceful." *Colorful* and *movable* do not relate to this definition. *Explosive* certainly conveys the meaning of strength and force, but *powerful*, choice B, is the best answer.

3. **A.** *Compelled* means "forced or coerced." The words *happy*, *honored*, and *free* do not relate to the definition of *compelled*. *Forced*, choice A, is the correct answer.

4. **D.** *Vivid* is based on a root word meaning "full of life or lively." *Dull*, *poor*, and *stagnant* do not share the meaning of the definition. *Colorful*, choice D, is the best answer.

5. **B.** *Futile* means "pointless or incapable of achieving success." *Successful*, *easy*, and *powerful* do not share the meaning of the word *futile*. Thus, choice B, *useless*, is the best answer.

6. **C.** *Solicited* means "asked, begged, or implored." *Ignored*, *rejected*, and *followed* do not share this meaning. Choice C, *requested*, is the best answer.

7. **A.** *Appease* means "to pacify, soothe, or give in to someone else's demands." It does not mean *hurt*, *scare*, or *wonder*. *Satisfy*, choice A, is the correct answer.

8. **C.** To *abhor* means "to feel disgust for or to detest." The words *adore* and *worship* are both opposites of this definition and are clearly wrong. If you detest something you might *reject* it, but the best answer is choice C, *hate*.

9. **B.** To be *prudent* means "to exercise good judgment; to be cautious and not rash." *Reckless* and *wasteful* seem to be near opposites of the definition and cannot be correct. Someone who

is prudent with money might be a little *stingy*, but *careful*, choice B, is the best answer.

10. **A.** To be *staunch* means "to be loyal, solid, dependable, unfaltering, and strong." A staunch supporter cannot be *part time* or *periodic*, so those words are clearly wrong. The word *talkative* does not relate to the meaning of the word, so *steadfast*, choice A, is the correct answer.

11. **D.** To be *zealous* means "to be fervent, obsessive, and extreme." The word *whimsical* does not relate to the definition. A person who is zealous might be *incorrect* and even *unusual*, but the word *passionate*, choice D, is the closest in meaning to *zealous*.

12. **C.** *Volatile* means "unstable, unpredictable, and hot-tempered." It is based on a root word meaning "to fly," as in "fly off the handle." The words *acceptable*, *motivated*, and *humorous* do not relate to the definition. The best answer is choice C, *explosive*.

13. **B.** To be *terse* is to be succinct and not use unneeded words to make a point. *Terse* means "brief and abrupt." The words *expansive* and *wordy* are opposite in meaning to *terse*. Something terse might possibly be a little *confusing*, but the best answer is choice B, *concise*.

14. **C.** To be *austere* means "to be stern and severe; plain and lacking in ornaments." *Austere* means "stark and unadorned." The words *fancy* and *extravagant* are opposite in meaning. *Austere* does not mean *crowded*. Choice C, *simple*, is the best answer.

15. **A.** To be *meticulous* means "to be extremely careful with details, thorough, and precise." *Careless* seems to be the opposite of the definition. The word *doubtful* is not related. One must be *persistent* if one is meticulous, but *painstaking*, choice A, is the best answer.

16. **D.** *Vital* means "essential to life; indispensable, critical, central, or crucial." *Unrelated* and *unimportant* seem almost opposite to the definition.

Harmful does not seem related to the word *vital*. *Essential*, choice D, is the best answer.

17. **B.** *Timid* means "lacking in self-confidence, fearful, and shy." *Talkative* seems to be the opposite in meaning. The words *upset* and *present* do not relate to the definition. Someone who is timid is *hesitant*. Choice B is the correct answer.

18. **A.** The prefix *in-* means "not." *Inaudible* means "not audible." *Audible* means "able to be heard." So *inaudible* means "not able to be heard." The word *loud* seems somewhat opposite in meaning. The words *funny* and *suspicious* are not related to the definition. The best answer is choice A, *unheard*.

19. **B.** *Dormant* means "resting or sleeping." *Growing* seems to be the opposite of that definition. The words *dry* and *seedless* are not related to the definition. *Inactive*, choice B, is the best answer.

20. **C.** *Relinquished* means "gave up, handed over, or let go." The words *saved*, *transferred*, and *enjoyed* do not match this definition. *Surrendered*, choice C, the best answer.

21. **B.** *Delegated* means "transferred or handed over power or responsibility." *Absolved* means "forgiven," and that is clearly wrong. The words *withheld* and *promoted* are also unrelated to the definition. *Assigned*, choice B, is the correct answer.

22. **C.** *Dispensable* means "not important, not necessary, or not needed." The prefix *in-* reverses the meaning, so *indispensable* means the reverse of "not important," which is "important." It is clear that the words *hurtful*, *annoying*, and *energizing* do not match the definition. *Crucial* is the best synonym, which makes choice C the correct answer.

23. **D.** *Plausible* means "possible, probable, and credible." It does not mean *ridiculous*, *lengthy*, or *lofty*. Choice D, *credible*, is the best answer.

24. **A.** To be *advantageous* means "to provide an advantage or a benefit." Something *advantageous* is helpful and useful. The words *harmful* and *useless* are opposite in meaning. The word *compromising* is unrelated. Choice A, *beneficial*, is the best answer.

25. **D.** *Aroused* means "stimulated, provoked, and stirred to action." The word *reduced* seems to be somewhat opposite that concept. The words *increased* and *appeased* are unrelated to the definition. *Awakened* is the best synonym, so choice D is the correct answer.

26. **C.** *Provisional* means "temporary, interim, and conditional." The words *lasting* and *final* are near opposites to that definition. A provisional decision might be *initial*, but choice C, *tentative*, is the best answer.

27. **B.** To be *unparalleled* means "to be unequaled, unmatched, and incomparable." *Equaled* is the opposite of that meaning and is incorrect. The words *appreciated* and *misunderstood* do not relate to the concept. Choice B, *unmatched*, is the correct answer.

28. **A.** *Intermittent* means "starting and stopping at intervals; blinking, alternating, and sporadic." *Constant* seems to be the opposite of that definition. The words *disruptive* and *annoying* are unrelated to the definition. *Periodic*, choice A, is the correct answer.

29. **D.** To be *obstinate* means "to insist on having your own way; to be determined, pigheaded, adamant, and inflexible." The words *positive*, *unsure*, and *hopeful* are unrelated to that definition. *Stubborn*, choice D, is the only clear choice.

30. **C.** *Tactful* means "able to say the right thing in a delicate situation without offending anyone." *Discreet*, *judicious*, *polite*, and *careful* are synonyms. *Thoughtless* seems nearly the opposite. The words *forceful* and *unruly* are somewhat opposite to the definition. *Diplomatic*, choice C, is the best answer.

31. **D.** To be *proficient* means "to be highly competent, talented, and gifted." The word *unschooled* seems unrelated to the definition. It does not necessarily follow that if you are *trained*, then you are competent or talented. *Superior* is close to the definition, but the best answer is *skilled*, choice D.

32. **B.** *Augment* means "make greater, add to, enlarge, or expand." *Reduce* seems to be the opposite of this idea. The words, *invent* and *forget* are unrelated to the meaning. To *supplement* means "to enhance" and is thus a good synonym. Choice B is the correct answer.

33. **D.** *Vindictive* means "revengeful, malicious, and cruel." *Gentle* is the opposite of the definition and cannot be correct. The words *accurate* and *mistaken* are not related in meaning. Choice D, *spiteful*, is the correct answer.

34. **C.** *Obsolete* means "out of use, out of date, or outmoded." The words *vibrant* and *state of the art* are nearly opposite of the definition and cannot be correct. *Broken* is not sufficiently accurate in meaning. The only answer that makes sense is choice C, *outdated*.

35. **B.** *Inundated* means "covered with, deluged, engulfed, or overwhelmed." The words *awakened*, *interrupted*, and *greeted* are not related to that definition. The closest answer is *flooded*, choice B.

PART 4. PARAGRAPH COMPREHENSION

1. **A.** The passage indicates that bombs fall close to the vent, so choice A is the correct answer.

2. **A.** There is no indication in the passage that airlines have gone bankrupt because of volcanoes (choice D) or that planes need to fly around volcanoes even when they are not erupting (choice C). The passage does suggest that crops affected by the ash can be destroyed, but no indication is given that the entire farming

industry has had problems from volcanoes; choice B is not correct. The passage does indicate that areas that are a substantial distance from the volcanoes can be affected by even minor ash falls, so choice A is the correct answer.

3. **B.** The passage states that volcanoes are found in Alaska, Hawaii, California, Oregon, and Washington. These are states in the western part of the United States. Choice B is the correct answer.

4. **D.** The passage directly states that sulfur dioxide mixed with water causes acid rain. Choice D is the correct answer.

5. **D.** The passage indicates that during an explosive eruption, volcanic glass, minerals, and ash rise high into the air, forming a huge, billowing, eruption column. The sentence gives you a sense that the eruption column is growing and expanding. The correct answer is choice D, *swelling*.

6. **C.** There is some truth to all the answers, but the basic thrust of the passage indicates that Hughes started from poor roots to become a praised and successful poet. The title that best represents that is choice C, "From Busboy to Poet."

7. **D.** The passage states that Lindsay was a famous literary critic, so choice D is the correct answer.

8. **C.** The passage indicates that "The Negro Speaks of Rivers" was Hughes's first great poem. Choice D is not the correct answer. There are no overall descriptors of the places Langston Hughes lived in Illinois or Missouri, so choices A and B cannot be correct. The passage indicates that the Harlem Renaissance was the African American artistic movement in the 1920s that celebrated black life and culture. Choice C is the correct answer.

9. **A.** Hughes's poems were praised by Lindsay (choice D), but the passage does not say that Lindsay's praise is the reason Hughes is considered a great poet, so choice D is not correct. It is also true that Hughes came from a humble background to become well known (choice B), but the passage does not suggest that this is what made him a great poet, so choice B is not correct. The passage does not say anything about Hughes's poems being published in many countries, so choice C is not correct. The passage does state that Hughes's literary works helped shape American politics and literature and that he was one of the most important writers and thinkers of his time. Choice A is the correct answer.

10. **D.** There is nothing in the passage to suggest that Hughes did not want to stay in Missouri or that he didn't want to move to Ohio. Choices A and B are not correct. Also there is nothing in the passage that indicates he was not allowed to write poems, so choice C is incorrect. The passage states that his father moved to Mexico and that his mother had to find steady work and was away for long periods. Thus, he was raised by his grandmother. Choice D is the correct answer.

11. **C.** Although Wilder's family planted crops, the passage does not suggest that she chose to become a farmer (choice A). Being a daughter is not a profession, so choice B is incorrect. Wilder must have studied to earn a teaching certificate, but she did not remain a student (choice D). The profession she chose was that of teacher (choice C).

12. **C.** The passage does not suggest that Wilder was unoccupied (choice A), that she was too old to become an author (choice B), or that she was driven by a desire for success (choice D). The passage states that she wanted to pass on her experiences to the next generation, so choice C is the correct answer.

13. **B.** The passage indicates that Laura and Almanzo settled on a farm in Missouri called "Rocky Ridge." Choice B, Missouri, is the correct answer.

14. **D.** The passage mentions many hardships that Wilder faced when she was growing up and when she was married. Volcanic eruptions are not one of them. Choice D, volcanic eruptions, is the correct answer.

15. **B.** *Little House in the Big Woods* (choice D) was the title of Wilder's book and is not the correct answer. Although there are descriptions of the hardships of farmers (choice C), that is not the basic intent of the passage. That Wilder was an author (choice A) is also not the major idea in the story. Choice B, "Pioneer Woman and Author," reflects the major aspects of the passage—that Wilder and her family were pioneers and that she developed writing skills for a newspaper and eventually authored a book. Choice B is the correct answer.

PART 5. MATHEMATICS KNOWLEDGE

1. **A.** Isolate the unknown on one side. Add 37 to both sides.

$$k = -10 + 37$$
$$k = 27$$

2. **C.** Perform operations in the parentheses first, then the exponents, and then the remaining operations.

$$-3(4)^2$$
$$= -3(16) = -48$$

3. **D.** $45 = x\%$ of 180. $45 = x180$. Solve for x.

$$x = \frac{45}{180}$$
$$x = \frac{1}{4} \text{ or } 25\%$$

4. **B.** To change a fraction to a percent, first transform the fraction into a decimal. In this problem the decimal equivalent is 0.125. Move the decimal point two places to the right and add the percent sign to obtain the correct answer of 12.5%.

5. **C.** To divide mixed number and fractions, change the mixed number to an improper fraction. As with other division problems, invert and multiply the divisor.

$$\frac{7}{2} \div \frac{3}{4}$$
$$= \frac{7}{2} \times \frac{4}{3}$$

Simplify by canceling.

$$\frac{7}{\cancel{2}} \times \frac{\cancel{4}^2}{3} = \frac{14}{3} = 4\frac{2}{3}$$

6. **C.** In this problem you can easily see that the factors will begin with y. The last terms in the factors need to multiply to 15. The middle term tells you that the numbers you multiply need to add to 8. The factors will have plus signs. $5 \times 3 = 15$ and $5 + 3 = 8$. This gives you the factors $(y + 3)(y + 5)$.

7. **D.** When subtracting, change the sign and add. In this problem, $4x^3 + 5x^2 - 12$ is changed to $-4x^3 - 5x^2 + 12$ and is added to the first term.

$$4x^3 - 2x^2 + 6$$
$$-(4x^3 - 5x^2 + 12)$$

Note that the first term drops out, leaving the correct answer $-7x^2 + 18$.

8. **B.** For this problem, you want to isolate the h on one side of the equals sign. Multiply each side by $\frac{1}{3}$ resulting in $\frac{1}{h} = \frac{1}{33}$. Invert the fractions, which results in $h = 33$. Another way to solve this problem is to cross multiply: $1h = 33$; $h = 33$.

9. **A.** Set each equation to equal zero. Next, arrange the equations so that when one is subtracted from or added to the other, one of the terms results in a zero and drops out of the result. Next, solve for the remaining unknown. Set the equations equal to zero:

$$3m + 3p - 24 = 0$$
$$2m + p - 13 = 0$$

In this problem, notice that if you multiply the second equation by −3 and add the two equations, you can eliminate the p term.

$$3m + 3p − 24 = 0$$
$$(−3)(2m + p − 13) = 0$$

$$3m + 3p − 24 = 0$$
$$−6m − 3p + 39 = 0$$

Add the equations:

$$−3m + 15 = 0$$

Solve for m:

$$−3m = −15$$
$$m = 5$$

Substitute the value of m into one of the original equations.

$$3(5) + 3p − 24 = 0$$
$$15 + 3p − 24 = 0$$
$$−9 + 3p = 0$$

Solve for p.

$$3p = 9$$
$$p = 3$$

The correct answer is $m = 5$, $p = 3$.

10. **C.** To divide numbers with exponents, subtract the exponents. In this problem, there are numbers to divide as well as exponents $\dfrac{3r^5 s^3}{12 r^3 s^2}$. Note that $\dfrac{3}{12}$ can be simplified to $\dfrac{1}{4}$. Next, subtract the 3 from 5 in the r term giving an exponent of 2 and the 2 from 3 in the s term giving an exponent of 1 or plain s. Putting all that together gives:

$\dfrac{1}{4} r^2 s$ or $\dfrac{r^2 s}{4}$, the correct answer.

11. **B.** Factor the equation as follows: $(x − 15)(x − 3)$. Set each factor to equal zero.

$$x − 15 = 0; \; x = 15$$
$$x − 3 = 0; \; x = 3$$

The correct answer is $x = 15$, $x = 3$.

12. **D.** To multiply fractions, multiply the numerators and denominators. Simplify where possible.

$$\frac{6g}{2} \times \frac{3}{4y} = \frac{18g}{8y}$$

Simplify.

$$\frac{9g}{4y}$$

13. **A.** To divide fractions, invert the second term and multiply. Simplify where possible.

$$\frac{x^2}{3y} \div \frac{3x}{2y} \text{ becomes } \frac{x^2}{3y} \times \frac{2y}{3x}.$$

Multiply the numerators and denominators.

$$\frac{2x^2 y}{9xy}$$

Simplify.

$$\frac{2x}{9}$$

14. **B.** The cube root is the number that, when multiplied by itself three times, results in the answer. In this instance, $10 \times 10 \times 10 = 1{,}000$, so the cube root of 1,000 is 10.

15. **A.** In this type of problem, where two fractions are separated by an equals sign, you can cross multiply and then solve for the variable of interest. Cross-multiply to get $4g = 1(8)$. In solving for g, $g = 2$.

16. **D.** A quadrilateral has four sides, so choice D is the correct answer.

17. **C.** An isosceles triangle has at least two equal sides. Multiply the length of one of the two equal sides by 2 and add the length of the base. In this problem, the result is $2(24) + 3 = 51$ meters, the correct answer.

18. **C.** Multiply 6.25 by 4.5 to obtain 28.13. When two negatives are multiplied, the result is positive. Thus, the answer is 28.13, not −28.13.

19. **D.** If the two angles add up to 75°, the third angle must be 105°.

20. **A.** Adding the measures of any two sides of a triangle gives a result that is greater than the measure of the third side. So in this problem, $AB + BC > AC$.

21. **B.** The right angle is 90°. The other angles have to add up to 90°. If this is an isosceles triangle, those two angles are equal. So 90° ÷ 2 = 45°.

22. **B.** To find the area of a parallelogram, multiply the base times the height. The height of a parallelogram is a line from one angle to the opposite side, making a right angle. In this problem, the area can be determined by multiplying the base 10 by the height 4, giving the correct answer of 40 units^2.

23. **D.** $\angle 2$ and $\angle 4$ are adjacent angles and thus form a straight angle or a straight line. If one angle is 25°, then the other is 180° − 25°, or 155°.

24. **C.** The volume of any rectangular solid is the product of its length, width, and height. For a cube, those measurements are the same, so the volume is the product of three edges. In this problem the edges are each 3 cm, so the volume = 3 × 3 × 3 or 27 cm^3.

25. **D.** The volume of a sphere is found by using the formula $\frac{4}{3}\pi r^3$. You are given the diameter of 10, so the radius is 5. Substituting the information into the formula gives $\frac{4}{3}(3.14)(5^3)$, or 4 × 3.14 × 125 ÷ 3, making the correct answer 523.33 cm^3.

PART 6. ELECTRONICS INFORMATION

1. **D.** Neoprene is an insulator. The other listed materials are all good conductors.

2. **B.** Volts measure the pressure in an electrical circuit (how hard the electrons are "trying" to get around the circuit).

3. **C.** Watts are the unit of electrical power; watts = amperes × volts.

4. **D.** Watts = volts × amperes = 12 × 4 = 48.

5. **A.** Transformers change voltage.

6. **C.** Depending on the chemical treatment, a semiconductor material can work as an insulator or a conductor.

7. **D.** Electric motors create magnetic fields that repel each other (exhibit repulsion).

8. **A.** A generator takes rotating motion and creates electric current.

9. **A.** Without a load, you have a short circuit, which will quickly heat up and burn out. Why are the other answers incorrect? Choices B and C: when a circuit is carrying electricity, it would have both current and voltage, but a "cold" circuit would not. Choice D: a switch is useful but optional.

10. **C.** In a rechargeable battery, the electrochemical reaction goes in one direction during charging and in the opposite direction while discharging.

11. **D.** In a series circuit, add the individual resistances to find the total resistance.

12. **B.** In a parallel circuit like this one, $1/R_{total} = 1/R_1 + 1/R_2 = 1/20 + 1/10 = 3/20$ so $R_{total} = 20/3$ ohms, or $6\frac{2}{3}$ ohms.

13. **C.** A series-parallel circuit has series and parallel sections.

14. **C.** Silicon is a semiconductor and is able to become a conductor or insulator as needed.

15. **A.** Diodes are devices that allow current to flow in one direction only.

16. **B.** The equipment grounding system is a safety, used only when the normal ground fails. Choice C is incorrect because the circuit may work correctly unless the normal ground fails.

17. **D.** A simple circuit tester simply detects whether a circuit is hot (carrying current) or cold (not carrying current).

18. **A.** 3-way switches are used in pairs to control a light.

19. **C.** An electromagnet "throws" a circuit breaker if too much current is present.

20. **D.** Volts = amperes × ohms = 10 × 40 = 400 volts.

PART 7. AUTO AND SHOP INFORMATION

1. **C.** A loose or worn fan belt squeals; the pitch rises along with engine speed.

2. **D.** A spark plug starts the explosion of the gasoline–air mixture.

3. **C.** This is the characteristic smell of leaking coolant.

4. **A.** Without a flywheel, the opposing compression and power strokes would create a jerking motion in the crankshaft.

5. **A.** Two rigid surfaces, like the mating faces of the engine block and the cylinder head, cannot be sealed unless something compressible (like a gasket) is between them.

6. **D.** The ignition coil supplies high-voltage current to the spark plugs.

7. **C.** The piston slides up and down, moving the connecting rod in a linear (straight-line) motion. It connects to the crankshaft, creating the rotary motion.

8. **D.** Brakes are a hydraulic system, where motion created in one place is transferred to another place through hydraulic fluid. The master cylinder pushes brake fluid through the brake lines, expanding the wheel cylinders, which push the brake pads against the disk or drum.

9. **C.** Cylinder heads must be firmly fastened to the engine block, without being warped during tightening. A torque wrench ensures proper tightness, and correct tightening order prevents warping.

10. **D.** Fan belts should be tight enough to grab the pulleys, but if they are too tight, bearings can be damaged.

11. **A.** The tube that supplies a fuel injector is full of high-pressure gasoline. A leak will likely cause a spray of gasoline in the engine compartment.

12. **C.** In a four-stroke engine, only one of the four strokes is a power stroke. Each cylinder fires (explodes) on every other stroke. Each revolution of a 4-cylinder engine has two explosions. Therefore, at 2,000 rpm, there would be 4,000 power strokes, or 4,000 explosions.

13. **B.** Why are the other answers incorrect? Choice A: a carriage bolt is stronger than a lag bolt, which is stronger than a screw. Choice C: a screw is stronger than a nail. Choice D: a carriage bolt is stronger than a nail.

14. **D.** The tool shown is a bar clamp.

15. **B.** Carpenter's glue does not hold unless it is clamped for 30 minutes.

16. **C.** Flux prevents oxidation during heating.

17. **A.** Tin snips shear sheet metal.

18. **A.** Unreinforced concrete is hard to compress, but easy to pull apart. Steel reinforcement adds tensile strength, allowing concrete to be used for beams, bridges, and the like.

19. **C.** The tool shown is a Phillips screwdriver.

20. **C.** Use a center punch to make a starting point when drilling metal. Otherwise, the bit will wander across the surface.

21. **D.** In plywood, actual size is the same as nominal (named) size. Why are the other answers incorrect? Choice A: a 2 × 4 is actually 1½ in. × 3½ in. Choice B: a 1 × 4 is actually ¾ in. × 3½ in. Choice C: an 8 in concrete block is actually 7⅝ in high (to allow for a ⅜ in mortar joint).

22. **C.** The tool shown is a wood chisel.

23. **B.** A lag-shield anchor must be inserted in a hole drilled in the concrete.

24. **B.** Why are the other answers incorrect? Choice A: a pipe wrench does not have two fixed jaws. Choice C: a pipe wrench can be used on pipe and pipe fittings. Choice D: length varies.

25. **D.** Pipe-joint compound seals a joint and prevents it from seizing up or rusting.

PART 8. MECHANICAL COMPREHENSION

1. **A.** Because 1 pound of pull on the rope produces 1 pound of lift, this pulley system has a mechanical advantage of 1.

2. **C.** When the shaft rotates faster, centrifugal force causes the balls to move outward. This enables a governor to regulate engine speed.

3. **D.** Lubricating the pavement with some sort of grease would reduce friction and cause the block to slide more easily.

4. **C.** In a class 3 lever, the load is between the fulcrum and the force.

5. **D.** When two sheaves are connected by a belt, the larger sheave will rotate more slowly than the smaller one. Sheaves A and B are connected to the same shaft, so they rotate at the same speed. Sheave C is larger than sheave B, so it will rotate more slowly. Sheave D is the largest of all, so it will rotate the slowest.

6. **B.** After ½ hour (30 minutes), the tank will have received 30 gallons. The engine will have used 1.5 gallons (½ hour × 3 gallons/hour = 1.5 gallons). 30 − 1.5 = 28.5.

7. **C.** The mechanical advantage is 2 (two supporting strands). MA = load/effort. 2 = 200/effort. Effort = 100 lb.

8. **B.** MA = load/effort. 400/100 = 4.

9. **C.** MA of an inclined plane = slope/vertical. The correct answer is 13/5.

10. **A.** When gear A rotates 3 times, it will move 72 teeth (3 × 24 = 72 teeth). 72/36 = 2, so gear B will rotate 2 times.

11. **C.** The handles pivot around a pin at the center, making them a lever. And the blades that shear the metal are just wedges.

12. **B.** MA = effort arm/load arm = 6/4 = 1.5.

13. **C.** Because three strands of rope must be shortened to raise the weight, the mechanical advantage is 3. MA = effort distance/load distance.

$$3 = 12/\text{load distance}$$
$$3 \times \text{load distance} = 12$$
$$\text{load distance} = 4$$

14. **D.** Why are the other answers incorrect? Choice A: either gear can drive (smaller drive gears increase the force; larger ones increase speed). Choices B and C: equal sized gears rotate at the same speed, while different sized gears rotate at different speeds.

15. **D.** Calculate the force needed to balance each weight separately, then add them. MA = effort arm/load arm, and MA = load/effort. For the 50-lb weight, the mechanical advantage is 12/6 = 2.

$$2 = \text{load/effort} = 50/\text{effort}$$

Effort = 25 lb. For the 30 lb weight, MA = 12/8 = 1.5. MA = load/effort. 1.5 = 30/effort. Effort = 20.

Total effort = 25 + 20 = 45 lb

16. **C.** The only object that is elastic (springy) is the tire.

17. **B.** Of the choices, only adding a post under the load would help the beam carry a heavier load. Choice D, convert one post into a fulcrum, would make the beam collapse.

18. **B.** Five turns tighter minus three turns looser equals two turns tighter. At 1/8 in per turn, the bolt will tighten by 2/8 in, or ¼ in.

19. **A.** With each turn of the screw, the 12-thread/inch screw will move the load a shorter distance, so it has a greater mechanical advantage. Choices C and D are incorrect because it is not possible to tell from this information whether the screw is stronger or weaker.

20. **D.** Applying the force shown would tip the beam, adding weight to B, but removing it from A.

21. **C.** Water pressure on the bottom depends on the depth of the water, so as water is added, pressure will increase.

22. **A.** MA = driven gear/drive gear. Mechanical advantage is reduced when the driven (rear) gear gets smaller. Therefore, more force is needed on the pedals.

23. **D.** In a closed hydraulic system, pressure must be the same at all points.

24. **A.** When the fulcrum is closer to the load, the effort arm gets longer and the load arm gets shorter. MA = effort arm/load arm. Therefore, the MA increases.

25. **D.** A lock nut and a lock washer are both designed to prevent a nut from moving on a bolt after it is tightened.

PART 9. ASSEMBLING OBJECTS

1. **C.** The original figure has three objects, while choice A has four. Look for the square. The square in choice B is too large, and there are two squares in choice D. Choice C is the correct answer.

2. **A.** The original figure has three objects, while choice C has four and choice D has only two. Look for the arch. Choice B does not have this object. Choice A is the correct answer.

3. **B.** The original figure has three objects, while choice C has four. Look for the five-sided object that looks like a house. Choice D does not have it. Look for the rectangle. Choice A does not have it. Choice B is the correct answer.

4. **C.** The original figure has five objects, while choice A has four and choice B has six. Look for the large rounded object at the top left. Choice D has a circle instead. Choice C is the correct answer.

5. **B.** The original figure has four objects, while choice C has five. The largest piece is the most distinctive. It looks like a rectangle with a corner carved out. Choice C does not have it and in choice A it is a mirror image (choice A also has two triangles). Look for the isosceles right triangle. Choice D does not have it. Choice B is the correct answer.

6. **D.** The original figure has three objects, while choice A has four and choice C has two. Look for the rectangle. Choice B does not have it. Choice D is the correct answer.

7. **A.** The original figure has three objects, while choice B has two and choice C actually only has one object. Look for the object shaped like a plus sign. Choice D has a similar object, but it is shaped like an X rather than a plus sign. Choice A is the correct answer.

8. **C.** The original figure has four objects, while choice A has three and choice D has five. Look for the long object at the top that looks like a rectangle with pointed ends. Choice B does not have this object. Choice C is the correct answer.

9. **B.** The original figure has three objects, while choice C has four. Look for the two identical rhombuses. Choice A has only one Look for the rectangle. Choice D does not have it. Choice B is the correct answer.

10. **C.** The original figure has three objects, while choices A and D have four. Look for the square. Choice B does not have a square. Choice C is the correct answer.

11. **C.** The original figure has three objects, while choice A has four. Look for the big curved object at the lower right. The curved objects in choices B and D are not the same shape. Choice C is the correct answer.

12. **B.** The original figure has four objects, while choice A has only three. Look for the large quadrilateral at the far right. Choice D does not have it. Choice C has two quadrilaterals, so eliminate it as well. Choice B is the correct answer.

13. **A.** The original figure has three objects, while choice C has four. Look for the triangle. Choice D does not have a triangle. Look for the two identical quadrilaterals. Choice B has rectangles instead. Choice A is the correct answer.

14. **D.** The original figure has four objects, while choice A has five. Look for the five-sided object at the top left that looks like a rectangle with one corner cut off. Choices B and C do not have this object. Choice D is the correct answer.

15. **D.** The original figure has three objects, while choice A has four and choice B has only two. Look for the most distinctive object: the five-sided object at the right that looks like a kite. Choice C does not have this object. Choice D is the correct answer.

16. **C.** The original figure has four objects, while choice D has five. Look for the most distinctive object: the object on the far left that looks like a chair. Choice B does not have this object. Look for the curved object at the upper right. Choice A does not have it. Choice C is the correct answer.

17. **B.** The original figure has four objects, while choice C has five. Look for the object that looks like a triangle with one concave side. Choice D does not have this object. Choice A has a similar object, but it is not the same shape. Choice B is the correct answer.

18. **D.** The second object has a more distinct point, so start with that object. Point B is on the inside corner of the right "leg" of the second object. Choices A and B do not connect there. Choice C connects at the outside corner of that "leg." Choice D is the correct answer.

19. **A.** Point A is at the bottom right corner of the first object in its original orientation. Choice D does not connect the first object at that point. Point B is on the large rounded corner of the second object. Choice B connects at a pointed end. Choice C also connects at a (different) pointed end. Choice A is the correct answer.

20. **D.** Point A is at the bottom right corner of the first object in its original orientation. Choices B and C do not connect the first object at that point. Point B is in the center of the top end of the second object in its original orientation. Choice A does not connect at this spot. Choice D is the correct answer.

21. **A.** For the first object, point A is at the tip of the "leg" on the left (in the object's original orientation). Choice B does not connect the first object at that point. Point B is at the tall point of the second object. Choices C and D do not connect at the pointed end. Choice A is the correct answer.

22. **A.** If you picture the first object sitting on its flat base, then point A is at the lower left corner. All the choices have point A in the correct spot. Point B is at the center of the tip of the teardrop-shaped second object. All the choices have point B in the correct spot as well. Point C is at the indented center of the right side of the third object, in its original orientation. Choices C and D do not connect at that spot. That leaves choices A and B. Since this figure has three objects, check to see if each choice connects the pieces in the same order as the original: A to B and B to C. Choice B connects C to A and A to B. Choice A is the correct answer.

23. **D.** Point A is at middle of the long flat base of the first object in its original orientation. Choices A and B do not connect the first object at that point (and choice B has a mirror image of the object, which would also eliminate it as a possibility). If you picture the second object sitting on its larger flat base, then point B is at the top right corner. Choice C does not connect at this spot. Choice D is the correct answer.

24. **B.** Point A is at center of the concave part of the curve on the first object. Choice A does not connect the first object at that point. If you picture the second object sitting on its larger flat base, then point B is at the upper right corner. Choices C and D do not connect at this spot. Choice B is the correct answer.

25. **A.** In this question, the second object is more distinctive, so start with that one. Point B is on the right "leg" at the lower left corner. Choices B and C do not connect at that point. If you picture the first object—the isosceles triangle—sitting on its base, then point A is at the lower left corner. Choice D connects at the lower right corner. Choice A is the correct answer.

Part 5

JOBS IN TODAY'S MILITARY

Jobs in Today's Military

So you want to enlist in the military, but you have to train for a job. What job or jobs will it be? This section of the book will provide some information about what kinds of jobs are available. This is not a full listing, but a good overview of the variety of jobs available.

In this part of the book, you will receive a general overview of the job, a description of what people do in that job, some helpful attributes, and the Services that have that job category.

This section of the book should give you a pretty good idea of the variety of military jobs available. For more detailed information, you should ask your recruiter. Don't be afraid to ask for more information from them. It's their job to inform you!

People in the military do just about every kind of work that exists in civilian life. They may be physical therapists, computer repair technicians, photographers and journalists, management analysts, auto mechanics, life scientists . . . you get the idea. And free job training is provided for every one of the over 400 individual types of military jobs. If a particular type of work interests you, you can most likely find it in the military. The range of job opportunities in today's military is extremely broad.

Military jobs fall into several categories:

- Accounting, Budget, and Finance
- Aviation
- Combat Operations
- Construction, Building, and Extraction
- Education and Training
- Environmental Health and Safety
- Human Resources Management and Services
- Intelligence
- Law Enforcement, Security, and Protective Services
- Mechanic and Repair Technologists and Technicians
- Naval and Maritime Operations
- Transportation, Supply, and Logistics
- Arts, Communications, Media, and Design
- Business Administration and Operations
- Communications Equipment Technologists and Technicians
- Counseling, Social Work, and Human Services
- Engineering and Scientific Research
- Health Care Practitioners
- Information Technology, Computer Science, and Mathematics
- International Relations, Linguistics, and Other Social Sciences
- Legal Professions and Support Services
- Medical and Clinical Technologists and Technicians
- Personal and Culinary Services

Here you will find many opportunities to serve our country and perhaps receive training that can easily transfer to a job when you leave the military.

High-quality job training is provided right from the start. After Basic Recruit Training, you probably will attend advanced training (AT), where you attend a school that teaches you the skills necessary to do a particular job. AT combines classroom and hands-on learning environments. Well-trained instructors and a low student/teacher ratio are the norm. Over 60 percent of the courses in advanced training are certified for college credit by the American Council on Education.

Can You Choose Your Job If You Enlist?

You *do* have an influence on the job training you receive, but your ASVAB scores and the needs of the military at the time you'd like to enlist also affect your job assignment. Ask your recruiter about all the specialties for which you are eligible based upon your ASVAB scores. If the job you want isn't open at the moment, ask about the Delayed Enlistment Program, or explore other services that may have an equivalent job. Remember that the higher your ASVAB scores, the more job options you have!

Are Jobs in the Military All Open to Women?

About 20 percent of the active duty forces are women. Opportunities for women in the military have never been better. All jobs are open to females who qualify.

At the current time, the percentage of women in each branch of the military is approximately as follows:

Percentage of Female Enlisted Service Members

Army	17%
Marine Corps	10%
Navy	24%
Air Force	24%
Coast Guard	21%

Given that all jobs are now open to women, these number are likely to rise.

What Do All Military Jobs Have in Common?

Responsibility. Excellence. Pride. You're expected to perform your assigned duties with a high level of competence. Others on your team and in your unit need to know that you are dependable and that they can count on you. More often than not, the military offers a surprising amount of responsibility early in a career. Leading a team or crew of six or seven people at age 20 is common, so is making tactical decisions on the spot—even at junior levels. Most military jobs offer the opportunity to mature quickly . . . through responsibility.

How Many Military Jobs Have Civilian Equivalents?

About 85 percent of military jobs have counterparts in the civilian world, and many military jobs are virtually identical to the "hot" jobs that are forecasted to grow the fastest over the next few years. Virtually all military jobs give you an edge in the civilian world. That's because employers prize the things military veterans have learned—among them dependability, focus on the task, extra effort, reliability, team play, and striving for excellence. Computer and cybersecurity skills are becoming more critical and plentiful in military life. People with those skills are much needed in the private sector and in government.

No matter what your job in the military, there's an opportunity for you to do it well and earn the respect of your peers (and superiors) for your skills. This nearly always leads to greater self-confidence and a sense of pride. Every job in the military is important because each contributes to military readiness and each helps protect the freedoms all Americans enjoy.

INFORMATION ON MILITARY OCCUPATIONS

The following pages provide you with some summary information on a sampling of enlisted occupations in the U.S. military. These pages give you brief information on the following characteristics:

- A brief description of the occupation
- What people in the occupation do
- Helpful attributes
- Services that offer this occupation

ADMINISTRATIVE SUPPORT SPECIALISTS

Summary: Administrative support specialists are responsible for a variety of duties in the office. They make sure that information is recorded, stored, and delivered to keep operations running as smoothly as possible. They supervise or perform administrative, clerical, and typing duties. They may also be in charge of scheduling meetings, making travel arrangements, and organizing any other work-related events.

Helpful Attributes

- Ability to organize and plan
- Interest in keeping organized and accurate records
- Interest in operating computers and other office machines
- Preference for office work

What They Do

- Arrange travel and lodging, coordinate itineraries, and prepare trip folders

- Assist in planning, preparing, arranging, and conducting official functions
- Coordinate, perform, and manage a variety of tasks and activities in direct support of senior leaders to include office management, human resources, executive staff support, postal, official mail, and a variety of other services and duties
- Manage process of planning, coordinating, managing, sharing, and controlling organization's data assets
- Manage databases for the storage, modification, and retrieval of information to produce reports, answer queries, and record transactions
- Manage timeliness, accuracy, and maintenance of published content
- Perform personnel, general, operational, and manpower management administration at all levels
- Perform clerical and personnel security and general administrative duties, including typing and filing
- Manage processes and activities to support organizational communications, including correspondence preparation, distribution, suspense tracking, workflow management, electronic mail management, content management, and other related duties
- Perform various administrative functions in support of military and civilian leaders, including calendar management and meeting support
- Ensure accuracy of information in personnel and manpower database systems
- Perform research, interview personnel, and prepare analytical historical publications; assemble and maintain historical document repositories for reference and research

Services

Marine Corps

Navy

Air Force

Coast Guard

AIR CREW MEMBERS

Summary: Air crew members perform in-flight duties to ensure the successful completion of combat, reconnaissance, transport, and search and rescue missions. They perform inspections to ensure equipment is in working order. They operate and monitor engine and aircraft systems controls, panels, indicators, and devices. Their responsibilities may vary by type of aircraft and include such tasks as operating mine sweeping, refueling, and electronic warfare systems.

Helpful Attributes

- Ability to work as a team member
- Ability to work under stress
- Interest in flying
- Ability to work with mechanical and electrical systems

What They Do

- Perform search and rescue (SAR) operations and Airborne Mine Countermeasure (AMCM) operations utilizing sonar, magnetic, mechanical, and acoustic mine sweeping systems and logistics support in support of tactical missions worldwide
- Perform aircrew operations administration, flight and ground training, internal and external cargo movement, medical evacuations (MEDEVAC), passenger transport, aerial gunnery, small arms handling, and observer duties for flight safety
- Conduct planning and execution of tactical missions such as air-to-air refueling, assault support missions, rapid ground refueling, low-level flight, transporting hazardous cargo, aerial delivery, and battlefield illumination
- Perform in-flight refueling aircrew duties; check forms for equipment status; perform visual and operational checks of air refueling and associated systems and equipment
- Supervise cargo/passenger loading and off-loading operations; direct the placement of material handling equipment to accomplish cargo on/off-loading operations; ensure cargo/passengers are placed according to load plans
- Compute and apply aircraft weight, balance, and performance data manually or electronically; determine and verify passenger, cargo, fuel, and emergency and special equipment distribution and weight; compute takeoff, climb, cruise, and landing data
- Provide passenger briefings to include the use of emergency equipment, evacuation procedures, and border clearance requirements
- Perform aircraft inspections; conduct pre-flight inspection of the aircraft, guns, defensive systems,

cargo/airdrop systems, aerospace ground equipment, and related aircraft equipment according to flight manual procedures

- Use night vision goggles (NVG) to perform scanner duties in relation to particular aircraft type and mission
- Regulate aircraft systems, such as electrical, communication, navigation, hydraulic, pneumatic, fuel, air conditioning and pressurization, ventilation, auxiliary power unit, and lubrication
- Observe warning indicators and lights for fire, overheat, depressurization, and system failure; report abnormal conditions to pilot, and recommend corrective action
- Conduct thorough airborne analysis/evaluation of weapons, defensive systems, and associated equipment
- Serve as members of presidential helicopter crews
- Operate mission equipment such as advanced imaging multi-spectral sensors, radar for safety of flight, and hand-held cameras
- Detect, analyze, classify, and track surface and subsurface contacts; operate an advanced sonar system utilizing sonobouys, radar, electronic support measures (ESM), Magnetic Anomaly Detector (MAD), Identification Friend or Foe/Selective Identification Feature (IFF/SIF), and Infrared Detector (IR)

Services

Marine Corps

Navy

Air Force

AIR TRAFFIC CONTROLLERS

Summary: Air traffic controllers direct the movement of aircraft into and out of military airfields. They track aircraft using navigational aids. They coordinate and communicate aircraft movement information, as well as weather and airfield conditions. They also provide critical information to direct the action of combat aircraft engaged in close air support and other offensive air operations.

Helpful Attributes

- Ability to make quick, decisive judgments
- Ability to work under stress
- Skill in math computation

What They Do

- Coordinate aircraft movement information with associated facilities or agencies, coordinate current weather and airfield conditions as required, and perform air traffic control duties in both tactical and non-tactical air traffic control organizations
- Initiate and issue air traffic control clearances, instructions, and advisories to ensure the safe, orderly, and expeditious flow of air traffic operating under instrument and visual flight rules
- Formulate and issue clearances and instructions to aircraft and vehicular traffic operating on runways, landing areas, and taxiways
- Maintain radar surveillance of airspace, processing and coordinating aircraft passing through the terminal control area, and entering arrival/departure zones
- Direct the action of combat aircraft engaged in close air support and other offensive air operations
- Direct pilots of aircraft in making departures and approaches using radio communications and surveillance and carrier controlled approach radar systems
- Manage assigned forces and air operations using voice and data communications and radar systems
- Respond to emergency air traffic situations
- Maintain current flight planning information and reference material
- Assist pilots in preparing and processing flight plans
- Provide functional expertise and input for activating, employing, deploying, or deactivating battle management systems
- Recognize various types of interference encountered on electronic equipment and recommend corrective action
- Supervise and participate in preparation of system equipment for movement and combat
- Participate in the coordination of surface-to-air weapons and interceptors in an anti-air warfare environment

Services

Marine Corps

Navy

Air Force

AIRCRAFT ENGINE MECHANICS

Summary: Some aircraft mechanics in the Military work solely on the engines, or powerplants, of the aircraft. These mechanics troubleshoot and perform organizational-, intermediate-, and depot-level maintenance on the engines and related components of the Military's airplanes and helicopters. Aircraft engine mechanics must have specialized knowledge of the mechanical, electrical, and hydraulics principles applying to jet and turboprop engines.

Helpful Attributes

- Ability to use hand and power tools
- Interest in engine mechanics
- Interest in work involving aircraft

What They Do

- Supervise, inspect, and perform aviation unit (AVUM), intermediate (AVIM), and depot maintenance on aircraft turbine engines and components
- Remove, replace, service, prepare, preserve, clean, and store engine assembles or components
- Disassemble, repair, reassemble, adjust, and diagnostically test turbine engine systems, subsystems, and components according to directives
- Assist in troubleshooting engines and rigging engine controls
- Requisition and maintain shop and bench stock for repair of aircraft engines
- Evaluate maintenance operations and facilities for compliance with directives, technical manuals, work standards, safety procedures, and operational policies
- Perform maintenance trend analysis and apply production control, quality control, and other maintenance management principles and procedures to aircraft engine maintenance and shop operations
- Instruct personnel and conduct technical training in aircraft engine system maintenance, supply, and safety techniques
- Inspect, maintain, test, repair, and perform complete repair of helicopter power plants and power plant systems
- Inspect, maintain, modify, test, and repair propellers, turboprop and turboshaft engines, jet engines, small gas turbine engines, and engine ground support equipment (SE)
- Determine resource requirements, including facilities, equipment, and supplies
- Advise, perform troubleshooting, perform engine health management, and determine repair procedures on aircraft engines
- Solve maintenance problems by studying drawings, wiring and schematic diagrams, technical instructions, and analyzing operating characteristics of aircraft engines and propellers
- Implement maintenance and safety policies for egress systems and integral egress system components to include personnel parachute assemblies and survival kits

Services

Marine Corps

Navy

Air Force

AIRCRAFT LAUNCH AND RECOVERY SPECIALISTS

Summary: Aircraft launch and recovery specialists ensure the safety of aircraft as they launch from and return to aircraft carriers. They perform this critical work by operating and maintaining catapults, arresting gear, and other equipment used in aircraft carrier takeoff and landing operations.

Helpful Attributes

- Ability to use hand tools and test equipment
- Interest in aircraft flight operations
- Interest in working on hydraulic and mechanical equipment

What They Do

- Responsible for the safe and expeditious movement of aircraft on flight decks, hangar decks, and ashore, which includes the installation and removal of chocks, tie downs, and tow bars
- Direct the movement and spotting of aircraft ashore and afloat
- Utilize visual hand and light signals required in landing recovery operations
- Follow published safety regulations relative to working on and near a runway and know the hazardous zones

- Prepare aircraft recovery equipment for storage and shipment
- Coordinate recovery operations with other airfield operations
- Perform crash rescue, firefighting, crash removal, and damage control duties
- Operate, inspect, and maintain emergency arresting gear systems for high performance military aircraft
- Install and perform preventive maintenance on visual landing aids
- Operate pressurized hydraulic and pneumatic equipment utilizing both basic and precision tools
- Perform preventative and corrective maintenance on equipment and maintain the material condition of all assigned spaces
- Operate, maintain, and perform maintenance on steam catapults, barricades, arresting gear, and associated equipment ashore and afloat
- Operate, maintain, and perform organizational maintenance on ground-handling equipment used for moving and hoisting of aircraft ashore and afloat

Services

Marine Corps

Navy

AIRCRAFT MECHANICS

Summary: Aircraft mechanics inspect, service, and repair the Military's fleet of helicopters and airplanes. These mechanics troubleshoot and maintain the aircraft's structure, engines, and other components, including mission-critical features such as weapons and electronic warfare, and coatings for stealth purposes.

Helpful Attributes

- Ability to use hand and power tools
- Interest in engine mechanics
- Interest in work involving aircraft

What They Do

- Inspect, service, maintain, troubleshoot, and repair aircraft engines, auxiliary power units, propellers, rotor systems, powertrain systems, and associated airframe and systems-specific electrical components

- Service, maintain, and repair aircraft fuselages, wings, rotor blades, fixed and movable flight control surfaces, and bleed aviation air, hydraulic, and fuel systems
- Disassemble, repair, reassemble, adjust, and diagnostically test turbine engine systems, subsystems, and components
- Inspect and identify aircraft corrosion for prevention and repair; apply corrosion control treatment to aircraft metals
- Evaluate maintenance operations and facilities for compliance with directives, technical manuals, work standards, safety procedures, and operational policies
- Perform ground handling and servicing of aircraft and conduct routine aircraft inspections
- Perform end-of-runway, postflight, preflight, thru-flight, special inspections, and phase inspection administrative duties
- Supervise and perform aircraft jacking, lifting, and towing operations
- Direct aircraft battle damage repair and crash recovery operations
- Operate and maintain electronic warfare systems
- Ensure that funds and resources are projected to support the maintenance effort and are managed to optimize mission accomplishment
- Solve maintenance problems by studying drawings, wiring and schematic diagrams, technical instructions, and analyzing operating characteristics of aircraft engines and propellers
- Design, repair, modify, and fabricate aircraft, metal, plastic, composite, advanced composite, low observables, and bonded structural parts and components
- Perform helicopter and propeller maintenance including rotors, gearboxes, and drive accessory repairs
- Advise on structural and low observable repair, modification, and corrosion protection treatment with respect to original strength, weight, and contour to maintain structural and low observable integrity

Services

Army

Marine Corps

Navy

Air Force

Coast Guard

ALL-SOURCE INTELLIGENCE SPECIALISTS

Summary: Intelligence specialists who specialize in all-source intelligence are specifically trained in the planning, collection, and implementation of all intelligence disciplines across the full spectrum of operations. All-source intelligence specialists consolidate imagery, signals, measurement, human, and open-source data into information to support national defense.

Helpful Attributes

- Ability to organize information
- Ability to think and write clearly
- Interest in gathering information and studying its meaning
- Interest in computers

What They Do

- Advise and assist in the planning, collection, and implementation of all intelligence disciplines across the full spectrum of intelligence operations
- Supervise analytical cells supporting intelligence preparation of the battle-space, intelligence targeting analysis, network engagement, and intelligence production where an analytical outcome is required
- Perform Tasking, Collection, Processing, Exploitation, and Dissemination (TCPED) of all-source tactical intelligence
- Perform tactical- and operational-level research and analysis
- Conduct analysis of threat forces disposition, capabilities, tactics, and training in support of strike mission planning
- Perform analysis and reporting of all-source information for tactical and strategic commanders in support of battle force and national intelligence objective
- Support all aspects of Air Force operations by discovering, collating, analyzing, evaluating, and disseminating intelligence information
- Produce all-source intelligence, situation estimates, threat studies of adversarial nations, terrorists, and insurgents, and other intelligence reports and studies
- Advise commanders on force protection and intelligence information for U.S. and Partner Nations

- Conduct intelligence debriefings of U.S. and allied military personnel involved in combat operations
- Instruct military personnel on collecting and reporting requirements and procedures, recognition techniques, and assessing offensive and defensive weapon system capabilities
- Collate intelligence and operations materials, and assemble final products for mission briefing, study, and use
- Use intelligence automated data systems to store, retrieve, display, and report intelligence information
- Perform support to mission planning and execution; provide tailored collections planning, threat analysis, and intelligence expertise necessary to develop detailed mission plans for air, space, cyberspace, and special operations
- Assess vulnerabilities of Department of Defense (DoD) cyberspace enterprise that could be exploited by adversaries

Services

Marine Corps

Navy

Air Force

Space Force

ARMORED ASSAULT VEHICLE CREW MEMBERS

Summary: Armored assault vehicle crew members work as a team to operate armored equipment and fire weapons to destroy enemy positions. They normally specialize by type of armor, such as tanks, light armor (cavalry), or amphibious assault vehicles.

Helpful Attributes

- Ability to follow directions and execute orders quickly and accurately
- Ability to work as a member of a team
- Ability to work well under stress
- Readiness to accept a challenge

What They Do

- Operate tracked and wheeled vehicles over varied terrain and roadways in combat formation and armor marches
- Drive the unit, operate the breaching/bridging systems, and provide protection

- Participate in reconnaissance, security, cordon/search, and other combat operations
- Serve as a member of an LP/OP (listening/observation post) while employing principles of cover and concealment
- Gather information on the size, activity, location, unit, time, and equipment of the enemy
- Perform or manage reconnaissance of fording sites, tunnels, and bridges, and report on information collected
- Maintain responsibility for individual weapons, crew-served weapons, and the maintenance of assigned vehicle and equipment
- Manage distribution and adjust direct and aerial fires in combat
- Employ operations security (OPSEC) and casualty evacuation (CASEVAC) measures
- Coordinate and conduct platoon resupply
- Provide input and oversight of the unit's ammunition allocation and semiannual gunnery qualification and certification programs
- Assist and perform target detection and identification, placing turret in operation and determining range to target
- Position vehicle in firing position and secure battle position
- Secure, prepare, and stow ammunition aboard tank; load, unload, clear, and perform misfire procedures on main gun; exercise safety precautions in ammunition handling
- Advise leaders in the tactical employment of the organic weapons systems of the unit; advise leaders on the condition, care, and economical use of unit resources and equipment
- Prepare amphibious combat vehicle (ACV) and associated equipment for movement and combat, locate and engage targets, drive the ACV in the water and ashore, and perform first echelon preventive and corrective maintenance

Services

Army

Marine Corps

ARTILLERY AND MISSILE CREW MEMBERS

Summary: Artillery and missile crew members target, fire, and maintain weapons used to destroy enemy positions, aircraft, and vessels. The specific duties they perform vary by the type of combat operations they support. Field artillery crew members predominantly use guns, cannons, and howitzers in ground combat operations; air defense artillery crew members predominantly use missiles and rockets; and naval artillery crew members predominantly use torpedoes and missiles launched from a ship or submarine.

Helpful Attributes

- Ability to think and remain calm in stressful situations
- Ability to work as part of a team
- Interest in cannon and rocket operations
- Willingness to face danger

What They Do

- Operate high technology cannon artillery weapon systems; load and fire howitzers; set fuse and charge on a variety of munitions, including high explosive artillery rounds, laser guided projectiles, scatterable mines, and rocket assisted projectiles
- Employ rifles, machine guns, and grenade and rocket launchers in offensive and defensive operations
- Drive and operate heavy and light wheeled trucks and tracked vehicles
- Participate in reconnaissance operations to include security operations and position preparation
- Operate in reduced visibility environments with infrared and starlight enhancing night vision devices
- Supervise handling, transportation, accountability, and distribution of ammunition
- Supervise and direct the construction, camouflage, and defense of the section position
- Perform computer operations, including fire mission processing, fire plan schedules, and database construction; compute and apply meteorological and muzzle velocity corrections
- Compile information for, and present briefings on, current operations, situations, and after-action reports
- Maintain fire capability charts, and friendly and enemy situation maps
- Lead, supervise, and participate in coordination and implementation of cannon, missile, rocket, or target acquisition operations

- Provide situational awareness (SA) of airspace and early warning; conduct current and future operations planning and execution of airspace management requirements for the supported echelon
- Inspect and prepare missile launcher system for employment to include movement to and from concealment positions and firing positions, operate the fire control systems, and handle multiple launch rocket systems
- Operate, perform, and coordinate organizational and intermediate maintenance on guided missile launching systems, missile launching groups, guns, gun mounts, small arms, and associated handling equipment

Services

Army

Marine Corps

Navy

AUDIO VISUAL AND BROADCAST TECHNICIANS

Summary: Audiovisual and broadcast technicians plan, supervise, and coordinate the operation, maintenance, deployment, and management of systems, facilities, and personnel engaged in visual information operations. They play a high-level, but hands-on, role in the production process.

Helpful Attributes

- Experience in school plays or making home movies
- Interest in creative and artistic work
- Preference for working as part of a team

What They Do

- Install, operate, maintain, and perform maintenance on visual information equipment and systems
- Perform maintenance on broadcast systems such as radio, television, satellite, and transmission equipment
- Act as crewmember for multimedia productions, including serving as the director of a production crew
- Manage, supervise, or perform as instrumentalist, music arranger, vocalist, or audio and lighting engineer in military band activities

- Direct action in the production of scenes and episodes, analyze existing scripts, and recommend revisions
- Determine scene composition, coordinate performing personnel, direct audio capture during recording, and approve set design and props
- Execute broadcast journalism coordination and train personnel in advanced multimedia techniques
- Maintain and operate broadcast control and production equipment such as receivers, monitors, cameras, audio consoles, generators, amplifiers, closed circuit systems, and visual imagery satellites
- Organize, plan, direct, and inspect production activities
- Schedule and coordinate rehearsals, ceremonies, and performances
- Advise higher authorities on status, equipment maintenance and requirements, personnel training, and operational efficiency

Services

Army

Marine Corps

Navy

Air Force

AUTOMOTIVE AND HEAVY EQUIPMENT MECHANICS

Summary: After ensuring adequate manpower, supplies, and workspace are available, automotive and heavy equipment mechanics maintain and repair various vehicles and systems. They determine the overall mechanical condition of vehicles and heavy equipment, diagnose malfunctions, and initiate restorative actions related to equipment in their area of responsibility.

Helpful Attributes

- Interest in automotive engines and how they work
- Interest in troubleshooting and repairing mechanical problems
- Preference for physical work

What They Do

- Perform preventive maintenance on diesel and gasoline equipment, such as tractors, power shovels, road machinery, and concrete mixers

- Repair fuel systems, electrical systems, diesel engines, cooling systems, transmissions, brake systems, steering systems, hydraulic systems, and auxiliary drives
- Supervise compliance with shop safety program and use, maintenance, and security of hand and shop power tools
- Maintain wheeled vehicles and their associated trailers and systems, including inspecting, servicing, maintaining, repairing, replacing, adjusting, and testing
- Conduct overhaul of engines, transmissions, and powertrain assemblies of hydraulic and fuel systems
- Diagnose mechanical and electronic circuitry malfunctions using visual and auditory senses, test equipment, and technical publications
- Remove, disassemble, repair, clean, treat for corrosion, assemble, and reinstall accessories and components
- Establish production goals, quality controls, operating instructions, annual budgets, and self-inspection programs
- Evaluate and replace brake actuators, batteries, starter motors, alternators, mechanical fuel pumps, hydraulic cylinders/pumps/control valves, drive shafts, universal joints, service brake shoes, disc brake pads, water pumps, turbochargers, cylinder heads, high-pressure fuel injection pumps, wheel bearings/seals, steering unit torque link, and road wheels

Services

Army

Marine Corps

Navy Air Force

AVIONICS TECHNICIANS

Summary: Avionics technicians inspect, service, maintain, troubleshoot, and repair avionics systems that perform communications, navigation, collision avoidance, target acquisition, and automatic flight-control functions. They perform operational tests on aircraft components to determine condition, analyze performance, and isolate malfunctions in the radar, sensors, weapons control, electronic warfare (EW), flight control, and engine control systems. They repair and replace systems and equipment when deficiencies are identified.

Helpful Attributes

- Ability to work with tools
- Interest in electronics and electrical equipment
- Interest in solving problems

What They Do

- Diagnose and troubleshoot malfunctions in electrical and electronic components, identifying location and extent of equipment faults
- Repair equipment by adjusting, aligning, repairing, or replacing defective components
- Clean, preserve, and store electrical/electronic components and aircraft instruments
- Use Test, Measurement, and Diagnostic Equipment (TMDE) Test Program Sets (TPS) and Interactive Electronic Technical Manuals (IETM) to determine the cause and location of malfunctions, extent of faults, and category of maintenance required
- Inspect, service, maintain, troubleshoot, and repair aircraft batteries, AC and DC power generation, conversion and distribution systems, as well as the electrical control and indication functions of all airframe systems
- Repair amplifier and logic circuits, microwave equipment, servomechanisms, radio frequency circuits, video displays, and power supply circuits
- Serve as aircrew members on flight
- Replace assembly components using hand tools, soldering devices, and electronic instruments
- Plan and organize integrated avionics equipment assembly, calibration, repair, modification, and maintenance activities
- Upload ground maintenance and operational software
- Coordinate with supply, operations, and other support activities to improve procedures and resolve problems
- Use automated maintenance systems; input, validate, and analyze data processed to automated systems; clear and close out completed maintenance discrepancies in automated maintenance systems
- Supervise and perform aircraft, engine, and component inspections; interpret inspection findings and determine adequacy of corrective actions
- Requisition and maintain stock for repair of aircraft avionics equipment, and maintain records
- Resolve technical problems and improve maintenance methods and techniques

Services

Army

Marine Corps

Navy

Air Force

Coast Guard

BOOKKEEPING, ACCOUNTING, AND AUDITING CLERKS

Summary: Bookkeeping, accounting, and auditing clerks perform duties related to budgeting, disbursing, and accounting of government funds. They are responsible for maintaining and supporting all financial management processes, including verification and submission of financial reports, payment for travel and commercial vendors, and auditing of pay transactions.

Helpful Attributes

- Ability to work with numbers
- Interest in using office machines such as computers, and calculators
- Interest in work requiring accuracy and attention to detail

What They Do

- Process, compute, and audit financial transactions
- Review and process travel claims related to military and civilian Temporary Additional Duty (TAD) and military Permanent Change of Station (PCS)
- Estimate travel costs and audit travel claims
- Maintain military pay records for officer and enlisted personnel
- Prepare vouchers related to disbursement and collection
- Determine fund availability and interpret financial directives
- Maintain, analyze, and verify payroll reports
- Perform audits and reviews
- Maintain accounting records, prepare accountability reports, and determine fund availability
- Review and analyze financial data for accuracy verification
- Identify possible inaccuracies and resolve financial data discrepancies

- Provide supervision and technical guidance to lower grade personnel, and support training efforts to ensure successful financial management practices

Services

Army

Marine Corps

Navy

Air Force

BUILDING ELECTRICIANS

Summary: Building electricians install, support, and troubleshoot electrical wiring systems. They ensure that electrically operated equipment and services are maintained in top running condition and in compliance with safety standards.

Helpful Attributes

- Ability to use hand tools
- Interest in electricity
- Preference for doing physical work

What They Do

- Supervise or perform installation, repair, and maintenance of interior electrical systems and equipment, such as service panels, switches, and electrical boxes
- Read and interpret drawings, plans, and specifications
- Employ the appropriate equipment to test the operational condition of circuits
- Climb utility poles and operate special purpose vehicles and equipment, including line maintenance and high-reach trucks to inspect, maintain, and repair overhead distribution systems
- Install, maintain, and repair interior, exterior, overhead, underground electrical power distribution systems and components such as capacitor banks, vacuum and air brake switches, breakers, transformers, fuses, lighting fixtures, receptacles, and motors
- Prepare cost estimates for in-service work; apply engineered performance standards to plan and estimate jobs
- Advise on problems encountered with the installation and repair of electrical power distribution and special purpose electrical systems

- Use meters, testing devices, indicators, and recorders to identify and isolate equipment, distribution, and motor controller malfunctions and faults
- Diagnose malfunctions and recommend appropriate repair procedures to correct defective equipment

Services

Army

Marine Corps

Air Force

CARDIOPULMONARY AND EEG TECHNICIANS

Summary: Cardiopulmonary and electroencephalogram (EEG) technicians assist physicians by performing a variety of diagnostic and therapeutic procedures that involve the heart, lungs, blood, and brain. They operate sophisticated equipment in performing this job and interact with patients on a regular basis.

Helpful Attributes

- Ability to follow strict standards and procedures
- Ability to keep accurate records
- Interest in electronic equipment
- Interest in learning how the heart, lungs, and blood work together

What They Do

- Perform and manage cardiopulmonary laboratory activities for diagnostic and interventional cardiac procedures, pulmonary function testing, diagnostic and therapeutic bronchoscopies, and respiratory therapy
- Prepare and instruct patients prior to procedure; take medical history and document results
- Perform and supervise cardiopulmonary functions such as electrocardiograms, exercise stress testing, and ambulatory electrocardiographic monitoring
- Assist physician in performing electroencephalography tests
- Operate electroencephalography equipment to examine patients for organic brain diseases
- Assist medical officers in treating patients, using controlled breathing apparatus that employs medical gases; perform medical gas therapy, nebulization therapy, mechanical ventilation, and pulmonary function testing

- Draw and analyze arterial blood gas samples
- Adhere to infection control and universal precaution procedures, including disposition of contaminated materials; practice safety and security measures
- Monitor data display on physiological equipment, and obtain and record vital signs
- Perform cardiopulmonary resuscitation on both adults and infants
- Place patient on continuous ventilator care, if ordered
- Provide instruction to cardiology and non-cardiology providers on advanced cardiac life support
- Perform laboratory administrative, maintenance, and support functions
- Determine inventory level of disposable supplies, stock accordingly, and return excess stock
- Perform user maintenance on equipment

Services

Air Force

CARDIOVASCULAR TECHNICIANS

Summary: Cardiovascular technicians specialize in diagnosing and treating heart conditions. They administer and document tests to obtain diagnostic data on the condition of the heart, to include electrocardiography, echocardiography, holter monitoring, and cardiac stress testing. They also assist physicians with diagnostic and interventional catheterizations, perform device implantations, and conduct electrophysiology studies.

Helpful Attributes

- Ability to follow strict standards and procedures
- Ability to keep accurate records
- Interest in electronic equipment
- Interest in learning how the heart, lungs, and blood work together
- Concern for others

What They Do

- Assist with the management of cardiac clinics
- Perform specialized invasive and noninvasive cardiac tests and examinations

- Administer and record tests to obtain diagnostic data on condition of heart
- Supervise operation of invasive and diagnostic laboratories
- Supervise daily operations for all cardiology service functions, to include all aspects of clinic and laboratory operations, including management of budget and staff
- Assist the physician in all invasive and non-invasive cardiac studies, to include diagnostic cardiac catheterization, interventional cardiology, pacemaker implantation, peripheral intervention, echocardiography, and stress testing
- Provide instruction to cardiology and non-cardiology providers on Advanced Cardiac Life Support
- Assist in the performance of all diagnostic and interventional cardiac procedures, to include cardiac and peripheral angiography, angioplasty, and stenting, as well as echocardiography and pacemaker technology
- Set and maintain the sterile field and care for all sterile instruments
- Possess knowledge of radiation safety and use of fluoroscopy

Services

Army

Navy

CAREER COUNSELING/RETENTION SPECIALISTS

Summary: Career counseling/retention specialists provide military personnel with consultation and guidance on career exploration by reviewing their interests, education, strengths, and abilities. They support the development and implementation of career information programs, and they are responsible for collecting and analyzing retention and attrition data. They provide service members with guidance and motivation in maximizing their career potential, and they provide counseling to transitioning or retiring personnel.

Helpful Attributes

- Ability to speak before groups
- Ability to work independently
- Interest in working with youths
- Listening and communication skills

What They Do

- Provide career information guidance and assistance to help service members explore and evaluate their education, training, interests, and capabilities
- Determine service member eligibility to immediately reenlist or extend; review reenlistment and extension documents for accuracy; coordinate retention ceremonies; prepare and process reenlistment, extension, or transition documents
- Brief leaders on matters relating to retention activities
- Manage objectives, statistics, and awards program
- Plan and conduct retention and transition training; evaluate subordinate training presentations and provide feedback
- Coordinate required tests, interviews, and physicals for lateral move candidates
- Develop, coordinate, and implement career information programs and policies
- Collect and examine retention and attrition data, and provide trend analysis; maintain reports required for systematic retention
- Provide executive guidance in all aspects of interviewing and counseling efforts regarding career opportunities and professional development
- Ensure individual and other family members are presented information concerning career opportunities, transition services and benefits, reenlistment incentives, rights and benefits, and advantages of a military career
- Advise and assist commands in organizing and implementing a career information program for reserve enlisted personnel
- Supervise and coordinate marketing, prospecting, interviewing, processing, classification, and counseling efforts
- Assist supervisors and commanders in counseling enlisted personnel on reenlistment opportunities, retraining, and benefits
- Provide management consultant services relating to career opportunities, progression, and planning
- Conduct and administer symposiums, workshops, or conference

Services

Army

Marine Corps

Navy

Air Force

CARGO SPECIALISTS

Summary: Cargo specialists ensure service members all over the world receive needed supplies and are themselves transferred safely and efficiently to their destinations. They are responsible for transferring or supervising the transfer of cargo to and from air, land, and water transport by manual and mechanical methods. They also plan and organize loading schedules.

Helpful Attributes

- Interest in working with forklifts and cranes
- Preference for physical work

What They Do

- Check, tally, and document cargo, utilizing both manual and automated data processing (ADP) systems
- Load and unload supplies and equipment from ships, docks, beaches, railheads, boxcars, warehouses, motor vehicles, and aircraft
- Operate and maintain all types and sizes of winches, cranes, and forklifts
- Inspect cargo, supervise cargo checking and hatch operations, control aircraft loading and unloading, oversee railhead tie-down crews, direct container stuffing and unstuffing, plan warehouse storage, and manage crane operations
- Supervise operator maintenance for cargo handling equipment such as cranes and forklifts
- Enforce safety practices and documentation procedures
- Provide staff supervision, policy, and guidance for personnel and cargo movement by air, rail, motor, and water transport
- Receive cargo/passenger load briefings, check placement of cargo/passengers against aircraft limitations/restrictions, and determine adequacy of cargo documentation
- Supervise cargo/passenger loading and offloading activities
- Determine cargo placement and restraint requirements, and direct and check the placement of restraint equipment
- Compute aircraft weight and balance
- Plan, supervise, train, and perform various duties pertaining to the preparation and packaging of various types of material for movement or shipment by common carriers

Services

Army

Navy

Air Force

CASE WORKERS

Summary: Caseworkers are responsible for overseeing and managing resources, programs, and functions regarding mental health services within the military community. They support services in psychiatry, social work, psychology, family advocacy, substance abuse awareness, and rehabilitation. They provide assistance to mental health professionals and advise in planning and administering treatment methods. They perform managerial duties in directing and coordinating resources, as well as developing, implementing, and evaluating training related to mental health programs.

Helpful Attributes

- Interest in working with people
- Patience in dealing with problems that take time and effort to overcome
- Sensitivity to the needs of others

What They Do

- Manage mental health service resources and activities
- Assist mental health professionals with developing and implementing treatment plans
- Perform specified mental health treatment
- Manage and direct personnel resource activities
- Interpret and enforce policy and applicable directives
- Establish control procedures to meet work goals and standards
- Recommend or initiate actions to improve functional operation efficiency
- Plan and program work commitments and schedules
- Develop plans regarding facilities, supplies, and equipment procurement and maintenance
- Participate in patient care conferences and substance abuse intervention

- Coordinate with appropriate agencies regarding treatment, prevention, rehabilitation, and administrative functions
- Maintain contact with military and community agencies to obtain collateral information
- Arrange and assist with patient referral to public, private, and military community agencies
- Manage preparation and maintenance of records and reports pertaining to specialty services
- Review procedures and requirements within specialty services to ensure efficiency of personnel
- Develop, maintain, and evaluate specific mental health training and evaluation programs

Services

Marine Corps

Navy

Air Force

COMMUNICATION EQUIPMENT OPERATORS

Summary: Communications equipment repairers install, sustain, troubleshoot, and repair standard voice, data, video network, communication security (COMSEC), and cryptographic devices in fixed and deployed environments, to ensure the ability to communicate and the Military's continued success. Communications equipment repairers usually work in repair shops and laboratories on land or aboard ships.

Helpful Attributes

- Ability to use tools
- Interest in solving problems
- Ability to apply electronic principles and concepts
- Interest working with electrical, electronic, and electrochemical equipment

What They Do

- Install, operate, and perform unit-level maintenance on cable and wire systems, to include Digital Group Multiplexes (DGM), Remote Multiplexing Combiners (RMC), repeaters, restorers, voltage protection devices, telephones, test stations, intermediate distribution frames, and related equipment
- Install, operate, perform strapping, restarting, preventive maintenance checks and services (PMCS), and unit-level maintenance on communication security (COMSEC) devices
- Perform tests on cable communications systems to ensure circuit and system quality
- Coordinate and supervise team member activities in the construction, installation, and recovery of cable and wire communications systems and auxiliary equipment
- Operate and perform PMCS on assigned communications equipment, vehicles, and power generators
- Inspect equipment for faults and completeness; test equipment to determine operational condition; troubleshoot to determine location and extent of equipment faults
- Repair equipment by adjusting, aligning, repairing, or replacing defective components; test repaired equipment to ensure compliance with technical specifications
- Maintain diagnostic and operational tools, support, and test equipment; maintain accountability of all parts of communication systems including software and spare parts
- Oversee requisition of supplies and spare parts; maintain necessary records of maintenance and compile data for reports
- Perform organizational-level planned and corrective maintenance actions on an air traffic control (ATC) digital communication switching systems
- Maintain and monitor the operational readiness of a surface ship's exterior communications suite on a system level; perform preventive and corrective maintenance on the surface ship's exterior communications system
- Deploy, sustain, troubleshoot, and repair standard radio frequency wireless, line-of-sight, beyond line-of-sight, wideband, and ground-based satellite and encryption transmission devices in a fixed and deployed environment
- Climb antenna support structures and wooden poles to various heights for maintenance and installation actions on cable and antenna systems
- Supervise, plan, organize, and direct cable and antenna installation and maintenance activities

Services

Army

Marine Corps

Navy

Air Force

Space Force

COMMUNICATIONS EQUIPMENT REPAIRERS

Summary: Communications equipment repairers install, sustain, troubleshoot, and repair standard voice, data, video network, communication security (COMSEC), and cryptographic devices in fixed and deployed environments, to ensure the ability to communicate and the Military's continued success. Communications equipment repairers usually work in repair shops and laboratories on land or aboard ships.

Helpful Attributes

- Ability to use tools
- Interest in solving problems
- Ability to apply electronic principles and concepts
- Interest working with electrical, electronic, and electrochemical equipment

What They Do

- Install, operate, and perform unit-level maintenance on cable and wire systems, to include Digital Group Multiplexes (DGM), Remote Multiplexing Combiners (RMC), repeaters, restorers, voltage protection devices, telephones, test stations, intermediate distribution frames, and related equipment
- Install, operate, perform strapping, restarting, preventive maintenance checks and services (PMCS), and unit-level maintenance on communication security (COMSEC) devices
- Perform tests on cable communications systems to ensure circuit and system quality
- Coordinate and supervise team member activities in the construction, installation, and recovery of cable and wire communications systems and auxiliary equipment
- Operate and perform PMCS on assigned communications equipment, vehicles, and power generators
- Inspect equipment for faults and completeness; test equipment to determine operational condition; troubleshoot to determine location and extent of equipment faults

- Repair equipment by adjusting, aligning, repairing, or replacing defective components; test repaired equipment to ensure compliance with technical specifications
- Maintain diagnostic and operational tools, support, and test equipment; maintain accountability of all parts of communication systems, including software and spare parts
- Oversee requisition of supplies and spare parts; maintain necessary records of maintenance and compile data for reports
- Perform organizational-level planned and corrective maintenance actions on an air traffic control (ATC) digital communication switching systems
- Maintain and monitor the operational readiness of a surface ship's exterior communications suite on a system level; perform preventive and corrective maintenance on the surface ship's exterior communications system
- Deploy, sustain, troubleshoot, and repair standard radio frequency wireless, line-of-sight, beyond line-of-sight, wideband, and ground-based satellite and encryption transmission devices in a fixed and deployed environment
- Climb antenna support structures and wooden poles to various heights for maintenance and installation actions on cable and antenna systems
- Supervise, plan, organize, and direct cable and antenna installation and maintenance activities

Services

Army

Marine Corps

Navy

Air Force

Space Force

COMMUNITY AND RECREATION SPECIALISTS

Summary: Community and recreation specialists provide support services to military personnel. These services include offerings such as community housing, food service, physical fitness facilities, laundry services, libraries, youth and outdoor recreation activities, arts and crafts, ceremonies, and unit-level sports. Community and recreation specialists are also available to help personnel and their families during wartime and other crises.

Helpful Attributes

- Interest in working with people
- Interest in physical fitness, athletics, arts and crafts, and other recreational activities
- Ability to plan and organize group and individual events, programs, and activities
- Enjoy helping, teaching, and caring for others
- Ability to work with minimal supervision

What They Do

- Establish and supervise bare base facilities that provide food, fitness, lodging, sports management, recreation, laundry, mortuary services, field resale operations, and protocol support to deployed personnel
- Ensure storage facilities and procedures are in place to adequately safeguard subsistence, equipment, and supplies
- Operate fixed, bare base, missile alert, and portable food service facilities and equipment; plan, prepare, and adjust menus; determine resource availability, pricing, and merchandise trends
- Administer fitness programs; explain and demonstrate concepts of fitness requirements, proper conditioning procedures, weight training, and aerobic equipment techniques
- Participate in arrival and departure ceremonies for the president, foreign heads of state, and other national or international dignitaries
- Develop and provide personal and family readiness services related to pre-deployment, deployment/sustainment, redeployment/reintegration, and post deployment
- Provide advocacy for the unique educational needs of military children and families during deployments and assist in ensuring school personnel are aware of the unique issues and stressors impacting military children
- Brief and assist military members, DoD civilians, and families during emergencies and natural disasters
- Supervise or perform laundry and shower operations; maintain facility upkeep, and provide customer service
- Provide haircuts that conform to military grooming standards; maintain barber equipment; meet sanitation requirements
- Administer daily housing operations to include assigning individuals to spaces

- Mentor residents and assist them in their adjustment to military life, the development of military attitude, and their enhancement of social skills; mediate resident disputes
- Supervise staff in maintaining barracks, buildings, and grounds in a clean, sanitary, and orderly condition
- Perform duties in both combat and noncombat environments pertaining to the search and recovery, processing, tentative identification, interment, disinterment, and transportation of human remains and personal effects

Services

Army

Marine Corps

Navy

Air Force

COMPUTER REPAIRERS

Summary: Some electrical instrument and equipment repairers focus on repairing the vast computer systems and networks operated in the Military. Computer repairers install, configure, test, troubleshoot, and repair computer systems (both hardware and software) that are used for both combat and noncombat missions. They may also provide technical assistance to system users.

Helpful Attributes

- Ability to use tools
- Interest in solving problems
- Ability to apply electronic principles and concepts
- Interest working with electrical, electronic, and electrochemical equipment

What They Do

- Maintain data link networks, tactical networks, computer-based systems, and peripheral computer equipment
- Analyze equipment operation, align computer and network configurations, and troubleshoot and repair computer-based equipment to the lowest replaceable unit
- Perform casualty control procedures, restoring operability for all assigned equipment, recognizing

redundancies within the systems, performing routine repairs, performing computer maintenance, and providing estimated repair times for corrective maintenance

- Perform administrative responsibilities which include maintaining technical manuals and equipment records
- Monitor test equipment calibration requirements
- Maintain copiers, audiovisual equipment, and fiber optics
- Perform micro-miniature repair, module test and repair, and test equipment calibration
- Supervise personnel who complete maintenance, maintain work spaces, conduct tools and test equipment inventory and logistics support, and implement equipment and system training programs
- Perform preventive and corrective maintenance on the ADS MK6 and UYQ-70 Display Systems on Baseline 7 DDG 91-102 AEGIS Destroyers
- Perform maintenance on weapon systems, nodes, and networks

Services

Navy

CONSTRUCTION EQUIPMENT OPERATORS

Summary: Construction equipment operators perform or manage the use of bulldozers, cranes, graders, and other heavy equipment in military construction. They manage daily activities devoted to the construction and maintenance of concrete and asphalt runways, structural systems and wooden, masonry, metal, and concrete buildings, aircraft parking aprons, and roads. They direct and coordinate the efforts of crews in the use of earth moving, quarrying, well, mixing, asphalt batching, and paving equipment. They also oversee demolition efforts and ensure adherence to environmental regulations.

Helpful Attributes

- Interest in operating heavy construction equipment
- Preference for working outdoors

What They Do

- Supervise or operate electric, pneumatic, and internal combustion-powered machines used in drilling, crushing, grading, and cleaning gravel and rock
- Supervise or operate all equipment used in concrete and asphalt production and paving; plan and direct layout of asphalt, and supervise production of hot mix asphalt
- Detonate explosives to blast rock in quarries and at construction sites; direct quarry rock evacuation to produce required sizes and tonnage of rock
- Operate air compressor and tools in support of horizontal construction projects
- Plan, construct, and repair airfield pavements, roads, streets, curbs, surface mats, membranes, and other improved areas using paving and surfacing procedures
- Inspect, lubricate, and perform operator maintenance on construction and snow removal equipment
- Solve basic mathematics problems related to earthwork and equipment effectiveness, and apply engineering performance standards to plan and estimate job specifications
- Interpret construction drawings and surveys using information such as subgrade contours and grade alignment
- Determine the geographical area most suitable for developing a water supply
- Investigate proposed work sites to determine resource requirements; prepare cost estimates for in-service work and oversee contractor work
- Determine type and application of equipment to use in various construction, maintenance, and repair operations
- Perform planning activities and conduct facility surveys
- Establish and maintain blaster qualification certification program
- Oversee stowage of blasting materials and maintain records of expended and stowed explosives

Services

Army

Marine Corps

Navy

Air Force

CONSTRUCTION SPECIALISTS

Summary: Construction specialists perform and manage the construction of buildings, bridges, foundations, utility systems, dams, and bunkers. They manage, construct, repair, and modify structural systems, and wooden, masonry, metal, and concrete buildings. They also oversee bridge building, rafting, and river-crossing operations to coordinate the building of structures during combat. Construction specialists must have a working knowledge of carpentry and masonry.

Helpful Attributes

- Ability to communicate effectively
- Ability to understand and apply math concepts
- Ability to coordinate and prioritize the work and activities of others
- Interest in work requiring accuracy and attention to detail

What They Do

- Lift and move heavy objects and equipment by setting up, bracing, and utilizing rigging devices and equipment
- Design and construct concrete forms and structures
- Erect and work from scaffolding, ladders, and mobile platforms
- Direct the construction of fighting positions and wire entanglements: camouflage positions, vehicles, or equipment with lightweight screening system
- Conduct ground reconnaissance for roads, routes, bridges, tunnels, fords, rivers, and ferries
- Perform basic demolition, mine warfare, and combat construction operations
- Direct minefield installation and removal, and submit minefield reports
- Provide conventional and powered bridge and rafting support for wet and dry gap crossing operations
- Interpret construction drawings and blueprints and prepare building layout
- Build and repair wood, concrete, and masonry structures; cut lumber to specific dimensions; perform rough and finished carpentry and masonry; install sheet rock, paneling, ceramic tile, millwork, trim
- Erect steel and lay out trusses and structures to specific dimensions
- Weld, cut, braze, and solder ferrous and nonferrous metals using various welding processes
- Operate and maintain shop tools and equipment, such as electric saws, sanders, planers, routers, drills, and other millwork tools

- Build, repair, and maintain technical security infrastructures in remote locations worldwide
- Inspect all phases of construction and installation, including civil, architectural and structural, electrical and mechanical, for compliance with drawings, specifications, and acceptable safe operating, installation and construction practices
- Plan and estimate material, manpower, and equipment requirements for various construction jobs; perform scheduling, procurement, production control, and management reporting of construction projects

Services

Army

Marine Corps

Navy

Air Force

CONTRACTS SPECIALISTS

Summary: Contracts specialists negotiate, procure, and process administrative actions necessary to acquire contracted resources. They understand pricing techniques, market trends, and supply sources of goods and services the Military needs to operate. Contracts specialists serve as business advisors, buyers, and administrators to support all functions of procurement actions.

Helpful Attributes

- Interest in planning and directing the work of others
- Interest in work requiring accuracy and attention to detail
- Ability to work with numbers

What They Do

- Manage, perform, and administer contracting functions for commodities, services, and construction using simplified acquisition procedures, negotiation, and other approved methods
- Use automated contracting systems to prepare, process, and analyze transactions and products
- Prepare memoranda, determinations and findings, justifications, and approvals to document contracting files

- Analyze statistical data and ensure compliance with bonding, insurance, and tax requirements
- Review requirements including descriptions, government furnished property, availability of funds, justifications for sole source, brand name purchasing, and delivery requirements
- Review government estimates, evaluate responsiveness of bids and offers, and conduct negotiations
- Conduct site visits to determine adequacy of contractor compliance and customer satisfaction
- Resolve claims, disputes, and appeals
- Evaluate methods and procedures used in purchasing commodities, services, and construction
- Ensure contractors' adherence to delivery schedules and prices

Services

Marine Corps

Air Force

CORRECTIONS CASE WORKERS

Summary: Corrections caseworkers provide individual and group counseling and assist awardees and prisoners in achieving successful transition back to military life. They perform initial and weekly reviews and assist in prisoner or awardee evaluation. They monitor and report significant behavioral changes and participate in treatment, clemency, parole, and work assignments. They also review and apply policies and directives, and support communication between unit and confined personnel.

Helpful Attributes

- Interest in working with people
- Patience in dealing with problems that take time and effort to overcome
- Sensitivity to the needs of others
- Listening and communication skills

What They Do

- Provide counseling, assist in evaluations, and recommend custody classification, treatment programs, and work assignments
- Make recommendations on clemency, parole, and/or restoration to duty requests

- Utilize operational control and mitigation methods
- Conduct prisoner interviews
- Conduct survival skills program, pre-release program, work program, and academic program
- Maintain prisoner sex offender registry program
- Conduct recreation and physical fitness activities
- Serve as a counselor role in correctional therapy
- Conduct incentive program
- Identify and use stress management techniques

Services

Marine Corps

Navy

CORRECTIONS SPECIALISTS

Summary: The corrections specialty in military law enforcement primarily involves guarding and supervising confined and restrained personnel. Law enforcement specialists that focus on corrections not only act as jailers but are also licensed military law enforcement specialists that ensure law and order. They prevent and quell riots and disturbances, prevent and suppress crimes against military personnel, and maintain order at military installations.

Helpful Attributes

- Ability to remain calm under pressure
- Interest in law enforcement and crime prevention
- Good judgment
- Negotiating skills

What They Do

- Provide safe, humane care and custody of enemy combatants
- Perform routine inspections, investigations, and searches
- Recognize, collect, and secure contraband
- Identify aspects of critical incidents in a correctional environment and participate in emergency response by providing first aid, when necessary
- Transport prisoners and process personnel for confinement and release
- Complete personal property inventory, oversee prisoner funds and personal property, and maintain appropriate records
- Facilitate corrections treatment programs such as counseling, management, training, and employment of military prisoners

- Plan and conduct security force drills, and test and monitor security systems
- Conduct formal and informal investigations by collecting intelligence for law enforcement and war crimes investigations
- Secure perimeter security posts, cell block dormitories, and outside places of confinement
- Employ unarmed self-defense techniques, riot control formations, and deadly force, when necessary

Services

Army

Marine Corps

Navy

CYBERSECURITY SPECIALISTS

Summary: Cybersecurity specialists are responsible for protecting military networks and the country against cyberattacks from enemy forces. These specialists monitor, analyze, detect, and respond to unauthorized activity in the cyberspace domain. They also perform deliberate actions to strengthen information systems and networks, perform vulnerability assessments, and respond to incidents. Cybersecurity specialists may focus on a specific type of information system, coordinating with network and system administrators, to ensure the security of DoD information networks.

Helpful Attributes

- Ability to communicate effectively
- Ability to understand and apply math concepts
- Interest in solving problems
- Interest in work requiring accuracy and attention to detail

What They Do

- Administer and manage cyber security programs to include Communications Security (COMSEC), Emissions Security (EMSEC), and Computer Security (COMPUSEC) programs
- Use defensive measures and information collected from a variety of sources (including intrusion detection system alerts, firewall logs, network traffic logs, and host system logs) to identify, analyze, and report events that occur or might occur within the network
- Conduct assessments of threats and vulnerabilities (through such tasks as authorized penetration testing, compliance audits, and risk assessments) to determine deviations from acceptable configurations and enterprise or local policies
- Protect information from, and recover information after, loss or damage using backups, virus detection, and recovery software procedures
- Perform deliberate actions to modify information systems or network configurations in response to alert or threat information
- Manage and validate network systems security using hardware, software, and established procedures
- Validate reported incidents, perform incident correlation and trending, conduct network damage assessments, and develop response actions
- Use mitigation, preparedness, and response and recovery approaches, as needed, to maximize network and information system confidentiality, integrity, and availability
- Implement and monitor security measures for communication information systems networks, and ensure that systems and personnel adhere to established security standards and governmental requirements for security on these systems
- Audit and enforce the compliance of cybersecurity procedures and investigate security-related incidents to include COMSEC incidents, classified message incidents, classified file incidents, classified data spillage, unauthorized device connections, and unauthorized network access
- Create a defensive security strategy that incorporates offensive protective measures to ensure the protection of information systems, networks, data, and the personnel that utilize this architecture
- Develop and supervise implementation of cybersecurity policies, plans, procedures, data network security measures, and defensive network intrusion detection and forensics systems
- Perform risk management framework security determinations of fixed, deployed, and mobile information systems (IS), and telecommunications resources to monitor, evaluate, and maintain systems, policy, and procedures to protect clients, networks, data/voice systems, and databases from unauthorized activity

- Supervise and perform as computer analyst, coder, tester, and manager in the design, development, maintenance, testing, configuration management, and documentation of application software systems, client-server, and web-enabled software and relational database systems critical to warfighting capabilities

Services

Army

Marine Corps

Navy

Air Force

Space Force

CYBER-OPERATIONS SPECIALISTS

Summary: Cyber-operations specialists conduct offensive and defensive cyberspace operations in support of the full range of military options. They utilize devices, computer programs, and techniques designed to create an effect across cyberspace. Offensive operations involve applying force to target enemy and hostile adversary activities and capabilities. Defensive operations are conducted to protect data, networks, net-centric capabilities, and other designated systems by detecting, identifying, and responding to attacks against friendly networks.

Helpful Attributes

- Ability to communicate effectively
- Ability to understand and apply math concepts
- Interest in solving problems
- Interest in work requiring accuracy and attention to detail

What They Do

- Perform cyber operational preparation of the environment
- Conduct cyber intelligence, surveillance, and reconnaissance activities on specified systems and networks
- Conduct network terrain audits, penetration testing, basic digital forensics data analysis, and software threat analysis, and recommend appropriate mitigation countermeasures
- React to cyberspace events, employ cyberspace defense infrastructure capabilities, collect basic digital forensics data, provide incident response impact assessments, and produce network security posture assessments
- Analyze computer system and network architectures, and determine and implement exploitation methods
- Perform initial cryptologic digital analysis to establish target identification and operational patterns
- Operate automated data processing (ADP) equipment for both remote and local collection, processing, and reporting
- Prepare technical products and time sensitive reports in support of cyber operations
- Evaluate operational effectiveness of communications, sensors, intrusion detection, and related support equipment
- Establish situational awareness of both friendly and adversary operations
- Partner with DoD, interagency, and Coalition Forces to detect, deny, disrupt, deceive, and mitigate adversarial access to sovereign national cyberspace systems
- Target adversary cyberspace functions or use first-order effects in cyberspace to initiate cascading effects into the physical domains to affect weapon systems, command and control processes, and critical infrastructure/key resources
- Provide threat indications and warnings, identify users, and, in some cases, identify the activities of the target
- Respond to crisis or urgent situations within the network to mitigate immediate and potential cyber threats
- Implement and monitor security measures for communication information systems networks, and ensure that systems and personnel adhere to established security standards and governmental requirements for security on these systems

Services

Army

Marine Corps

Navy

Air Force

Space Force

DENTAL HYGIENISTS AND ASSISTANTS

Summary: Dental hygienists and assistants have many responsibilities, including teeth cleaning, examining patients for signs of oral diseases such as gingivitis, and providing other preventive dental care. They also educate patients on ways to improve and maintain good oral health and perform other miscellaneous tasks, such as equipment maintenance and supplies inventory.

Helpful Attributes

- Ability to follow spoken instructions and detailed procedures
- Good eye–hand coordination
- Interest in working with people

What They Do

- Perform paraprofessional tasks and oral hygiene duties
- Supervise dental assistant functions
- Assist the dentist in the delivery of dental care
- Receive patient, examine dental health record, and prepare patient for treatment
- Adjust the dental chair and select and arrange instruments, materials and medicaments for use
- Measure and record blood pressure
- Prepare syringe for injection of anesthetics
- Retract tissues and maintain clear operating field
- Prepare materials for making impressions and restoring defective teeth
- Apply anticariogenic agents and place sealants
- Assist in planning, developing, and conducting comprehensive dental health programs
- Expose and process dental radiographs/images
- Clean, sterilize, and sharpen dental instruments; conduct sterilization equipment monitoring; perform daily inspection and user maintenance of dental equipment
- Perform dental administrative duties: coordinate patient appointments, maintain dental health records, filing systems, and publications, and review correspondence, reports, and records for accuracy
- Perform dental material functions related to budgeting procurement, custodial responsibilities, and maintenance and disposition of dental supplies and equipment
- Administer unit self-inspection program, and inspect and evaluate administrative and paraprofessional practices employed in the dental service
- Manage dental clinic and laboratory activities

Services

Army

Navy

Air Force

Coast Guard

DENTAL LABORATORY TECHNICIANS

Summary: Dental laboratory technicians construct and repair dentures and other dental appliances, including crowns, bridges, partial dentures, pre- and post-oral and maxillofacial surgical devices, and orthodontic appliances in a laboratory. Although dental technicians seldom work directly with patients, they provide a valued health care service by working with dentists to improve patient health and appearance.

Helpful Attributes

- Ability to follow detailed instructions and work procedures
- Ability to use precision tools and instruments
- Interest in work requiring attention to detail
- Interest in working with one's hands

What They Do

- Perform basic and intermediate-level prosthetic laboratory procedures
- Fabricate and finish dental prostheses: complete dentures, removable partial dentures, and other prescribed protective and restorative intraoral appliances
- Repair, reconstruct, and reline dental prostheses
- Conduct routine and prescribed equipment maintenance
- Coordinate technical and clinical application and dental technology training
- Assist the maxillofacial prosthodontist in the clinical and technical procedures required to fabricate prostheses and appliances for oral, craniofacial, and other anatomical defects
- Construct and finish ocular, extraoral, intraoral, and somato prostheses of silicone and other related materials
- Design and construct stone, metal, and/or silicone molds for prosthetic rehabilitation procedures

- Liaise with and assist other medical/dental specialists in related disciplines
- Perform and supervise procedures and techniques required in the construction of complex and precision dental prostheses: fixed partial dentures, porcelain fused to metal systems, dental ceramic arts, precision attachment prostheses, and the arrangements of artificial teeth for aesthetic, phonetic, and functional requirements

Services

Navy

Air Force

DIETETIC SUPPORT SPECIALISTS

Summary: Dietetic support specialists perform basic clinical dietetic functions in the dietary management and treatment of patients in nutrition clinics, clinical dietetics branches, health promotion, and wellness clinics. They perform nutritional assessments and screening of individual patients for nutritional risk, followed by procuring, storing, preparing, and serving regular and therapeutic diets and nourishment. They establish production controls and standards for quantity and quality of foods. Additionally, they complete administrative tasks related to medical dietary cases such as screening patients, obtaining histories, and recording data.

Helpful Attributes

- Interest in cooking
- Interest in working with the hands
- Interest in work that helps others
- Interest in work requiring accuracy and attention to detail

What They Do

- Provide medical nutrition care therapy to patients and staff in support of humanitarian missions in field and fixed hospitals
- Perform and supervise the nutritional assessments and screening of individual patients for nutritional risk
- Prepare and serve modified and regular food items in the management of the normal nutrition

needs of individuals across the lifespan (i.e., infants through older adults), and a diversity of people, cultures, and religions in support of the mission, under the supervision of a dietitian or noncommissioned officer (NCO)
- Perform and supervise basic clinical dietetics functions in the dietary management and treatment of patients in nutrition clinics, clinical dietetics branches, health promotion, and wellness clinics
- Practice timely, quality fundamentals of food preparation for regular and therapeutic diets, tube feedings, and therapeutic in-flight and box lunches
- Prepare and cook food items included in regular and therapeutic diets to conform with menus, recipes, and food production worksheets
- Assemble and disassemble patient trays on food carts in patient tray assembly area
- Establish production controls and standards for quantity and quality of foods
- Plan menus according to established patterns
- Provide field feeding, accountability, sanitation, and layout during disasters or contingencies
- Conduct dietary rounds to interview patients on regular and therapeutic diets to determine satisfaction and food preferences
- Assist in determining requirements, preparing requisitions, and making local purchase orders
- Receive, verify, store, and issue foods and supplies from the prime vendor, commissary, and medical logistics
- Monitor food quality, quantity, sanitation, safety, and security standards
- Advise dietitian on equipment status, maintenance, adequacy, personnel training, and operational efficiency and economy

Services

Army

Air Force

DIVERS

Summary: They perform such tasks as reconnaissance, demolition, ship repair, search for missing persons, salvage in underwater conditions, or construction. They usually specialize in either scuba diving or deep-sea diving.

Helpful Attributes

- A high degree of self-reliance
- Ability to stay calm under stress
- Interest in underwater diving

What They Do

- Perform basic explosive ordnance disposal (EOD) related diving procedures, including bottom and hull searches, day or night, using SCUBA, surface supplied, and mixed or gas diving equipment
- Test, repair, and adjust scuba and associated underwater equipment
- Recognize and assist in the treatment of diving-related injuries, such as decompression sickness and other forms of barotrauma
- Perform diving, demolition, ordnance location and identification, EOD detachment operational support, and staff and logistics support to EOD commands
- Supervise deep-sea diving missions and dives conducted deeper than 100 feet in salt water
- Formulate demolition plan for operational and training missions
- Perform underwater work such as taking measurements, making templates and fittings, placing shores, pouring cement, using excavating nozzles, and removing and repairing ships appendages
- Manage preventive and corrective maintenance on diving equipment, support systems, salvage machinery, handling systems, and submarine rescue systems
- Conduct day and general underwater search, detailed ship-bottom search, and routine inspections to a depth of 60 feet using underwater compass, depth indicators, and associated underwater equipment
- Perform adjustments and field shop maintenance on SCUBA and underwater accessories
- Dive and perform underwater demolition for the purpose of open ocean salvage, ship husbandry, or underwater construction operations

Services

Army

Marine Corps

Navy

Coast Guard

DOCUMENT SECURITY SPECIALISTS

Summary: Document security specialists safeguard sensitive material and arrange for delivery by courier. They provide adequate protection for material from receipt through delivery or to storage, prepare classified correspondence, and manage administrative functions and security procedures for various security programs (e.g., Special Security Program, Sensitive Compartmented Information Security Program).

Helpful Attributes

- Ability to organize and plan
- Interest in keeping organized and accurate records
- Interest in operating computers and other office machines
- Preference for office work
- Reliability and discretion

What They Do

- Conduct and supervise administrative functions and security procedures governing the Special Security Program
- Retain knowledge of security for classified matter safeguards and procedures
- Assist command personnel in the completion and submission of Single Scope Background Investigation (SSBI) packages and periodic reviews
- Prepare and review the checklist for Sensitive Compartmented Information Facility (SCIF) accreditation and periodic inspection
- Manage administrative functions and security procedures governing the Special Security Program
- Prepare classified correspondence
- Perform multiple security disciplines in the following areas: Personnel Security (initiate and update security clearance/access eligibility); Security Access Eligibility Reports (SAER); Single Scope Background Investigation (SSBI); Periodic Re-Investigation (PR); Eligibility; polygraph management; Special Security Office (SSO) files; Special Access/Compartmented Programs; SCI Facility oversight; Information Security (violation reporting or investigations, classification management, Freedom of Information Act (FOIA), and privacy act); and Security Education & Awareness

- Ready SCI material for electronic transmission and physical transfer by the Defense Courier Service (DEFCOS)
- Provide and maintain accreditation of SCI facilities
- Assist leadership with managing and implementing Service-specific security programs
- Safeguard and deliver Armed Forces Courier Service material
- Provide adequate protection for material from receipt through delivery or to storage, and caution handlers to exercise care in storing material
- Maintain constant surveillance over material in custody on the courier route
- Make positive identification of Armed Forces couriers, addresses, and Top Secret control officers before releasing material

Services

Navy

Air Force

Space Force

ELECTRICAL INSTRUMENT AND EQUIPMENT REPAIRERS

Summary: Electrical instrument and equipment repairers install, maintain, and repair instruments and equipment, including communications equipment, radar and sonar systems, tactical data systems, and computers. They use a variety of approaches to troubleshoot and replace faulty components, subassemblies, and assemblies to restore instruments and equipment to optimum operating condition.

Helpful Attributes

- Ability to use tools
- Interest in solving problems
- Ability to apply electronic principles and concepts
- Interest working with electrical, electronic, and electrochemical equipment

What They Do

- Install cable systems
- Use tools, meters, and special test equipment necessary to correctly test, align, troubleshoot, and repair electrical and electronic components

- Detect, diagnose, and isolate equipment failures using computer aided Preventative Maintenance/Fault Location (PM/FL) software, Fault Logic Diagrams (FLD), schematics, wiring, and logic diagrams
- Repair electric motors, electronic modules, motor control circuits, and electric power-generation equipment
- Repair circuits and control devices
- Repair failed electrical and electronic components by removal and replacement to the lowest replaceable unit (LRU)
- Perform computer configuration, management, and troubleshooting
- Use computers and software tools to perform system/network operations
- Interpret system and equipment error codes to correct system faults
- Prepare system and equipment related forms and reports
- Ensure availability of backup equipment, spares, and repair parts to sustain system operations
- Compile system and network statistics for reports
- Plan, schedule, and implement installation and maintenance functions associated with electrical systems
- Provide technical assistance and training for user owned and operated automated telecommunication computer systems, local area networks and routers, signal communications and support electronic equipment, and satellite radio and communications equipment

Services

Army

Marine Corps

Navy

Air Force

Space Force

Coast Guard

EMERGENCY MANAGEMENT SPECIALISTS

Summary: Whether it is a natural disaster or man-made as a result of a chemical, biological, radiological, or nuclear incident, emergency management specialists are trained for response and recovery operations anywhere in the world. They prepare emergency plans, coordinate emergency response teams, and train other people to meet mission needs and to minimize casualties and damage in the event of any disaster situation, including floods, earthquakes, hurricanes, or enemy attack.

Helpful Attributes

- Ability to communicate effectively
- Ability to plan and organize
- Ability to work calmly under stress

What They Do

- Advise command on how to integrate into the normal command organization the functions necessary to prepare for, defend against, and recover from major accidents and natural and man-made disasters
- Assist in the coordination with local, civic authorities on disaster response operations
- Conduct training for nuclear, biological, and chemical (NBC) warfare defense to include hazard awareness, individual protection, decontamination, and mission restoration
- Coordinate actions to ensure prompt response during emergency management (EM) operations, including immediate mobilization of resources and participation of agencies and organizations
- Receive, process, and disseminate emergency action messages via voice and record copy systems
- Coordinate and execute search and rescue activities
- Prepare and submit operational, defense readiness, international treaty, and aerospace asset reports
- Operate and monitor voice, data, and alerting systems
- Develop, maintain, and initiate quick reaction checklists that support situations such as suspected or actual sabotage, nuclear incidents, natural disasters, aircraft accidents or incidents, evacuations, dispersal, and aerospace anomalies
- Coordinate actions to ensure prompt response during disaster operations (pre, trans, and post), including immediate activation and recall of all resources and participating agencies and organizations
- Receive and disseminate time-critical information to and from the commander to internal and external agencies during daily operation, natural disasters, and wartime and contingency operations to affect positive control of assigned forces and weapons systems

- Prepare, plan, train, educate, and equip personnel and installation leaders on ways to prepare for, prevent, respond to, maintain mission capability, and recover from threat events, including major accidents, natural disasters, weapons of mass destruction, and wartime attacks

Services

Army

Marine Corps

Navy

Ari Force

Coast Guard

EMERGENCY MEDICAL TECHNICIANS

Summary: EMTs are health professionals trained to respond quickly to emergency situations. Military EMTs provide emergency medical treatment, limited primary care, force health protection, and evacuation in a variety of operational and clinical settings from point of injury or illness through the continuum of military health care. They may work in military health facilities or in the field.

Helpful Attributes

- Ability to communicate effectively
- Ability to work under stressful conditions
- Interest in helping others

What They Do

- Perform paramedical skills, basic life support, triage, minor surgical procedures, and other routine and emergency medical care
- Serve as member of primary emergency medical response to in-flight emergencies and potential mass casualty scenarios for on- and off-base incidents
- Monitor and record physiological measurements
- Operate emergency medical and other vehicles; load and unload litter patients
- Prepare patient with special equipment for transfers
- Perform and assist with examinations and special procedures, including mechanical ventilation, cardiovascular, and neurovascular procedures and dialysis
- Participate in contingency or disaster field training, exercises, and deployments

- Administer trauma stabilization to include initial treatment in a combat environment
- Perform aeromedical evacuation (AE) ground and/or flight duties; perform preflight/inflight patient care
- Observe, report, and record observations in patient progress notes and at team conferences
- Perform portions of medical treatment, diagnostic, and therapeutic procedures
- Provide emergency care in the event of diving medical or hyperbaric chamber emergencies
- Render emergency dental treatment

Services

Army

Navy

Air Force

ENVIRONMENTAL HEALTH AND SAFETY SPECIALISTS

Summary: Environmental health and safety specialists protect the military community through programs to ensure that military facilities and food supplies are free of disease, germs, and other hazardous conditions. These specialists identify occupational and environmental health hazards and risks that may negatively impact health, human performance, and environmental health quality. They manage programs related to communicable disease control and prevention, food safety and defense, sanitary compliance, occupational health and safety, hazardous material control, and public health contingency response.

Helpful Attributes

- Interest in gathering information
- Interest in protecting the environment
- Preference for work requiring attention to detail

What They Do

- Perform inspections and surveys of food and food service facilities, berthing spaces, childcare facilities, recreational facilities, potable water systems, solid waste and wastewater disposal sites/systems, vehicles, and transport containers
- Conduct epidemiological investigations and reporting; interview and counsel patients with communicable diseases; administer mass immunization programs
- Organize and assist in communicable disease prevention and control programs
- Perform duties of radiation monitor, radiation surveys, gas and liquid analyses with knowledge of the medical aspects relating to personnel exposed to ionizing radiation
- Identify hazards, unsafe work practices, and health hazardous conditions
- Provide safety indoctrination and education, and maintain safety equipment and material
- Inspect meat, poultry, water, food, eggs, dairy products, operational rations, fresh fruits, and vegetables in depots, supply points, and on military installations
- Conduct inspections in food handling establishments; evaluate and recommend corrective actions for unsanitary conditions
- Collect, prepare, and transmit samples to laboratory for testing; review laboratory test results
- Provide care to animals, with a focus on prevention and control of disease transmitted from animal to human
- Collect data to ascertain accident trends; investigate accidents when they occur
- Respond to accidents, natural disaster, and attack by hostile forces that may result in exposure(s) to occupational and environmental health threats
- Provide control recommendations to mitigate or eliminate occupational and chemical, biological, radiological, or nuclear health threats
- Provide consultation to supervisors and workers in personal hygiene, occupational hazards, hazard communications, and personal protective equipment

Services

Army

Navy

Air Force

Coast Guard

EQUAL OPPORTUNITY SPECIALISTS

Summary: Equal opportunity specialists perform administrative functions, including preparation and maintenance of equal opportunity (EO) case files. They are responsible for the maintenance and analysis of pertinent data, and the development of relevant reports. They control budget operations, assess EO education program activities, counsel military personnel and civilian employees on EO policies, and provide recommendations and advice to leaders in resolving problems related to EO matters. They promote effective organizational climate by implementing seminars, focus groups, and other techniques.

Helpful Attributes

- Ability to compose clear instructions or correspondence
- Ability to follow detailed procedures and instructions
- Interest in working closely with others
- Interest in equal opportunity policies

What They Do

- Provide instruction, assistance, and advice on all equal opportunity (EO) matters to installation and tenant commanders
- Maintain administrative control of the Discrimination and Sexual Harassment (DASH) reporting system
- Assist commands with investigations into allegations of discrimination, to include sexual harassment, as directed by cognizant authority
- Advise commanders and military personnel on complaint resolution procedures both formal and informal
- Conduct inspections of command equal opportunity program, as directed by cognizant authority
- Assist the command's EO representatives with unit EO training, cultural events, or celebrations on those days set aside for recognition of contributions of various groups
- Provide input into EO policies and programs for both installation and tenant commands
- Assist commanders with monitoring organization EO climate, identifying trends and areas of concern, and suggesting methods for improving command EO climate
- Emphasize the use of the Informal Resolution System (IRS) to resolve conflicts at the lowest level
- Provide briefings on all aspects of EO
- Participate in EO meetings, conferences, and seminars

Services

Marine Crops

Navy

Air Force

EXPLOSIVE ORDINANCE DISPOSAL SPECIALISTS

Summary: EOD specialists detect, locate, render safe, and dispose of explosive threats all over the world. These threats include chemical, biological, and nuclear weapons, as well as improvised explosive devices (IED).

Helpful Attributes

- Ability to remain calm under stress
- Interest in working with guns and explosives
- Attention to detail

What They Do

- Perform or supervise render safe and disposal procedures on conventional, chemical, and improvised explosive devices
- Read and interpret x-rays, diagrams, drawings, and other technical information on explosive ordnance
- Operate and maintain specialized EOD equipment, such as robotics, x-ray, landmine, and chemical, biological, radiological, and nuclear (CBRN) detection equipment
- Locate and identify buried and underwater ordnance
- Perform toxic chemical agent tests using chemical detector kits
- Select disposal site and transport demolition explosives and equipment to authorized disposal areas
- Fabricate explosive demolition charges, and dispose of hazardous devices, ordnance, and explosives
- Perform and supervise open and closed-circuit SCUBA diving, explosive demolitions, parachuting, tactical delivery, and extraction by unconventional insertion methods
- Support and conduct ordnance-related intelligence collection and counterterrorism operations
- Assist in operating an Emergency Contamination Control Station (ECCS) and Emergency Personnel Decontamination Station (EPDS)

- Prepare technical intelligence and incident reports
- Advise military and civilian agencies on disposition of hazardous materials and searching techniques for improvised explosive devices
- Perform administrative and supervisory functions in maintenance, security, supply, and training management
- Supervise coordination and deployment of EOD response teams

Services

Army

Marine Corps

Air Force

FIELD COMBAT MEDICS

Summary: Some Emergency Medical Technicians (EMTs) are trained to provide medical care in an operational or combat environment. These EMTs, sometimes called field/combat medics, provide frontline trauma and medical care to deployed personnel. They care for those suffering from disease as well as those injured in combat. In addition to being prepared to work in combat, these health professionals are often trained in health concerns related to specific conditions, such as those encountered during diving or flight operations.

Helpful Attributes

- Ability to communicate effectively
- Ability to work under stressful conditions
- Interest in helping others

What They Do

- Perform patient care, medical administration, logistical duties, diagnostic procedures, advanced first aid, basic life support, nursing procedures, minor surgery, basic laboratory procedures, and industrial hygiene surveillance programs
- Assist medical officers in the prevention and treatment of illnesses associated with diving and high-pressure conditions
- Conduct technical and administrative medical assistance supporting the mission and functions of field units

- Maintain field treatment facilities rendering routine medical and emergency care to unit personnel and combatants
- Coordinate and perform medical and casualty evacuation procedures
- Ensure the observance of field sanitary and preventive medicine measures supporting force health protection
- Assist with the procurement and distribution of related supplies and equipment for peacetime use and in combat areas
- Conduct first aid, health, dental, and medical education training programs
- Provide dental treatment in the field
- Advise operational personnel in measures for the prevention of illness and treatment of injuries associated with swimming, open and closed-circuit SCUBA diving, military freefall, and amphibious operations
- Manage preventive medicine and industrial hygiene surveillance programs
- Perform aircrew and emergency medical care functions assigned in support of search and rescue missions
- Perform clinical diagnostics, advanced paramedical skills, Advanced Cardiac Life Support (ALCS), basic surgical anesthesia, basic dental exams, and other routine and emergency medical health care procedures as required
- Order, store, catalog, safeguard, and distribute medical supplies, equipment, and pharmaceutical supplies

Services

Army

Navy

Air Force

FIRE FIGHTERS

Summary: Military firefighters do much more than fight fires. They also perform inspections to minimize fire dangers, provide first aid to accident victims, and respond to hazardous materials spills. They assist civilian fire departments when needed.

Helpful Attributes

- Ability to remain calm under stress
- Ability to think and act decisively
- Willingness to risk injury to help others

What They Do

- Supervise or perform firefighting, rescue, salvage, and fire protection operations
- Perform rescue and firefighting operations during structural fires, aircraft emergencies, vehicle emergencies, and wildland fires
- Perform emergency response duties during hazardous materials incidents
- Operate and maintain firefighting equipment and vehicles during emergency and non-emergency operations
- Conduct fire prevention operations to include determining building classification and installation-level inspections; plan, organize, and direct all fire protection activities
- Analyze fire protection operations, determine trends and problems, and formulate corrective measures
- Conduct and evaluate training on specialized fire protection equipment and procedures
- Perform inspections and organizational maintenance on fire protection vehicles, equipment, and protective clothing
- Inspect facilities, and identify fire hazards and deficiencies; determine fire extinguisher distribution requirements and perform inspections and maintenance
- Conduct public relations and fire prevention awareness and educational training
- Drive and operate fire apparatus, specialized tools, and equipment
- Preserve and protect emergency scene evidence; investigate fires to determine origin and cause
- Administer emergency first aid

Services

Army

Marine Corps

Navy Air Force

FLIGHT ATTENDANTS

Summary: Flight attendants are responsible for passenger safety during aircraft operations. They manage cabin duties such as briefing passengers on aircraft systems, amenities, and equipment. Flight attendants are responsible for orderly, expeditious evacuation of passengers and crew, and providing emergency medical assistance when necessary.

Additionally, they perform pre-flight, through-flight, and post-flight inspections. They also operate a variety of aircraft systems and equipment.

Helpful Attributes

- Ability to work as a team member
- Ability to work under stress
- Interest in flying
- Excellent communication skills

What They Do

- Provide for safety and comfort of passengers during aircraft operations
- Demonstrate and maintain proficiency in emergency equipment use, emergency procedures, and egress
- Brief passengers on normal and non-normal use of aircraft systems and equipment
- Perform preflight, through-flight, and postflight inspections of aircraft emergency, cabin, and galley equipment
- Operate aircraft systems and equipment such as electrical, environmental, water, interphone, doors, and exits
- Plan all menus and coordinate meals on normal and non-normal use of aircraft systems
- Provide highest level of service, etiquette, and protocol as the direct contact between the Military and passengers
- Prepare meals utilizing the fundamentals of culinary arts, including knife skills, basic cooking methods (baking, braising, sautéing, etc.), and making sauces and emulsions
- Perform loading and off-loading of aircraft
- Coordinate with military and civilian airfield agencies to acquire supplies and transportation
- Perform passenger and baggage inspections
- Ensure access to escape exits; direct safety, security, and fire prevention procedures
- Develop and direct instruction in equipment operation and flight attendant activities
- Evaluate flight attendant activities and compliance with technical manuals, regulations, and work standards

Services

Air Force

FLIGHT ENGINEER TECHNICIANS

Summary: Flight engineering technicians perform aircraft inspections, including aircrew visual inspection, aircraft maintenance, and preflight, through-flight, and postflight inspections. They perform aircrew administration, flight/ground training, internal/external cargo movement, medical evacuations, and passenger transport. They also serve as crew members aboard military aircraft.

Helpful Attributes

- Ability to work as a member of a team
- Interest in working with mechanical systems and equipment
- Skill in using wiring diagrams and maintenance manuals
- Strong desire to fly

What They Do

- Operate and monitor engine and aircraft systems controls, panels, indicators, and devices
- Manage flight engineer functions and activities
- Maintain aircraft forms and records during flight and while aircraft is away from home station
- Compute and apply aircraft weight, balance, and performance data
- Determine and verify passenger, cargo, fuel, and emergency distribution weight
- Compute takeoff, climb, cruise, and landing data
- Calculate engine fuel consumption using airspeed, atmospheric data, charts, and computers
- Assist pilot by performing engine starts, flight operation, and engine shutdown
- Control, monitor, and regulate aircraft systems such as electronic, communication, navigation, hydraulic, fuel, air conditioning, pressurization, and ventilation systems
- Recommend corrective action related to overheating, depressurization, and system failure
- Organize flight engineering standardization, qualification, and other requirements of flight engineer logs, reports, and records for accuracy, completeness, format, and compliance

Services

Marine Corps

Navy

Air Force

FLIGHT OPERATIONS SPECIALISTS

Summary: Flight operations specialists help with administrative functions that are necessary to keep military aircraft up and running. They prepare flight schedules and authorizations, administer aircrew training and qualification testing, perform airfield inspections, and maintain flight logs and other records. These specialists help ensure that both the aircraft personnel and equipment are prepared for mission accomplishment.

Helpful Attributes

- Ability to keep accurate records
- Interest in work involving computers
- Interest in work that helps others
- Ability to work as team member
- Interest in aircraft

What They Do

- Schedule and dispatch tactical aircraft missions and perform associated operational administrative duties
- Process cross-country and local flight clearances, including examination for conformance with flight rules and regulations
- Prepare flight authorizations and monitor individual flight requirements and unit flying hours
- Maintain individual flight records, files of aviation operations publications, records, and correspondence in accordance with current directives
- Coordinate scheduling, standardization and evaluation, flight and ground training, flight records, and squadron operations
- Prepare and process aeronautical orders and military pay orders
- Monitor flight physicals, physiological training, aircrew qualifications, and other aircrew and parachutist-related programs
- Perform airfield inspections and checks to include runways, taxiways, aprons, pavements, arresting systems, signs, lighting, and airfield clearance areas
- Respond to wildlife, foreign object debris, and other flight safety hazards affecting the airfield environment
- Provide aircrews with preflight briefings; brief pilots current airfield status, arresting system configuration, runway surface conditions, correct taxi routes, and any hazards to operations in person and via air-to-ground radios

- Procure, maintain, and produce information regarding the safe operation of aircraft on the airfield and through the national and international airspace systems
- Decode sequence reports and synoptic charts, and disseminate weather information to pilots
- Initiate lost plane procedures
- Maintain situation and operation maps
- Assist in the preparation of visual and instrument flight plans

Services

Army

Marine Corps

Air Force

FOOD SERVICE SPECIALISTS

Summary: Food service specialists function in every aspect of food preparation, including administration, procurement, storage, and distribution. They plan menus, purchase supplies, cook food, operate food service equipment, and provide customer service.

Helpful Attributes

- Interest in cooking
- Interest in working with the hands

What They Do

- Operate, maintain, and clean kitchen equipment
- Develop and analyze menus and coordinate menu substitutions
- Season, bake, fry, braise, boil, simmer, steam, and sauté food
- Prepare fruits, vegetables, meat, fish, and poultry
- Ensure adherence to proper procedure and food temperature guidelines during food preparation
- Inspect dining facility, food preparation/storage areas, and dining personnel
- Survey individual preferences, food preparation, and food conservation
- Portion and select nourishment for regular and therapeutic diets
- Ensure safety and sanitation of equipment and dining facility
- Calculate funding, requisitioning, purchasing, receiving, and accounting for food supply for unit
- Oversee inventory management and maintain financial transaction records

Services

Army

Marine Corps

Navy

Coast Guars

FUEL SUPPLY SPECIALISTS

Summary: Fuel supply specialists receive, store, account and care for, dispense, issue, and ship various fuel products, including petroleum, alternate fuel, and cryogenic products. They manage, maintain, and operate fuel support equipment used for base and tactical operations. They are responsible for ensuring compliance with all safety and environmental regulations.

Helpful Attributes

- Interest in working with machines and equipment
- Preference for physical work

What They Do

- Install, operate, maintain, and repair fuel handling units and accessory equipment, and test petroleum products
- Assure adherence to safety procedures and ensure an efficient, clean, and safe work environment
- Supervise aircraft refueling and defueling operations
- Direct receipt, storage, transfer, and issue operations for petroleum, alternate fuel, and cryogenic products
- Forecast product requirements, place orders for products, and perform product receipt operations
- Maintain inspection and maintenance records for facilities, report facility deficiencies to appropriate maintenance activity, and initiate facility upgrades and construction projects
- Process computer transactions to ensure proper billing and payment and ensure accuracy of receipt, inventory, transfer, and issue records
- Monitor inventory levels, ensure adequate stocks are on-hand, and reconcile information systems
- Oversee unit personnel readiness by allocating personnel to authorized positions, participating in mobility planning, and submitting resource and training system data

- Conduct preventative maintenance inspections on vehicle and equipment, analyze malfunctions, document deficiencies, and coordinate repairs
- Adjust and repair valves, manifolds, pumps, filters, and meters

Services

Army

Marine Corps

Navy

Air Force

GEOSPATIAL IMAGING SPECIALISTS

Summary: Geospatial imaging specialists perform the collection, analysis, and dissemination of geospatial imagery to support war fighting operations and other activities. They operate geographic information systems (GIS) that store and analyze maps and other geographic information. Intelligence derived by the geospatial imaging specialist is critical for mission success at the national, theater, and tactical levels. As such, they disseminate multi-sensor geospatial intelligence products to appropriate parties.

Helpful Attributes

- Interest in computers
- Interest in earth science
- Strong critical thinking skills

What They Do

- Collect, analyze, and process geophysical data and geographic information to aid in the production of geographic intelligence products
- Utilize global satellite communication networks to support geospatial intelligence
- Utilize survey and mapping instrumentation, such as theodolites, electronic and satellite positioning equipment, and microcomputer-based mapping equipment
- Operate imagery exploitation equipment, including computer-assisted exploitation, geospatial analysis manipulation, and automated database systems; construct queries and retrieve historical files to conduct comparative analysis
- Exploit and analyze multi-sensor imagery and geospatial data and products in conjunction with all-source intelligence information

- Extract geospatial data from remote sensed imagery, field reconnaissance, digital data, existing topographic products, and other collateral data sources
- Analyze still, motion, radar, infrared, spectral imagery, and geospatial data
- Detect and report on observed image activities that are of significant military, civilian, industrial, infrastructural, and environmental importance to decision makers and warfighters
- Determine type, function, status, location, significance of military facilities and activities, industrial installations, and surface transportation networks
- Perform targeting functions to include target development, weaponeering, force application, execution planning, and combat assessment
- Validate collection requirements for strategic and tactical intelligence, surveillance, and reconnaissance (ISR) platforms
- Utilize basic drafting techniques to tailor terrain products and revise planimetric and topographic maps; perform digital manipulation of topographic information by querying, viewing, evaluating, and downloading digital data
- Compile geospatial data into a printable map/product and prints (maps, overlays, and special products)

Services

Army

Marine Corps

Navy

Air Force

Space Force

GRAPHIC DESIGNERS AND DESKTOP PUBLISHERS

Summary: Graphic designers and desktop publishers use their creativity in the areas of production, maintenance, and coordination related to visual products and presentations. They use computer software or manually develop the overall layout and production design for various media, such as advertisements, brochures, magazines, and reports.

Helpful Attributes

- Ability to convert ideas into visual presentations
- Interest in artwork or lettering

- Neatness and an eye for detail
- Understanding of visual communication

What They Do

- Create illustrations, layouts, map overlays, posters, graphs, charts, and internet web pages in support of battlefield operations, psychological operations, military intelligence, medical, public affairs, and training functions
- Determine media, style, design, and technical requirements to create visual information presentation products
- Choose and develop sequence and manner of presentation of products, and operate assigned equipment to carry out visual information presentations
- Prepare graphic products using manual, mechanical, and electronic equipment
- Demonstrate skill in the principles of design and layout, fundamentals of color theory, realistic drawing, color media, lettering, printing reproduction, image editing, and desktop publishing
- Generate, import, export, and edit artwork files
- Utilize web page design software, animation software, and digitized audio and video software
- Manage, transmit, and archive imagery
- Prepare originals for reproduction and complete tasks such as collating, binding, stapling, and hole-punching
- Coordinate graphic requirements, applications, and capabilities for leaders and customers

Services

Army

Marine Corps

Navy

HEATING AND COOLING MECHANICS

Summary: Heating and cooling mechanics are responsible for providing the Military with technical and mechanical services regarding heating, cooling, and related systems. They install, operate and perform inspections, and conduct testing, troubleshooting, and repairs on malfunctioning systems. They perform maintenance and quality control functions, ensuring compliance with safety and environmental regulations. They supervise and provide guidance to subordinates and provide recommendations regarding installations and repairments of HVAC/R systems.

Helpful Attributes

- Ability to use hand and power tools
- Interest in solving problems
- Interest in working on machines

What They Do

- Perform inspection, testing, adjustment and repair of gasoline engine systems, bottle cleaning/charging stations, fire extinguisher rechargers, and fire extinguishers/valves
- Perform inspection, testing, adjustment and repair of air conditioner electrical systems, air conditioner vapor systems, refrigeration unit electrical systems, and portable heater fuel/electrical systems
- Install and operate HVAC/R systems and equipment
- Interpret drawings and schematics and install HVAC/R components
- Install, repair, fabricate, and test piping and tubing systems
- Install, operate, and maintain combustion equipment and industrial air compressors
- Evaluate water treatment and balance air and water in HVAC/R systems
- Monitor operation of systems and modify, repair, or replace components or equipment to ensure efficiency
- Ensure compliance with safety and environmental regulations for fuels, refrigerants, and hazardous materials
- Perform recurring maintenance and seasonal overhaul on systems and components
- Perform management activities and quality control functions
- Conduct facility surveys, prepare cost-estimates, and determine resource requirements
- Provide technical guidance and supervision to subordinate personnel
- Provide recommendations and advice on installations and repairs of HVAC/R systems
- Troubleshoot and suggest solutions related to maintenance issues through studying layout drawings and by analyzing construction and operating characteristics

Services

Army

Marine Corps

Navy

Air Force

HEAVY EQUIPMENT MECHANICS

Summary: Heavy equipment mechanics specialize in the maintenance and repair of heavy equipment, such as tanks and other combat vehicles. They maintain and repair construction equipment including that used for earthmoving, grading and compaction, lifting and loading, quarrying and rock crushing, asphalt/concrete mixing and surfacing, and water pumping, as well as special purpose equipment, including power-generation equipment and air conditioning/refrigeration systems. They also supervise and perform diagnostic troubleshooting to determine maintenance repair criteria.

Helpful Attributes

- Interest in automotive engines and how they work
- Interest in troubleshooting and repairing mechanical problems
- Preference for physical work

What They Do

- Supervise and perform field maintenance
- Diagnose and troubleshoot malfunctions
- Maintain and repair electrical/fuel heater systems, pumps, water purification systems, decontamination systems, protective filter systems, and smoke generator systems.
- Perform organizational maintenance and on-board direct support tasks on the suspension systems, steering systems, hydraulic systems, auxiliary power units, fire extinguisher/suppression systems, gas particulate systems, vehicular mounted armament, gun turret drive system, and the fire control system on combat equipment
- Maintain wheeled vehicles, their associated trailers, and material handling equipment (MHE) systems
- Conduct in-process inspection/troubleshooting procedures during repairs and overhaul of engines, transmissions and power train major assemblies

and components, hydraulic system, and fuel system components

- Supervise recovery team performance of wheeled vehicle recovery operations
- Supervise and perform diagnostic troubleshooting to determine maintenance repair criteria using Test Measurement Diagnostic Equipment (TMDE)
- Perform equipment classification inspections, and annotate and submit appropriate forms and documents
- Perform battlefield damage and assessment and repair (BDAR)
- Service, inspect, maintain, and repair motor transport equipment at the field level
- Supervise the maintenance, repair, and inspection of motor transport vehicles, and direct the activities of assigned enlisted personnel in a motor transport repair shop or facility
- Perform tasks involved in maintenance, repair, and overhaul of automotive, materials handling, and construction equipment
- Assign and supervise activities of assistants who locate, analyze, and correct malfunctions in equipment, and issue repair parts
- Train assistants in repair procedures and techniques

Services

Army

Marine Corps

Navy

HUMAN INTELLIGENCE SPECIALISTS

Summary: Human intelligence specialists identify adversarial elements, strengths, dispositions, tactics, equipment, personnel, and capabilities through collecting information from people. HUMINT specialists screen documents and other materials to identify potential source leads. They conduct and oversee interviews, interrogations, screenings and debriefings in English and work with translators when necessary. They also exploit information found in a variety of media.

Helpful Attributes

- Ability to organize information
- Ability to think and write clearly

- Interest in gathering information and studying its meaning
- Interest in foreign cultures

What They Do

- Identify adversarial elements, intentions, compositions, strength, dispositions, tactics, equipment, personnel, and capabilities through the use of military source operations (MOS), interrogations, screenings, debriefings
- Assist in the translation and exploitation of documents and media
- Conduct debriefings and interrogations of HUMINT sources, under supervision; assist in screening HUMINT sources and documents; participate in HUMINT source operations
- Perform HUMINT analysis, as required, and prepare appropriate intelligence reports
- Plan, analyze, develop, design, distribute, disseminate, and evaluate psychological operations (PSYOP) across the range of military operations
- Plan and organize work schedules and assign specific tasks in support of PSYOP missions; conduct liaison with the supported unit staff; coordinate resource requirements for the development, production, and dissemination of PSYOP products
- Use foreign language skills and/or interpreters to conduct Counterintelligence/Human Intelligence (CI/HUMINT) activities
- Acquire intelligence information from human sources in response to military and national requirements by supervising and conducting tactical HUMINT collection operations
- Use interpreters and manage interpreter/translator operations; conduct liaison and coordination in foreign language with host nation agencies
- Screen documents and open-source materials to identify source leads; contact and assess leads to determine value and validity of source information
- Prepare Intelligence Information Reports (IIRs) and summaries from collected data citing specific requirements
- Publish knowledge-level briefs, notices of intelligence potential, and requests for requirements to alert the intelligence community on source availability and information
- Maintain familiarity with validated requirements and apply them to screenings, assessments, debriefings, interrogations, and any documents resulting from these activities

- Interrogate prisoners of war, enemy deserters, and civilian detainees
- Assess potential sources by examining biographical records and personal documents, and by assessing subject's demeanor, grade, apparent status, and other pertinent data; evaluate source reliability and make applicable IIR field comments

Services

Army

Marine Corps

Navy

Air Force

HUMAN RESOURCE SPECIALISTS

Summary: Human resources specialists oversee the maintenance and processing of personnel records, including performance monitoring, training, and evaluation, as well as all human resource related actions such as military pay, transfers, leaves, and promotions. They are responsible for managing service member records, assessing personnel performance, and providing training and evaluation reports. They verify personnel readiness and compliance with military policies, and they provide support to leaders by accompanying them on inspections and providing them with recommendations. They promote organizational effectiveness and performance improvement by planning and implementing training and establishing performance measures and standards.

Helpful Attributes

- Ability to compose clear instructions or correspondence
- Ability to follow detailed procedures and instructions
- Interest in working closely with others

What They Do

- Consult with leadership on all matters related to enlisted personnel
- Verify readiness and knowledge of command regulations and policies of newly assigned enlisted personnel

- Monitor, implement, and assess training of enlisted personnel
- Consult with and provide guidance to enlisted personnel
- Assess duties of subordinate staff, report performance deficiencies, and proceed with plans for correction
- Oversee processing of noncommissioned officer performance evaluation reports
- Review evaluation reports and provide recommendations to commanders and related officials
- Process requests for promotions, reductions, transfers, reenlistments, leaves, retirements, and discharges
- Monitor and process requests for military and special pay programs, meal cards, personnel security clearances, and orders for temporary duty and travel
- Prepare and maintain personnel records for officers and enlisted members
- Manage personnel administration activities and support programs related to formal training, promotions, classification, performance reports, and career progression
- Prepare casualty documents and provide next of kin with assistance on casualty-related procedures
- Oversee personnel activities and functions and ensure personnel compliance with policies
- Counsel and guide personnel to promote efficiency and ensure adaptation to the military environment

Services

Army

Marine Corps

Navy

Air Force

Coast Guard

INFANTRY

Summary: Members of the infantry are ground troops that engage with the enemy in close-range combat. They operate weapons and equipment to engage and destroy enemy ground forces. This job is typically considered to be the job in the Military that is more physically demanding and psychologically stressful than any other job.

Helpful Attributes

- Ability to stay in top physical condition
- Interest in working as a member of a team
- Readiness to accept a challenge

What They Do

- Operate and maintain mounted and dismounted weapons such as rifles, machine guns, mortars, and hand grenades
- Participate in reconnaissance operations
- Employ, fire, and recover antipersonnel and antitank mines
- Operate, mount/dismount, zero, and engage targets using night vision sights
- Locate, construct, and camouflage infantry positions and equipment
- Evaluate terrain and record topographical information
- Operate and maintain field communications equipment
- Assess need for and direct supporting fire
- Place explosives and perform minesweeping activities on land
- Deliver long-range precision fire from concealed positions
- Support security, stability, transition, and reconstruction operations

Services

Army

Marine Corps

Navy

INSTRUMENTALISTS

Summary: Instrumentalists play one or more musical instruments in recital, in accompaniment, and as members of an orchestra, band, or other musical group. For instrumentalists, knowledge of musical conducting, rehearsal techniques, transpositions, and the ability to read printed music are mandatory.

Helpful Attributes

- Ability to play more than one instrument
- Dedication, patience, and persistence
- Poise when performing in public

What They Do

- Perform on musical instruments in a band in concerts, parades, band drills, honor guards, guard mounts, and various other official functions and ceremonies
- Train or assist in the training of musician personnel, and perform other additional duties as required, such as section leader, supply non-commissioned officer (NCO), music librarian, storeroom keeper, and arranger
- Perform duties as instrumentalists on primary and/or secondary instrument(s) as members of authorized military bands
- Conduct preventive maintenance and make minor repairs to assigned musical instruments
- Perform on a musical instrument in a variety of ensembles, ranging from solo performance to full concert band
- Tune an instrument to a given pitch
- Perform duties in unit administration, training, supply, and/or operations, as required
- Organize, instruct, train, counsel, and evaluate musicians and senior musicians of the Musical Performance Team (MPT)
- Advise commanders on all aspects of band operations and serve as the band commander in their absence
- Assist with the inspections of instruments and accessories, arrange and transcribe music, and supervise the necessary administrative duties within the musical unit
- Drill with the marching band, and execute commands based on drum major verbal or nonverbal direction while playing and carrying a musical instrument

Services

Marine Corps

Navy

Air Force

INTELLIGENCE SPECIALISTS

Summary: Intelligence specialists play a key role in ensuring that military operations are planned using the most accurate, current information about enemy forces and capabilities. They oversee efforts to collect, exploit, develop, analyze, and produce intelligence information for dissemination to key military leaders and consumers worldwide. The results of their work are used to develop targets and provide situational awareness to operations personnel and key leadership. Intelligence specialists may focus on one type of intelligence information, such as signals or human intelligence, or they may consolidate all sources into usable products.

Helpful Attributes

- Ability to organize information
- Ability to think and write clearly
- Interest in gathering information and studying its meaning
- Interest in foreign cultures
- Interest in computers

What They Do

- Supervise, perform, or coordinate collection management, analysis, production, processing, and dissemination of strategic and tactical intelligence
- Compile intelligence information and disseminate data through media, such as plots, briefings, messages, reports, and publications; maintain intelligence libraries
- Perform advanced measurements on communication and noncommunications signals using various tactical data processing equipment
- Capture, correlate, and fuse technical, geographical, and operational intelligence information
- Conduct in-depth analysis of communications characteristics and target tactics, techniques, and procedures
- Assess electronic warfare (EW) risks and vulnerabilities and recommend countermeasures; assess friendly capabilities and missions in EW terms
- Conduct and supervise investigations of individuals, organizations, installations, and activities to detect, identify, assess, counter, exploit, and neutralize threats to national security
- Review, edit, translate, and exploit foreign documents; may act as an interpreter/translator for intelligence matters and materials
- Perform initial analysis of enemy foreign propaganda products and other media
- Assist in determining significance and reliability of incoming information; assist in integrating incoming information with current intelligence holdings and prepare and maintain the situation map

- Disseminate threat warning information to affected entities via established channels
- Oversee, prepare, and deliver intelligence briefings and other reports
- Assist in establishing and maintaining systematic, cross-referenced intelligence records and files
- Determine and manage required intelligence collection procedures and priorities; safeguard classified and sensitive material

Services

Army

Marine Corps

Navy

Air Force

Space Force

Coast Guard

INTERNATIONAL AND CIVIL AFFAIRS SPECIALISTS

Summary: International and civil affairs specialists perform various duties incident to the planning, coordination, and conduct of civil–military operations. They support the operation of civil affairs planning and coordination centers. They also perform civil–military assessments of their assigned operational area and coordinate with a wide variety of civilian populations, organizations, and agencies.

Helpful Attributes

- Interest in living and working in a foreign country
- Interest in working closely with people
- Ability to express ideas clearly and concisely

What They Do

- Assist in conducting assessments, coordination, analysis, production, and processing of civil affairs' products
- Serve as key advisor and consultant to commanders and supervisors on issues pertaining to foreign language and regional culture
- Function as interpreter or translator, and provide language skills, regional expertise, and cultural

capabilities to commanders throughout the phases of military operations
- Analyze and apply operational culture
- Build and maintain relationships and rapport with counterparts to further mission objectives
- Recognize and mitigate cultural stress
- Develop and implement training plans and events
- Participate in security cooperation planning
- Maintain and operate various voice and data communication devices
- Operate data processing equipment and computer programs

Services

Army

Marine Corps

Navy

Air Force

INTERPRETERS AND TRANSLATORS

Summary: Interpreters and translators are responsible for training military personnel in foreign language familiarization and foreign cultural awareness. They perform written translations, and they identify, translate, and summarize communications. They utilize foreign language skills, including knowledge of grammar and vocabulary, to collect and analyze intelligence information.

Helpful Attributes

- Interest in reading and writing
- Interest in working with people
- Talent for foreign languages

What They Do

- Identify, translate, and summarize communications effectively conveying the appropriate information to understand the meaning of an activity or a situation
- Conduct low- and mid-level escort interpretation
- Perform written translations from a foreign language into English
- Brief supported element on interpreter/translator utilization
- Provide and implement language familiarization and cultural awareness training

- Apply critical thinking and analytic skills to collect, evaluate, and combine data from multiple sources using language processing tools
- Provide guidance to subordinate personnel
- Oversee and support the development of English and foreign language skills
- Perform proficient utilization of foreign language grammar, vocabulary, specialized technical, and military vocabularies
- Operate electronic equipment relating to audio digital files, computerized databases, and analytical systems
- Perform quality control on language-derived materials
- Utilize foreign language skills to collect, transcribe, translate, analyze, and report intelligence information
- Use foreign language skills to search for, monitor, identify, and process communications involving activities of interest

Services

Army

Marine Corps

Navy

Air Force

INVESTIGATIVE SPECIALISTS

Summary: Investigations specialists are primarily responsible for investigating any criminal allegations and offenses that threaten the safety of military personnel, property, resources, or facilities. Specialties may include narcotics, economic crimes, cybercrimes, armed robbery, and death, among others.

Helpful Attributes

- Ability to remain calm under pressure
- Interest in law enforcement and crime prevention
- Listening and communication skills
- Critical thinking and problem-solving skills

What They Do

- Investigate felony and other significant crimes of military interest as defined by regulations, military, and federal law

- Plan, organize, conduct, and supervise overt and covert investigations
- Examine and process crime scenes
- Collect, preserve, and evaluate physical evidence for scientific examination by laboratories and use in judicial proceedings
- Obtain and execute arrest and search warrants
- Conduct raids and task force operations
- Perform hostage negotiations
- Interview and interrogate victims, witnesses, suspects, and subjects, and obtain written statements executed under oath
- Develop, evaluate, and manage informants and other sources of criminal intelligence
- Write, review, and approve technical investigative reports
- Analyze and detect ongoing crime, and recommend actions to prevent crime that could result in significant economic loss and reduced combat effectiveness
- Conduct personnel security vulnerability assessments for designated military officials
- Formulate special investigations policy governing investigative and related programs in counterintelligence, cyber threats, counterthreat, criminal, fraud, and technical services areas

Services

Army

Navy

Air Force

LAW ENFORCEMENT SPECIALISTS

Summary: Law enforcement specialists are equipped with the latest law enforcement tools and techniques for keeping the peace. They are responsible for investigating crimes that are committed within military bases. They also control traffic, respond to emergencies, conduct lawful arrests, interview the arrested suspects, guard military installations and correctional facilities, and conduct patrolling activities.

Helpful Attributes

- Ability to remain calm under pressure
- Interest in law enforcement and crime prevention

What They Do

- Lead military police squads and sections, operate police desks, plan crime prevention measures, operate evidence rooms, and prepare operations plans and orders in military police detachments
- Supervise or conduct investigations of incidents and offenses or allegations of criminality affecting personnel, property, facilities, or activities
- Perform mounted and dismounted individual and team patrol movements, tactical drills, battle procedures, convoys, military operations other than war, antiterrorism duties, and other special duties
- Perform specialized duties in law enforcement, physical security, antiterrorism operations, and detection of explosives and/or illicit drugs in the military community utilizing assigned military working dogs
- Operate communications equipment, vehicles, intrusion detection equipment, individual and crew-served weapons, and other special purpose equipment
- Apply self-aid, buddy-care, and lifesaving procedures as first responders to accident and disaster scenes
- Detect and report presence of unauthorized personnel and activities and implement security reporting and alerting system
- Direct vehicle and pedestrian traffic; investigate motor vehicle accidents, minor crimes, and incidents; and operate speed measuring, drug and alcohol, and breath test devices
- Secure crime and incident scenes; apprehend and detain suspects; search persons and property; and collect, seize, and preserve evidence
- Conduct interviews of witnesses and suspects, obtain statements, and testify in official judicial proceedings
- Defend personnel, equipment, and resources from hostile forces throughout the base security zone of military installations
- Conduct covert operations, personal protective services, hostage negotiations, polygraph examinations, laboratory examinations, and liaison with other senior military, civil, and federal law enforcement agencies
- Guard and provide 24-hour supervision of the daily activities of confined and restrained personnel to prevent disturbances and escapes
- Supervise or provide support to the battlefield by conducting maneuver and mobility support, area security, internment resettlement operations, police intelligence operations, prisoner of war operations, civilian internee operations, law and order operations on the battlefield, and to the peacetime military community

Services

Army

Marine Corps

Navy

Air Force

Coast Guard

LOGISTICS SPECIALISTS

Summary: Logistics specialists manage the particulars surrounding procurement, maintenance, and transportation of military materiel, facilities, and personnel. They develop, evaluate, monitor, and supervise logistics plans and programs including war readiness materiel, deployment, employment, and support planning. These individuals must be able to integrate the separate functions of planning and implementing a logistics management program.

Helpful Attributes

- Ability to keep accurate records
- Interest in operating forklifts and other warehouse equipment
- Preference for physical work
- Preference for work requiring attention to detail

What They Do

- Direct, develop, or perform logistics management operations that involve planning, coordinating, or evaluating the logistical actions required to support a specified mission, weapons system, or other designated program
- Supervise preparation of logistics annexes for operations plans and orders, programming, general support, contingency, and exercise plans
- Manage preparation and maintenance of workforce records and reports
- Prepare, evaluate, and oversee all aspects of deployment planning, dispersal, sustainment, recovery, reconstitution, and support

- Compile, coordinate, publish, distribute, maintain, and implement base support plans
- Activate and manage command and control centers during peacetime, wartime, and contingency operations
- Identify limiting factors, shortfalls, and alternate support methods to enhance supportability of transiting forces
- Monitor deployment of personnel and equipment products
- Review planning documents to determine deployment taskings
- Input, extract, and interpret data in automated information systems
- Develop crisis action procedures
- Create plans regarding facilities, supplies, and equipment procurement and maintenance
- Analyze work activities to ensure quality and compliance with policies, current directions, and other publications

Services

Marine Corps

Navy

Air Force

MACHINISTS

Summary: Machinists perform various duties incident to fabrication, repair or modification, and motor transport. They are experienced machine tool operators who use a variety of equipment and devices, such as lathes, drill presses, grinders, and other machine shop equipment.

Helpful Attributes

- Ability to apply mathematical formulas
- Interest in making things and finding solutions to mechanical problems
- Preference for working with the hands

What They Do

- Perform work using lathes, shapes, milling machines, internal and external grinders, drill presses, saws, and cylinder or line-boring machines
- Interpret sketches, diagrams, blueprints, and written specifications

- Coordinate with other repair shops and report of work complete
- Compute calculations using basic algebra and trigonometry
- Manufacture and repair precision parts from blueprints
- Utilize tool and cutter grinders, surface grinders, cylinder grinders, universal milling machines, optical comparators, surface finish analyzer, disintegrators, vertical turret lathes, and horizontal boring mills
- Understand the nature and physical properties of metal and alloys with demonstration and practice in metal testing, identification, and heat treating
- Utilize portable machinery, hand tools, and measuring instruments found in a machine shop to perform work outside the shop
- Supervise recovery operations and inspect completed work
- Fabricate, repair, and modify metallic and nonmetallic parts

Services

Army

Marine Corps

Navy

MAGNETIC RESONANCE IMAGING TECHNICIANS

Summary: Magnetic resonance imaging (MRI) technicians in the Military operate sophisticated equipment to produce high-definition, three-dimensional diagnostic images, and assist physicians with diagnostic imaging functions and activities. These expert technicians have knowledge of magnetism, magnetic safety, radio frequency, and magnetic physics; techniques of operating MRI equipment; and advanced knowledge of cross-sectional anatomy applicable to MRI. They are an important part of the Military health care team.

Helpful Attributes

- Ability to follow strict standards and procedures
- Interest in activities requiring accuracy and attention to detail
- Interest in helping others
- Interest in working in a medical environment
- Listening and communication skills

What They Do

- Select imaging protocols and required accessories, and make adjustments based on the specific examination requirements
- Operate fixed and portable radiographic equipment to produce routine diagnostic medical images
- Compute techniques and adjust control panel settings such as kilovoltage, milliamperage, exposure time, and focal spot size
- Position patient to image desired anatomic structures
- Select image recording media, adjust table or image receptor, align distance and angle, and restrict radiation beam for maximum patient protection
- Expose and process images
- Manipulate the recorded images using computer applications
- Assist physicians with fluoroscopic, interventional, and special examinations
- Prepare and assist with contrast media administration
- Maintain emergency response cart, sterile supplies, and equipment
- Operate accessory equipment, such as automatic pressure injectors, digital imagers, stereotactic biopsy devices, and vital signs monitoring equipment
- Perform image subtraction and manipulation techniques
- Clean and inspect equipment and perform preventive maintenance
- Receive patients, schedule appointments, and prepare and process examination requests and related records

Services

Air Force

MANPOWER SPECIALISTS

Summary: Manpower specialists provide support on manpower, which is the composition of the force. They oversee manpower resources and manage manpower requirements. They determine how many and which capabilities are required to execute a mission, and they advise on the allocation of military and civilian resources. They support accession planning, reenlistment, and force development programs. They operate manpower data systems, prepare relevant reports, and supervise efforts focused on continuous improvement.

Helpful Attributes

- Ability to compose clear instructions or correspondence
- Ability to follow detailed procedures and instructions
- Interest in working closely with others
- Analytical skills

What They Do

- Update military personnel data system (PDS) records
- Ensure compliance with personnel policies, directives, and procedures
- Conduct interviews to determine individual interests, qualifications, and personnel data
- Help commanders develop career information and motivation programs
- Monitor retention programs and provide reports and statistics
- Create, maintain, and audit personnel records
- Update computerized personnel data
- Schedule individuals for processing personnel actions, such as reenlistment, promotion, separation, retirement, or reassignment
- Administer standard tests, act as test monitor, score tests, and record results
- Advise members on official and personal obligations incident to relocation, training, and promotion
- Maintain files of correspondence, directives, instructions, and other publications
- Administer casualty program, prepare related reports and documents, assist next of kin of deceased and missing personnel to apply for death gratuity pay, arrears of pay, veterans' affairs, social security, government and commercial life insurance, and other benefits
- Ensure proper counseling of individuals on personnel programs, procedures, and benefits
- Maintain and monitor duty status changes

Services

Marine Corps

Air Force

MARINE ENGINE MECHANICS

Summary: Marine engine mechanics repair, inspect, test, and maintain gasoline and diesel engines on ships, boats, and other watercraft. They also repair shipboard mechanical and electrical equipment, including refrigeration and air conditioning systems. Specialized tools help them make accurate diagnoses and estimate how long a part will be reliable before it should be replaced.

Helpful Attributes

- Ability to use hand and power tools
- Interest in fixing engines, machinery, and auxiliary equipment
- Preference for doing physical work

What They Do

- Perform daily systems checks and post all instrument and gage readings to the engineer logbook
- Position fuel control racks and adjust throttle controls to maximize engine efficiency
- Inspect, troubleshoot, test, service, adjust, repair, and replace batteries, electrical system components, fuel system elements, propellers and propeller shafts, pumping assemblies and parts, and other marine engine equipment
- Start, operate, troubleshoot, and secure vessel engines
- Perform preventive maintenance and make repairs to small craft, including riverine assault craft; metal, fiberglass, and rubber hulled small craft; diesel motors; propeller and hydrojet propulsion systems; and small craft trailers and cradle systems
- Take readings and make adjustments required for the proper operation and repair of engines
- Instruct and supervise marine engine department personnel in all systems maintenance
- Test, inspect, and perform organizational maintenance on shipboard weapons and cargo handling elevators on system and component levels
- Operate, maintain, and repair internal combustion engines, main propulsion machinery, refrigeration, air conditioning, gas turbine engines, and assigned auxiliary equipment
- Inspect, service, adjust, replace, repair, and overhaul engine components, throttle controls, accessory drives, boiler and piping systems, vessel steering mechanisms, electrical and wiring assemblies, cooling and lubrication systems, and vessel hulls for general upkeep
- Perform preventive, organizational-, and/or intermediate-level maintenance on submarine diesel engines and components, including pumps, blowers, governors, injectors, cylinders, liners, pistons, connection rods, bearings, crankshaft, vertical drive assembly, lubrication oil, fuel oil, scavenging air, exhaust, starting air, and cooling water systems

Services

Army

Marine Corps

Navy

Coast Guard

MEDICAL ASSISTANTS

Summary: Medical assistants serve as critical team members in providing medical care to service members and their families. They give patients the care and treatment required to help them recover from illness or injury, including injuries suffered in combat. Medical assistants help prepare equipment for diagnostic tests and other procedures for patients.

Helpful Attributes

- Ability to follow directions precisely
- Ability to work under stressful or emergency conditions
- Interest in helping others

What They Do

- Assist physicians and other health professionals in the prevention and treatment of disease and injury
- Provide routine and emergency medical care in health care facilities or in the field
- Perform first aid and preventive medicine procedures
- Assist with physical examinations and provide instructions to patients
- Administer medications and perform general pharmacy services

- Perform supply and accounting record procedures in a variety of environments, such as aboard ships
- Conduct diagnostic testing, x-rays, and clinical lab tests
- Monitor, operate, and maintain equipment used for tests and treatments
- Support physicians in performing invasive procedures, such as biopsies and dialysis
- Assist in the treatment of patients with orthopedic conditions and injuries, including those sustained in combat
- Assist in minor operative procedures, such as arthrocentesis, suture removal, dressing changes, and insertion and removal of skeletal traction devices
- Support flight surgeon or medical officer in providing special examinations and treatments for naval aviators and flight personnel
- Operate and maintain training devices peculiar to the aviation physiology and aviation water survival training programs
- Assist medical officer in examination and treatment of urological patients

Services

Army

Navy

Air Force

Coast Guard

MEDICAL EQUIPMENT REPAIRERS

Summary: Medical equipment repairers test, adjust, and repair biomedical equipment. They service and maintain medical equipment with mechanical, hydraulic, pneumatic, electronic, digital, optical, and radiological principles. They may work on patient monitors, defibrillators, ventilators, anesthesia machines, and other life-supporting equipment, as well as medical imaging equipment (x-rays, CAT scanners, and ultrasound equipment), voice-controlled operating tables, and electric wheelchairs. To do their work, medical equipment repairers use a variety of tools, including specialized test-equipment software.

Helpful Attributes

- Ability to use tools

- Interest in electric motors and appliances
- Interest in solving problems

What They Do

- Perform Preventive Maintenance Checks and Services (PMCS), Calibration, Verification, Certification (CVC), and electrical safety tests on medical and medically related equipment
- Assemble, maintain, troubleshoot, align, and calibrate medical equipment
- Troubleshoot to isolate malfunctioning or defective parts and/or boards on medical and medically related equipment
- Replace malfunctioning or defective parts and/or boards on medically related equipment
- Fabricate field expedient repair parts for field medical equipment
- Compute power requirements for field medical equipment
- Perform pre-issue inspections and installations of medical equipment
- Inspect, service, and modify biomedical equipment and support systems
- Evaluate user maintenance procedures and ensure safe medical equipment practices are exercised
- Instruct and advise personnel in the care and safe, effective use of medical equipment
- Calibrate medical equipment according to manufacturer's technical literature, pertinent federal regulations, national standards, and state and local laws
- Apply electrical, electronic, optical, mechanical, pneumatic, hydraulic, and physiological principles to diagnose and locate system malfunctions

Services

Army

Nave

Air Force

MEDICAL IMAGING TECHNICIANS

Summary: Medical imaging technicians are primarily responsible for operating the equipment used in collecting, diagnosing, and treating medical conditions. They specialize in collection techniques of different types of medical imagery such as sonography, magnetic resonance imaging (MRI), x-ray, and nuclear medicine.

Helpful Attributes

- Ability to follow strict standards and procedures
- Interest in activities requiring accuracy and attention to detail
- Interest in helping others
- Interest in working in a medical environment

What They Do

- Operate fixed and portable specialized diagnostic equipment to produce images and assist physicians with special procedures; operate other equipment such as vital signs monitoring equipment
- Prepare sterile supplies and equipment for diagnostic studies and therapeutic procedures
- Instruct patients preparing for procedures and position patients to image desired anatomic structures
- Establish and employ health protective measures, such as transmission-based precautions and radiation protection
- Develop plans regarding facilities, supplies, and equipment procurement and maintenance
- Clean and inspect equipment and perform preventive maintenance
- Compute techniques based on the specific examination requirements and adjust control panel settings such as kilovoltage, milliamperage, exposure time, and focal spot size
- Expose, process, and record images
- Manipulate the recorded image using computer applications
- Perform administrative and managerial tasks as required, including receiving patients, scheduling appointments, and preparing and processing examination requests and related records
- Select image recording media, and adjust table or image receptor (cassette holder); align x-ray tube for correct distance and angle, and restrict radiation beam for maximum patient protection

Services

Navy

Air Force

MEDICAL LABORATORY TECHNICIANS

Summary: Medical laboratory technicians perform and/or supervise the application of basic and advanced laboratory procedures, such as collecting, processing and analyzing biological specimens and other substances; performing clinical diagnostic laboratory tests; and monitoring and reporting results. They conduct these laboratory tests to aid in the diagnosis, treatment, and prevention of disease and other medical disorders.

Helpful Attributes

- Ability to follow detailed procedures precisely
- Interest in scientific and technical work

What They Do

- Perform elementary blood banking and clinical laboratory procedures in hematology, immunohematology, clinical chemistry, serology, bacteriology, parasitology, and urinalysis
- Collect blood specimens by venipuncture and capillary puncture
- Prepare tissue for electron microscopy
- Assemble, disassemble, and perform preventive maintenance on laboratory equipment
- Train lower-level staff in all elementary laboratory procedures; provide technical and administrative management, coordination, control, and operational duties in senior positions
- Participate in a wide variety of studies and reviews
- Coordinate with other activities, agencies, and organizations
- Inspect and evaluate medical laboratory activities
- Advise superiors regarding status and adequacy of equipment, supplies, personnel training, and operating efficiency
- Assist the health care provider in providing diagnoses to patients
- Assist pathologists in the preparation of surgically obtained tissue of human, animal, or plant origin
- Perform accessioning, grossing, freezing, cutting, mounting tissue on slides, and staining with special dyes

Services

Army

Navy

Air Force

MEDICAL RECORD TECHNICIANS

Summary: Medical record technicians perform a variety of medical administrative duties that provide support to health-related services and medical departments. They are responsible for maintaining patient medical records and preparing requests for diagnostic tests and referrals. They monitor long-term patient data and audit completed records and forms for administrative and technical accuracy. They process patient admission to medical facilities, and they schedule patients for medical board action.

Helpful Attributes

- Ability to communicate well
- Interest in using computers and other office machines
- Interest in work requiring accuracy and attention to detail

What They Do

- Prepare, consolidate, and file medical documents to ensure proper sequencing of forms
- Utilize knowledge of medical terminology related to anatomy and physiology
- Transcribe physicians' orders and prepare requests for diagnostic tests, consultations, and referrals
- Maintain a medical record tracking system within the medical treatment facility
- Prepare medical records for retirement
- Compile data and prepare required statistical reports on outpatient visits, inpatient visits, admissions, dispositions, and other selected workload areas
- Verify eligibility for care and process patients for admission to medical facilities
- Account for and safeguard funds received for deposit in medical services account
- Provide technical guidance and assign tasks to lower grade personnel to improve flow of workload and promote efficiency of services
- Counsel eligible beneficiaries concerning military health care benefits
- Counsel patients on physical disability processing procedures
- Release medical information, answer inquiries, and provide medical record information to requesting parties, as authorized
- Perform quality control checks on medical records regarding effectiveness of care and treatment

Services

Army

Air Force

MENTAL HEALTH CASEWORKER

Summary: Mental health caseworkers assess, diagnose, and treat psychological conditions. They implement initial mental health assessment procedures, including standardized psychological testing, clinical interviewing, and substance abuse evaluations. They monitor, record, and report on patient progress. Mental health caseworkers conduct and/or provide individual and group counseling, and they establish preventative measures related to patient injuries, including those sustained in combat. They supervise and support caring for individuals with acute and post-traumatic stress reactions.

Helpful Attributes

- Interest in working with people
- Patience in dealing with problems that take time and effort to overcome
- Sensitivity to the needs of others
- Listening and communication skills

What They Do

- Perform evaluations on mental health service tasks and activities
- Support mental health services in psychiatry, psychology, social work, family advocacy, substance abuse prevention, treatment and aftercare, integrated operational support, and other mental health programs
- Manage mental health service resources and activities
- Assist mental health professional staff with developing and implementing treatment plans
- Perform general and diagnosis-specific mental health counseling
- Maintain thorough and accurate reports on patient care and treatment
- Assist psychiatrists and psychologists in performing assessments, crisis triage and management,

co-facilitation of therapy groups, short-term counseling, and psychological testing

- Provide intervention for persons affected by psychological trauma, mental illness, and crisis
- Complete observations and documentation in the care and treatment of patients in inpatient and outpatient hospital settings and field environments
- Identify patients' strengths, weaknesses, problems, and needs
- Assist individuals, families, and groups in achieving treatment goals through exploration of problems, establishing treatment plans, maintaining therapeutic relationships, examining attitudes and affective responses, providing alternative solutions, and assisting in decision-making

Services

Navy

Air Force

METEOROLOGICAL SPECIALISTS

Summary: Meteorological specialists predict the weather and study the causes of particular weather conditions using information obtained from the land, sea, and upper atmosphere. They use the latest forecast technology to predict severe weather, tornadoes, flooding, hurricanes, strong winds, and excessive heat and cold.

Helpful Attributes

- Ability to communicate effectively
- Interest in gathering and organizing information
- Interest in learning how weather changes
- Interest in working with formulas, tables, and graphs

What They Do

- Operate atmospheric and space-sensing instruments and computer workstations to gather data from weather radars, meteorological satellites, and products provided by military, national, and international weather agencies
- Collect, analyze, tailor, and integrate meteorological, oceanographic, and space environmental information into military decision-making processes
- Help develop weather-related policies and plans

- Manage weather operations, ensure quality, and adapt resources to meet mission requirements
- Utilize weather tactics, techniques, and procedures to integrate weather information into the decision-making process at all levels to mitigate and exploit weather impact on operations
- Observe, record, and transmit surface, upper air, and space environment observations
- Issue advisories, watches, and warnings to alert users of dangerous, inclement, or operationally significant terrestrial and space weather events
- Evaluate, record, and transmit surface weather, oceanographic, geographic, riverine, and space environment observations
- Prepare and operate weather observation equipment, such as theodolites and surface sensors, that measures atmospheric readings; and prepare and operate survey equipment, such as inertial navigation systems, global positioning system receivers, theodolites, and electronic distance measuring equipment

Services

Army

Marine Corps

Navy

Air Force

MILITARY DOG HANDLERS

Summary: Military dog handlers are in charge of the basic care and training of military working dogs, which are generally used for drug interdiction, locating lost or wanted persons, or bomb-sniffing missions. They perform specialized duties in law enforcement, physical security, antiterrorism operations, and detection of explosives and/or illicit drugs in the military community, utilizing an assigned military working dog.

Helpful Attributes

- Ability to remain calm under pressure
- Interest in law enforcement and crime prevention
- Ability to judge a situation accurately and react instantly
- Respect for dogs

What They Do

- Meet and maintain training and certification standards associated with the use and maintenance of assigned military working dog
- Execute or oversee the daily care and grooming of assigned military working dog
- Conduct searches of open areas, buildings, and vehicles for detection of explosives or illegal drugs
- Detect and locate weapons, ammunition, explosives, and Improvised Explosive Devices
- Store and maintain accountability of narcotic and explosive training aids, as well as all other assigned equipment
- Locate lost or wanted persons by scouting and tracking
- Provide oversight, guidance, and assistance to commanders with the application of physical security and force protections, to include detecting and reporting presence of unauthorized personnel and activities
- Plan, coordinate, and supervise activities pertaining to organization, training, and combat operations
- Edit and prepare tactical plans and training material
- Participate in emergency planning and recovery operations, such as response to disaster relief operations
- Defend personnel, equipment, and resources from hostile forces throughout the base security zone of military installations
- Perform individual and team patrol movements, tactical drills, battle procedures, convoys, military operations, and antiterrorism duties
- Investigate motor vehicle accidents, minor crimes, and incidents

Services

Army

Marine Corps

Navy

Air Force

MILITARY-SPECIFIC/TACTICAL TRAINING SPECIALISTS

Summary: Training specialists that focus on military-specific/tactical training may conduct training under conditions that closely approximate combat, such as Survival, Evasion, Resistance, and Escape (SERE) situations. They may be required to develop and conduct training related to aerospace and operational physiology, antiterrorism, weapons and tactics, or marksmanship, to name a few. For example, small-arms marksmanship instructors conduct training in all phases of basic marksmanship, both ashore and afloat. This training includes firearms safety, mechanical training on small arms, instructional and qualification firing, basic range operations, and reporting.

Helpful Attributes

- Oral and written communication skills
- Interest in counseling and promoting human relations
- Interest in teaching
- Military knowledge

What They Do

- Serve as SERE subject matter expert in the design and development of curriculum, functional structure, and procedures for SERE courses and programs
- Determine requirements for training, facilities, space, equipment, visual aids, and supplies to support military training requirements
- Plan training schedules according to course control documents, directives, policies, and instructional principles
- Assess readiness and efficacy of equipment, supplies, and training aids, to include conducting Developmental Testing and Evaluation (D&TE) and Operational Testing and Evaluation (O&TE) on SERE equipment
- Ensure standardization and compliance with policies, procedures, directives, course control documents, and operational guidance
- Conduct classroom, laboratory, and operational recruit training for newly enlisted personnel using lecture, demonstration and performance, guided discussion, case study, and time and circumstance instructional methodology
- Utilize training environments, including but not limited to global environmental conditions, combat situations, and the full spectrum of captivity environments

- Guide recruits in the fundamentals of service life and the development of the recruit's discipline, physical fitness, pride, and love of military service and country
- Instruct in general orders for sentinels, interior guard duty, personal hygiene, first aid, military bearing and neatness, and care of clothing and equipment
- Instruct in nomenclature, disassembly, assembly, and functioning of small arms, and assist in marksmanship instruction
- Assist in conduct of parades, reviews, and drills
- Execute remedial training as necessary

Services

Army

Marine Corps

Navy

Air Force

MORTUARY AFFAIRS SPECIALISTS

Summary: Mortuary affairs specialists are there for personnel and their families during times of grief and sadness. They work in both combat and noncombat situations to search for, and identify, deceased personnel. They help with the inventory and safeguarding of personal effects and provide internment. These specialists also assist the next of kin and provide funeral services.

Helpful Attributes

- Solid code of ethics
- Listening and communication skills
- Knowledge of customs and traditions
- Enjoy helping, teaching, and caring for others
- Ability to work with minimal supervision

What They Do

- Inventory, safeguard, and evacuate personal effects of deceased personnel
- Disinter remains, record personal effects, and evacuate remains and personal effects to designated points
- Determine and record recovery locations on maps, sketches, and overlays

- Establish and record tentative identification
- Assist in preparation, preservation, and shipment of remains
- Instruct in special handling, marking, and shipping of contagious disease cases and processing of contaminated remains
- Accompany remains and personal effects to designated locations, and assist with arrangements for military honors at place of burial
- Advise commanders and headquarters staff on mortuary affairs activities and coordinate activities of subordinate units
- Advise on temporary cemetery locations, emergency burials, and security and disposition of remains and personal effects
- Coordinate with non-U.S. service authorities, other service authorities, and civilian officials on mortuary affairs matters
- Supervise mortuary affairs activities, and provide technical and administrative support on graves registration matters, acquisition of land for temporary cemeteries, equipment requirements for mortuaries, and recovery/evacuation procedures
- Perform duties in both combat and non-combat environments pertaining to the search and recovery, processing, tentative identification, interment, disinterment, and transportation of human remains and personal effects
- Coordinate the transfer of remains and personal effects of remains in the area of operations
- Inspect and/or process remains for transportation to the United States overseas locations, or local burial; ensure proper documentation accompanies remains
- Conduct liaison with next of kin and foreign government officials for disposition of remains

Services

Army

Marine Corps

Navy

MUSIC COMPOSERS AND ARRANGERS

Summary: Music composers and arrangers write and transcribe musical scores. They adapt music into customized musical arrangements and create original music for various musical ensembles. Knowledge of music theory, transpositions, capabilities of instruments, and the ability to arrange and compose music using industry standard software are mandatory.

Helpful Attributes

- Knowledge of multiple musical instruments
- Creativity and originality
- Poise when performing in public

What They Do

- Develop and write musical scores/arrangements to meet various performance requirements
- Manage, supervise, and perform as an instrumentalist, music arranger, vocalist, or audio and lighting engineer in military band activities
- Adapt music into customized musical arrangements, and create original music for various musical ensembles
- Plan, organize, direct, and inspect band activities
- Analyze musical requirements and provide appropriate musical support
- Plan, schedule, and coordinate rehearsals, ceremonies, master classes, and performances
- Advise higher authority on band status, equipment maintenance and requirements, personnel training, and operational efficiency
- Manage unit administrative and support functions such as Operations, Administration, Publicity, Resources, and Auditions

Services

Navy

Air Force

MUSICIANS

Summary: Musicians in the Military play not only for members of their own service branch but also to engage the public. They play indoors in theaters and concert halls, at dances, and outdoors at parades and open-air concerts. They perform many types of music, including marches, classical, jazz, and popular music. They also travel regularly.

Helpful Attributes

- Ability to play more than one instrument
- Ability to sing
- Poise when performing in public

What They Do

- Perform all applicable styles of music, including marching band, ceremonial band, concert band, classical, jazz, ethnic, and popular music compositions
- Provide technical guidance to junior grade personnel
- Advise higher authority on band status, equipment maintenance and requirements, personnel training, and operational efficiency
- Perform in a variety of ensembles, ranging from solo performance to full concert band
- Tune an instrument to a given pitch
- Transpose moderately easy music
- Organize, instruct, train, counsel, and evaluate musicians and senior musicians of the Music Performance Team (MPT)
- Supervise MPT operator maintenance
- Manage, supervise, and perform as an instrumentalist, music arranger, vocalist, or audio and lighting engineer in military band activities
- Analyze musical requirements and provide appropriate musical support
- Plan, schedule, and coordinate rehearsals, ceremonies, master class, and performances

Services

Army

Marine Corps

Navy

Air Force

Coast Guard

NETWORK AND DATABASE ADMINISTRATORS

Summary: Network and database administrators develop, install, operate, and maintain the Military's computer networks and databases. They monitor system performance and make sure the appropriate personnel have access to data. They perform administrative duties, including providing user support to military personnel. They also provide information security to protect the Military's computer systems against cyberattacks. Network and database administrators may work in military facilities or aboard ships and submarines.

Helpful Attributes

- Ability to communicate effectively
- Ability to understand and apply math concepts
- Interest in solving problems
- Interest in work requiring accuracy and attention to detail

What They Do

- Operate and maintain computer systems and networks; perform system administration and maintain computers
- Install, test, maintain, and upgrade operating systems software and hardware to comply with information assurance requirements
- Manage accounts, network rights, and access to network systems and equipment
- Establish and maintain computer systems, analog and digital voice systems (telephones and voicemail), and install and maintain the physical network infrastructure that ties the systems together
- Plan, coordinate, share, and control the Military's data and information assets; manage technologies to capture, organize, and store tacit and explicit knowledge
- Manage databases for the storage, modification, and retrieval of information to produce reports, answer queries, and record transactions
- Install, support, and maintain server operating systems or other computer systems and the software applications pertinent to its operation, while also ensuring current defensive mechanisms are in place
- Create, administer, and audit system accounts; perform system-wide backups and data recovery; ensure continuing systems operability by providing ongoing optimization and problem-solving support
- Protect operating systems, application software, files, and databases from unauthorized access to sensitive information, or misuse of communication-computer resources
- Perform proactive security functions to deter, detect, isolate, contain, and recover from information system and network security intrusions
- Determine, analyze, and develop requirements for software systems through interpreting standards, specifications, and user needs
- Translate system specifications and requirements into program code and database structures; implement designed functionality as software coders
- Perform client-level information technology support functions; manage hardware and software; perform configuration, management, and troubleshooting
- Deploy, sustain, troubleshoot, and repair standard voice, data, and video network infrastructure systems, internet protocol (IP) detection systems, and cryptographic equipment
- Apply communications security programs to include physical, cryptographic, transmission, and emission security

Services

Army

Marine Corps

Navy

Air Force

Space Force

Coast Guard

NEURODIAGNOSTIC (EEG) TECHNICIANS

Summary: Neurodiagnostic technicians in the Military focus on diagnostic tests related to brain functions. They assist physicians in performing electroencephalography (EEG) tests. They prepare patients for examination and operate EEG equipment to identify organic brain diseases. These specialists also maintain the EEG equipment.

Helpful Attributes

- Ability to follow strict standards and procedures
- Ability to keep accurate records
- Interest in electronic equipment
- Interest in learning about the brain and disease of the brain
- Concern for others

What They Do

- Assist the neurologist in identifying patients with neurological disorders
- Perform electro-neurodiagnostic studies to include EEGs, evoked potentials, polysomnography (sleep studies), and nerve conduction studies

- Assist medical officer in performing electroencephalography tests
- Prepare patients for examinations
- Assist in operating electroencephalography equipment to examine patients for organic brain diseases
- Record test results
- Maintain electroencephalography equipment
- Record and study the electrical activity in the brain and the nervous system to diagnose neurological issues
- Ensure patient safety
- Talk with patients and prepare them for testing by explaining what is going to occur

Services

Navy

NON-DESTRUCTIVE TESTERS

Summary: Non-destructive testers find the smallest imperfections and take necessary corrective measures to keep the Military's equipment working safely. They utilize everything from x-rays to ultrasound to perform the job.

Helpful Attributes

- Interest in machines and how they work
- Thoroughness and dependability
- Interest in operating test equipment

What They Do

- Advise on metals machining, welding, designing, and production problems
- Design, manufacture, or modify special precision tools, gauges, dies, and fixtures to facilitate metal working operations
- Calculate cutting speeds and settings, welding processes, and pre-heat and post-heat requirements
- Use manual and computer numerical controlled metal working machines, mills, and lathes to manufacture and repair cams, gears, slots, and keyways for aircraft components
- Assemble, disassemble, and fit component parts using machine screws, bolts, rivets, press fits, and welding techniques
- Inspect and maintain hand tools and metalworking machinery

- Determine test method and prepare fluids and parts for non-destructive inspection
- Compute and monitor personal exposure areas for radiographic operations, and monitor personal exposure data
- Perform inspection on structures, components, and systems to detect flaws such as cracks, delaminations, voids, processing defects, and heat damage
- Execute operator maintenance and service inspections on shop equipment and tools

Services

Navy

Air Force

NUCLEAR MEDICINE TECHNICIANS

Summary: Nuclear medicine technicians in the Military administer radioactive isotope via injection, inhalation, and oral administration to create images of organs and organ systems, study body functions and flow, analyze biological specimens, and treat disease. They operate and maintain Gamma camera imaging equipment, devices, and probes.

Helpful Attributes

- Ability to follow strict standards and procedures
- Interest in activities requiring accuracy and attention to detail
- Interest in helping others
- Interest in working in a medical environment
- Listening and communication skills

What They Do

- Operate and maintain diagnostic equipment
- Sterilize the room in which the procedure is to take place
- Assist medical officers in preparing and conducting radioactive isotope therapy
- Perform diagnostic imaging and radiopharmaceuticals administration
- Support physician or health care team in preparing and conducting various radioactive therapies
- Monitor patient for unusual reactions to any drugs administered

- Coordinate communication between multiple health care providers
- Explain procedures to patient and answer process-related questions
- Maintain and update patient records
- Ensure compliance with radiation disposal and safety procedures

Services

Navy

OCCUPATIONAL THERAPY ASSISTANTS

Summary: Occupational therapy assistants support occupational therapists and/or medical officers in administering occupational therapy, a form of therapy for those recuperating from physical or mental illness that encourages rehabilitation through the performance of activities required in daily life. These specialists find new and easy ways for people going through physical, mental, and emotional trauma so that they can overcome the problems and lead a satisfying and happy life. Some of their responsibilities include examining patients, planning treatment, reviewing progress, updating prescribed treatment, examining the patients' homes, helping children participate in school activities, and helping their patients with doing daily tasks like washing, dressing, cooking, and eating.

Helpful Attributes

- Ability to communicate effectively
- Interest in working with and helping people
- Patience to work with people whose injuries heal slowly

What They Do

- Interview, test, plan treatment programs, and assist patients in activities of daily living with outpatient care and treatments, or with inpatient care and treatment, under the supervision of an Occupational therapist (OTR) or a Certified Occupational Therapy Assistant (COTA)
- Administer emergency and routine combat stress/orthopedic treatment to battlefield casualties
- Promote lifestyle modification to improve fitness

- Supervise activities of combat stress unit, medium-sized occupational therapy clinic, and mobile treatment facilities
- Assist occupational therapist and/or medical officer in administering occupational therapy
- Assist with carrying out established treatment plans for acute and chronic rehabilitation services
- Assist with the fabrication and fitting of upper and lower extremity, high and low temperature orthotics, and with splinting and rehabilitation for tendon lacerations, fractures, burns, and various neuromusculoskeletal disorders
- Conduct range of motion and strengthening programs using simulators and workstations
- Assist in the management of hypertrophic scarring, edema, hypersensitivity, and wound care
- Assist in the development and application of psychosocial rehabilitation for psychiatric patients
- Assist in the management of combat stress-related disorders in wartime

Services

Army

Navy

OPTICIANS

Summary: Opticians are highly skilled eye care professionals who undergo rigorous and extensive training. They design and dispense eyeglasses, contact lenses, low vision aids, and prosthetic ocular devices for customers. They also assist with frame selection and manage clinic offices.

Helpful Attributes

- Ability to communicate effectively
- Interest in work requiring accuracy and attention to detail

What They Do

- Assemble spectacles utilizing pre-surfaced single vision lens
- Supervise optical laboratory personnel
- Fabricate, repair, and assemble prescription spectacles
- Maintain tools and equipment
- Perform and manage optometry and ophthalmology clinic activities, including personnel, supplies, equipment, and programs

- Record patient case history, conduct visual screening tests such as visual acuity, eye cover, pupillary, color vision, depth perception, visual field charting, corneal topography, and tonometry for analysis and interpretation
- Take ophthalmic photographs
- Administer ophthalmic drops and ointments and apply ocular dressings
- Order, fit, dispense, and process prescriptions for military eyewear
- Instruct patients on contact lens procedures
- Assist aircrew members in aviator contact lens and night vision goggle programs
- Perform as an ophthalmic surgical assistant, prepare preoperative patients, and provide care for postoperative patients
- Manage ophthalmic resources, including determining requirements for supplies, equipment, and personnel
- Perform all phases of fabrication of single vision and multifocal spectacles from prescriptions, including marking, cutting, edging, and inserting single vision and multifocal lenses into appropriate frames
- Maintain and repair optical laboratory equipment

Services

Army

Navy

Air Force

ORDINANCE SPECIALISTS

Summary: Ordnance specialists are responsible for the safety, security, and accountability of the Military's weapons and ammunition. They perform a wide variety of duties, including the safe receipt, storage, and transport of ordnance. They also inspect, prepare, and dispose of weapons and ammunition. Some of these specialists deal solely with the destruction and demilitarization of explosive items.

Helpful Attributes

- Ability to remain calm under stress
- Interest in working with guns and explosives

What They Do

- Identify munitions and equipment requirements; operate and maintain automated data processing equipment (ADPE) to perform inspection, testing, and stockpile management activities
- Store, maintain, test, assemble, issue, and deliver assembled munitions; prepare munitions for loading; check safety and arm mechanisms
- Install warheads, guidance units, fuses, arming wires, squibs, strakes, wings, fins, control surfaces, and tracking flares
- Analyze stockpile requirements to determine type and quantity of munitions facilities needed to safely store, inspect, maintain, and secure munitions assets
- Supervise the maintenance, repair, and inspection of all small-arms weapons and towed artillery howitzers and direct the activities of assigned enlisted personnel in a small arms or towed artillery repair shop or facility
- Operate and perform operator maintenance on armament weapons support equipment and aircraft armament equipment
- Inspect ammunition, components, and containers for defects; prepare documentation indicating identification and quantity of assets
- Conduct functional tests of racks, launchers, adapters, electrical components, and aircraft armament circuits
- Locate, identify, render safe, and dispose of foreign and domestic conventional, chemical, or nuclear ordnance and improvised explosive devices (IED)
- Utilize and maintain advanced equipment, such as, robotics, x-ray, landmine, and chemical, biological, radiation, and nuclear detection equipment
- Provide immediate initial support to nuclear weapon accidents or incidents to evaluate nuclear weapon/delivery status, mitigate risk, and provide site stabilization and situational awareness
- Identify munitions by type, nomenclature, and explosive hazard; identify the hazard present and appropriate emergency response procedures for each type of munitions handled
- Educate personnel of the hazards, safety principles, and practices related to the use of ammunition and explosives
- Perform tactical and operational mine warfare planning, coordination, and execution; plan tactical

employment of airborne and underwater mine countermeasure systems

Services

Army

Marine Corps

Navy

Air Force

PARALEGALS AND LEGAL ASSISTANTS

Summary: Paralegals and legal assistants in the Military do everything from research and interviews to processing cases and discovery management to ensure law and order. Legal assistants and paralegals perform many of the same duties as an attorney but only under the supervision of an attorney.

Helpful Attributes

- Ability to keep organized and accurate records
- Ability to listen carefully
- Interest in the law and legal proceedings

What They Do

- Administer and supervise the provision of legal services and court reporting and assist judge advocates/attorneys
- Conduct extensive legal research
- Perform paralegal tasks including, but not limited to, legal research, writing, analysis, interviewing, and discovery management in the areas of administrative law, military justice, operational law, claims, and office management
- Prepare written communications, process correspondence, and maintain suspense files
- Compile, input, update, retrieve, and interpret statistical data; prepare and present statistical reports on legal activities in various forums
- Act as a trial team member by assisting attorneys with investigating leads, conducting witness interviews, reviewing case status, and developing case strategy
- Examine all actions and records of legal proceedings to ensure accuracy and completeness
- Transcribe verbatim records of legal proceedings
- Investigate the facts of cases and ensure that all relevant information is considered

- Identify appropriate laws, judicial decisions, legal articles, and other materials that are relevant to assigned cases
- Help prepare legal arguments, draft pleadings and motions to be filed with the court, obtain affidavits, and assist attorneys during trials
- Organize and track files of important case documents and make them available and easily accessible to attorneys
- Prepare and process legal documents in support of courts-martial, nonjudicial punishment, and other military justice matters, line of duty determinations, separation board proceedings, and other administrative law matters, legal assistance services, and claims processing and investigations

Services

Army

Marine Corps

Navy

Air Force

PEST CONTROL SPECIALISTS

Summary: Pest control specialists use their expertise to protect military members and their families from diseases borne of insects. They conduct assessments and determine what pest management actions are needed to control and prevent infestations in military facilities and the surrounding environment. They interact with other military medical providers and follow environmental compliance guidelines.

Helpful Attributes

- Interest in gathering information
- Interest in protecting the environment
- Preference for work requiring attention to detail
- Customer relations skills

What They Do

- Manage, evaluate, and execute pest management techniques
- Conduct pest management surveys
- Determine pest management approaches needed to control and prevent infestations of plant and animal pests

- Select chemicals and operate pesticide dispersal equipment
- Interact and coordinate with other medical personnel to control health hazards
- Provide maximum benefits consistent with environmental protection parameters
- Ensure compliance with applicable laws and directives
- Maintain tools, equipment, facilities, and storage areas
- Ensure correct use and maintenance of personal protective equipment and tools
- Evaluate proposed work, determine resource requirements, and prepare cost estimates
- Identify, budget for, and acquire specialized equipment
- Inspect facilities and provide assistance to building managers on pest prevention and control practices
- Maintain historical databases and tracking systems
- Perform quality assurance and evaluation of contracted pest management functions

Services

Air Force

PHARMACY TECHNICIANS

Summary: Pharmacy technicians perform all functions related to preparing, controlling, and dispensing pharmaceutical products. They verify prescription orders for dosage and provide instructions to patients regarding medication and possible side effects. They perform administrative functions, safeguard and account for inventories, and submit required reports. They maintain pharmacy system databases and pharmacy reference files.

Helpful Attributes

- Ability to follow strict procedures and directions
- Ability to work using precise measurements and standards
- Interest in body chemistry

What They Do

- Prepare, control, compound, and dispense pharmaceutical products

- Receive, interpret, fill, label, and file prescriptions manually or using computerized systems
- Verify orders for dosage, dosage regimen, and quantity to be dispensed
- Maintain and operate pharmacy information systems
- Perform and verify pharmaceutical calculations; calculate and annotate proper dosage
- Provide pharmacists or physicians with information regarding availability, strength, and composition of medications
- Ensure patient eligibility for receipt of medication
- Confer with prescribers or patients on questions to assure desired therapeutic outcome
- Provide patients with instructions regarding medication consumption and side effects
- Perform quality control checks on medications
- Assign and record prescription numbers
- Deliver unit dose, sterile products, bulk drug, and controlled drug orders
- Prepare prescription labels and affix auxiliary labels
- Perform inventory control functions
- Prepare and maintain files of controlled substances, stock cards, records, and work units
- Clean and disinfect pharmacy equipment and work areas

Services

Army

Navy

Air Force

PHOTOGRAPHERS AND VIDEOGRAPHERS

Summary: Military photographers and videographers have an important job in communicating and recording military activity. Photographers photograph people, landscapes, merchandise, and other subjects, using digital or film cameras and other equipment. Videographers use a variety of electronic media to record videos for television, internet, and film.

Helpful Attributes

- Ability to recognize and arrange interesting photo subjects
- Accuracy and attention to detail

What They Do

- Use video cameras and editing equipment in a variety of environments to support information operations, operational imagery, civil affairs, intelligence imagery, investigations, research, development, test and evaluation, recruiting, etc.
- Perform video equipment inspections, preventive maintenance, visual and statistical quality control, captioning, archive accessioning, and video dubbing
- Draft reports, official correspondence, and budgets; supervise and instruct personnel in the operation of all equipment and software, as well as supervise and organize all aspects of video/photographic operations
- Supervise, plan, and operate electronic and film-based still, video, and audio acquisition equipment to document combat and noncombat operations
- Operate equipment, such as broadcast, collection, television production, battlefield video teleconferencing, and distribution equipment
- Create visual information products in support of combat documentation, psychological operations, military intelligence, medical, public affairs, training, and other functions
- Take photographs using all types of cameras and accessories in a variety of environments (including in low-light and underwater conditions)
- Perform chemical mixing, equipment inspections, preventive maintenance, visual and statistical quality control, captioning, archive accessioning, and photographic finishing
- Produce news pictures, feature pictures, picture stories, and multimedia packages that lend balance and impact to visual media communications
- Coordinate and direct personnel and operational requirements to produce audiovisual, audio, and television productions in both fixed and tactical environments
- Take still photographs to illustrate written articles, provide coverage of special events, and prepare layouts for base newspapers
- Train personnel in photojournalism and advanced multimedia techniques

Services

Army

Marine Corps

Navy

Air Force

PHYSICAL THERAPY ASSISTANTS

Summary: Physical therapy assistants support physical therapists in providing physical therapy treatments and procedures. Under the direction of a physical therapist and following medical referral, these specialists administer physical therapy to decrease physical disabilities and promote physical fitness of service members. They treat disease, injury, or deformity by physical methods such as massage, heat treatment, and therapeutic exercise, rather than by drugs or surgery.

Helpful Attributes

- Ability to communicate effectively
- Interest in working with and helping people
- Patience to work with people whose injuries heal slowly

What They Do

- Supervise or administer physical therapy to decrease physical disabilities and promote physical fitness of patients, under the direction of a physical therapist and following medical referral
- Assist physical therapists and/or medical officers in administering physical therapy
- Assist with the development, teaching, and supervision of exercise programs and activities of daily living to enhance patient's strength, endurance, coordination, and mobility
- Assist with assessment of range of motion, strength, and ambulation skills
- Teach the use of ambulation aids
- Apply physical modalities of whirlpool, paraffin bath, ultrasound, diathermy, infrared, hot and cold packs, and electrical stimulation
- Assist with the management of a small clinic or medium- or large-sized physical therapy section

Services

Army

Navy

PHYSICAL AND OCCUPATIONAL THERAPY SPECIALISTS

Summary: Physical and occupational therapy specialists assist physical and occupational therapists in the development of treatment plans, carry out routine functions, document the progress of treatment, and modify specific treatments in accordance with patient status and within the scope of treatment plans established by the therapist, with the goal of helping their patients regain strength and mobility.

Helpful Attributes

- Ability to communicate effectively
- Interest in working with and helping people
- Patience to work with people whose injuries heal slowly

What They Do

- Supervise or administer physical therapy to decrease physical disabilities and promote physical fitness of patients, under the direction of a physical therapist
- Assist with the management of a small clinic
- Administer emergency and routine combat stress/orthopedic treatment to battlefield casualties
- Perform interviews, test, and assist patients in activities of daily living with outpatient care and treatments or with inpatient care and treatment under the supervision of an Occupational Therapist (OTR) or a Certified Occupational Therapy Assistant
- Administer patient care activities in physical therapy, occupational therapy, and orthotic services
- Participate in planning, providing, and evaluating patient care interventions
- Utilize therapeutic principles to restore function and support activities of daily living
- Conduct treatments by utilizing special equipment, modalities, and other treatment procedures
- Fabricate splints and aid devices to protect or assist patient in achieving optimal independent physical function
- Assist with the development, teaching, and supervision of exercise programs and activities of daily living to enhance patients' strength, endurance, coordination, and mobility
- Apply physical modalities of whirlpool, paraffin bath, ultrasound, diathermy, infrared, hot and cold packs, and electrical stimulation
- Assist with carrying out established treatment plans for acute and chronic rehabilitation services
- Assist with the fabrication and fitting of upper and lower extremity, high and low-temperature

orthotics, and splinting and rehabilitation for tendon lacerations, fractures, burns, and various neuromusculoskeletal disorders
- Conduct range of motion and strengthening programs using simulators and workstations
- Supervise treatment programs, and teach and assist patients to facilitate maximum recovery by decreasing physical and mental disabilities resulting from illness or trauma, prevent injury, and promote lifestyle modification to improve fitness, under the direction of a registered military occupational therapist

Services

Air Force

PLUMBERS AND PIPEFITTERS

Summary: Plumbers and pipefitters assemble, install, maintain, and repair many different types of pipe systems.

Helpful Attributes

- Ability to work with detailed plans
- Preference for doing physical work

What They Do

- Install and repair pipe systems, plumbing fixtures, and equipment
- Read and interpret drawings, plans, and specifications to determine layout, and identify types and quantities of materials required
- Connect pipe sections using appropriate valves, couplings, reducers, and other fittings by threading, bolting, soldering, and other established joining procedures
- Conduct inspection of plumbing facilities and ensure compliance with proper safety procedures
- Perform maintenance on marine propulsion boilers
- Review and maintain records and reports pertinent to boiler repair
- Demonstrate knowledge of repair procedures, repair and maintenance standards, quality assurance specifications, and proper use of maintenance industrial machinery

Services

Army

Navy

POSTAL ADMINISTRATIVE SPECIALISTS

Summary: Postal administrative specialists perform all duties necessary to the efficient running of military postal operations. They perform postal finance, mail handling, locator service, and mail distribution functions. Mail handling duties can include, but are not limited to, accepting, sorting, manifesting, and dispatching all types of mail. They may also issue and cash U.S. Postal Service (USPS) money orders, sell stamps, apply postage to, and mail out parcels.

Helpful Attributes

- Ability to organize and plan
- Interest in keeping organized and accurate records
- Interest in operating computers and other office machines
- Preference for office work

What They Do

- Advise customers on proper postal claims procedures for loss and damaged articles
- Conduct fleet audits and postal inspections
- Prepare mail routing instructions
- Prepare mail/PAX/Cargo and mail on hand reports
- Prepare and release postal incident reporting messages, including Postal Net Alert (PNA)/postal offense and monthly updates
- Prepare mail transportation forms, including international mail and customs requirements
- Review and issue receipt for capital equipment
- Operate postage meter equipment and maintain records
- Prepare financial reports for correctness prior to submission
- Provide directory service on all undeliverable mail
- Train department mail orderlies on safeguarding mail

- Prepare and maintain standard operating procedures for ships' postal operations
- Ensure post office security is compliant with Department of Defense regulations
- Manage accountable mail processes and procedures

Services

Marine Corps

Navy

Air Force

POWER PLANT ELECTRICIANS

Summary: Power plant electricians in the Military maintain and repair electricity-generating equipment in mobile and stationary power plants. They may work in power plants on land or aboard ships and submarines. They provide maintenance for different types of power-generating equipment, including nuclear power plants.

Helpful Attributes

- Ability to use hand and power tools
- Interest in electricity
- Interest in working with machinery

What They Do

- Test circuits and components to isolate malfunctions and repair defects
- Perform scheduled and emergency inspections and repairs of distribution systems and equipment
- Supervise or perform maintenance, repair, calibration, and tests of internal substation equipment
- Operate and perform advanced organizational and/or intermediate maintenance to gas turbine main propulsion- and electric-generating plants control systems
- Perform intermediate-level maintenance on electrical components, including generators, constant speed drive units, control units, voltage regulators, inverters, converters, transformer rectifiers, and the generator test stand
- Perform on-equipment repair and intermediate maintenance on hydraulic test stands, hydraulic and electrical control circuits and purifiers, hydraulic power supplies, flow dividers, hydraulic pumps, and aircraft jacks

- Maintain, modify, and repair electrical power-generating and control systems, automatic transfer switches, aircraft arresting systems, and associated equipment
- Install, maintain, and repair energized and de-energized electrical distribution systems and components
- Inspect, maintain, and repair fixed and portable airfield lighting systems, including runway, threshold, approach, taxiway, visual glide slope, obstruction, and distance marker lights
- Install, maintain, and repair cathodic protection and grounding systems and voltage and current regulators
- Climb utility poles and operate special purpose vehicles and equipment, including line maintenance and high reach trucks to inspect, maintain, and repair overhead distribution systems
- Review layout drawings and wiring diagrams
- Observe and interpret instruments such as ammeters, voltmeters, frequency meters, synchroscopes, automatic temperature and pressure recorders, and engine oil, fuel, and coolant gauges
- Advise on problems installing and repairing electrical power distribution and special purpose electrical systems
- Perform operations and basic preventive maintenance of electronic equipment used for reactor control, rod control, protection and alarm system, primary plant instrumentation, nuclear instrumentation, primary plant control, steam generator water-level control, and other electrical and electronic support equipment

Services

Navy

Air Force

POWER PLANT OPERATORS

Summary: Powerhouse mechanics install, maintain, operate, and repair electrical and mechanical equipment in power-generating stations. They maintain various types of power-generating equipment on land, on ships, and on submarines. They also support the mobile utility equipment that the Military provides as portable, temporary sources of electricity.

Helpful Attributes

- Interest in nuclear power
- Interest in repairing machines and equipment
- Preference for doing physical work

What They Do

- Operate, install, repair, and perform organizational and intermediate maintenance on mechanical components of gas turbine engines, main propulsion machinery, assigned auxiliary equipment, and propulsion control systems
- Perform organizational-level maintenance on electronic automatic boiler controls systems, steam plant control systems (chameleon), boiler combustion monitoring systems, and boiler ignitor systems
- Perform mechanical, electrical, and instrumentation functions necessary to install and prepare power station equipment for initial start-up
- Perform on-equipment repair and intermediate maintenance on Mobile Electric Power Plants (MEPP)
- Troubleshoot and repair electric motors, generators, voltage regulator systems, over and under voltage systems, frequency control systems, fault indicator systems, and other control circuits
- Perform electrical assessments, facilities maintenance, and quality control of electrical distribution systems and facilities
- Analyze plant equipment and system operating characteristics to determine operational condition
- Repair or overhaul power-generating equipment and associated systems components
- Determine and isolate complex malfunctions, utilizing diagnostic tests and troubleshooting techniques
- Perform field- or sustainment-level maintenance on tactical utility, precise power-generation sets, internal combustion engines, and associated equipment
- Apply the laws and concepts of advanced mathematics, physics, thermal dynamics, rotational kinematics, kinetic energy, energy conservation, electrical/electronic engineering, and mechanical engineering to provide technical solutions to complex problems
- Operate and perform organizational-level maintenance on mechanical systems for surface ship nuclear propulsion plants and support equipment

- Perform maintenance of hydraulic power plants and hydraulic system components, emergency diesel engines, compressed gas systems such as air, oxygen, hydrogen, nitrogen, sanitation systems, seawater systems, freshwater systems, oxygen generation equipment, and atmosphere control equipment
- Inspect and assess material condition, monitor general readiness, provide on-board maintenance training, and diagnose improper operating procedures and equipment casualties/failures of marine main propulsion diesel engines, diesel generators, and component systems

Services

Navy

Air Force

POWERHOUSE MECHANICS

Summary: Powerhouse mechanics install, maintain, operate, and repair electrical and mechanical equipment in power-generating stations. They maintain various types of power-generating equipment on land, on ships, and on submarines. They also support the mobile utility equipment that the Military provides as portable, temporary sources of electricity.

Helpful Attributes

- Interest in nuclear power
- Interest in repairing machines and equipment
- Preference for doing physical work

What They Do

- Operate, install, repair, and perform organizational- and intermediate-level maintenance on mechanical components of gas turbine engines, main propulsion machinery, assigned auxiliary equipment, and propulsion control systems
- Perform organizational-level maintenance on electronic automatic boiler controls systems, steam plant control systems (chameleon), boiler combustion monitoring systems, and boiler ignitor systems
- Perform mechanical, electrical, and instrumentation functions necessary to install and prepare power station equipment for initial start-up

- Perform on-equipment repair and intermediate maintenance on Mobile Electric Power Plants (MEPP)
- Troubleshoot and repair electric motors, generators, voltage regulator systems, over and under voltage systems, frequency control systems, fault indicator systems, and other control circuits
- Perform electrical assessments, facilities maintenance, and quality control of electrical distribution systems and facilities
- Analyze plant equipment and system operating characteristics to determine operational condition
- Repair or overhaul power-generating equipment and associated systems components
- Determine and isolate complex malfunctions, utilizing diagnostic tests and troubleshooting techniques
- Perform field- or sustainment-level maintenance on tactical utility, precise power-generation sets, internal combustion engines, and associated equipment
- Apply the laws and concepts of advanced mathematics, physics, thermal dynamics, rotational kinematics, kinetic energy, energy conservation, electrical/electronic engineering, and mechanical engineering to provide technical solutions to complex problems
- Operate and perform organizational-level maintenance on mechanical systems for surface ship nuclear propulsion plants and support equipment
- Perform maintenance of hydraulic power plants and hydraulic system components, emergency diesel engines, compressed gas systems such as air, oxygen, hydrogen, nitrogen, sanitation systems, seawater systems, freshwater systems, oxygen generation equipment, and atmosphere control equipment
- Inspect and assess material condition, monitor general readiness, provide on-board maintenance training, and diagnose improper operating procedures and equipment casualties/failures of marine main propulsion diesel engines, diesel generators, and component systems

Services

Army

Navy

POWERHOUSE MECHANICS

Summary: Powerhouse mechanics install, maintain, operate, and repair electrical and mechanical equipment in power-generating stations. They maintain various types of power-generating equipment on land, on ships, and on submarines. They also support the mobile utility equipment that the Military provides as portable, temporary sources of electricity.

Helpful Attributes

- Interest in nuclear power
- Interest in repairing machines and equipment
- Preference for doing physical work

What They Do

- Operate, install, repair, and perform organizational- and intermediate-level maintenance on mechanical components of gas turbine engines, main propulsion machinery, assigned auxiliary equipment, and propulsion control systems
- Perform organizational-level maintenance on electronic automatic boiler controls systems, steam plant control systems (chameleon), boiler combustion monitoring systems, and boiler ignitor systems
- Perform mechanical, electrical, and instrumentation functions necessary to install and prepare power station equipment for initial start-up
- Perform on-equipment repair and intermediate maintenance on Mobile Electric Power Plants (MEPP)
- Troubleshoot and repair electric motors, generators, voltage regulator systems, over and under voltage systems, frequency control systems, fault indicator systems, and other control circuits
- Perform electrical assessments, facilities maintenance, and quality control of electrical distribution systems and facilities
- Analyze plant equipment and system operating characteristics to determine operational condition
- Repair or overhaul power-generating equipment and associated systems components
- Determine and isolate complex malfunctions, utilizing diagnostic tests and troubleshooting techniques
- Perform field- or sustainment-level maintenance on tactical utility, precise power-generation sets, internal combustion engines, and associated equipment
- Apply the laws and concepts of advanced mathematics, physics, thermal dynamics, rotational kinematics, kinetic energy, energy conservation, electrical/electronic engineering, and mechanical engineering to provide technical solutions to complex problems
- Operate and perform organizational-level maintenance on mechanical systems for surface ship nuclear propulsion plants and support equipment
- Perform maintenance of hydraulic power plants and hydraulic system components, emergency diesel engines, compressed gas systems such as air, oxygen, hydrogen, nitrogen, sanitation systems, seawater systems, freshwater systems, oxygen generation equipment, and atmosphere control equipment
- Inspect and assess material condition, monitor general readiness, provide on-board maintenance training, and diagnose improper operating procedures and equipment casualties/failures of marine main propulsion diesel engines, diesel generators, and component systems

Services

Army

Navy

PRECISION INSTRUMENT AND EQUIPMENT REPAIRERS

Summary: Precision instrument and equipment repairers calibrate, maintain, and adjust instrumentation that is used for precise functions. These instruments include measurement, hazard detection, communication, laser, testing, diagnostic, and laboratory equipment. These individuals must have excellent fine motor skills and attention to detail. They often specialize by the type of equipment they repair.

Helpful Attributes

- Ability to solve mechanical problems
- Ability to use repair tools
- Interest in electronics, communications equipment, and digital services

What They Do

- Use Test, Measurement, and Diagnostic Equipment (TMDE), Test Program Sets (TPS), and

Interactive Electronic Technical Manuals (IETM) to determine the cause and location of malfunctions, extent of faults, and category of maintenance required

- Supervise the process and use of TMDE to perform voltage, current, power, impedance, frequency, microwave, temperature, physical-dimensional, and optical measurements
- Remove and replace microminiature components such as surface mounted, highly static sensitive and multi-lead devices; utilize the microminiature component repair station, the appropriate electronic maintenance kit, and automated test equipment
- Disassemble, repair, clean, and reassemble digital encoders
- Repair flexible conductors, multilayer conductors and laminates, illuminated panels, and remove and replace welded lead components and surface mounted devices
- Repair locks using various hand tools, including screwdrivers, cold chisels, and hammers; repair or replace tumblers, springs, and other parts
- Perform duties involving physical, mechanical, plane, and angular measurements and calibration, including flow and temperature measurement and calibration
- Perform corrective maintenance on circuit cards and electronic modules by interpreting schematics, troubleshooting, and repairing the faulty cards and modules
- Perform and manage repair, calibration, and modification of TMDE, including precision measurement equipment laboratory (PMEL) standards and automatic test equipment
- Perform mechanical, mechanical-optical, and electrical repairs to include inspecting, troubleshooting, and adjusting fire control instruments and systems
- Inspect, maintain, and overhaul all musical equipment assigned to a band
- Maintain and calibrate combat critical systems such as fire control devices, Global Positioning System (GPS) receivers, night vision devices/equipment, laser and fiber optic systems, mine detection and dispensing systems, battlefield illumination devices, electronic azimuth determining devices, and nuclear, biological, and chemical (NBC) warning and measuring devices

Services

Army

Marine Corps

Navy

Air Force

PREVENTIVE MAINTENANCE ANALYSTS

Summary: Preventive maintenance analysts develop schedules to ensure the Military's fleet and equipment is regularly inspected and maintained. They plan and control work methods, maintenance, and production schedules, operating procedures, and performance standards. They establish priorities and allocate resources to support mission requirements.

Helpful Attributes

- Ability to use mathematical formulas
- Interest in working with computers
- Interest in working with numbers and statistics
- Preference for work requiring attention to detail
- Interest in mechanical operation

What They Do

- Perform managerial functions in the areas of operational commitments, aircraft flight schedules, workload requirements, component induction, bench status, and personnel assets
- Perform inspections and monitor maintenance quality control
- Manage activities related to inspection, diagnostics, repair, modification, refinishing, and data collection for the vehicle and equipment fleet
- Analyze maintenance reports, past and current performance, and inspection reports to ensure cost-effective operations, timely preventive maintenance, repairs, and rebuilding of vehicular equipment
- Schedule maintenance jobs, help establish work priorities, and monitor completion
- Oversee the daily implementation and scheduling of required maintenance items
- Investigate proposed work sites to determine resource requirements
- Prepare cost estimates for in-service work requirements

- Solve maintenance problems by studying layout drawings, studying wiring and schematic drawings, and analyzing construction and operating characteristics
- Analyze deficiencies in areas such as equipment performance, materiel consumption, scheduling, management, and resources, their impact on the maintenance mission, and results of corrective actions
- Identify problems and recommend and apply corrective actions related to maintenance information systems (MIS) operation and maintenance
- Plan, schedule, and organize use and maintenance of aerospace vehicles, engines, munitions, missiles, space systems, aerospace ground equipment (AGE), and associated support systems
- Ensure vehicles, equipment, tools, parts, and manpower are available to support mission requirements
- Ensure maintenance and supply documentation is complete and accurate
- Maintain records and publications/files, and prepare reports, logs, directives, and correspondence within other maintenance and repair activities

Services

Marine Corps

Navy

Air Force

PROTECTIVE SERVICES SPECIALISTS

Summary: In the Military, protective services specialists perform, plan, coordinate, and execute protective service missions for personnel in high-risk and high-profile positions who are potential targets of terrorism. They are responsible for protective service tactics, anti-ambush operations, counter surveillance operations, evasive driving techniques, and physical security.

Helpful Attributes

- Ability to remain calm under pressure
- Interest in law enforcement and crime prevention
- Listening and communication skills

What They Do

- Provide physical security for assigned individuals and/or establishments during field exercises and in combat

- Conduct risk/threat validity assessments
- Monitor early-warning detection systems, security alarms, and cameras to identify and address security risks
- Transport, escort, and monitor assigned subjects and locations
- Serve as first responder in emergency situations and provide first aid
- Patrol and secure designated perimeters
- Execute traffic and parking enforcement
- Conduct security and threat audits, and make recommendations for improvement of security protocol
- Ensure safe and effective movement and monitoring in a wide variety of environments
- Conduct surveys and assessments, and prepare evaluation reports
- Coordinate operational and logistical details with counterparts at other law enforcement agencies
- Document safety and security related incidents

Services

Navy

PSYCHIATRIC ASSISTANTS

Summary: Medical assistants specializing in the field of psychiatry assist psychiatrists or other health care professionals with diagnosing, preventing, and treating mental disorders. Psychiatric assistants may work in outpatient or inpatient settings, and they are available to help service members or their families suffering from combat-related disorders. They interview and observe patients and provide them with counseling. Oftentimes, these professionals represent the front lines of treatment for service members suffering with psychological conditions.

Helpful Attributes

- Ability to follow directions precisely
- Ability to work under stressful or emergency conditions
- Interest in helping others
- Social perceptiveness
- Listening and communication skills

What They Do

- Assist with the management and treatment of in/outpatient mental health activities, during peacetime or mobilization, under the supervision of a psychiatrist, social worker, psychiatric nurse, or psychologist

- Collect and record psychosocial and physical data, under supervision
- Counsel and treat clients/patients with personal, behavioral, or mental health problems, under supervision
- Assist with care and treatment of psychiatric, and drug and alcohol patients, under close supervision
- Assist professional staff in the supervision of patient treatment programs, personnel matters, supply economy procedures, and fiscal, technical, and administrative matters
- Assist in the care and treatment of neuropsychiatric patients
- Aid medical officers in the administration of special neuropsychiatric therapy procedures
- Care for patients in accordance with nursing methods
- Observe and report symptoms and psychotic manifestation of mental patients
- Maintain neuropsychiatric ward and clinic equipment

Services

Army

Navy

PUBLIC AFFAIRS SPECIALISTS AND BROADCAST JOURNALISTS

Summary: Public affairs specialists and broadcast journalists publish communication products through radio, television, video, and web content for worldwide distribution. They are responsible for advising leadership, public affairs planning, developing and executing information strategies, and community relations. Some are trained to operate audio and video equipment in both tactical and non-tactical environments.

Helpful Attributes

- Ability to keep detailed and accurate records
- Ability to write clearly and concisely
- Interest in researching facts and issues for news stories
- Strong, clear speaking voice

What They Do

- Research, prepare, and disseminate news releases, articles, web-based material, and photographs of military personnel and activities
- Conduct media awareness training for personnel and facilities
- Advise on accuracy, propriety, timing, and relative importance of information for release to the public, and recommend methods of communicating information
- Perform as writer, reporter, editor, videographer, producer, or program host in radio and television productions
- Compose news releases and feature articles, shoot still and video imagery, serve as spokesperson, and maintain websites to raise public awareness of important issues and news stories
- Determine equipment requirements, research subject matter, and establish outlines
- Perform operational equipment checks and preventative maintenance
- Determine camera angle, lighting, and special effects, and operate camera, lighting, microphones, and related equipment
- Maintain liaison with media by receiving queries for news media, obtaining information, coordinating answers, and giving responses to news media
- Engage in digital and print photojournalism, to include photographing, writing, editing, and managing content for websites and other social media platforms, periodicals, guides, pamphlets, and fact sheets

Services

Army

Marine Corps

Navy

Air Force

Coast Guard

QUARTERMASTERS AD BOAT OPERATORS

Summary: Quartermasters and boat operators focus on small boat safety, handling, operation, navigation, communications, and maintenance. They supervise the embarking and disembarking of troops from the vessel, establish and enforce safety procedures, oversee deck maintenance, and supervise maintenance of life saving equipment.

Helpful Attributes

- Ability to follow detailed instructions and read maps
- Ability to work with mathematical formulas
- Interest in sailing and navigation

What They Do

- Provide supervision and technical guidance for subordinates
- Employ the onboard weapon system
- Process operations and intelligence information
- Conduct crew drills and supervise training
- Administer the vessel mess functions to include all money exchanges, headcount records, daily cook worksheets, and food utilization reports
- Maintain vessel charts, publications, orders, and vessel logbook
- Conduct Riverine operations, including patrolling, observation/lessening posts, ambushes, fire support, and vessel search and seizure

Services

Army

Marine Corps

Navy

RADAR AND SONAR OPERATORS

Summary: Radar and sonar operators set up, operate, and perform preventive maintenance on sophisticated radar and sonar equipment. They use radar and sonar to detect, track, recognize, analyze, and identify objects. The radar and sonar systems are also used to direct artillery fires and forecast the weather.

Helpful Attributes

- Ability to concentrate for long periods
- Ability to work under stress
- Interest in advanced communications and electronic equipment
- Detail-oriented

What They Do

- Operate radar systems, interior communications circuits, fathometers, and related equipment
- Operate submarine sonar, oceanographic equipment, and submarine auxiliary sonar
- Detect, analyze, classify, localize, and report acoustic transmissions using sophisticated sensors
- Create realistic acoustic models for optimized sensor performance
- Review, analyze, and summarize data collected from radar and sonar systems resulting in usable intelligence
- Operate surface ship underwater fire control systems
- Operate firefinder radar and assist in engagement planning; conduct reconnaissance of general position areas for relocation of weapons locating radar
- Recognize and respond to environmental and tactical changes, and provide continuous tactical advice
- Construct field fortifications and camouflage to protect the equipment position
- Prepare radar equipment, power generator, and associated equipment for movement and operation
- Recognize and interpret systems alerts and degraded/casualty indications
- Perform system reconfiguration, clear faults, or correct faults by removal and replacement of failed modules and components
- Maintain system documentation

Services

Army

Marine Corps

Navy

Air Force

RADAR AND SONAR SYSTEM REPAIRERS

Summary: Radar and sonar system repairers troubleshoot and repair radar and associated equipment assemblies, subassemblies, or modular and circuit elements, using test equipment to isolate and fix malfunctions. Radar and sonar system repairers in the Military monitor sophisticated equipment, typically in security-controlled areas, to ensure that the equipment utilized by service members is in perfect working order.

Helpful Attributes

- Ability to use tools
- Interest in solving problems
- Ability to apply electronic principles and concepts
- Interest working with electrical, electronic, and electrochemical equipment

What They Do

- Troubleshoot radar and associated equipment assemblies, subassemblies, modular and circuit elements with common and system peculiar test equipment, to identify deficiencies and malfunctions
- Test repaired system to ensure compliance with technical specifications
- Maintain, troubleshoot, and repair electronic equipment used for detection, tracking, recognition, and identification of maritime vessels
- Operate and maintain radar systems, interior communications circuits, Voyage Management Systems (VMS), alarm systems, ship's control stations, entertainment systems, gyrocompasses, Central Atmosphere Monitoring Systems (CAMS), liquid-level detection and tank-level indicator circuits, flow meters, pressure and temperature sensing circuits, valve position indicators, hovering systems, and missile compensation systems
- Maintain shipboard and airfield radar systems, radar systems switchboards, Identification Friend or Foe (IFF) equipment, meteorological equipment, navigation equipment, associated cabling, cooling water systems, and dry air systems
- Operate radar and sonar sub-systems and equipment in all modes and perform all preventive and corrective maintenance on the sonar equipment including advanced corrective maintenance
- Perform preventive and corrective maintenance at the organizational level on radar using ordnance publications, circuit diagrams, and other appropriate documentation
- Perform casualty analysis and fault isolation, and operate, test, align, and repair individual equipment, subsystems, and interfaces

- Operate the consoles and associated equipment, as applicable, in a tactical situation and during test and evaluation
- Install, maintain, and repair fixed or mobile air traffic control, weather, ground aircraft control, and warning radar systems, related radar operator training devices, aircraft identification equipment, remoting systems, video mappers, computerized processors, and communications subsystems
- Plan, organize, and schedule work assignments, workloads, and maintenance procedures for ground radar activities
- Evaluate and resolve problems encountered during siting, installing, repairing, and overhauling ground radar systems
- Install ground radar systems; assemble, connect, modify, and adjust ground radar subassemblies, such as antennas, transmitters, receivers, processors, indicator groups, and ancillary systems such as beacon equipment and video mappers
- Establish requirements for tools, support equipment, personnel, supplies, and technical documents
- Identify maintenance problem areas and initiate corrective action; develop methods for improving maintenance effectiveness and efficiency

Services

Army

Marine Corps

Navy

Air Force

RECRUITERS

Summary: Recruiters perform all tasks related to the application process. They participate in promotional events, interviews, and consultations related to potential enlistments. They provide applicants with all the necessary information related to enlisting and they prepare enlistment reports. They maintain statistics on recruiting programs and maintain administrative records. They serve as liaison with high schools, colleges, and industry officials, and they participate in community activities, such as state ceremonies and fundraising drives.

Helpful Attributes

- Ability to speak before groups
- Ability to work independently

- Interest in working with youths
- Listening and communication skills

What They Do

- Recruit, interview, counsel, and qualify applicants for enlistment
- Explain benefits; prepare enlistment forms and documents
- Assist in market research and analysis, and make appropriate recommendations to the chain of command
- Maintain statistics on recruiting
- Plan, organize, and coordinate recruiting activities
- Develop and direct training programs to assist subordinates
- Establish liaison with educational authorities and other civilian agencies
- Screen each applicant to determine eligibility relative to physical defects, moral character, criminal involvement, age, drug abuse, satisfactory prior service, citizenship, education, and dependency
- Schedule working applicants to take the Armed Services Vocational Aptitude Battery test at the Military Entrance Processing Station (MEPS)
- Arrange for physical examinations of mentally qualified applicants at the MEPS
- Arrange for publication and broadcasting of recruiting programs and provide publicity material; identify and cultivate community centers of influence
- Develop marketing information sources, such as employment agencies, driver's license and job advertisement lists, high school and college student lists, and separation reports
- Develop publicity programs; plan, direct, and evaluate sales promotional projects using media, such as direct mail, press, radio, and television presentations
- Develop community relations programs; plan, organize, and provide support for recruiter special events, such as state and municipal ceremonies, exhibits, fairs, parades, centennials, and sporting events
- Develop and maintain market data and allocate recruiting goals

Services

Army

Marine Corps

Navy

Air Force

RECRUITING AND RETENTION SPECIALISTS

Summary: Recruiting and retention specialists provide prospective candidates with all the appropriate information and guidance prior to and throughout the military enlistment process. They are responsible for interviewing, orienting, and screening potential candidates, and they participate in recruiting actions by attending promotional events and maintaining liaison with high schools and college officials. They plan and implement interviews with military personnel, and they provide unit leaders with feedback and guidance for improving and enhancing retention. They maintain personnel records and prepare enlistment packets.

Helpful Attributes

- Ability to speak before groups
- Ability to work independently
- Interest in working with youths

What They Do

- Provide civilians with information about future military careers and potential career opportunities
- Inform high school students and school counselors regarding the Armed Services Vocational Aptitude Battery (ASVAB) and explain its role in the selection process
- Attend job fairs and provide potential applicants with appropriate information regarding career opportunities in the Military
- Maintain records in Recruiter Zone
- Plan and implement interviews for enlistments and retention of current enlistments, as well as re-enlistments
- Provide consultation regarding all enlistment requirements, including enlistment contracts, applicant test results, incentives, and military benefits
- Prepare Prior-Service (PS) and Non-Prior Service (NPS) enlistment packets
- Conduct Military Entrance Processing Station (MEPS) pre-enlistment screening briefings
- Provide training and consultation to first line leaders (FLL) regarding available programs and career development options

- Monitor the effectiveness of programs and policies by maintaining regular contact and conducting interviews with personnel, families, and employers
- Consult with unit leaders and provide suggestions for improvement based on personnel input regarding current processes and policies
- Support retention programs and provide assistance to the Recruiting and Retention Manager (RRM) by ensuring successful implementation of retention interviews and exit surveys
- Perform sales promotional projects utilizing various media resources

Services

Army

Navy

Air Force

RELIGIOUS PROGRAM SPECIALISTS

Summary: Religious program specialists assist military chaplains in building a culture of spiritual care and facilitating the free exercise of religion for military service members and their families. As experts in religious diversity, they conduct worship services, provide pastoral counseling, and offer crisis intervention. They also advise leaders at all levels on religious accommodation, ethical, and moral issues.

Helpful Attributes

- Ability to express ideas clearly
- Interest in administrative work
- Interest in religious guidance
- Knowledge of various religious customs and beliefs
- Sensitivity to the needs of others

What They Do

- Provide and manage support of religious observances to include worship services for all faiths, liturgies, rites, ceremonies, and memorial services
- Identify and support battle fatigued soldiers and their families; perform emergency religious ministrations on the battlefield and grief support sessions
- Provide counseling, safeguard privileged communications and offerings, screen and refer prospective counselees
- Support the combat stress team, conduct critical incident stress debriefing, and train family support group leaders on family ministry issues; assist soldiers in developing moral values and resolving conflict
- Conduct staff assistance visits at all levels of assigned command
- Execute crisis intervention counseling to include suicide prevention and intervention, and traumatic stress response
- Offer religious support to hospitals and mortuaries
- Manage, develop, and implement ministry needs assessment and identify available resources to meet spiritual, religious, ethical, and moral needs
- Recruit, train, and organize volunteers for specific religious ministries
- Prepare and present religious customs and culture briefings
- Partner with chaplains in responding to emergency and stressful situations, including air crashes, mass casualty sites, hostage situations, casualty collection points, evacuation and deployment processing points, and work centers
- Manage financial resources

Services

Army

Navy

Air Force

RESPIRATORY TECHNICIANS

Summary: Respiratory technicians focus on testing and treatment related to lung functions. They administer respiratory care using controlled breathing equipment. They perform pulmonary function testing, and breathing therapies such as medical gas therapy, nebulization therapy, and mechanical ventilation. They also assist physicians with intubation and extubation of patients.

Helpful Attributes

- Ability to follow strict standards and procedures
- Ability to keep accurate records
- Interest in electronic equipment

- Interest in learning how the heart, lungs, and blood work together
- Concern for others

What They Do

- Assist with the management of a respiratory unit
- Administer respiratory therapy and perform pulmonary function tests under the supervision of a physician or nurse anesthetist
- Gather and check equipment, take history, and assess the patient
- Document results on treatment, administer drugs as ordered, and troubleshoot any problems that occur with patient during treatment
- Perform pulmonary functions, additional assessments as required, and transport patients needing ventilator support by air and/or land
- Perform cardiopulmonary resuscitation on both adults and infants
- Provide mechanical guidance, management, and training to junior personnel
- Perform and assist in advanced medical procedures, such as thoracentesis, bronchoscopy, and tracheal intubation
- Operate sterilizing equipment and patient transport apparatus
- Prepare budget, train other medical care personnel on respiratory support, and assist in research on practical teams
- Monitor departmental procedures to assure compliance with government regulations
- Supervise home care programs
- Assist medical officers in treating patients using a variety of controlled breathing apparatus that employ medical gases
- Perform medical gas therapy, nebulization therapy, mechanical ventilation, blood gas analysis, and pulmonary function testing
- Perform minor repair and preventive maintenance on tracheostomy and endotracheal apparatus, ventilators, blood gas equipment, and electronic monitoring and resuscitation equipment

Services

Army

Navy

SALES AND STOCK SPECIALISTS

Summary: Military sales and stock specialists work to manage the operation of military retail stores to keep stock of merchandise, record transactions and deposits, and provide military personnel with a friendly, convenient option that allows them to get their necessities on base or onboard ships without difficulty. Other duties include pricing of merchandise, accounting and bookkeeping, and making deposits of cash.

Helpful Attributes

- Ability to use cash registers and calculators
- Interest in working with people

What They Do

- Use a variety of modern e-commerce and civilian automated systems and tools, including computer-based cash registers, laser scanning, cashless debit, and electronic data interchange (EDI) systems
- Understand best business practices of perpetual inventories and cost-based accounting
- Maintain financial and material accountability in a variety of operational environments through production and analysis of management reports produced by the Resale Operations Management (ROM) II system
- Provide diverse logistics and accounting support in a global setting to aviation, surface, subsurface, and expeditionary forces
- Order, receive, inspect, stow, preserve, package, ship, and issue materials and cargo
- Prepare and maintain required forms, records, correspondence, reports, and files
- Provide direct personal services by operating and managing retail and service activities to include: ship stores, vending and cash collection machines, shipboard barbershops, and laundry operations
- Perform administrative and automated stock control functions for all activities operated
- Provide and account for the constant stream of supplies, clothing, commissary items, and spare parts necessary
- Procure, store, preserve, and package supplies, spare parts, provisions, technical items, and all other necessary supplies and services

- Maintain inventories, prepare requisitions, and check incoming supplies
- Handle all logistical functions and maintain the accounting system, preparing financial accounts and reports
- Utilize all types of office equipment and use computers extensively
- Operate all types of material handling equipment, including forklifts

Services

Marine Corps

Navy

Coast Guard

SEAMAN

Summary: Seamen are responsible for the smooth operation of the majority of occurrences above deck. They are capable of performing almost any task in connection with deck maintenance, small boat operations, navigation, and supervising all personnel assigned to a ship's deck force or shore unit. They have a working knowledge of all programs performed in, or related to, the marine environment. They must establish and maintain knowledge of ropes and cables, including different uses, stresses, strains, and proper stowing. They facilitate upkeep of ships external structures, riggings, deck equipment, and lifeboats.

Helpful Attributes

- Ability to work closely with others
- Interest in sailing and being at sea
- Preference for physical work

What They Do

- Execute deck watches, such as helmsman, lookout, and messenger
- Maintain ship's compartments, decks, deck machinery, equipment, external structure, lines, and riggings
- Serve as petty officer-in-charge of picket boats, self-propelled barges, tugs, and other yard and district crafts
- Perform intermediate-level repair of rigging equipment, provide general rigging and manufacturing

services, and the weight testing of equipment, fixtures, ladders, handrails, and elevators
- Run small boats while conducting search and rescue aid to law enforcement and security efforts
- Train, direct, and supervise personnel in military duties and all activities relating to seamanship, in painting and upkeep of ship's structure, deck equipment, boat seamanship, and lifeboats
- Operate hoists, cranes, and winches to load cargo or set gangplanks; stand watch for security, navigation, or communications; and have a general knowledge of ropes and cables, including different uses, stresses, strains, and proper stowing
- Perform and supervise rigging, streaming, and recovery of all minesweeping gear

Services

Navy

Coast Guard

SECURITY SPECIALISTS

Summary: Security specialists in the Military protect and defend. They conduct risk/vulnerability assessments, analyze crime, and recommend appropriate courses of action to eliminate conditions conducive to terrorism, espionage, sabotage, wrongful destruction, malicious damage, theft, and pilferage. Law enforcement specialists that focus on security require a breadth of knowledge in weapon systems, antiterrorism, law enforcement, defense, and combat arms.

Helpful Attributes

- Ability to remain calm under pressure
- Interest in law enforcement and crime prevention
- Listening and communication skills

What They Do

- Plan, evaluate, and supervise the implementation of site-specific security plans to protect assets designated as vital to national security
- Restore and/or maintain security to prevent access, damage, or removal of vital assets
- Demonstrate skill in land navigation and patrolling
- Execute interior guard procedures, and be familiar with antiterrorist tactics and techniques

- Conduct offensive infantry tactics in confined spaces, ashore, and/or in the air, to restore security when breached
- Provide armed internal security to designated United States diplomatic and consular facilities to prevent the compromise of classified information and equipment integral to national security
- Perform law enforcement techniques, small arms handling and employment, and emergency first aid
- Conduct inspections, preliminary investigations, and vulnerability assessments
- Control entry and exit of military and civilian personnel, vehicles, and other equipment at military installation access points, on United States diplomatic missions, or at other designated establishments
- Execute convoy escort and defense operations

Services

Marine Corps

Navy

SHIP ELECTRICIANS

Summary: Ship electricians operate and repair electrical systems on ships. They keep electrical power plants, wiring, and machinery in working order, maintain and repair shipboard elevator systems, and interpret electrical sketches, diagrams, and blueprints.

Helpful Attributes

- Ability to use tools
- Interest in electricity and how electrical devices operate
- Interest in solving problems
- Interest in electronics and communication devices

What They Do

- Test, inspect, and perform organizational-level maintenance on shipboard weapons and cargo handling elevators on a system and component level
- Troubleshoot and repair electronic/electrical systems and equipment, such as controllers, sensors, switches, and the electrical components of hydraulic/mechanical interfaces
- Operate and perform preventive and corrective maintenance and all authorized fault isolation and repair procedures on engineering control system equipment (ESCE)
- Perform operational, preventive, and corrective maintenance on unique electrical components, 400HZ static frequency converters, and the SSM Degaussing System
- Perform intermediate-level electrical maintenance procedures on various shipboard equipment and systems in support of fleet-directed requirements
- Perform intermediate-level maintenance procedures, inspecting, testing, and servicing motors/generators and correcting vibration on rotating machinery in support of fleet-directed requirements
- Perform required electrical maintenance on the low-pressure electrolyzer onboard submarines
- Perform organizational-level preventive and corrective maintenance, system alignment, troubleshooting, and fault isolation to the lowest replaceable unit (LRU) on the SPCS
- Perform organizational-level preventive and corrective maintenance, troubleshooting, and fault isolation procedures on main propulsion and auxiliary control consoles
- Perform maintenance on catapult and arresting gear systems, including synchros, repeaters, 110/220/440V electrical systems, contacts, relays, micro switches, series and parallel switches, pressure and temperature sensing switches, torque and gear limit switches, chronograph, and control console electrical functions
- Operate and perform maintenance on power and lighting circuits, electrical fixtures, motors, generators, voltage and frequency regulators, controllers, distribution switchboards, and other electrical equipment; test for short circuits and grounds; and rebuild electrical equipment, including solid state circuitry elements

Services

Navy

Coast Guard

SIGNALS INTELLIGENCE SPECIALISTS

Summary: Signals intelligence specialists oversee the collection and exploitation of electromagnetic signals, including communication and non-communication signals. They operate sophisticated equipment to gather, sort, and scan intercepted foreign communications and non-communications. They identify and process the intercepted signals and perform analysis to establish target identification and operational patterns. They then take this information and produce combat, strategic, and tactical intelligence reports.

Helpful Attributes

- Ability to organize information
- Ability to think and write clearly
- Interest in gathering information and studying its meaning
- Interest in computers and related technology

What They Do

- Supervise and perform analysis and reporting of intercepted foreign communications and non-communications at all echelons
- Create combat, strategic, and tactical intelligence reports
- Gather, sort, and scan intercepted messages to isolate valid intelligence
- Operate automated data processing (ADP) equipment for SIGINT collection, processing, and reporting
- Perform operator maintenance on surveillance systems, organic communications equipment, light wheeled vehicles, and power sources
- Assist in the emplacement, camouflage, and recovery of surveillance systems, and/or associated equipment
- Search the radio frequency (RF) spectrum to collect, identify, and record target communications and selected categories or classes of electro-optic or foreign instrumentation signals (FIS)
- Perform and supervise acquisition, collection, collection resource management, analysis, and exploitation of foreign communications and radar signals at all echelons
- Plan, coordinate, and execute Signals Intelligence (SIGINT) and Electronic Warfare (EW) operations from start to finish, while utilizing both program of record equipment, as well as commercially acquired equipment
- Prepare and issue reports to include translation summaries, intelligence reports, and technical reports as needed

- Perform signals analysis of Radio Detection and Ranging (RADAR) signals, other signals derived from worldwide Technical/Operational Electronic Intelligence (TECHELINT/OPELINT), (EW) collection, and other multiple intelligence production resources
- Acquire, process, identify, analyze, and report on electromagnetic emissions
- Utilize a wide range of complex analysis hardware and software to process signals, including receivers, demodulators, spectrum analyzers, and other associated computer equipment
- Use advanced computer software programs to manipulate and extract intelligence data from electromagnetic emissions
- Operate computer terminals for data entry, query, data restructuring, and signals development

Services

Army

Marine Corps

Navy

Air Force

Space Force

SONOGRAPHY TECHNICIANS

Summary: Sonography technicians must be knowledgeable about ultrasound physics. They operate medical radiology equipment while performing ultrasound examinations, which are commonly used to study a developing fetus, abdominal and pelvic organs, muscles and tendons, and the heart and blood vessels. Sonography technicians produce ultrasound diagnostic images while providing patient care for appropriate study and diagnosis. Sonography technicians balance patient interaction and technological performance while working cohesively with a health care team.

Helpful Attributes

- Ability to follow strict standards and procedures
- Interest in activities requiring accuracy and attention to detail
- Interest in helping others
- Interest in working in a medical environment
- Technical skills and eye–hand coordination

What They Do

- Maintain ultrasound equipment and sterilize the room in which the procedure is to take place
- Assess patient and determine ability to undergo requested procedure
- Utilize imaging equipment that non-invasively emits sound waves directed toward internal organs, blood vessels, tissues, and other structures
- Demonstrate knowledge of vascular and abdominal anatomy (topical and cross-sectional), including normal variant anatomy, abnormal anatomy, and obstetric anatomy, as well as abnormal anatomy transducer characteristics, differences, and uses
- Participate in imaging of specialty areas, such as breast, obstetric, neurologic, or vascular sonography
- Explain procedures to patients and answer process-related questions
- Assist with interventional or minimally invasive procedures for either diagnostic or therapeutic purposes, from ultrasound-guided surgeries to biopsies
- Ensure probe captures all angles and sections of anatomy that need to be assessed
- Evaluate images for technical quality
- Present images and preliminary findings to physician or health care team
- Maintain patient records with medical notes related to the ultrasound procedure

Services

Navy

SPACE OPERATIONS SPECIALISTS

Summary: Space operations specialists handle space warning and control systems, orbital mechanics, data analysis, and transmission. These specialists operate space systems to perform space control missions.

Helpful Attributes

- Ability to work with formulas to solve math problems
- Interest in operating electronic equipment and systems
- Interest in space exploration
- Interest in working as part of a team

What They Do

- Detect, identify, and maintain surveillance on low orbiting and deep space satellite vehicles, using optical radar sensors
- Track sea-launched and intercontinental ballistic missiles using a variety of system peculiar sensors
- Plan satellite contacts, resolve emergencies, and perform satellite commanding during launch, early orbit, daily operations, and end-of-life testing
- Perform launch and on-orbit operations for military communications and global positioning system satellites
- Provide on-orbit telemetry, tracking, and commanding for satellite systems
- Support spacelift and aeronautical testing to fulfill warfighting and commercial users
- Validate and enhance warfighter capabilities through testing and evaluation of space systems
- Replicate adversary's space capabilities to improve combat training and increase awareness of threats from space
- Oversee space surveillance, space lift, space warning, and satellite command and control
- Develop future plans for systems, facilities, and personnel

Services

Air Force

Space Force

SPECIAL FORCES

Summary: Special forces members implement unconventional operations by air, land, or sea during combat or peacetime as members of elite teams. These activities include offensive raids, demolitions, reconnaissance, search and rescue, and counterterrorism. In addition to their combat training, special forces members often have specialized training in swimming, diving, parachuting, survival, emergency medicine, and foreign languages.

Helpful Attributes

- Ability to remain calm in stressful situations
- Ability to work as a team member
- Readiness to accept a challenge

What They Do

- Conduct advanced reconnaissance operations and collect intelligence information
- Recruit, train, and equip friendly forces; plan, organize, train, advise, assist, and supervise indigenous and allied personnel on collection and processing of intelligence information
- Conduct raids and invasions on enemy territories; lay and detonate explosives for demolition targets
- Locate, identify, defuse, and dispose of ordnance
- Evaluate terrain, select weapons emplacement sites, and assign targets and areas of fire
- Process prisoners of war, write, and establish security plans, and perform security duties
- Maintain all classified documents in the operational area, and establish destruction and evacuation plans
- Perform, plan, lead, supervise, instruct, and evaluate pararescue activities
- Supervise and perform insertion, infiltration, exfiltration, and extraction functions
- Perform basic and advanced trauma casualty management and paramedic-level skills, basic emergent surgical procedures, treatment of basic traumatic dental emergencies, and other routine and emergency medical health care procedures
- Perform overt, low-visibility, or clandestine movement in friendly, hostile, denied, or sensitive land and water areas
- Conduct parachuting operations, such as day and night jumps, to include the use of oxygen and combat equipment
- Demolish underwater and land objects and obstacles using explosives
- Operate, maintain, and repair specially configured combatant craft
- Conduct Chemical, Biological, Radiological, and Nuclear Explosive (CBRNE) defense measures

Services

Army

Marine Corps

Navy

Air Force

SUBSTANCE ABUSE CASEWORKERS

Summary: Substance abuse caseworkers provide individual and group counseling, perform crisis intervention, and offer education and counseling related to substance abuse. They perform urinalysis screenings and conduct assessments and interviews. They develop treatment plans and provide assistance to leaders in addressing issues related to substance abuse. They prepare summary reports and maintain patient progress data, and they are responsible for implementing prevention, outreach, and educational programs related to alcohol and drug abuse for members of the Military and their families.

Helpful Attributes

- Interest in working with people
- Patience in dealing with problems that take time and effort to overcome
- Sensitivity to the needs of others
- Listening and communication skills

What They Do

- Perform intake, which includes gathering demographics and completing required admission forms
- Provide program orientation to new patients
- Conduct assessments, which includes clinical interviewing to gather biopsychosocial, mental health, and substance abuse histories
- Document assessment results, treatment plans, reports, progress notes, discharge summaries, and other patient care records
- Administer urinalysis screenings, ensuring proper control of specimen chain of custody
- Assist patients with nutritional needs, hygiene, and comfort measures
- Provide training and education on substance abuse prevention
- Integrate quality communication, foster partnership, ensure flexible and responsive treatment options, optimize force management and development, and empower continuous process improvement
- Evaluate program resource needs, conduct program evaluations, and coordinate appropriate support
- Collect, update, and brief statistical data

Services

Marine Corps

Navy

SURGICAL ASSISTANTS

Summary: Surgical assistants provide care, safety, and support to service members and their families before, during, and after surgery. They assist physicians and other health care providers by preparing the operating room, sterilizing the surgical instruments, and assisting with anesthesia. These professionals may need to assist with surgery for injuries resulting from combat or other disasters.

Helpful Attributes

- Ability to follow directions precisely
- Ability to work under stressful or emergency conditions
- Interest in helping others

What They Do

- Assist the nursing staff in preparing the patient and the operating room (OR) environment for surgery and for providing assistance to the medical staff during surgical procedures
- Operate the centralized material service (CMS), and prepare and maintain sterile medical supplies and special equipment for medical treatment facilities
- Assist in preparing patients and the operating room environment for surgery and provide assistance to the medical staff during surgical procedures, to include the creation and maintenance of sterile fields for surgical procedures; the draping of patients; the preparation, manipulation, and delivery of basic surgical instruments and equipment; and the accountability for all instruments, needles, sponges, and medications placed within the sterile field
- Perform various preoperative and postoperative procedures as directed
- Receive, clean, decontaminate, sterilize, store, and issue various medical supplies and equipment used during surgical procedures
- Monitor quality of sterilization techniques to ensure adherence to established standards
- Prepare and maintain various reports and files
- Assist with instruction, supervision, and evaluation of students and other personnel-assigned duties relating to surgery
- Assist the executive management team with developing, interpreting, and evaluating instructions, regulations, policies, and procedures
- Accomplish routine safety checks and preventive maintenance on fixed and movable medical equipment and fixtures
- Schedule and prepare patients, and set up instruments, supplies, and equipment for specialized procedures in the OR and specialty clinics
- Perform management and training functions within surgical services
- Analyze requirements and supervise requisition, storage, maintenance, and issue of equipment and supplies

Services

Army

Navy

Air Force

SURVEYING, MAPPING AND DRAFTING TECHNICIANS

Summary: Surveying, mapping, and drafting technicians help determine, describe, and record geographic areas or features. They conduct land surveys, take measurements, make maps, and prepare detailed plans and drawings for construction projects. They are also sometimes called upon to provide maps and surveys that locate military targets and help plot troop movements. These technicians play key roles in the field of geospatial information.

Helpful Attributes

- Ability to convert ideas into drawings
- Interest in maps and charts
- Interest in working with drafting equipment and computers
- Interest in working with surveying equipment

What They Do

- Record field data, prepare schematic sketches, and mark survey stations
- Perform astronomic observation, and measure azimuths and angles with angular measuring equipment
- Direct and perform civil engineering design, drafting, surveying, and contract surveillance to support military facility construction and maintenance programs

- Prepare computer aided design (CAD) drawings, building information modeling (BIM) solutions, construction contract specifications, and cost estimates
- Evaluate potential construction sites and perform field tests on soils, asphalt, and concrete
- Perform drafting duties; interpret rough engineering sketches to produce working drawings using CAD/BIM techniques; produce architectural, structural, civil, mechanical, and electrical drawings
- Conduct reconnaissance, site location, construction, and mapping surveys
- Utilize auto-levels, electronic total stations, resource and survey grade global positioning system (GPS) equipment and related instruments to complete surveys
- Produce installation maps using a geographic information systems (GIS) interface
- Manage and inspect construction and maintenance contracts; interpret plans, specifications, and other contract documents
- Prepare timekeeping records, construction schedules, and planning and estimating, quality control and progress reports

Services

Army

Marine Corps

Navy

Air Force

SURVIVAL EQUIPMENT SPECIALISTS

Summary: Survival equipment specialists in the Military inspect, fit, maintain, and repair survival equipment, such as parachutes, aircraft life support equipment, search and rescue equipment, and air–sea rescue equipment, along with survival kits, medical kits, flight clothing, protective wear, night vision equipment, aircrew oxygen systems, liquid oxygen converters, anti-exposure suits, and G-suits.

Helpful Attributes

- Ability to do work requiring accuracy and attention to detail
- Interest in working for the safety of others

What They Do

- Inventory, clean, receive, store, and issue all equipment used in airdrop operations
- Install, inspect, and test extraction and release systems
- Perform technical, routine, and in-storage rigger-type inspection on cargo, extraction, and personnel parachutes, as well as other airdrop equipment before, during, and after each use
- Use and maintain machines and tools for fabrication, modification, and repair to parachute and other airdrop equipment
- Inspect, maintain, and repair parachutes, seat pans, survival equipment, and flight and protective clothing and equipment
- Troubleshoot oxygen systems, repair and test oxygen regulators and liquid oxygen converters removed from aircraft
- Operate and repair sewing machines
- Plan, direct, organize, and evaluate AFE operational aspects, such as equipment accountability, personnel reliability, mobility readiness, and other activities necessary to meet operational readiness
- Inspect, service, maintain, troubleshoot, and repair: cargo aerial delivery systems, drag parachute systems, aircraft oxygen systems, helicopter emergency flotation systems, portable dewatering pumps, air–sea rescue kits, and special-purpose protective clothing
- Inspect, maintain, pack, and adjust aircrew flight equipment, such as flight helmets, oxygen masks, parachutes, flotation devices, survival kits, helmet-mounted devices, aircrew night vision and other ocular systems, anti-G garments, aircrew vision and respiratory protective equipment, chemical biological protective oxygen masks and coveralls, other types of aircrew flight equipment, and aircrew chemical defense systems

Services

Army

Marine Corps

Navy

Air Force

TACTICAL DATA SYSTEM REPAIRERS

Summary: Electrical instrument and equipment repairers in the tactical data system specialty may work in the field or aboard ships to troubleshoot and maintain these vital systems. Tactical data system repairers perform maintenance on tactical data systems, common hardware/software suites, and other computerized information processing systems. They are responsible for installation, configuration management, system administration, and maintenance.

Helpful Attributes

- Ability to use tools
- Interest in solving problems
- Ability to apply electronic principles and concepts
- Interest working with electrical, electronic, and electrochemical equipment

What They Do

- Inspect, maintain, and repair tactical remote sensor system (TRSS) equipment
- Provide technical assistance during the installation and operation of the TRSS
- Plan and perform operational management and maintenance on command and control, tactical data, and common hardware/software systems
- Perform and supervise installation, alignment, inspection, testing, maintenance, and repair of electronic assemblies and subassemblies, fiber optic cables, and system software of all Tactical Air Operations and Air Defense Systems
- Make periodic inspections and perform preventive maintenance
- Use proper safety procedures in systems maintenance and operation
- Diagnose and isolate malfunctions
- Remove and replace hardware LRUs (lowest replaceable unit), perform corrective measures of software, and verify that any malfunction has been corrected
- Assist in maintaining accountability of all parts of the system including software, spare parts, and in requisitioning supplies and spare parts
- Maintain necessary records of maintenance and compile data for reports
- Provide organizational- and/or intermediate-level maintenance on a system level for Shipboard Tactical Data systems
- Maintain diagnostic and operational tools, support, and test equipment

Services

Marine Corps

Navy

TRAINING SPECIALISTS

Summary: Training specialists plan, conduct, and administer programs that train and continually improve the knowledge, skills, and abilities of military personnel. Training programs may be conducted in a classroom, in a laboratory, in the field, or online. Some specialize by functional area, such as military-specific/tactical.

Helpful Attributes

- Ability to communicate effectively, in writing and speaking
- Interest in counseling and promoting human relations
- Interest in teaching

What They Do

- Conduct classroom, laboratory, and operational training; use lecture, demonstration and performance, guided discussion, case study, and time and circumstance instructional methodology
- Design and develop curriculum, functional structure, and procedures
- Determine adequacy of existing courses and programs
- Collect and analyze job performance data and conduct occupational analysis surveys
- Sequence objectives, select instructional design, media, and delivery mode, and identify resource needs
- Create materials to achieve objectives
- Maintain education and training data, and provide statistical reports on programs and operations; monitor progress, identify problem areas, determine causes, recommend corrective action, and provide counsel
- Organize and control facilities, supplies, and equipment to support education and training needs
- Forecast education and training requirements, determine validity, and assess costs
- Participate in utilization and training workshops, training planning teams, and training planning groups

Services

Marine Corps

Navy

Air Force

TRANSPORTATION SPECIALISTS

Summary: Transportation specialists coordinate, monitor, control, and supervise the movement of personnel, equipment, and cargo by air, land, and sea. Along with managing technical and logistical support for all facets of transportation operations, they initiate, research, and propose necessary changes to the traffic management system for cost effectiveness and to meet mission requirements.

Helpful Attributes

- Interest in arranging travel schedules
- Interest in serving people
- Interest in using computers

What They Do

- Coordinate, monitor, control, and supervise the movement of personnel, equipment, and cargo by air, rail, highway, and water
- Determine the most efficient mode of transport that accomplishes mission requirements
- Research, interpret, prepare, and coordinate actions pertaining to travel entitlements
- Identify and report problem areas within the traffic management system to prevent additional costs, losses, and damage
- Establish budget estimates for materials, equipment, and transportation services
- Evaluate appropriate transport capability and prepare movement schedules for all modes of transportation
- Ensure transport capability is appropriate, cost effective, and meets mission requirements
- Maintain liaison with air, rail, highway, and water transportation facilities
- Prepare, consolidate, and review technical, personnel, and administrative reports and forms covering transportation matters (e.g., unit movement, personal property, passenger travel, freight/cargo, and material movement reports)
- Complete network capability and infrastructure assessments

- Coordinate movements and ensure compliance with operating timetables, rule book, and any other specific instructions

Services

Army

Marine Corps

Navy

Air Force

UNMANNED VEHICLE (UV) OPERATIONS SPECIALISTS

Summary: Unmanned vehicle operations specialists operate, maintain, and control the Military's fleet of unmanned vehicles. These vehicles include unmanned aerial vehicles (UAV), aerial systems (UAS), ground vehicles (UGV), surface vehicles (USV), and underwater vehicles (UUV). The specialists control these vehicles to perform reconnaissance and surveillance missions. They also operate sensors for target detection, deploy ground and air systems, and perform basic maintenance on systems as needed. Personnel normally specialize by the type of vehicle they operate.

Helpful Attributes

- Enjoy working with tools
- Knowledge of electronic theory and schematic drawing
- Superior adaptability to three dimensional spatial relationships

What They Do

- Supervise or operate the UAV, to include mission planning, mission sensor/payload operations, launching, remotely piloting, and recovering the aerial vehicle
- Perform mission planning, preflight, inflight, and postflight duties in accordance with aircraft technical orders
- Plan and prepare for mission by reviewing mission tasking, intelligence, and weather information
- Supervise mission planning, equipment configuration, and crew briefing
- Ensure ground station and aircraft are preflighted, inspected, loaded, and equipped for mission

- Operate aircraft controls and equipment; perform, supervise, or direct navigation, Intelligence, Surveillance, and Reconnaissance (ISR), and weapons employment operations
- Employ airborne sensors in manual or computer-assisted modes to actively and/or passively acquire, track, and monitor airborne, maritime, and ground objects
- Deploy and redeploy the UAV ground and air system
- Operate mission equipment, systems, and electronic protection (EP) equipment
- Monitor aircraft and weapons systems status to ensure lethal and non-lethal application of airpower
- Operate mission planning ancillary equipment to initialize information for download to airborne mission systems
- Operate radar, electronic support measures, and advanced electro-optics and infrared sensor systems to track surface and underwater contacts
- Operate and perform operator-level maintenance, assembly, and disassembly on communication equipment, power sources, light wheeled vehicles, ground control stations, ground data terminals, portable ground control stations, portable ground data terminals, transport, launch and recovery trailers, and tactical landing systems

Services

Army

Marine Corps

Navy

Air Force

VEHICLE DRIVERS

Summary: Vehicle drivers in the Military operate a multitude of vehicles in direct support of mission requirements. They operate all-wheeled vehicles and equipment over varied terrain to transport personnel and cargo. Military vehicle drivers operate and perform preventive maintenance on light and heavy-duty vehicles such as buses, truck and semi-trailer combinations, forklifts, and wrecker/recovery vehicles.

Helpful Attributes

- Interest in driving
- Interest in mechanics

What They Do

- Oversee loading and unloading of personnel and cargo on vehicles
- Employ land navigation techniques using radios and other navigation equipment
- Perform vehicle recovery to include towing, when necessary
- Perform preventive maintenance on government motor vehicles; correct or report vehicle deficiencies, supporting mechanics, as necessary
- Prepare vehicles for movement or shipment by air, rail, or vessel
- Organize and participate in convoys
- Conduct dispatch operations through planning and scheduling of vehicle operations resources to meet transportation support requirements
- Receive and fill requests from authorized persons for motor transport
- Coordinate, schedule, and document cargo movement using automated and non-automated tracking processes
- Provide transportation services for distinguished visitors and special events
- Administer driver qualification and licensing program
- Review contingency, mobility, and natural disaster plans to determine support requirements
- Conduct night operations to include use of night vision goggles and global positioning systems
- Refuel and defuel vehicles

Services

Army

Marine Corps

Air Force

VOCALISTS

Summary: Vocalists read, sing, and memorize vocal parts for public performances. They perform in a variety of musical performance units, such as chorus, popular musical ensemble, jazz band, country band, quartet, and soloist. Vocalists perform to boost morale, entertain, and assist with military ceremonies. They are part of a group of talented musicians who support the Military's mission around the world.

Helpful Attributes

- Ability to sing
- Poise when performing in public
- Clear enunciation

What They Do

- Read, sing, and memorize vocal parts for public performance
- Perform in a variety of musical performing units to include (but not limited to) chorus, popular musical ensemble, jazz band, country band, quartet, and soloist
- Perform duties as solo/lead vocalist for authorized military bands
- Perform all applicable styles of music
- Organize, instruct, train, counsel, and evaluate musicians and senior musicians
- Perform as vocalist in field music element or special band
- Serve in administrative, training, supply, or operations positions
- Perform as a musician or in direct support of the mission of military bands, including The U.S. Army Band, The U.S. Army Field Band, U.S. Military Academy Band, or the 3rd Infantry (The Old Guard) Fife and Drum Corps
- Provide ceremonial and entertainment services afloat and ashore

Services

Navy

Air Force

WAREHOUSING AND DISTRIBUTION SPECIALISTS

Summary: Warehousing and distribution specialists oversee the receipt, storage, documentation, transport, and quality assurance of military supplies. They conduct inventory control using automated or manual methods, and they keep detailed records of the number and condition of supplies. They perform the critical job of maintaining and safeguarding items that are important to the success of military missions and to the health and well-being of military personnel.

Helpful Attributes

- Ability to keep accurate records
- Interest in operating forklifts and other warehouse equipment
- Preference for physical work
- Preference for work requiring attention to detail

What They Do

- Receive, inspect, inventory, load, unload, segregate, store, issue, deliver, and turn in organization and installation supplies and equipment
- Establish and maintain stock records and other documents such as inventory, materiel control, accounting, and supply reports
- Prepare reports on personnel and equipment availability, storage space, relocation of materiel, and warehouse denials
- Conduct warehousing surveillance and inspections
- Inspect and identify property; determine condition of property received; perform shelf-life inspections of stock
- Develop methods and improve procedures for storing property
- Ensure proper supply flow under various issue methods; ensure effective internal controls for processing property documents and maintaining accountable records
- Apply special handling procedures for controlled medical items, gases, precious metals, dangerous and hazardous materials, and refrigerated or frozen materials
- Comply with fire and safety regulations; use protective measures for items in storage, including open storage lots and hazardous materials storage areas; establish field supply support areas
- Perform or supervise requisitioning, receipt, inventory management, storage, preservation, issue, salvage, destruction stock control, quality control, property management, repair parts management, inspection, packing and shipping, care, segregation, and accounting of medical supplies and equipment

Services

Army

Marine Corps

Navy

Air Force

Coast Guard

WATER AND SEWAGE TREATMENT PLANT OPERATORS

Summary: Military bases operate their own water treatment plants, which provide drinking water and safely dispose of sewage when public facilities cannot be used. Water and sewage treatment plant operators control processes and equipment that remove pollutants from the water so that it is safe to drink and can be returned safely to natural areas or be reused. Water and sewage treatment plant operators run the equipment, control the processes, and monitor the plants that treat the water.

Helpful Attributes

- Interest in chemistry and pollution control
- Interest in working with mechanical equipment

What They Do

- Assist in water reconnaissance, site preparation, and setup of water treatment activity
- Operate and maintain water treatment equipment
- Receive, issue, and store potable water
- Perform water quality analysis testing and verification
- Determine treatment method and treat water for purification
- Prepare water treatment reports
- Manage the maintenance of non-electric kitchen equipment, such as grease traps and other miscellaneous collection systems
- Install, inspect, maintain, troubleshoot, modify, repair, and manage plumbing, water distribution, wastewater collection systems, water and wastewater treatment systems, fire suppression, backflow prevention systems, natural gas distribution systems, liquid fuel storage, distribution, and dispensing systems
- Manage and direct facility and infrastructure systems and daily activities devoted to fuel, heating, cooling, ventilation, combustion equipment, industrial air compressors, natural gas, refrigeration, interior plumbing, sprinkler, irrigation systems, pest management, chemical application processes, and associated operations

Services

Army

Marine Corps

Navy

Air Force

WEAPONS MAINTENANCE TECHNICIANS

Summary: Weapons maintenance technicians are responsible for servicing and maintaining the Military's weapons. They ensure these assets are fully operational at all times. These experts plan, organize, and perform every service necessary to keep these systems in pristine working condition so they are ready when needed.

Helpful Attributes

- Ability to do work requiring accuracy and attention to detail
- Interest in working with electronic or electrical equipment
- Interest in working with weapon systems

What They Do

- Perform routine equipment inspections and emergency operating procedures
- Conduct component removal, repair, and replacement procedures, often using special tools and test equipment
- Supervise and perform unit maintenance and recovery of all self-propelled field artillery cannon weapon systems, including automotive, turret, carriage-mounted armament, associated fire control systems, and chemical protection subsystems
- Prepare and maintain equipment logs, equipment modification and utilization records, exchange logs, and calibration data cards; complete maintenance and supply forms and records
- Maintain, operate, and supervise maintenance on ground and air missiles, spacelift boosters, payloads, guidance and control systems, and subsystems
- Troubleshoot, repair, and service missile weapon systems, spacelift and research and development (R&D) equipment, facilities, and support equipment (SE)
- Load and unload nuclear and non-nuclear munitions, explosives, and propellant devices on aircraft
- Inspect, assemble, disassemble, maintain, and modify nuclear weapons, bombs, missiles, reentry vehicles and systems, launchers, pylons, penetration aids, and associated test and handling equipment

- Perform preventive and corrective maintenance
- Perform and manage maintenance, inspection, storage, handling, modification, accountability, custody, and repair of nuclear weapons, weapon components, associated equipment, and general or specialized test and handling equipment
- Perform operations and basic-level maintenance of submarine anchor systems, arms ammunition and explosives, weapons shipping and handling systems, submarine launched mines, missiles and torpedoes, weapons delivery systems, vertical launch systems, countermeasure systems, and participate in weapon handling functions
- Perform quality control measures; inspect, test, adjust components to specific tolerances; determine shortcomings and malfunctions in electronic, electrical, mechanical, pneumatic, optical, and electromechanical assemblies, subassemblies, modules and circuit elements, with common and system special design test equipment
- Manage maintenance, processing, acquisition, and operation of ground and air launched missiles, aircraft missile rotary launchers and pylons, spacelift boosters, payloads, related subsystems, test, calibration, support and handling equipment, and facilities

Services

Army

Marine Corps

Navy

Air Force

Coast Guard

WELDERS AND METAL WORKERS

Summary: Welders and metal workers deal with shaping, brazing, soldering, and forming metals to aid in different construction efforts that serve military missions worldwide. They create custom parts to repair the structural components of ships, tanks, submarines, landing craft, buildings, and equipment. They also install sheet metal products, such as roofs, air ducts, gutters, and vents. These enlisted workers use a wide range of skills, processes, and tools.

Helpful Attributes

- Interest in working with repair tools
- Preference for physical work

What They Do

- Provide expert advice on metals machining, welding, designing, and production problems
- Design, manufacture, or modify special precision tools, gauges, dies, and fixtures to facilitate metal working operations
- Perform metals technology shop calculations, such as determining cutting speeds and settings, welding processes, and pre-heat and post-heat requirements
- Weld, braze, solder, and heat treat metals
- Use manual and computer numerical controlled (CNC) metalworking machines, mills, and lathes to manufacture and repair cams, gears, slots, and keyways
- Check completed components and determine serviceability in accordance with drawings and specifications
- Disassemble, assemble, and fit component parts using machine screws, bolts, rivets, press fits, and welding technique
- Maintain and inspect hand tools and metalworking machinery
- Solve fabrication, airframe, maintenance, local manufacture, and support equipment repair problems
- Inspect and evaluate fabrication maintenance activities
- Perform installation, operation, maintenance, and repair of metalworking and welding equipment and material
- Perform tasks directly related to fabrication and erection of pre-engineered structures, including steel reinforcement; control job site deployment of materials and equipment

Services

Marine Corps

Navy

Air Force

Coast Guard

X-RAY TECHNICIANS

Summary: X-ray technicians utilize medical radiology equipment and assist radiologists in performing radiographic procedures including mammographic examinations, fluoroscopic examinations, vascular procedures, and computerized axial tomographic examinations (CAT scans), to name a few. They also help administer medical dyes, accurately target body structures for imaging, and work with health care professionals to deliver therapy to wounded or sick members of the Military.

Helpful Attributes

- Ability to follow strict standards and procedures
- Interest in activities requiring accuracy and attention to detail
- Interest in helping others
- Interest in working in a medical environment
- Listening and communication skills

What They Do

- Operate fixed and portable radiology equipment
- Pack, unpack, load, unload, assemble, and disassemble radiology equipment
- Read and interpret radiographic requests and physician orders
- Prepare, assemble, and adjust instruments, materials, and equipment
- Conduct radiographic examinations of the upper and lower extremities, vertebral column, trunk, and skull
- Perform soft tissue radiographic examinations and bone surveys
- Execute body section radiography, foreign body localization, prenatal, pediatric, urogenital, and radiographic examinations of the digestive, respiratory, vascular, and nervous systems
- Develop radiographic images using digital and manual processing
- Utilize hospital information systems to maintain patient locator files, radiographic files, and report files
- Inspect and perform operator maintenance on radiology equipment, ensuring compliance with radiation safety procedures

Services

Army

Navy